Pulmonary Drug Delivery Systems: Material and Technological Advances

Piyush Pradeep Mehta
Vividha Dhapte-Pawar
Editors

Pulmonary Drug Delivery Systems: Material and Technological Advances

Springer

Editors
Piyush Pradeep Mehta
Global Respiratory R&D
Cipla R&D Center
Mumbai, India

Vividha Dhapte-Pawar
Department of Pharmaceutics, Poona
College of Pharmacy
Bharati Vidyapeeth (Deemed to be
University)
Pune, India

ISBN 978-981-99-1922-2 ISBN 978-981-99-1923-9 (eBook)
https://doi.org/10.1007/978-981-99-1923-9

This Springer imprint is published by the registered company Springer Nature Singapore Pte Ltd.
The registered company address is: 152 Beach Road, #21-01/04 Gateway East, Singapore 189721, Singapore

Dedicated to our bookstores that witness every event in life.

Piyush Pradeep Mehta

Vividha Sagar Dhapte-Pawar

Foreword

In 1861, Thomas Graham [1] coined the term colloid (which means "glue" in Greek) to describe dispersions where particles have low diffusion rates (particle diameter at least 1 nm), yet do not settle under the influence of gravity (upper size limit of about 1 mm). As such, colloids provide a critical interface between the molecular and macroscopic worlds. Colloids play a major role in many biological processes, and understanding these connections may hold the key to development of dosage forms for gene therapy, gene editing, and other novel biotherapeutics.

Hakän Wennerström and Fennell Evans first suggested that the colloidal domain is where physics, chemistry, biology, and technology meet [2]. Indeed, the disruptive innovations occurring in biology are driving the need for innovation into delivery systems. This is certainly true for pulmonary drug delivery. This book provides an excellent primer on new technologies being advanced to deliver these new classes of APIs, as well as small molecule therapeutics.

A common theme within many of the chapters in this book is the development of nanoparticles or nanostructured materials in the colloidal domain. These methods include top-down processes for reducing the size of crystalline drug particles into microparticles and nanoparticles (e.g., jet milling, media milling). They also include bottom-up methods where particles are created from solution or dispersions in a liquid medium. These include technologies such as spray drying, spray freeze drying, flash nanoprecipitation with microfluidics systems, thin film freezing, printing technologies, and various formats of supercritical fluid manufacturing processes. Many of the particle engineering principles (e.g., spray drying) used to provide high lung delivery efficiencies and room temperature stability for proteins are being applied to the new biologics and vaccines [3]. New techniques such as thin film freezing that results formation of a brittle matrix of nanoparticle agglomerates have shown promise in the delivery of biologics including adjuvanted vaccines [4]. Discussion is provided around how to ensure delivery of nanoparticles into the lungs while preventing exhalation of the extra-fine particles. This may be accomplished by embedding the nanoparticles in microparticles. Upon contact with airway surface liquid, the microparticle matrix dissolves releasing the nanoparticles into the lungs.

Another theme throughout the book is controlling lung targeting and dose consistency through particle design. Optimal targeting to the lungs requires high delivery efficiencies and slow clearance of engineered particles from the lungs. As an industry, we are at the point where 10–30% delivery to the lungs with portable inhalers is unacceptable. It leads to high variability associated with oropharyngeal filtering of particles, and dramatic differences in deposition with inspiratory flow rate. Today, achieving 40–60% delivery efficiencies with the various bottom-up processes described herein is easily achieved. More recently, the envelope has been pushed to 90–95% delivery for both carrier-based and carrier-free formulations [5, 6]. Sustained delivery of inhaled therapeutics in the lungs is also critical as this leads to improved efficacy and safety for many drugs. As discussed herein, controlled clearance of therapeutics may be achieved with neutral forms of drugs, via encapsulation within liposomes or polymeric microspheres, or via use of cyclodextrins or mucoadhesive polymers.

In the 20 years since the human genome was mapped, breakthroughs in molecular biology have resulted in tremendous opportunity. Unlocking the potential of these drugs, be they siRNA, mRNA, antisense oligonucleotides, gene editing therapeutics, DNA, peptides, proteins, antibodies, or even small molecules, requires corresponding breakthroughs in the science for delivering these molecules to their targets. This book provides a framework to build engineered particles for delivery of novel therapeutics to the lungs.

References

1. Graham T. Liquid diffusion applied to analysis. Philos Trans R Soc. 1861;151:183–224.
2. Wennerström H, Evans DF. The colloidal domain: where physics, chemistry, biology and technology meet. New York: Wiley-VCH; 1999.
3. Heida R, Hinrichs WLJ, Frijlink HW. Inhaled vaccine delivery in the combat against respiratory viruses: a 2021 overview of recent developments and implications for COVID-19. Expert Rev Vaccines. 2021;7:957–74.
4. Alzhrani RF, Xu H, Moon C, Suggs LJ, Williams RO, Cui Z. Thin film freeze drying is a viable method to convert vaccines containing aluminum salts from liquid to dry powder. Methods Mol Biol. 2021;2183:489–98.
5. Weers JG, Son YJ, Glusker M, Haynes A, Huang D, Kadrichu N, Le J, Li X, Malcolmson R, Miller DP, Tarara TE, Ung K, Clark A. Idealhalers versus realhalers: is it possible to bypass deposition in the upper respiratory tract? J Aerosol Med Pulm Drug Deliv. 2019;32:55–69.
6. Miller DP, Tarara TE, Weers JG. Targeting of inhaled corticosteroids to the small airways: nanoleucine carrier formulations. Pharmaceutics. 2021;13:1855.

Jeffry Weers
Cystetic Medicines, Inc., Burlingame, CA, USA

Preface

The book *Pulmonary Drug Delivery Systems: Material and Technological Advances* provides a succinct coverage of the promising technologies and novel materials, which have been explored as carriers for drug delivery to treat pulmonary complications. The book brings together new technologies and material advances in the area of pulmonary drug delivery. This book focuses on a wider audience, including early career researchers, students, academicians, formulation scientists, materials engineers, clinicians, interdisciplinary professionals, and experienced researchers. It is a must-have for those who wish to explore the new dimensions while working for effective drug delivery approaches to treat pulmonary ailments.

The book initially elucidates the role of thin film freezing, supercritical fluid technology, nano-in-microparticles, crystal-engineered microstructures, and porous particles in pulmonary drug delivery. The subsequent book chapters elaborate on various functional excipients such as chitosan, cyclodextrins, and Vitamin E-TPGS to attain local and systemic therapeutic action. There are dedicated book chapters on diverse novel carrier systems such as hydrogels, quantum dots, metal-organic framework, as well as prodrug approach. Also, the book contains chapters, exclusively dedicated for biological and numerical simulation for pulmonary therapeutics.

Every book chapter focuses on the pulmonary relevance of technology or polymeric materials, carrier synthesis schemes, current technical state-of-the-art, along with clinical, industrial, as well as regulatory aspects. Also, every chapter contains a future perspective section that systematically reflects the current state of advances in pulmonary drug delivery. Additionally, it offers a practical basis for audience to comprehend the design and function of the pulmonary delivery systems for better therapeutic outcomes.

In concise, this book provides balanced views by considering the investigations from various scientific domains and industrial knowledge. The book enriches the understanding of the readers about the upcoming challenges, evolving novel applications, and future strategies for better pulmonary drug delivery.

We hope that this book will enlighten readers, academicians, and researchers understanding about material and technological advances in pulmonary drug delivery systems and will be able to contribute to the field of pulmonary drug delivery in the best possible way. The book would be valuable resource for budding scientists to pursue their careers in this field towards sustainable healthcare and well-being of future generations.

Mumbai, India Piyush Pradeep Mehta
Pune, India Vividha Dhapte-Pawar

Contents

Contributors

Nasser Alhajj Pharmaceutical and Molecular Biotechnology Research Centre (PMBRC), Waterford Campus, South East Technological University, Waterford, Ireland

Helen Cathcart Pharmaceutical and Molecular Biotechnology Research Centre (PMBRC), Waterford Campus, South East Technological University, Waterford, Ireland

Narumon Changsan College of Pharmacy, Rangsit University, Pathum Thani, Thailand

Vividha Dhapte-Pawar Department of Pharmaceutics, Poona College of Pharmacy, Bharati Vidyapeeth (Deemed to be University), Pune, Maharashtra, India

Alireza Ebrahimi Biotechnology Research Center, Tabriz University of Medical Sciences, Tabriz, Iran

Student Research Committee, and Department of Food Science and Technology, Faculty of Nutrition and Food Sciences, Tabriz University of Medical Sciences, Tabriz, Iran

Diana A. Fernandes OnResp, Fruebjergvej 3, Copenhagen Ø, Denmark

Hamed Hamishehkar Drug Applied Research Center, Tabriz University of Medical Sciences, Tabriz, Iran

Piyush Pradeep Mehta Global Respiratory R&D, Cipla R&D Center, Mumbai, Maharashtra, India

Basavaraj Nanjwade Surya College of Pharmacy and MRMH Pharma Pvt Ltd., Bangalore, Karnataka, India

Ali Nokhodchi Lupin Inhalation Research Center, Coral Springs, FL, USA

School of Life Sciences, University of Sussex, Brighton, UK

Niall J. O'Reilly Pharmaceutical and Molecular Biotechnology Research Centre (PMBRC), South East Technological University, Cork Road Waterford, Ireland

SSPC – The Science Foundation Ireland Research Centre for Pharmaceuticals, Limerick, Ireland

Satish Polshettiwar Department of Pharmaceutical Sciences, School of Health Science and Technology, Dr. Vishwanath Karad MIT-World Peace University, Pune, India

Somchai Sawatdee Drug and Cosmetics Excellence Centre and School of Pharmacy, Walailak University, Nakhon Si Thammarat, Thailand

Teerapol Srichana Drug Delivery System Excellence Centre and Department of Pharmaceutical Technology, Faculty of Pharmaceutical Sciences, Prince of Songkla University, Songkhla, Thailand

Tan Suwandecha Drug and Cosmetics Excellence Centre and School of Pharmacy, Walailak University, Nakhon Si Thammarat, Thailand

About the Editors

Piyush Pradeep Mehta is currently working as deputy manager in global respiratory R&D at Cipla R&D Centre, Mumbai, India. Dr. Piyush has rich experience in the field of inhaled drug products. He has worked in reputed pharmaceutical companies like Cipla R&D and Zydus Cadila. He has been instrumental in completing a few dry powder inhaler (DPIs) products in India. He has been elected as prestigious Newton-Bhabha Fund, Royal Society of Chemistry (RSC) Researcher (2019–2020) owing to his contribution to the field of pulmonary drug delivery. He has been selected for On-Demand Presentations on pulmonary drug delivery from The Drug Delivery to the Lungs (DDL 2020), UK because of his contribution to modified throat research. He has also received appreciation for the contribution of Institution's Innovation Council by MHRD's Innovation Cell, Government of India (2018–2019). He was a guest editor for a special issue "Advances in Orally Inhaled and Nasal Drug Products" in "Frontiers in Pharmacology" journal. He has filed four device patents and published more than 45 articles, book chapters, and scientific communications related to his proficiency in several peer-reviewed journals. He has delivered > 20 presentations at several international summits.

Vividha Dhapte-Pawar is working as Associate Professor in the department of Pharmaceutics at Poona College of Pharmacy, Pune. Dr. Vividha has worked in novel drug delivery systems, especially nano-formulations, pulmonary medicines, and chemotherapeutics. Dr. Vividha has 02 patent published, 40 international papers, and 10 book chapters to her credit. She is a recipient of 6 Gold Medals during her academics as well as at various national competitions along with awards for best oral and poster presentations at national and international conferences. Dr. Vividha is a recipient of AICTE fellowship, Dhirubhai Ambani Scholarship, and Ratan Tata Trust Fellowship. Also, she has received research grants, travel grants from UGC, Bharati Vidyapeeth University, and Pune University. She serves as a consultant for microbial testing, student start-ups, and generic medicines in Hospitals.

Crystal Engineering: A Versatile Platform for Pulmonary Drug Delivery

1

Piyush Pradeep Mehta and Vividha Dhapte-Pawar

Abstract

Crystal engineering is a potential discipline of materials science that involves sound design and development of different molecular solids. At present, crystal engineering is receiving more attention in the pharmaceutical field due to its ability to tailor drug substances' physicochemical, microstructural, and biopharmaceutical properties. It aids in designing multiple solid forms such as co-crystals, polymorphs, and solvate of the drug substance. The key objective of the present chapter is to summarize the applications of crystal engineering in pulmonary drug delivery. The chapter addresses vital background information regarding crystal engineering, synthesis schemes, and their applications in the pharmaceutical domain. Also, various crystal engineering-based inhaled medicines are reviewed, focusing on their biophysical properties, and aerodynamic performance. In addition, the chapter will carefully discuss the biological outcomes of intratracheal administered crystalline solids, followed by a detailed discussion section dealing with formulation aspects and clinical hurdles. Briefly, this chapter is an exhaustive account of crystal engineering in inhaled formulations.

P. P. Mehta (✉)
Global Respiratory R&D, Cipla R&D Center, Mumbai, Maharashtra, India
e-mail: piyu053@gmail.com

V. Dhapte-Pawar
Department of Pharmaceutics, Poona College of Pharmacy, Bharati Vidyapeeth (Deemed to be University), Pune, Maharashtra, India
e-mail: dhapte.vv@bharatividyapeeth.edu

P. P. Mehta, V. Dhapte-Pawar (eds.), *Pulmonary Drug Delivery Systems: Material and Technological Advances*,
https://doi.org/10.1007/978-981-99-1923-9_1

Keywords

Crystal engineering · Co-crystals · Solid-state synthesis · Crystalline solids and AmphiCrys

1.1 Introduction

The "Crystal engineering" phrase was first introduced by Schmidt G. and thoroughly used by Pepinsky R. in 1955 [1]. Afterward, Desiraju and Etter systematically recognized, characterized, and explored the supramolecular interactions to determine the crystal engineering phenomenon [2, 3]. Today, crystal engineering is a diverse field used to develop and modify several solid dosage forms [4–6]. Crystal engineering is a scheme to design and develop novel molecular solids with the judicious use and understanding of intermolecular interactions [7]. Crystal engineering using supramolecular synthon chemistry allows us to develop different molecular solids (e.g., co-crystals, polymorphs, and solvate) with preferred physicochemical and mechanical properties [8]. Certainly, crystal engineering is a discipline of enormous opportunity and applications ranging from electrical [9], metallurgy, explosives, pharmaceuticals [10], biomaterial synthesis [11], and dyestuffs [12, 13] to the production of optic tools [14] and catalytic substances [15]. Likewise, it is also extensively studied and explored in the design of novel pharmaceutical solids [16, 17]. Specifically, in the pharmaceutical industry, crystal engineering is explored for improving the physicochemical (e.g., solubility [18], dissolution rate [19], and stability [20]), mechanical [21] (e.g., surface morphology and surface energetics), biological (e.g., bioavailability [22, 23] and permeability [24, 25]) as well as formulation (e.g., tablet ability [26], taste masking [27, 28], and powder flow [29]) properties of the drug substance [30, 31]. Additionally, photostability (e.g., levofloxacin [32]) and microbial stability (e.g., nitrofurantoin) of drugs have been altered by crystal engineering techniques [30]. Moreover, crystal engineering presents the unique potential to extend the intellectual property of the existing molecules so as to preserve the drug lifecycle and product market value [33]. Thus, crystal engineering holds substantial commercial potential in the pharmaceutical industry [34].

1.2 Crystal Engineering Techniques

To date, many synthesis schemes and techniques have been judiciously used for the design and development of novel crystalline materials [35]. The crystal engineering techniques can be classified as solid-state [36] or solution-based [37] synthesis techniques. Solid-state synthesis requires very minute or no solvent. In contrast, solution-based techniques involve a large amount of solvent for isolation, purification, and separation of crystalline product from the mother liquor [33, 38]. Contact formation, solid-state grinding [39] (neat grinding or mechanochemical synthesis), extrusion (e.g., hot-melt extrusion and twin-screw extrusion) [40], and high shear wet granulation [41] are the well-studied solid-state crystal synthesis methods. Solid-state crystal synthesis is an ecological synthesis scheme due to the absence of organic solvents during manufacturing. Most importantly, it can be utilized for continuous crystal synthesis. There are several instances where energy (e.g., temperature and solvent) is needed to accomplish the co-crystallization process smoothly [35]. One of the ways to address this issue is using a small amount of organic solvent instead of energy to assist the co-crystallization process as a solution-based synthesis technique [35]. Liquid-assisted grinding (solvent-drop grinding or wet granulation), evaporative co-crystallization, anti-solvent co-crystallization [42], ultrasound-assisted crystallization [43], cooling crystallization [6], reaction co-crystallization, isothermal slurry conversion, and supercritical fluid methods (e.g., co-crystallization with supercritical solvent, the rapid expansion of supercritical solvents, supercritical anti-solvent co-crystallization, and supercritical CO_2-assisted spray drying) are the well-studied and explored solution-based crystal synthesis methods [44–46]. Additionally, a few researchers also tried and tested typical formulation techniques, i.e., spray drying [47] and freeze drying [48] for crystal engineering [49]. Each type of synthesis scheme has its own unique merits and demerits considering the nature of the molecule, type of co-former, solvents used, and processing conditions [49]. So, many active chemists and formulation scientists are now dynamically involved in exploring new crystal engineering methods to improve crystalline solids' physicochemical quality and yield. Microfluidics with jet dispensing, a microfluidic system equipped with ultrasonication [50, 51], matrix-assisted co-crystallization [52], laser irradiation [53], electrospray technology [54], electrochemically induced co-crystallization [55], and resonant acoustic mixing [56] are the few newly investigated alternative approaches for the synthesis of crystalline solids. Furthermore, researchers also used freezing-induced precipitation technology, precisely the thin film freezing method, to synthesize crystalline solids [57]. Most of the newly developed methods are studied to fabricate crystalline solids with high purity and yield [58]. Various crystal engineering techniques with their merits and demerits are discussed in Table 1.1.

Table 1.1 Crystal engineering techniques with their comparative merits and demerits

Techniques	Merits	Demerits	References
Solid-state grinding	Eco-friendly process Good purity co-crystals	Low yield Difficult to scale-up Time-consuming process	[58]
Liquid-assisted grinding	High possibility of obtaining co-crystals that cannot be synthesized by solid-state grinding Solvent variation assists in isolating different polymorphs	Difficult to scale-up Involve the use of organic solvents High risk of solvates formation	[58]
Anti-solvent co-crystallization	High-purity co-crystals Relevant for a wide range of drug-co-former pairs Easy of in-line process control	Involve the use of organic solvents Time-consuming process	[58]
Ultrasound-assisted crystallization	Good purity co-crystals Yielding fine crystals with uniform size Gain nucleation at low supersaturation levels	Difficult to scale-up Involve the use of organic solvents High risk of solvates formation	[58, 59]
Spray drying	High-purity co-crystals Continuous process and easy to scale-up Ease of drying solution, suspension, and emulsions	Various process parameters to optimize Involve the use of organic solvents	[58]
Freeze drying	Easy to scale-up Relative high yields	Time-consuming process Involve the use of organic solvents	[58]
Supercritical fluid crystallization	High-purity co-crystals Involve low processing temperatures Continuous process and yield solvent-free co-crystals	Involve the use of organic solvents Time-consuming process Low process yield	[58, 60]
Extrusion (e.g., hot-melt extrusion)	Solvent-free and eco-friendly process Continuous process and easy to scale-up Single-step method and high-purity co-crystals Ease of process understanding via PAT (process analytical technology) application	Not suitable for thermolabile drugs Cost of equipment	[58]

Table 1.1 (continued)

Techniques	Merits	Demerits	References
Microfluidic crystallization	High-purity co-crystals and continuous process Applicable for high throughput screening Suitable to use different stoichiometric ratios Use moderate temperatures Easy to isolate polymorphs of the same co-crystal Accessible of in/online process monitoring	Involve the use of organic solvents Difficult to scale-up Cost of equipment	[50, 58]
Electrospray technology	Relevant for a wide range of drug-co-former pairs Suitable to use different anti-solvent	Low yield, difficult to scale-up Time-consuming process and various process parameters to optimize	[54, 58]
Thin film freezing	Ease to scale-up Relative high yields Suitable for thermolabile drugs Ease of drying and energy-efficient process	Time-consuming process Involve the use of organic solvents	[58]
Resonant acoustic mixing	Easy to scale-up One-pot synthesis approach Suitable to use different stoichiometric ratios	Involve the use of organic solvents Various process parameters to optimize	[56]

1.3 Crystal Engineering and Pulmonary Drug Delivery

Majority of the commercial inhalation powders are usually manufactured using top-down methods, specifically high-energy jet milling [61, 62], ball milling [63], or dry powder coating [64, 65] as these methods permit reproducibility at a manufacturing level, and any modifications during scale-up are much easier to execute [66, 67]. However, jet-milled powders are usually highly cohesive owing to the high surface energy produced during the milling progression [68]. Therefore, the jet-milled powder habitually acquires poor powder flow, fluidization, and aerosolization behavior [69]. Additionally, high-energy milling tends to produce partially amorphous solids that involve post-processing to enhance the stability of the final formulation [70, 71]. Other material processing techniques (e.g., bottom-up technique) are based on homogenization or precipitation of supersaturated solutions and involve surfactants to stabilize the end product [70]. Furthermore, it may generate different unstable polymorphs, solvates and hydrates during material processing. Besides, a small quantity of residual solvent can initiate major physicochemical stability issues during storage [72]. Therefore, the direct crystallization method offers a better alternative to tackle the issues associated with high-energy milling and bottom-up processing due to its ability to modify various physicochemical and mechanical properties of a drug substance. Additionally, most crystallization techniques are capable enough to develop particles (0.5–5 μm) suitable for pulmonary drug delivery [73]. Briefly, the crystal engineering process symbolizes a promising alternative to traditional high-energy milling and bottom-up processing method. Consequently, active efforts taken by several dynamic researchers and chemists to develop inhalable particles using crystal engineering methods are thoughtfully summarized in the following section (Fig. 1.1).

1.3.1 NanoCrySP—Novel Technology for Lung Delivery

Nanocrystal is a fresh and easily adaptable strategy to delivery of "difficult-to-deliver" drug substances. Nanocrystals are nanoscopic (100–1000 nm) crystals of a parent drug which enhance saturation solubility of drug owing to nanometer particle size and improve dissolution kinetics due to higher specific surface area. For pharmaceutical applications nanocrystals are fabricated using top-down and bottom-up techniques. The high-energy top-down methods involve application of high-energy procedure or mechanical attrition to transform coarse micron size drug particles to fine nano size drug particles. Top-down methods involve the use of high-pressure homogenization, pearl milling, media milling, and micro-fluidization. Commercial top-down methods are Nanomill®, NanoCrystals®, Nanopure®, and Dissocubes®. The bottom-up method comprises precipitation of drug nanocrystals using non-aqueous drug solution and aqueous surfactant/stabilizer solution. It involves formation of nanocrystals via molecular aggregation using methods like solvent precipitation and co-precipitation method. Pre-treatment process of nanocrystal synthesis may include extra steps such as wet media milling, controlled precipitation, lyophilization, spray drying, and spray freeze drying (SFD) [74, 75]. Additionally, combination techniques such as Nanoedge™ (Baxter) and smartCrystal® (Abbott Laboratories) are commercially viable methods to accelerate the development of nanocrystals. In view of benefits and challenges allied with existing top-down and bottom-up nanocrystal synthesis methods recently Shete and co-workers developed unique bottom-up, spray drying-based method "NanoCrySP" to formulate drug nanocrystals for several pharmaceutical applications.

The new bottom-up spray drying-based technology; NanoCrySP produced nanocrystalline solid dispersions in which drug nanocrystals are submerged in the milieu of a small molecule excipient. Fundamentally, a solution comprising small molecule excipient (e.g., mannitol) and drug is spray dried to attain powder particles (2–50 μm) containing drug crystals (< 1000 nm). NanoCrySP technology contains three key mechanisms for production of drug nanocrystals, i.e., (1) physical barrier to crystal growth, (2) plasticization, and (3) primary heterogeneous nucleation. NanoCrySP fabricates the nanocrystalline solid dispersion as a solid powder which can be easily transformed to appropriate finished dosage form (e.g., tablet or capsule) with slight downstream treatment. The nature and concentration of small molecule excipient, crystallization characteristics of drug substance, and spray drying

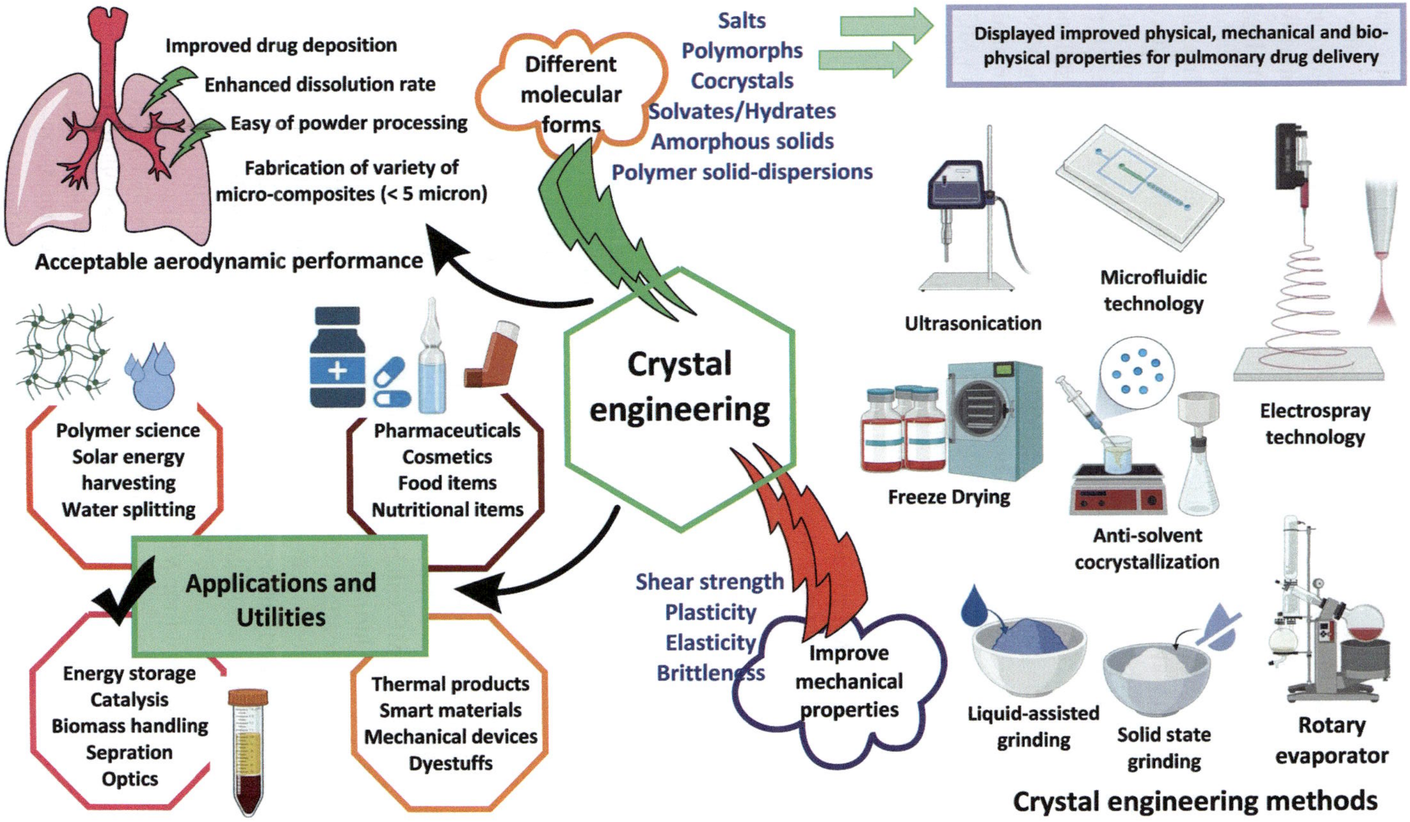

Fig. 1.1 Overview of crystal engineering methods, advantages, applications, for pulmonary drug delivery

procedure eventually direct the construction of nanocrystals. Furthermore, formulated drug nanocrystals appear under "Class I" based on the Nanotoxicological Classification System [76, 77]. This indicates low toxicity potential by drug nanocrystals, therefore, the NanoCrySP technology act as a value-added bottom-up spray drying-based technology in the pharmaceutical field [76, 77]. Additionally, authors established and presented a proof-of-concept for NanoCrySP technology using several natural (hesperidin and naringenin) [78, 79] and synthetic (e.g., celecoxib) [80] drug substances. Very interestingly same research group well investigated the one-step bottom-up spray drying-based NanoCrySP technology to produce ready-to-use respirable nanocrystalline solid dispersion dry powders which are discussed in the below segment.

Kaur et al. (2022) formulated high-dose nanocrystalline and microcrystalline solid dispersion powder of voriconazole (VRC) for pulmonary application. VRC nanocrystalline (VNC) system was fabricated using VRC, Pearlitol® 25C (mannitol), and soya lecithin in ratio 45:53.9:1.1% while VRC microcrystalline (VMC) system was fabricated using VRC, mannitol, and soya lecithin in ratio 60:38.5:1.5%. VNC and VMC were spray dried using methanol and water (70:30) mixture. Irregular, dense VNC and VMC particles displayed VRC loading of 45% and 60%, respectively. During the particle size assessment, VNC powder and VRC crystals embedded within VNC powder showed mean particle size of 2.7 μm and 645.86 nm, respectively, with a specific surface area of 5.19 m^2/g whereas VMC powder and VRC crystals embedded within VMC powder showed mean particle size of 3.1 μm and 4800 nm, respectively, with specific surface area of 4.17 m^2/g. Thermal analysis of VNC and VMC exhibited no glass transition incident and discrete melting endotherms of mannitol and VRC. X-ray diffraction (XRD) analysis presented a typical peak for VRC Form II while mixture of Form α, β, and δ for mannitol. During processing, mannitol generated physical barrier to crystal growth by intermingling with triazole nitrogen and pyrimidine ring fluorine of the VRC. During in vitro aerodynamic analysis, VNC (44.45 mg) and VMC (33.33 mg) equivalent to 20 mg of VRC were poured into size 3 HPMC capsules and actuated at 60 L/min using Rotahaler® device. VNC presented emitted dose (ED), mass median aerodynamic diameter (MMAD) and fine particle fraction (FPF) of 90.89%, 3.5 μm, and 76.69%, correspondingly whereas VMC exhibited ED, MMAD and FPF of 89.89%, 3.4 μm, and 73.60%, individually. During the long-term 24 months stability study at 25 °C/60% RH, VNC presented a slight change in mean particle size (991.66 nm), assay (99.29%), and moisture content (0.33%) of the VNC compared to initial data. In in vitro dissolution study using phosphate buffer saline (PBS; pH 7.4, 5 mL) by Franz cell (1000 rpm), >80% of VRC was dissolved quickly in first 10 min from VNC while 100% dissolution of VRC was accomplished in 30 min from VMC. Better dissolution performance from VNC was mainly credited to high surface area nanocrystalline solid, greater concentration of mannitol in VNC, and rapid wetting assisted by hydrophilic microenvironment of mannitol compared to VMC. In in vitro cellular uptake study using adenocarcinomic human alveolar basal epithelial cells (A549), VRC uptake from VNC was higher than VMC. VNC and VMC were taken up by fluid phase pinocytosis into the A549 intracellular compartment. VNC showed higher uptake due to the size-dependent quicker dissolution and absorption into the A549 cells. In in vivo pharmacokinetic study using mice after insufflation of VNC and VMC (5 mg) displayed a significant 2.94 and 1.92-fold higher area under curve (AUC) compared to Vorier® (10 mg/mL) intravenous administration. The systematic exposure of VRC from VNC (12.11-fold) and VMC (tenfold) were substantially lower compared to Vorier®. In vivo pharmacokinetic results were in good co-relationship with in vitro dissolution and cell line outcomes. The investigation verified the potential of one-step bottom-up spray drying-based NanoCrySP technology for generating high-dose dry powder inhalers (DPIs) [81].

Same research group also explored Aerosil® 200 (hydrophilic fumed silica) as the quaternary ingredient to optimize the ternary VNC. VNC system was fabricated using VRC, mannitol, and soya lecithin in ratio 45:53.9:1.1% by spray drying in presence of methanol and water (70:30) mixture. During the particle size estimation, VNC powder and VRC crystals embedded within VNC powder showed mean particle sizes of 2.7 μm and 587.90 nm, respectively, with 100% assay. Moreover, Aerosil® 200 (5%) was mixed with VNC during spray drying and physically mixed with the VNC. Aerosil® 200 treated VNC particles displayed mean particle size of 3.0 μm and crystal size of 702.53 nm with 97.24% assay. Aerosil® 200 presented halo-pattern in XRD analysis and did not affect crystalline nature and polymorphic forms of VRC and mannitol during spray drying. Spray-dried Aerosil® 200 / VNC particles showed specific surface area, porosity, and bulk density of 12.36 m^2/g, 96.25%, and 0.09 g/mL, respectively. During in vitro aerodynamic study, spray-dried Aerosil® 200 / VNC equivalent to 40 mg of VRC were poured into size 3 HPMC capsules and actuated at 60 L/min using Rotahaler® device. Spray-dried Aerosil® 200 / VNC offered ED, MMAD and FPF of 93.49%, 3.1 μm, and 94.37%, correspondingly. The use of quaternary ingredients in spray drying presented a marked impact on aerosolization performance of dry powder [82]. Therefore, NanoCrySP technology with different quaternary ingredient such as lactose fines, magnesium stearate, and leucine needs to be analyzed in the coming time to probe the full potential of this new crystal engineering system.

1.3.2 AmphiCrys: Novel Crystal Engineering Platform

Precipitation crystallization using traditional anti-solvent systems has gathered huge interest in recent times for crystal engineering [83]. Anti-solvent crystallization systems produced crystalline inhalable drug particles of various sizes (1–10 μm) and shapes (e.g., rod, needle, plate, and smooth surface) successfully [84]. However, particles do not frequently meet the required level of crystallinity and ecologically sensitive or complex mixture organic solvents (e.g., methanol, acetone, acetonitrile, and combination of two or more) are needed to dissolve hydrophobic drug substances [85]. Besides, poor control over particle size distribution owing to non-uniform mixing, dispersity in nucleation, irregular crystal growth kinetics during processing, and few post-crystallization events such as washing, agglomeration, and aging are the key challenges during scale-up [86–88]. Use of crystal growth inhibitors (e.g., hydroxypropyl cellulose, hydroxypropyl methylcellulose, and polyvinylpyrrolidone) [89, 90], controlled admixing, usage of ultrasound energy [91] and supercritical fluid were judiciously scrutinized to control agglomeration, aging and nucleation but, presence of stabilizing ingredients in the final crystalized drug particles is unfavorable [92]. Additionally, existence of stabilizing ingredients results in amorphous content and application is highly concentration dependent with more specific for the drug substance's chemical nature thus requiring cautious optimization during process validation.

Application of poly (ethylene glycol) (PEG) in anti-solvent crystallization denotes a suitable option instead of conventional anti-solvents since PEG displays good solubility in a variety of polar and nonpolar solvents [93]. Moreover, PEG is a valuable ecological solvent due to its non-volatility, low flammability, low toxicity, and biodegradable nature. In institutional and industrial research environment PEG is a choice of solvent for various chemical reactions, organic conversions, recovery of small molecules or biologicals, purification, and crystallization [94]. It is a highly capable and fascinating solvent system to solubilize low to high molecular weight hydrophobic pharmaceuticals. Waxy solid or low-melting solid PEG is available in various grades according to molecular weight and is listed in the US FDA "generally regarded as safe (GRAS)" list [95]. By knowing this physicochemical background of PEG and anti-solvent crystallization prerequisites Murnane and co-workers recently developed an amphiphilic crystallization (AmphiCrys) process using PEG and water to obtain crystalline drug particles suitable for pulmonary delivery.

Aqueous anti-solvent crystallization of salmeterol xinafoate (SX) from various organic solvents, i.e., PEG, acetone, propan-2-ol, and methanol, was thoroughly investigated to understand the mechanical and morphological properties of the resulting microcrystals. In thermal analysis, PEG-crystallized SX showed the absence of metastable crystal phases compared to the propan-2-ol-crystallized SX. PEG-crystallized SX followed the classical nucleation theory and developed the SX II form. SX microcrystals formulated from PEG 400 showed a narrow particle size compared to the other organic solvents. The SX microcrystals formulated from PEG 400 and 6000 showed mean particle sizes of 4.5 and 0.92 μm, respectively. PEG 6000 showed almost fivefold lower mean particle size, however, SX microcrystals formulated from PEG 400 (2.49) demonstrated lower span value compared to the PEG 6000 (10.42). Despite the initial organic solvent used, the SX microcrystals showed a typical plate-like morphology with crystal growth limited to two dimensions. SX microcrystals synthesized using anti-solvent crystallization showed thin and flakier morphology than the micronized SX. The SX microcrystals formulated from acetone showed fused agglomerates. Briefly, crystallization using PEG represents an excellent choice to formulate the thermodynamically stable microcrystal form at room temperature [96]. A similar research group also optimized the PEG 400-based anti-solvent crystallization process using the 2^4-factorial design. The high anti-solvent addition rate (200 g/min) with low stirrer speed (400 rpm) formulated SX microcrystals with a median diameter of 2.54 μm. The four parameters, i.e., SX concentration, PEG concentration, stirring speed, and anti-solvent addition rate were responsible for supersaturation of SX, supersaturation generation rate and hydrodynamic conditions, correspondingly, that influence particle size distribution. The factorial study displayed that the mean particle size decreases with increasing supersaturation. The size distribution of crystals is highly dependent on the balance of micro, meso-, and macro-mixing phases in the turbulent mixing process. During PEG-based anti-solvent crystallization, the sufficient micro- and meso-mixing balance is mainly obtained using a jet-reactor device compared to the stirred beaker. The use of PEG as a solvent for the anti-solvent crystallization of pulmonary drug substances revealed its capability to harvest particles with a median diameter in the low micron size range [97, 98].

Direct crystallization of fluticasone propionate (FP)/SX particles in the inhalable size range (0.5–5 μm) surmount the surface energization issues obtained from the traditional high-energy jet micronization. Thus, same research team used anti-solvent crystallization technique to formulate the inhalable SX and FP microcrystals. SX and FP solutions were prepared using PEG 400, while water was used as an anti-solvent to induce the crystallization phase. SX microcrystals (6.59 μm) showed thin plate-like morphology, whereas FP microcrystals (6.14 μm) displayed needle-like morphology. Additionally, SX and FP microcrystals were blended with lactose monohydrate (Lactohale® 200) and subjected to aerodynamic analysis using Cyclohaler® device at 60 L/min. SX microcrystals (26.26%) showed a 1.20-fold improvement in FPF as compared to the micronized SX (21.81%), whereas FP microcrystals (25.66%) showed 1.80-fold improvement in FPF as compared to the micronized FP (14.21%). The amphiphilic micro-crystallization of SX and FP using PEG represents the excellent potential to control particles' mechanical properties. Anti-solvent crystallization with PEG developed FP and SX microcrystals suitable for pulmonary applications. Thus, it is a feasible alternative to jet milling or ball milling-based micronization in developing DPIs. However, crystal engineering attributes such as surface area, surface energy, shape, and solid thermal state properties of microcrystal alone are needed to fine-tune to achieve dose to dose uniformity and consistent respirable fraction at scale-up level [99]. Briefly, PEG-based amphiphilic crystallization (AmphiCrys) proved an eco-friendly method to fabricate SX microcrystals with a particle size distribution suitable for effective pulmonary delivery.

1.3.3 Drug–Drug Co-crystal for Respiratory Applications

As pulmonary diseases are treated by several drug substances, combination remedy helps to enhance the pharmacological outcomes, reduce dosing frequency, costs of therapy, and ultimately improve patient compliance in the long-term treatment [100–102]. Drug-drug co-crystal is a modern approach for combination therapy against respiratory ailments. It is receiving more and more attention from numerous researchers and material chemists due to its synergistic pharmaceutical outcomes. It is a low-cost, low-risk, however, high-reward formulation strategy to fabricate therapeutically effective combination drug products. Compared to the traditional co-crystals approach drug-drug co-crystals use a drug substance as a co-former to develop a co-crystal. Accordingly, drug-drug co-crystals are more challenging and need a strong therapeutic rational before designing. For drug-drug co-crystal synthesis, drug substances were picked primarily based on therapeutic class and afterward characteristic supramolecular synthon evaluation such as host–guest interactions, hydrogen bonding, van der Waals forces, electrostatic interactions, π-π stacking, and metal–ligand coordination was performed. Drug-drug co-crystal synthesis received very good acceptance from the pharmaceutical industry. Few marketed co-crystal examples are Entresto® (sacubitril/valsartan), Lexapro® (escitalopram oxalate/oxalic acid), Steglatro® (ertugliflozin/L-pyroglutamic acid), and Suglat® (ipragliflozin/L-proline). Undoubtedly drug-drug co-crystal holds many biopharmaceutical advantages and great pharmaceutical market potential [103–105]. Accordingly, a few well-studied anti-asthmatic and antitubercular drug-drug co-crystal with important conclusions are listed in Table 1.2 whereas selected drug-drug co-crystal case studies are discussed in the following section.

Favipiravir (FAV) is a substituted pyrazine analog used against resistant cases of influenza virus. It exhibits an antiviral action by inhibiting RNA-dependent RNA polymerase. FAV is approved for the therapy of influenza in Japan and well researched for the management of several life-threatening organisms such as Lassa virus, Ebola virus, and currently severe acute respiratory syndrome coronavirus 2 (COVID-19). FAV is one of the most vigorously investigated drug molecule for respiratory disorders like COVID-19 [124]. However, ultra-high oral dose, short half-life, fast renal clearance of hydroxylated form and poor water solubility (2.29 mg/mL) at pH 2.0–6.1 at 20 °C reduced its pharmacological activity [125]. Moreover, ultra-high oral dose increases toxicity issues and poor compliance with multi-comorbidity patients. Therefore, inhaled FAV seems like a balanced approach to overcome the difficulties stated earlier. THP is a non-selective phosphodiesterase (PDE) inhibitor used in treatment of asthma and chronic obstructive pulmonary disease (COPD). Supramolecular chemistry presented that THP crystallize efficiently with various co-formers comprising primary amides such as formamide, pyrazinamide, and urea through developing an amide pseudo-amide synthon. Henceforth, pyrazine structure of FAV possesses high propensity to form a co-crystal with THP. Moreover, it offers a good clinical rational to develop the antiviral and PDE inhibitors in a single inhaled dosage form. Thus, Wong et al. (2022) developed a carrier-free DPI containing FAV: THP (1:1) co-crystal via spray drying and optimized the same using Quality-by-Design tool. FAV: THP co-crystal displayed a sharp melting endotherm at 194.9 °C and unique XRD pattern compared to individual component. Co-crystals presented elongated rod-shaped structures with smoother surfaces. Co-crystal powder also revealed porous nature and shaped bunches with different degrees of agglomeration. During the aerosolization study, 5 mg of powder was filled into size 3 HPMC capsules and dispersed using Breezhaler® at 60 L/min. Co-crystals disclosed MMAD and FPF of <5 μm and 79.3%, respectively. During moisture sorption analysis at 25 °C/80% RH co-crystals showed 3% mass change and measured as non-hygroscopic. In stability testing co-crystals showed no major changes in thermal properties when kept in screw-capped glass bottles at 60 °C/30% RH for 1 month. In dissolution study 10 mg of powder was filled into size 3 HPMC capsules and dispersed using Breezhaler® at 60 L/min by Fast Screening Impactor. Powder with aerodynamic diameters less than 5 μm was collected onto a glass fiber filter and dissolution was assessed using

Table 1.2 List of anti-asthmatic and antitubercular drug-drug co-crystal with vital conclusions

Drug-drug	Therapeutic category	Key outcome	References
Aspirin: Theophylline (THP)	NSAID and anti-asthmatic	Co-crystal showed good compatibility and offered mechanistic insights into co-crystal system	[106]
Flufenamic acid: THP	NSAID and anti-asthmatic	Co-crystal displayed improved hygroscopic stability, dissolution rate, and solubility of system	[107]
Diflunisal: THP	NSAID and anti-asthmatic	Co-crystal showed increased solubility, physical stability, and a typical "spring and parachute" shape dissolution kinetics	[108]
Diclofenac: THP	NSAID and anti-asthmatic	Co-crystal disclosed increased solubility, physical stability displayed a 1.3-fold higher dissolution rate	[108]
Entacapone: THP	Antiparkinsonian and anti-asthmatic	Co-crystal showed good thermodynamic stability with high solubility and permeability	[109]
Famotidine: THP	Histamine-2-blocker and bronchodilator	Co-crystal displayed improved stability at lower pH (1.2) setting	[110]
Sulfathiazole: THP	Antibacterial and bronchodilator	Co-crystal showed pH-dependent solubility and stayed stable under accelerated settings 40 °C/75%	[111]
Acetazolamide: THP	Diuretic and bronchodilator	Co-crystal presented good crystal packing and present systematic inputs for co-crystal formation	[112]
4-Aminosalicylic acid: Isoniazid	Antitubercular and antitubercular	Co-crystal presented good compatibility with insights into co-crystal formation	[113]
Pyrazinamide: Isoniazid	Antitubercular and antitubercular	Co-crystal displayed good drug-bridge-drug with optimized formulation and in vitro/vivo synergistic outcomes	[114]
5-fluorocytosine: Isoniazid	Antineoplastic and antitubercular	Co-crystal showed superior phase stability against moisture with potential for treatment of tuberculosis and associated infections	[115]
Entacapone: Isoniazid	Antiparkinsonian and antitubercular	Co-crystal displayed enhanced solubility	[109]

(continued)

Table 1.2 (continued)

Drug-drug	Therapeutic category	Key outcome	References
Furosemide: Isoniazid	Diuretic and antitubercular	Co-crystal displayed good equilibrium between solubility and diffusion. Highly suitable to improve absorption kinetics	[116]
Hydrochlorothiazide: Isoniazid	Diuretic and antitubercular	Co-crystal displayed enhancement in solubility and membrane permeability than parent drug	[117]
Nicotinamide: Isoniazid	Skin restorative agent and antitubercular	Co-crystal exhibited good solubility and dissolution kinetics	[118]
Entacapone: Pyrazinamide	Antiparkinsonian and antitubercular	Co-crystal offered enhanced solubility	[109]
Hydrochlorothiazide: Pyrazinamide	Diuretic and antitubercular	Co-crystal showed improved dissolution kinetics	[119]
Temozolomide: Pyrazinamide	Anticancer and antitubercular	Co-crystal presented mechanistic insights into co-crystal development	[120]
Nicotinamide: 4-aminosalicylic acid	Skin restorative agent and antitubercular	Co-crystal presented good compatibility and present physical insights on co-crystal development	[121]
Carbamazepine: 4-aminosalicylic acid	Antiepileptic and antitubercular	Co-crystal presented enhanced solubility	[122]
Salicylic acid: Ethionamide	NSAID and antitubercular	Co-crystal exhibited improved dissolution kinetics	[123]

simulated lung fluid. Co-crystals exhibited superior dissolution with total release within 20 min compared to physical mixture. In cell viability assay using A549 cells by MTT method at concentration range 1.6–1000 μM co-crystals showed 100% cell viability [126]. Briefly, robust single-step technique developed carrier-free antiviral drug-drug co-crystal powder with suitable aerosol functioning.

Colistin is a cyclic polypeptide antibiotic obtained from *Bacillus colistinus*. It is a cationic detergent with broad-spectrum bactericidal activity against Gram-negative organisms by damaging the cytoplasmic membrane [127]. It is available in parenteral form and high systemic exposure cause severe side effects such as nephrotoxicity which is a major dose-limiting parameter [128]. But high doses (2.5–5 mg/kg) are required to manage lung infections as only a small dose of drug reaches the airways after systemic delivery. Moreover, the increasing use of colistin as monotherapy accelerates resistance development and few cases were reported. Thus, co-administration of colistin with other antibiotics such as rifapentine is a valuable policy to improve therapeutic outcomes and diminish colistin resistance. Rifapentine is a long-acting, cyclopentyl-switched resultant substance of rifamycin. It is prescribed against leprosy and mycobacterium infections. Thus, Zhou et al. (2015) developed a respirable combination DPI of amorphous colistin and crystalline rifapentine by spray drying rifapentine particles (16 mg/mL) suspended in an aqueous colistin (16 mg/mL) solution for the treatment of respiratory infections. Spray-dried powder showed a particle size of 3.8 μm (d_{90}) with spherical colistin MPs adhered on the surface of elongated rod-shaped rifapentine MPs. The combined spray-dried powder exhibited lower moisture content. The moisture protection effect was mainly attributed to the rod-shaped elongated crystalline rifapentine (3.2 μm; d_{90}) MPs which reduced the contact between hygroscopic spherical-shaped colistin (2.7 μm; d_{90}) MPs. The combination powder proved high FPF for colistin (76.9%) and rifapentine (78.5%) when analyzed using the Aerolizer® device at 100 L/min. Moreover, combined spray-dried powder showed homogeneous distribution across each stage of the impinger. This combination spray-dried powder exhibited improved antimicrobial performance against the biofilm cultures and planktonic cells of *Pseudomonas aeruginosa* as indicated by reduced minimum biofilm inhibitory concentration (MBIC) and minimum inhibitory concentration (MIC) values for rifapentine (4 mg/L) and colistin (2 mg/L) as compared to pure rifapentine (MBIC 64 mg/L and MIC 32 mg/L) and pure colistin (MBIC 8 mg/L and MIC 4 mg/L), respectively. Concisely, the combined spray-dried powder approach not only showed an improved moisture protection effect but also demonstrated enhanced aerosol dispersion efficiency with superior antimicrobial performance [129].

Verapamil is a first generation, phenylalkylamine calcium channel blocker prescribed for therapy of hypertension and angina pectoris. Recent laboratory and preclinical investigations showed various activities of verapamil such as improvement in anti-mycobacterial activity of many tuberculosis molecules (e.g., isoniazid, rifampicin, clofazimine, streptomycin, and bedaquiline), reinforces the intracellular activity of antibiotics (e.g., rifampicin and isoniazid) microbial infections by enriching the macrophage uptake, replenish mycobacterium sensitivity to first-line antibiotics and avert the progress of drug resistance [130]. Moreover, verapamil displayed a significant reduction in bacillary burden, relapse rates, and accelerate bacillary clearance in the airways. However, oral delivery of verapamil showed high protein binding (90%) and high susceptibility to metabolizing enzyme CYP3A4 which meticulously diminishes its clinical effectiveness [131]. Accordingly, Parumasivam et al. (2016) tried to surpass the protein binding and metabolic limits by combining verapamil with rifapentine in inhaled delivery. Particularly, Parumasivam and team designed DPI of amorphous verapamil (1% w/w) and crystalline rifapentine (3% w/w) in the presence of anti-adherent leucine (20% w/w) by spray drying method for the effective treatment of pulmonary tuberculosis. Spray-dried powder showed a particle size of 1.9 μm (d_{50}) with nearly spherical and elongated particles. During aerodynamic assessment using an Osmohaler® device at 100 L/min flow rate, spray-dried powder showed FPF of 77.4 and 71.5% for verapamil and rifapentine, respectively. Improved aerodynamic performance was evenly attributed to lower intermolecular interaction between particles and control over moisture-induced recrystallization of amorphous verapamil due to the amino acid leucine. During microbiological analysis using *Mycobacterium tuberculosis,* i.e., H37Ra and H37Rv strains, spray-dried powder exhibited a MIC_{90} comparable to rifapentine alone. In contrast, spray-dried powder improved the killing of the H37Ra strain within the infected macrophage cells compared to verapamil and rifapentine alone. In a cellular toxicity study, spray-dried powder showed the acceptable half-maximal inhibitory concentration (IC_{50}; 62.5 μg/mL) value against both A549 and human leukemia monocytic cell line (THP-1). Results also showed that the presence of verapamil effectively improved the intracellular movement of rifapentine. Additionally, the spray-dried powder was stable in aerosolization performance for 3 months at 20 °C [132].

FP is a synthetic tri-fluorinated glucocorticoid used in various aerosol systems for anti-inflammatory and anti-asthmatic action. SX is a synthetic long-acting β-agonist frequently used for the treatment of COPD and asthma. Additionally, FP/SX is available in combination (e.g., Advair Diskus®, AirDuo Respiclick®, and Seroflo Ciphaler®) for the treatment of various respiratory conditions. Likewise, formoterol fumarate dihydrate (FFD) and Budesonide (BUD) are also used individually and in combination (e.g., Symbicort Turbohaler®, Foracort Rotacaps®, and DuoResp® Spiromax®) for management of several lung diseases. In most of above-listed combination products, two low-dose micronized drug substances were mixed with a lactose carrier using different techniques to formulate final formulation. Thus, to ease the formulation and handling process during manufacturing Kumon et al. (2010) fabricated novel crystalline inhalable particles using crystalline precursor mannitol by co-spray drying technique. FP/SX and mannitol (FP/SX-M) and BUD/FFD and mannitol (BUD/FFD-M) were dissolved in ethanol/water mixtures and subjected to spray drying. Particularly, FP/SX-M (0.8/0.08/7.12 w/v) was dissolved in ethanol/water (75:25) mixtures while BUD/FFD-M (1.6/0.048/18.352 w/v) was dissolved in ethanol/water (50:50) mixtures to obtained inhalable powders. Spray-dried FP/SX-M powder showed a satisfactory 69% yield and ~ 100% drug content while BUD/FFD-M powder presented a satisfactory 58% yield and ~ 100% drug content. Partially hollow FP/SX-M particles showed irregular mosaic structure with a mean particle size of 2.02 μm and a cohesion force of 18.7 nN. In contrast, partially hollow BUD/FFD-M particles exhibited smaller prismatic projections with the mean particle size of 2.01 μm and cohesion force of 15.2 nN. During aerosol performance analysis using Aerolizer® at 60 and 100 L/min, co-spray-dried particles showed satisfactory FPF in the range of 54–62%. At 60 L/min, FP/SX-M particles showed FPF of 53.8 (SX) and 54 (FP) % whereas BUD/FFD-M particles displayed FPF of 56.8 (FFD) and 57.7 (BUD) %. At higher flow rate, i.e., 100 L/min, FP/SX-M particles exhibited FPF of 58 (SX) and 58 (FP) % whereas BUD/FFD-M particles presented FPF of 61.6 (FFD) and 60.5 (BUD) %. The partially hollow particles with drug-enriched surfaces displayed superior aerosol performance. Improved FPF values were also attributed to the lower inter-particulate force resulting from a mixture of the MPs' crystalline nature and rough facade. The co-spray drying technique offers an alternative option for effective processing of combined formulations for pulmonary delivery [133].

Moxifloxacin is a fourth-generation fluoroquinolone with antibacterial action and most frequently prescribed for community-acquired pneumonia, acute sinus infections, abdominal infections, and acute worsening of chronic bronchitis. It also used for treatment of tuberculosis [134]. Moxifloxacin has high molecular weight

(401.4 g/mol), low water solubility (13.9 mg/mL), high log P value (2.9), and two different ionizable functional groups, i.e., carboxylic acid (acidic- 6.3) and secondary amine (basic- 9.3). For treating tuberculosis typically prescribed oral daily dose of moxifloxacin is 400 mg, however, only a small fraction (~ 4%) of oral daily dose reaches to the airways resulting in a suboptimal level at the disease site. Pulmonary delivery of moxifloxacin straight to the airways improves drug concentration at the site of primary infection and logically it is most feasible approach. Trans-Cinnamic acid is a naturally occurring flavoring agent and has a rich history of human use. It is listed in the US FDA GRAS list. It is mostly used in food products and fragrances to improve qualitative properties of the products. Chemically it is a monocarboxylic acid obtained from *Cinnamomum cassia*. It was applied as an antituberculosis agent and exhibited synergistic action when used with other antituberculosis drugs. Therefore, Eedara et al. (2019) combined moxifloxacin and trans-cinnamic acid to modify biopharmaceutical properties of the moxifloxacin and improve its therapeutic outcome of same. Basically, the crystalline adduct of moxifloxacin and trans-cinnamic acid (1:1) was synthesized to improve the pharmacokinetic profile of moxifloxacin. Crystalline adduct was formulated using dichloromethane by cooling crystallization method. XRD showed sharp diffraction peaks for adduct representing crystalline nature. Long needle-shaped moxifloxacin crystalline adduct showed a significant decrease of 2.89 and 3.35-fold in equilibrium solubility and intrinsic dissolution rate, respectively, compared to the moxifloxacin alone in phosphate-buffered saline (PBS; pH 7.4). The low solubility and dissolution rate of crystalline adduct is owing to stable hydrogen bond formation between trans-cinnamic acid and moxifloxacin. It also denotes the stability of the crystalline adduct in the aqueous milieu. Moxifloxacin crystalline adduct showed FPF of 30.4% using Aerolizer® inhaler device. Additionally, during dissolution study using custom-made dissolution apparatus, the respirable fraction of moxifloxacin crystalline adduct displayed slower dissolution kinetics and permeation than the moxifloxacin alone. The lower equilibrium solubility, intrinsic dissolution rate, and permeation kinetics helped to enhance the local mean residence time and pharmacological action of moxifloxacin in the pulmonary airways [135].

1.3.4 Theophylline Crystal Engineering

THP (Fig. 1.2) is a well-studied alkaloid obtained from tea and cocoa plants. Chemically it is a xanthine derivative, i.e., dimethylxanthine, and appeared as an odorless white crystalline powder. THP possesses bronchodilator action (causes relaxation of smooth muscle) and has a rich history of lung disease management [136]. It is registered in the US FDA GRAS record. It is one of the most widely utilized coformer in the synthesis of pharmaceutical co-crystals due to its ability to configure hydrogen bond with a variety of natural and synthetic drug substances. The nitrogen atom and imino group of THP facilitate the hydrogen bonding with drug substances [137]. Most common bonds involved in THP-based co-crystals are iminazole-carboxide (THP-nicotinamide) [138], iminazole-hydroxyl (THP-salicylic

Fig. 1.2 Molecular structure and details of theophylline

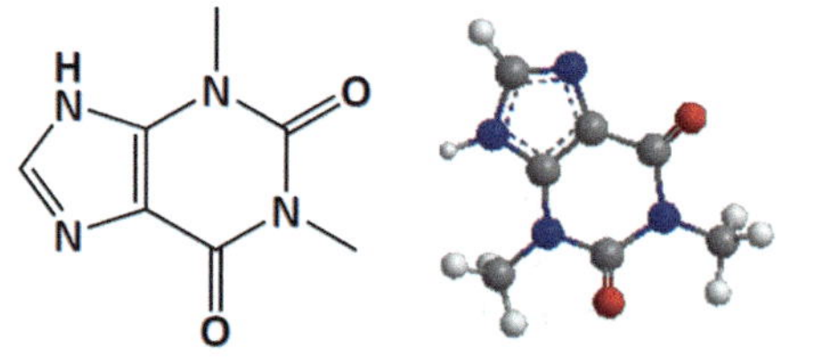

Theophylline **Theophylline - 3D**

1,3-dimethyl-3,9-dihydro-1*H*-purine-2,6-dione

Chemical Formula: $C_7H_8N_4O_2$

Exact Mass: 180.06

Molecular Weight: 180.17

m/z: 180.06 (100.0%), 181.07 (7.6%)

Elemental Analysis: C, 46.67; H, 4.48; N, 31.10; O, 17.76

acid) [139], iminazole-amino (THP-*p*-hydroxybenzoic acid) [140], carboxide-carboxide (THP-sulfacetamide) [141], carboxide-hydroxyl (THP-gentisic acid) [142], and carboxide-amino bonding (THP-baicalein) [143]. Synthesis of co-crystal with THP offers a better prospective in attaining apt physicochemical properties with synergistic biological actions. With these facts in cognizance several researchers explore THP to fabricate co-crystals and few such case studies are discussed below.

Tanaka et al. (2020) formulated co-crystal comprising THP anhydrate and oxalic acid (2:1; TA: OA) using SFD for better pulmonary application. The porous particles showed aerodynamic and geometric particle sizes of 3.03 and 7.20 μm, respectively. TA: OA co-crystals showed a much lower pseudo-polymorphic transformation rate and hydration capacity than theophylline anhydrate particles under a setting of 40 °C temperature and 96.4% relative humidity due to the formation of distinctive crystalline structure. The lower energy and intermolecular distance (1.16-fold lower) in the TA: OA co-crystals were accountable for the co-crystal's stability compared to the theophylline anhydrate particles alone. During processing of THP anhydrate alone, translation of THP anhydrate to monohydrate occurred due to bonding between nitrogen and hydrogen atoms of THP and water, respectively. Whereas during processing THP anhydrate with OA, co-crystal was formed due to bonding between nitrogen and hydrogen atoms of THP and OA, respectively. Moreover, aerodynamic particle sizer spectrometer analysis at 60 L/min demonstrated MMAD (3.03 μm) suitable for effective pulmonary delivery. SFD is suitable for THP co-crystallization and enable the preparation of DPI products under environmentally friendly conditions. In brief, SFD technology with THP is suitable for the preparation of co-crystallization-based DPIs [144]. Alhalaweh et al. (2013) synthesized THP co-crystals (1:1) using urea (THP-URE), saccharin (THP-SAC), and nicotinamide (THP-NIC) co-formers by spray drying and thoroughly investigated for the inhalation performance. The spray-dried THP particles showed elongated, acicular nature while THP-URE (2 μm) co-crystals displayed plate (flakes) shaped, agglomerated, and fractured morphology. Moreover, THP-SAC (3 μm) and THP-NIC (1.5 μm) co-crystals showed irregular and deformed particles. Variations in the solid form and crystal habitat of a material showed remarkable impact on its surface chemistry. The dispersive surface energies of spray-dried co-crystals studied using inverse gas chromatography were ranked in the following order THP-NIC > THP-SAC > THP-URE > spray-dried THP. Particularly, THP-URE, THP-SAC, and THP-NIC showed surface energy of 43.23, 45.70, and 69.98 mJ/m^2, respectively. During cascade analysis using Aerolizer® device at 60 L/min THP co-crystals alone showed an effective drug deposition pattern as compared to lactose-loaded THP co-crystals formulations. The aerodynamic outcome of different THP co-crystals was in the following order: THP-NIC > THP-URE>THP-SAC. Results demonstrated that spray-dried co-crystals were capable of pulmonary delivery of drug actives [145].

1.3.5 Itraconazole Crystal Engineering

Itraconazole (ITZ) is a triazole antifungal substance prescribed for the treatment of local and systemic fungal infections. ITZ is approved by the US FDA for the management of fungal infections, such as histoplasmosis and mucocutaneous candidiasis, i.e., oropharyngeal and esophageal candidiasis. ITZ displayed highest efficiency against respiratory aspergillosis. However, ITZ is a BCS class II drug with low aqueous solubility (1 ng/mL) and poor bioavailability. It has high molecular weight (705.6 g/mol), high log P value (5.66), and systemic use of ITZ is restricted owing to its hepatotoxicity. Long-term ITZ therapy is allied with transient, mild-to-moderate serum elevations and can cause clinically evident acute drug-stimulated liver injury. The pulmonary delivery of ITZ at a lower dose surmounts restrictions allied with oral and systemic delivery [146–148]. Consequently, the pulmonary delivery of ITZ has fascinated substantial consideration compared to oral and parenteral dosage forms. Few such investigations are discussed below.

Lin et al. (2017) thoroughly explored the hot-melt extrusion-based solvent-free fine solid crystal suspension (FSCS) method for the design and development of inhalable ITZ microparticles (MPs). Initially, jet-milled ITZ (2.86 μm) was mixed homogeneously with mannitol carrier (20:80) and subjected to a conical co-rotating, twin-screw extruder for fabricating extrudates. Subsequently, extrudates were subjected to jet milling. The irregular-shaped non-spherical, jet-milled MPs showed a mean particle size (2.19 μm) suitable for pulmonary drug delivery. In aerodynamic assessment using the Turbospin® inhaler at 60 L/min, jet-milled extrudates showed a superior FPF (50.69%). The improved aerodynamic outcome is mainly accredited to the lower level of particle agglomeration. Additionally, FSCS method-based extrudates showed 145-fold superior solubility compared to raw ITZ, and the crystalline form remained unaffected after storage under the accelerated conditions for 6 months (40 °C and 75% relative humidity). Thus, the mannitol-based FSCS technique is an appealing platform technology to improve water-insoluble molecules' solubility and aerodynamic efficacy [149].

Weng et al. (2019) modified the crystal structure of ITZ with a dicarboxylic acid, i.e., suberic acid (SUB), using the rotary evaporation technique. Moreover, a spray drying system was applied to obtain co-crystals with appropriate mean particle size and remove the residual solvent. Specifically, the rotary evaporation method effectively synthesized ITZ: SUB (1:1) co-crystal using a mixture of ethanol and chloroform (4:1). During the in vitro release study, ITZ-SUB co-crystal showed 39-folds superior intrinsic dissolution kinetics in 0.1 N HCL compared to unprocessed ITZ. In the in vitro aerodynamic study, ITZ-SUB co-crystal showed higher FPF (64.10%) with acceptable MMAD (2.56 μm) using the Breezhaler® device (90 L/min). As per thermal stress analysis, ITZ-SUB co-crystals remained stable for 1 month at 60°C [150]. ITZ respirable micronized co-crystals and spray-dried powders were formulated and thoroughly studied for in vitro and in vivo pulmonary properties. ITZ co-crystals (1:1) were synthesized using succinic acid (ITZ-SA) and l-tartaric acid (ITZ-TA) co-formers by anti-solvent crystallization scheme and subsequently micronized to a particle size of lesser than 2 μm. In contrast, ethanol, and water (4:1 v/v) containing ITZ and mannitol (1:4 w/w) were spray dried to formulate amorphous ITZ powder (< 5 μm). During microscopic analysis, the crystalline ITZ-SA, ITZ-TA, and spray-dried powder showed elongated, angular blocks, bulky agglomeration, and smooth spherical surface, respectively. The micronized ITZ-SA and ITZ-TA co-crystals powders exhibited an enhanced intrinsic dissolution rate compared to the amorphous spray-dried ITZ formulation in potassium dihydrogen phosphate buffer (pH 7.2) containing polysorbate 80 (5% w/v) and dipalmitoyl phosphatidylcholine (1% w/v). Following pulmonary administration using Dry Powder Insufflator™ in 8-week-old male Sprague-Dawley rats, ITZ-SA and ITZ-TA co-crystals displayed a significant improvement of 24 and 19-fold in ITZ bioavailability compared to the amorphous spray-dried ITZ formulation, respectively, at an equivalent dose (0.8 mg/kg). The poor intrinsic dissolution rate and pulmonary bioavailability of amorphous spray-dried ITZ formulation are mainly attributed to ITZ's rapid crystallization during processing. Briefly, the co-crystals formulated using the spray drying method denoted a suitable scheme for enhancing the pulmonary bioavailability of poorly soluble molecules [151].

1.3.6 Budesonide Crystal Engineering

BUD is a well-known glucocorticoid which showed a therapeutic effect on various human organs. BUD is mainly applied to treat inflammatory forms of the intestines and lungs. It is approved by US FDA for treatment of ulcerative colitis, Crohn's disease, COPD, and asthma. It is mostly delivered directly to the respiratory airways as aerosol (e.g., Pulmicort Turbuhaler®) for topical therapeutic action. However, due to low aqueous solubility (28 μg/mL) and high hydrophobic nature (log p of 3.2) it follows slow dissolution kinetics which eventually leads to inadequate BUD concentration at the target sites [152, 153]. Therefore, various crystal engineering efforts were used to surpass solubility, dissolution-related problems and deliver adequate concentration of the BUD to the inflamed lung tissues.

Microfluidics is a bottom-up scheme used to fabricate versatile MPs and nanoparticles (NPs) with various functionalities. It is one of the types of synchronized homogenization and emulsification systems utilized to develop various pharmaceutical, cosmetic, and food products. It is highly apt to produce homogenous and stable aqueous systems even with nonpolar components owing to its unique combined mechanism of ultra-high pressure, intense velocity impact, uniform cavitation, and precise shear kinetics. Microfluidics combined mechanism help to transmute material conformation and structure which eventually cause adaptations in the material attributes and act as a base for new versatile application [154–156]. Based on this preliminary understanding, Saboti et al. (2017) fabricated BUD crystalline inhalable particles using the microfluidic reactor integrated with an ultrasonic SFD system. Specifically, the T-junction chip microfluidic reactor (Dolomite Centre Ltd., UK) was utilized to formulate BUD suspension (using methanol–water, ethanol–water, and acetone–water mixture), which was immediately placed into an ultrasonic atomization probe. The mixture was subjected to liquid nitrogen, and finally, the freeze drying process was used to produce BUD dry powder. Rectangular-shaped multimodal and plate-like unimodal particles were obtained for acetone–water and methanol–water, ethanol–water, respectively. BUD crystalline particles showed mean particle size (D_{50}), i.e., acetone–water (5.85 μm), methanol–water (8.83 μm), and ethanol–water (7.33 μm) suitable for pulmonary drug delivery. XRD exhibited lower peak intensity due to decrease in particle size of BUD crystals. However, microfluidic method retained the crystal structure of BUD and thus confirms its suitability for the development of narrow-sized crystalline formulations. During dynamic vapor sorption analysis at 90% relative humidity acetone–water (2.79%), methanol–water (1.34%), and ethanol–water (2.12%) showed lower moisture uptake. Processed crystalline BUD particles were mixed with Respitose® SV010 lactose carrier (1: 67.5) geometrically using Turbula® low shear mixer (46 rpm for 30 min) for aerodynamic investigation. Formulated blend displayed a satisfactory aerodynamic outcome using Aerolizer® at 100 L/min with FPF of 54.9% (acetone–water) and 47.6% (ethanol–water), while the performance of the methanol–water was amid 54.9–47.6%. Briefly, microfluidics technology was successfully applied without stabilizers for formulating inhalable MPs of BUD. Briefly, the present investigation displayed that microfluidics can be effectively used without additional mechanical systems such as high-speed homogenizers, cryo-crushing, and micro-grinding or polymeric stabilizers for formulating ultra-fine inhalable particles for DPI application [157]. The anti-solvent precipitation method formulated the ultra-fine inhalable BUD MPs. The BUD solution was formulated using organic solvents, i.e., methanol and acetone, while water was selected as the anti-solvent to induce the precipitation phase. Moreover, stabilizers like hydroxypropyl cellulose (0.07% g/mL) and tyloxapol (0.16% g/mL) were added to the methanol–water system to regulate crystal growth in three dimensions. The acetone–water system developed rectangular-shaped crystals, while the methanol–water system formulated elliptic particles. Both systems showed flaky morphology with agglomerated MPs of size <5 μm and thickness of 300 nm. Interestingly, consistent spindle-shaped and uniform ellipsoidal MPs (<5 μm) were obtained in the presence of stabilizers.

During aerodynamic analysis using Aerolizer® at 60 L/min crystalline BUD MPs showed aerosol performance (emitted FPF) in the following order: methanol–water system (68%) > acetone–water system (60%) > methanol–water with tyloxapol (26%) > methanol–water with hydroxypropyl cellulose (23%) [158].

1.3.7 Dynamic Methods for Respirable Crystals

1.3.7.1 Plug Flow Crystallizer

Continuous crystallization is a rapidly growing segment in institutional and industry research agenda owing to its capacity to synthesize high-quality crystalline products with reproducibility and inexpensive manufacturing process. Continuous crystallization is highly apt for high-volume manufacturing processes of polymers, pharmaceutical process, and products such as tablets, capsules, and inhalers. Stirred-tank and tubular-flow (e.g., slug flow, segmented plug-flow, oscillatory-baffled, and unbaffled-tubular) were used for continuous crystallization process. Plug flow crystallizer (PFC) is one of the most explored continuous crystallizers. PFC is a compact assembly with a shaft having distinctively organized blades which revolve within a casing to generate plug flow and low attrition for continuous crystallization. In PFC materials flow in homogenous manner and gentle agitations prevent crystal disruption. It has several heating /cooling zones with opening slots to offer temperature control and addition of anti-solvents, crystallization seeds, respectively [159–161]. With this prior knowledge, Hadiwinoto et al. (2019) judiciously used single-step continuous PFC to fabricate rifapentine and beclomethasone dipropionate (BDP) crystals for pulmonary applications. Specially, for both drug substances continuous crystallization, spray drying, and integrated continuous crystallization and spray drying were used to fabricate inhalable crystalline particles. For development of tubular crystallizer, a rigid low-density polyethylene tube with inner diameter of 1/8 (0.125 inches) was used. The acetone–water and ethanol–water solvent systems were used for anti-solvent crystallization of rifapentine and BDP, respectively. Continuous crystallization using a tubular crystallizer presented spherical amorphous and needle-like crystalline rifapentine particles under various tested conditions while needle-shaped crystals were found for BDP in all verified settings. Spray drying showed needle-shaped crystals for rifapentine with a small portion of irregular-shaped particles whereas long needle and plate-shaped crystals were found for BDP. In last integrated continuous crystallization and spray drying process displayed needle-like shape particles with some small amorphous particles and a hollow morphology for rifapentine while small spike-shaped (monohydrate form) particles for BDP. The anti-solvent crystallization in continuous plug-flow facilitated adequate mixing of drug solution with an anti-solvent to achieve the high supersaturation and nucleation rate. Moreover, the narrow residence time distribution of the integrated process helped in crystal traceability. During aerodynamic analysis using Breezhaler® rifapentine crystals (99.5 L/min) and BDP crystals (90 L/min), showed a satisfactory FPF of 60.2% and 54%, respectively. Briefly, studied novel integrated continuous crystallization and spray drying process effectively fabricated rifapentine and BDP crystals with suitable aerodynamic characteristics for deep lung delivery [162].

1.3.7.2 Multiphase Static Mixer

Anti-solvent precipitation process is time consuming and in most cases during large-scale production heterogenous mixing leads to generation of wide particle size distribution [163, 164]. However, some novel mixing techniques such as membrane contactor, confined impinging jet mixers, microfluidics, T-mixer, and static mixers were available to achieve rapid and homogeneous mixing to control particle size distribution and other physical properties of the processed materials. Among these techniques, static mixer is an effective mixing tool for nonstop large-scale activity. Static mixers are used in various industries for key unit operations such as multiphase applications, heat transfer, blending solid materials, and uniform mixing of two or more fluid components. Static mixers contain stationary mixer gears with crossbars fixed in a suitable casing. Mixer gears are assembled in tortuous design to obtain a rapid and homogeneous mixing of two or more solvents. Static mixers can achieve turbulent and laminar flow during mixing through dividing and reuniting solvent streams [165]. Static mixers were also used for design and development of nanoparticles of various drug substances such as spironolactone [166], Cyclosporine A [167], progesterone, and carbamazepine [168]. Sheng et al. (2019) adopted static mixer-based multiphase mixing process to fabricate hollow crystals of poorly water-soluble BCS class II model drug spironolactone. Micron-sized, hollow crystals of spironolactone were fabricated using a static mixer system and povidone stabilizer. Specifically, deionized water containing povidone stabilizer (0.1 mg/mL) was added to spironolactone ethanolic solution (10 mg/mL) in the presence of SMV DN25 static mixer. During static mixing, povidone assisted in achieving the diffusion-limited growth phase, which is highly important for the effective growth of hollow crystals. After receiving the equilibrium stage, drug suspension was subjected to spray drying to obtain an inhalable dried powder. Elongated and smooth surface, spray-dried hollow crystals showed mean particle size and surface area of 16.1 μm and 2.79 m^2/g, respectively. During aerodynamic assessment using Aerolizer® device at 60 L/min, spray-dried spironolactone showed improved FPF of 23.1% compared to cohesive raw spironolactone (7.8%). Improved aerodynamic performance was mainly attributed to the high aspect ratio of hollow crystals and low level of particle aggregation. Static mixer followed by spray drying was found to be a robust method for formulating crystalline and hollow particles with good aerodynamic properties. Briefly, static mixing is a low-cost, less time consuming, and completely scalable method for continuous synthesis of micron-sized inhalable crystals [169].

1.3.7.3 Combined Crystallization Approach

The key crystal attributes such as shape, size, and nature in the crystallization process are greatly impacted by supersaturation kinetics because of its association with nucleation and crystal growth profile. Thus, controlled supersaturation kinetics is a key to acquiring anticipated crystal qualities. The supersaturation kinetics are regulated via seeding properties, optimal cooling conditions, and anti-solvent addition rate. Thus, in the traditional crystal engineering process cooling crystallization is preferred when solubility of the dissolved solute is mainly governed by temperature whereas anti-solvent crystallization is utilized when dissolved solutes have low thermal stability. But, in many instances the solubility of a solute is vastly affected by both addition of an anti-solvent and temperature. In such instances, combining anti-solvent and cooling process is valuable as it enhances percentage yield and presents two working parameters to exert better regulation on supersaturation trajectory and thus on the end results of crystallization development [170, 171]. By knowing combined cooling and anti-solvent crystallization technique benefits Ragab et al. (2010) used combined experimental methods to synthesize inhalable progesterone (naturally occurring chiral steroid) microcrystals. The anti-solvent combination of isopropyl alcohol and water was used to fabricate progesterone microcrystals and whole process was optimized using 2^4 factorial designs. The controlled cooling bath and linear cooling protocol were used to employ cooling crystallization. Progesterone (1 g/L) solution was formulated using a mixture of solvents, i.e., water–isopropanol, while water (50%) was utilized as an anti-solvent to incite the crystallization point. During the 2^4 factorial screening drug concentration, organic solvent composition and anti-solvent addition rate strongly impacted the physicochemical properties of progesterone microcrystals. Platelet-shaped progesterone microcrystals showed a higher surface-to-volume ratio than typical spherical particles with particle size and span of 3.22 (D_{90}) and 1.24, respectively. In the present case, isopropanol as a solvent hydrophobic agent exhibited the excellent capability to shape particles with a mean diameter apt for pulmonary delivery. During cascade analysis using Turbuhaler® at 30 L/min, progesterone microcrystals showed satisfactory FPF and MMAD of 50% and 4.52%, respectively. Additionally, progesterone microcrystals showed a lower aggregation index (0.50). Briefly, the factorial screening and in vitro aerodynamic assessment clearly disclosed the benefits of the combined crystallization approach [172].

1.3.7.4 Acidic Titration with Vertically Oriented Jet Mill

Diclofenac (DF) is a well-investigated non-steroidal anti-inflammatory agent usually formulated as potassium and sodium salts for treatment of various inflammatory conditions, osteoarthritis, arthritis, and migraine. Yazdi and Smyth (2016) studied the acidic titration followed by a jet milling (Jet-O-Mizer™) method to design inhalable crystalline DF sodium MPs. In the acidic titration method, aqueous DF sodium solution was treated with 2.5 N HCl to obtain DF. The acidic titrated soft white cake was dried to obtained DF powder. Finally, DF sodium and DF were subjected to jet milling using Jet-O-Mizer™ to obtain respirable fine crystals. Jet-O-Mizer™ is a vertically oriented jet mill with tangential situated grinding jets at bottom and static distribution happening at the top. Due to its vertical orientation, it offers few key benefits, i.e., ease of material feed, consumes very less energy, and generates narrow particle size distribution. Jet-milled DF (2.36 μm) and DF sodium (3.25 μm) displayed 1.38 and 8.2-fold higher specific surface area compared to unmicronized DF (6.98 μm) and DF sodium (19.5 μm), respectively. The higher surface area of jet-milled DF and DF sodium crystals is mainly attributed to the formation of a multilayer adsorbate and its condensation on top of the jet-milled fine particles. DF crystals exhibited smooth, hollow needle-shaped morphology with thicknesses of 0.5 μm and a length of 20 μm. Additionally, jet-milled DF and DF sodium presented lower moisture content of 0.45 and 2.99%, respectively. During aerosolization study of carrier-free materials (10 mg) using the high resistance Monodose RS01® inhaler at 58.5 L/min, neat DF, micronized DF, and milled DF sodium showed % FPF of 37.7, 60.4, and 75.6%, individually. Aerosolization study showed 8.71 and 3.95-fold higher % FPF $_{(1\ \mu m)}$ for milled DF sodium compared to neat DF and micronized DF, correspondingly. The superior aerosolization performance of jet-milled DF sodium crystals was significantly attributed to a large aspect ratio, i.e., elongated needle-shaped morphology with hollow structure. Concisely, acidic titration method-based carrier-free pulmonary formulation can reach deep airways to act systemically or locally with reduced adverse reactions by decreasing the total dose requirements [173].

1.3.7.5 Slow Solvent Evaporation with Spray Drying

Dapsone (DAP) is a synthetic sulfonamide with antibacterial and antiinflammatory actions. DAP is also known as 4,4′-sulfonyldianiline and is an interesting drug candidate used in treatment of many diseases such as leprosy, tuberculosis, malaria, and acute ischemic stroke-associated pneumonia. DAP is commonly co-administered with other antibiotics like rifampicin and clofazimine for the management of leprosy. But the remedial potential of DAP is significantly restricted by its low water solubility (0.16 mg/mL) [174, 175]. Thus, to improve its solubility, dissolution kinetics, and other physicochemical properties Amaral and co-workers fabricated inhalable multicomponent solid forms of DAP for pulmonary delivery. DAP/caffeine (1:1; DAP: CAF) co-crystals were synthesized and methodically studied for toxicity and permeability potential using lung cells. Initially, slow solvent evaporation investigations and X-ray diffraction screening presented acetone (needle-shaped crystals), ethanol (no-definite crystal habit), and ethyl acetate (orthorhombic crystals) as the ideal solvent system to develop co-crystals via liquid-assisted grinding and/or spray drying methods. Liquid-assisted grinding with acetone (40.60 μm), ethanol (39.10 μm), and ethyl acetate (31.20 μm) develop micron-sized coarse (D_{50}) crystals whereas spray drying with acetone (5.40 μm), ethanol (5.20 μm), and ethyl acetate (5.10 μm) developed micron-sized fine (D_{50}) crystals suitable for inhalation. Spray-dried smooth surface, non-porous spheres of acetone, ethanol, and ethyl acetate-based crystals showed surface areas of 2.5, 2.6 and 2.5 m^2/g, respectively. During in vitro cytotoxicity study using human lung cancer cell line (Calu-3), DAP: CAF acetone co-crystals showed 99.6 and 96.5% Calu-3 cells viability at 0.1- and 1.0-mM concentrations, respectively. Additionally, during in vitro permeability study using Calu-3 monolayer DAP: CAF acetone co-crystals showed a 2.25-fold improved apparent permeability coefficient for DAP compared to DAP alone. The improved permeation (basolateral to apical) of DAP: CAF acetone co-crystals was mainly accredited to the CAF as the efflux rate of DAP is tenfold lesser in the presence of CAF. Briefly, the spray drying method with judiciously selected co-formers denoted a suitable scheme for inhalation powders with applications in respiratory pathologies. Enhancement of surface morphology and analysis of aerosolization of DAP: CAF co-crystals is essential to study in near future to approve DAP co-crystals use for pulmonary delivery by means of a DPI [176].

Meloxicam is a long-acting, oxicam derivative prescribed mainly for the treatment of rheumatoid arthritis. It is classified under non-steroidal anti-inflammatory drugs and shows important therapeutic actions such as analgesic, anti-inflammatory, and antipyretic activities. In addition, this meloxicam also presented anti-angiogenetic and antioxidant effects and recent literature indicated its potential in therapy of cystic fibrosis or non-small cell lung cancer. But the chronic oral use of meloxicam (7.5 to 15 mg/day) lead to instances of severe liver injury, gastrointestinal ulcers and immune system-associated issues [177]. Thus, Chvatal et al. (2017) fabricated inhalable meloxicam particles for better therapeutic outcomes. Particularly, Chvatal and co-workers utilized high-purity specially synthesized salt form of meloxicam, i.e., meloxicam potassium monohydrate (MPM) to surpass meloxicam low water solubility (0.4 mg/mL at 37 °C) issues, formulation processing, and handling difficulties. MPM: leucine: polyvinyl alcohol (2:4:0.2%) combination with one step co-spray drying method was used to develop the carrier-free DPI. The wrinkled-shaped micro-composites showed particle size ($d_{0.5}$), surface area, and density of 3.21 μm, 2.29 m/g, and 0.153 g/cm^3, respectively. During aerodynamic assessment at 80 L/min flow rate using Breezhaler® inhaler device, micro-composites showed FPF and MMAD of 49.83% and 2.33 μm, respectively. During in silico assessment using the Stochastic Lung Deposition Model (80.22 L/min), micro-composite particles showed superior lung acinar region deposition (48%) with minimum extra-thoracic deposition compared to unprocessed MPM. The carrier-free micro-composites demonstrated a clear 7.12 and twofold improvement in % FPF and lung deposition during the in vitro and in silico pulmonary assessment, respectively, compared to the unprocessed MPM particles. Based on the in vitro and in silico aerodynamic assessment, the developed novel spray-dried micro-composites can be used in the treatment of several pulmonary diseases such as cystic fibrosis and non-small cell lung cancer [178].

1.3.7.6 Crystal Designing with High Shear Agitator

Novel steroid KSR-592 acicular crystals were formulated to improve the aerodynamic properties. Notably, the original KSR-592 (α-form) with plate-like morphology was transformed to β-form with needle-like morphology by agitating (8000 RPM for 15 min) the KSR-592 (α-form) in hexane containing ethanol (5%). The 2D nuclear growth equation denoted the polymorphic transformation kinetics from α- to β-form due to formation of β-form nuclei on the exterior of α-form and successive growth of β-form nuclei to needle-like morphology. Collected β-form crystals were de-aggregated using the chemical blender operational with two blades. The β-form needle-shaped crystals showed particle size (1.8 μm width and 41 μm length) suitable for pulmonary delivery. The Coulter Counter method showed volume mean diameter of 6.2 μm. Inhalation blend was formulated by mixing original α-form crystals and coarse needle-like β-form crystals (0.6 g) with Pharmatose® 325 M

(11.4 g; 60 μm) lactose at 500 RPM for 2 min. Moreover, fine needle-like β-form crystals were mixed with lactose at 1500 RPM for 2 min to obtain a respirable powder blend. Particle size analysis of respirable powder blend using Coulter Counter showed a lower mean diameter (5.8 μm) of drug crystals due to slight disintegration of crystals during the high-speed mixing. For aerodynamic assessment, powder blend (10 mg) was actuated using two different inhaler devices, i.e., Jethaler® (cartridge-based inhaler device equipped with milling chamber) and Diskhaler® (blister-based inhaler device equipped with a unique puncturing mechanism) at the same airflow rate. During the aerodynamic analysis, KSR-592 α- and β-forms alone presented FPF of 4.73 and 39.34%, respectively using Jethaler® while β-form alone with Diskhaler® displayed FPF of 23.62%. In aerosolization study using twin stage impinger lactose blended α- and β-form exhibited FPF of 5.80 and 43.80%, respectively, using Jethaler® whereas 5.40 and 32.90%, respectively, using Diskhaler®. During aerodynamic study using cascade impactor lactose blended β-form exposed superior aerosolization performance compared to α-form. The cascade deposition profile of β-form crystals was less affected by the airflow rate and exhibited more consistent inhalation behavior. Outcomes from this work indicated that β-form adhered to lactose surface were detached more effectively compared to the α-form crystals. In cascade impactor analysis Jethaler® disclosed good aerosolization outcome compared to Diskhaler®. Moreover, during device comparison study, Jethaler® showed efficient disintegration of powder aggregates into primary particles compared to Diskhaler® due to well-integrated milling system of device. Furthermore, lactose-loaded fine and coarse β-form crystals showed FPF of 21.87 and 39.34%, respectively, using Jethaler® which indicated that fine β-form crystals were not effectively detached from the lactose surface due to higher adhesive force between the fine β-form crystals and the lactose. Briefly, the inhalation behavior of KSR-592 was enhanced noticeably by varying the crystal shape from plate to needle [179].

1.3.7.7 Unidirectional Crystal Engineering

Unidirectional crystal growth is highly advantageous due to uniform mean particle size and unidirectional crystals such as fibers, needle-like, or elongated crystals showed good yield during batch or large-scale processing [180]. Unidirectional crystals with narrow aerodynamic size and low geometric standard deviation showed site-specific deposition within the airways. One-directional crystals reached the alveoli region via interception mechanism while compact crystals of similar aerodynamic attributes were easily exhaled unless extended breath holding is applied. Additionally, one-directional crystals hold a larger drug mass compared to compact crystals. Therefore, with one-directional crystals significant dose reduction is also feasible and such crystal habitats offer greater colloidal stability because of low interfacial energy. Chan and Gonda (1988) developed unidirectional cromoglycic acid crystal using precipitation method. Particularly, inhalable elongated cromoglycic acid crystals were fabricated using hydrochloric acid and aqueous cromolyn sodium by precipitation reaction. The obtained mixture was immediately recrystallized using different solvents such as hot water, dimethyl sulfoxide (DMSO), and mixture of water/DMSO. Hot water, methanol, DMSO: water mixture and nonpolar solvents (cyclohexane and benzene) presented flat fibers (242–247 °C), amorphous material, well-shaped rod-shaped crystals (238 to 244 °C), and deformation of crystals, respectively. XRD analysis presented that DMSO attaches loosely to the crystal surface while precipitation with hydrochloric acid exhibited a lower degree of crystallinity. As found in thermal and diffraction analysis cromoglycic acid crystal recrystallization using hot water fulfills all size and shape criteria with purity. Long, slender, needle-shaped cromoglycic acid crystals showed aerodynamic characteristics (0.7 μm; MMAD) suitable for effective pulmonary drug delivery. Briefly, cromoglycic acid crystals with a suitable inhalation device can be delivered into the selected parts of the lung. Future fine tuning in this development process is going to be useful even in clinical practice [181].

1.4 Future Perspective

Crystal engineering has exhibited considerable potential in designing and developing inhalable particles to treat lung diseases effectively. As carefully reviewed and summarized in previous sections, active efforts are invested by material chemists and formulators for the development of different inhalable crystalline solids (Table 1.3). The supersaturation phase is crucial in the crystal engineering process because it ultimately controls the nucleation kinetics and final crystal size distribution [182]. Solution-based crystal engineering techniques are mainly challenged by the liability of the less soluble constituent to supersaturate and crystallize primarily. Also, the high mechanical pressure and complexity of attaining a uniform mixture indicate the main problem for the grinding technique [183]. Similar issues were also found by a few pioneering methods such as hot-melt extrusion, thin film freezing, supercritical fluid technology, spray drying, and resonant acoustic mixing. Thus, various solvents, combinations of solvents, co-formers, polymers, and processing techniques were judiciously explored to obtain inhalable (< 5 μm) crystalline particles with reasonable control over the supersaturation phase and ultimately crystallization processes. Virtual computational screening and phase diagram methods are available to know the experimental supersaturation and crystallization points [184]. Process analytical technologies (PAT) tools can be adapted to define the critical process parameters in robust processes designed to obtain crystalline particles [30, 185]. Undoubtedly, PAT tools also assist in monitoring and maintaining co-crystal integrity throughout the production phases [184]. Furthermore, a novel crystallization process informatics system, i.e., CryPRIN (Loughborough University, UK), can be adopted for better crystallization process understanding. CryPRIN offers real-time monitoring of the crystallization process with temperature and periodic flow of the matrix [186]. CryPRIN can be easily integrated with Raman, Malvern laser diffraction, and XRD to better understand the microstructural properties and crystallization process [187–189]. Such screening strategies are very productive at lab scale and scale-up routes for continuous crystal manufacturing.

Table 1.3 Summary of inhaled crystalline platforms for treating pulmonary diseases

Formulation	Method	Particle properties	Outcome	References
VRC nanocrystalline system	Spray drying	Irregular, dense with 2.7 μm and VRC crystal of 645.86 nm	VNC presented ED, MMAD, and FPF of 90.89%, 3.5 μm, and 76.69%, correspondingly	[81]
VRC microcrystalline system	Spray drying	Irregular, dense with 3.17 μm and VRC crystal of 4800 nm	VMC exhibited ED, MMAD, and FPF of 89.89%, 3.4 μm, and 73.60%, individually	[81]
VRC nanocrystalline system with Aerosil® 200	Spray drying	Irregular, dense with 2.7 μm and VRC crystal of 587.90 nm	Hydrophilic fumed silica treated system offered ED, MMAD, and FPF of 93.49%, 3.1 μm, and 94.37%, correspondingly.	[82]
SX microcrystals	Anti-solvent crystallization (PEG 400 and water)	Thin plate-shaped and 6.59 μm	Microcrystals showed 1.20-fold higher % FPF compared to micronized drug using Cyclohaler® at 60 L/min	[99]
FP microcrystals	Anti-solvent crystallization (PEG 400 and water)	Needle shaped and 6.14 μm	Microcrystals showed more satisfactory % FPF compared to micronized drugs using Cyclohaler® at 60 L/min	[99]
FAV: THP (1:1)	Spray drying	Elongated rod-shaped structures with smoother surfaces	During the aerosolization study using Breezhaler® (60 L/min) satisfactory MMAD (< 5 μm) and FPF (79.3%) were obtained	[126]
Amorphous colistin: Crystalline rifapentine (1:1)	Spray drying	Spherical colistin MPs (2.7 μm) adhered to a surface of elongated rod-shaped rifapentine MPs (3.2 μm)	Combination MPs demonstrated FPF of 76.9% (colistin) and 78.5% (rifapentine) using Aerolizer® at 100 L/min with superior antimicrobial performance	[129]

(continued)

Table 1.3 (continued)

Formulation	Method	Particle properties	Outcome	References
Amorphous verapamil: Crystalline rifapentine (1:3)	Spray drying	Spherical, elongated particles and 1.9 μm	Combined MPs demonstrated FPF of 77.40% (verapamil) and 71.50% (rifapentine) using Osmohaler® device at 100 L/min with superior antimicrobial performance against *Mycobacterium tuberculosis* viz. H37Ra and H37Rv strains	[132]
FP-SX-M	Spray drying (ethanol/water)	Hollow irregular particles with mosaic structure and 2.02 μm	Respirable crystals showed FPF of 54.0% (FP) and 53.8% (SX) at 60 L/min	[133]
BUD-FFD-M	Spray drying (ethanol/water)	Smaller prismatic projections and 2.01 μm	Respirable crystals showed FPF of 56.7% (BUD) and 56.8% (FFD) at 60 L/min	[133]
Moxifloxacin: Trans-cinnamic acid (1:1)	Solution co-crystallization using dichloromethane	Needle-shaped crystalline adduct	Adduct showed a higher FPF of 30.40% using Aerolizer® and slower dissolution and permeation kinetics compared to the moxifloxacin alone	[135]
THP: OA (2:1)	SFD	Porous crystalline particles and 3.03 μm	Stable crystalline particles showed MMAD of 3.03 μmat 60 L/min	[144]
THP-URE (1:1)	Spray drying	Flakes shaped agglomerated and 2 μm	Co-crystals showed FPF of 43.23% using Aerolizer® at 60 L/min	[145]
THP-SAC (1:1)	Spray drying	Irregular and deformed particles with 3 μm	Co-crystals showed FPF of 45.70% using Aerolizer® at 60 L/min	[145]

Table 1.3 (continued)

Formulation	Method	Particle properties	Outcome	References
THP-NIC (1:1)	Spray drying	Irregular and deformed particles with 1.5 μm	Co-crystals showed FPF of 69.98% using Aerolizer® at 60 L/min	[145]
ITZ: Mannitol (1:4)	Spray drying	Spherical, smooth surface and < 5 μm	Amorphous ITZ showed a marked of 12-fold higher bioavailability for itraconazole as compared to raw itraconazole	[151]
ITZ: SA (1:1)	Anti-solvent crystallization	Elongated angular blocks and < 2 μm	Co-crystals showed a marked 24-fold higher bioavailability for ITZ as compared to spray-dried formulation	[151]
ITZ: TA (1:1)	Anti-solvent crystallization	Bulky agglomerates and < 2 μm	Co-crystals showed a marked 19-fold higher bioavailability for ITZ as compared to spray-dried formulation	[151]
ITZ: SUB (1:1)	Rotary evaporation followed by spray drying	–	Inhalable co-crystals showed FPF and MMAD of 64.10% and 2.56 μm using Breezhaler® at 90 L/min	[150]
ITZ: Mannitol (20:80)	HME-based solvent-free fine solid crystal suspension	Irregular-shaped non-spherical and 2.19 μm	Respirable extrudates displayed FPF of 50.69% using Turbospin® at 60 L/min	[149]
BUD crystals	Microfluidic reactor integrated with ultrasonic spray freeze drying (methanol-water)	Plate-like unimodal particles and 8.83 μm	Respirable crystals displayed FPF of ~50% using Aerolizer ® at 100 L/min	[157]
BUD crystals	Microfluidic reactor integrated with ultrasonic spray freeze drying (ethanol–water)	Plate-like unimodal particles and 7.33 μm	Respirable crystals displayed FPF of 47.60% using Aerolizer ® at 100 L/min	[157]

(continued)

Table 1.3 (continued)

Formulation	Method	Particle properties	Outcome	References
BUD crystals	Microfluidic reactor integrated with ultrasonic spray freeze drying (acetone–water)	Rectangular-shaped multimodal and 5.85 μm	Respirable crystals displayed FPF of 54.90% using Aerolizer ® at 100 L/min	[157]
BUD crystalline MPs	Anti-solvent precipitation (acetone–water)	Rectangular crystals and < 5 μm	Respirable crystals displayed FPF of 60% using Aerolizer ® at 60 L/min	[158]
BUD crystalline MPs	Anti-solvent precipitation (methanol–water)	Elliptic particles and < 5 μm	Respirable crystals displayed FPF of 68% using Aerolizer ® at 60 L/min	[158]
Rifapentine crystals	Continuous PFC integrated with spray drying	Small needle-shape	Inhalable crystals showed FPF of 60.20% using Breezhaler® at 99.5 L/min	[162]
BDP crystals	Continuous PFC integrated with spray drying	Small spike-shape (monohydrate form) particles	Inhalable crystals showed FPF of 54% using Breezhaler® at 90 L/min	[162]
Spironolactone: PVP (10:0.1)	Static mixer followed by spray drying	Elongated, smooth surface hollow crystals and 16.10 μm	Crystals showed threefold higher % FPF values as compared to drug alone using Aerolizer® at 60 L/min	[169]
Progesterone microcrystals	Anti-solvent crystallization	Spherical and 3.22 μm	Respirable microcrystals showed FPF and MMAD of 50% and 4.54 μm using Turbuhaler® at 30 L/min	[172]
Diclofenac crystals	Acidic titration followed by jet milling	Smooth, hollow crystals and 3.25 μm	Needle-shaped crystals demonstrated FPF of 75.6% and lowered MMAD (< 2 μm) using high resistance monodose RS01® inhaler at 58.5 L/min	[173]

Table 1.3 (continued)

Formulation	Method	Particle properties	Outcome	References
DAP: CAF (1:1)	Spray drying	Spherical-shaped and 5.4 μm	Co-crystals showed a 2.5-fold improvement in apparent permeability coefficient compared to a drug alone using Calu-3 monolayer.	[176]
Meloxicam potassium: Leucine: Polyvinyl alcohol (2:4:0.2)	Spray drying	Wrinkled-shaped micro-composites and 3.21 μm	Respirable micro-composites showed FPF and MMAD of 49.83% and 2.33 μm using Breezhaler® at 80 L/min	[178]
Steroid KSR-592 (β-form)	Agitating with hexane containing ethanol (5%)	Needle-like crystals and 1.8 μm width with 41 μm length	β-Form crystals showed consistent aerosolization profile and were less affected by the airflow rate as compared to α-form	[179]
Cromoglycic acid crystals	Precipitation reaction	Long, slender, and needle-shaped crystals	Unidirectional crystal growth showed suitable aerodynamic characteristics with MMAD of 0.7 μm	[181]

Among the reported methods, anti-solvent crystallization is one of the most used techniques to formulate inhalable crystalline particles. In anti-solvent crystallization, crystal size distribution is highly influenced by the origin of supersaturation and the nucleation time upon adding anti-solvent to the solution [190]. Additionally, Murnane et al. experimentally showed that high molecular weight PEG could be utilized effectively as solvents for the anti-solvent crystallization of small molecular weight drugs [99, 190]. Briefly, the adequate balance between solvents and co-formers with a selection of apt processing techniques is crucial for continuous crystallization. To date, various drug substances such as BUD, THP, ITZ, and BDP were explored to modify crystalline properties and aerosol outcomes. Additionally, dual combinations of drug substances such as FP/SX and BUD/FFD were also processed through crystal engineering techniques to revise the deagglomeration and dispersion of dry powders. Moreover, a few novel combinations such as meloxicam potassium/leucine/polyvinyl alcohol and SP/PVP were also used to transform the morphology of crystalline micro-composites and the percentage of respirable fractions. It is well known that the particle shape and size of inhaled drug substances significantly influenced the aerosol performance [191]. So, crystal engineering is an apt platform to control the particle morphology of the drug substance. Particularly for pulmonary delivery, it assists in producing several different morphologies such as flakes, plate, elongated, needle, spherical, wrinkled, hollow, prismatic, and rectangular-shaped particles to improve powder deagglomeration profile. In contrast, aggregates (< 5 μm), nanostructured aggregate, unimodal particles, and micro-composite particles can tune the drug deposition pattern.

In the above-summarized research articles, most authors have thoroughly discussed anti-solvent variety, co-former selection, particle morphology influence, and formulation development strategies; however, merely touched on the facts related to inhaler device techniques, delivered dose uniformity and in vivo pulmonary consequences. To explore the absolute potential of crystal engineering, it is also essential to study inhaler device selection, aerosolization assessment techniques, and pharmacokinetic/pharmacodynamic investigations. Furthermore, it is imperative to comprehend whether the co-former selection exerts any negative or positive impacts on the in vivo application of drug substances. Besides, a deep understanding of toxicological facts, pulmonary clearance rate, and the mechanism is essential for establishing the safety profile of crystalline drug products. Concisely, the carefully validated crystal engineering process aligned with rationalized manufacturing options, aerodynamic assessment, and biological measurement will positively contribute to the successful development of inhalable crystalline drug particles.

1.5 Conclusion

The crystal engineering strategies show good prospects to design inhaled powder and particles for deep lung delivery. Crystal engineering can easily tune the solid-state morphology, surface energetics, and mean particle size of a drug substance. It assists in designing co-crystals, polymorph, and solvate with different microstructural properties and aerodynamic diameter for better aerosolization outcomes. Overall, the solid-state synthesis schemes for crystal engineering seem appealing compared to solvent-based crystallization methods due to solvent free, few processing steps, reduced production time, and cost-efficient nature. Continuous crystallization techniques such as hot-melt extrusion and microfluidics with green synthesis principles are promising strategies for industrial applications. Even though numerous approaches have revealed hopeful outcomes, issues linked to large-scale crystal synthesis and long-term stability in the presence of carriers need to be addressed for better product performance. Additionally, active efforts are required for in vitro dissolution techniques, in vitro aerosolization techniques such as delivered dose uniformity and drug detachments process, and the in vivo pharmacokinetics models to design robust therapy regimens in the management of respiratory disorders.

References

1. Braga D, Grepioni F, Gavezzotti A, Re Bernstein J. Crystal engineering in the regulatory and patent literature of pharmaceutical solid forms. Cryst Growth Des. 2017;17(3):933–9. (Editorial)
2. Desiraju GR. Crystal engineering: the design of organic solids. Amsterdam; New York: Elsevier; 1989.
3. Pepinsky R. Crystal engineering-new concept in crystallography. Phys Rev. 1955;100:952.
4. Kavanagh ON, Croker DM, Walker GM, Zaworotko MJ. Pharmaceutical co-crystals: from serendipity to design to application. Drug Discov Today. 2019;24(3):796–804.
5. Kamble RN, Bothiraja C, Mehta PP, Varghese V. Synthesis, solid state characterization and antifungal activity of ketoconazole cocrystals. J Pharm Investig. 2018;48(5):541–9.
6. Mohite R, Mehta P, Arulmozhi S, Kamble R, Pawar A, Bothiraja C. Synthesis of fisetin co-crystals with caffeine and nicotinamide using the cooling crystallization technique: biopharmaceutical studies. New J Chem. 2019;43(34):13471–9.
7. Desiraju GR. Crystal engineering: a holistic view. Angew Chem Int Ed Engl. 2007;46(44):8342–56.
8. Aakeröy CB, Sinha AS. Chapter 1: Co-crystals: introduction and scope. In: Co-crystals: preparation, characterization and applications; 2018. p. 1–32.
9. Nwambaekwe KC, John-Denk V, Douman SF, Mathumba P, Yussuf ST, Uhuo OV, Ekwere PI, Iwuoha EI. Crystal engineering and thin-film deposition strategies towards improving the performance of kesterite photovoltaic cell. J Mater Res Technol. 2021;12:1252–87.
10. Jia S, Gao Z, Tian N, Li Z, Gong J, Wang J, Rohani S. Review of melt crystallization in the pharmaceutical field, towards crystal engineering and continuous process development. Chem Eng Res Des. 2020;166:268–80.
11. Saha S, Mishra MK, Reddy CM, Desiraju GR. From molecules to interactions to crystal engineering: mechanical properties of organic solids. Acc Chem Res. 2018;51(11):2957–67.
12. Dhir A, Rajkumar M, Desiraju GR. Co-crystal hydrate of Bandrowski's base and clotrimazole: a prospective ingredient for hair dye formulations. CrystEngComm. 2021;23(32):5421–4.
13. Delori A, Urquhart AJ, Oswald ID. Supramolecular hair dyes: a new application of co-crystallization. CrystEngComm. 2016;18(28):5360–4.
14. Yu P, Zhen Y, Dong H, Hu W. Crystal engineering of organic optoelectronic materials. Chem. 2019;5(11):2814–53.
15. Nangia AK, Desiraju GR. Crystal engineering: an outlook for the future. Angew Chem Int Ed Engl. 2019;58(13):4100–7.
16. Bolla G, Sarma B, Nangia AK. Crystal engineering and pharmaceutical crystallization. In: Hot topics in crystal engineering. Elsevier; 2021. p. 157–229.
17. Weng J, Huang Y, Hao D, Ji Y. Recent advances of pharmaceutical crystallization theories. Chin J Chem Eng. 2020;28(4):935–48.
18. Nugrahani I, Fisandra F, Horikawa A, Uekusa H. New sodium Mefenamate - nicotinamide multicomponent crystal development to modulate solubility and dissolution: preparation, structural, and performance study. J Pharm Sci. 2021;110(9):3246–60.
19. Zaini E, Fitriani L, Sari RY, Rosaini H, Horikawa A, Uekusa H. Multicomponent crystal of mefenamic acid and N-methyl-d-Glucamine: crystal structures and dissolution study. J Pharm Sci. 2019 Jul;108(7):2341–8.
20. Duggirala NK, Smith AJ, Wojtas Ł, Shytle RD, Zaworotko MJ. Physical stability enhancement and pharmacokinetics of a lithium ionic co-crystal with glucose. Cryst Growth Des. 2014;14(11):6135–42.
21. Reddy CM, Krishna GR, Ghosh S. Mechanical properties of molecular crystals-applications to crystal engineering. CrystEngComm. 2010;12(8):2296–314.
22. Luo Y, Chen S, Zhou J, Chen J, Tian L, Gao W, Zhang Y, Ma A, Li L, Zhou Z. Luteolin co-crystals: characterization, evaluation of solubility, oral bioavailability and theoretical calculation. J Drug Deliv Sci Technol. 2019;50:248–54.

23. Pi J, Wang S, Li W, Kebebe D, Zhang Y, Zhang B, Qi D, Guo P, Li N, Liu Z. A nano-cocrystal strategy to improve the dissolution rate and oral bioavailability of baicalein. Asian J Pharm Sci. 2019;14(2):154–64.
24. Zhang YX, Wang LY, Dai JK, Liu F, Li YT, Wu ZY, Yan CW. The comparative study of co-crystal/salt in simultaneously improving solubility and permeability of acetazolamide. J Mol Struct. 2019;1184:225–32.
25. Liu L, Zou D, Zhang Y, Zhang Q, Feng Y, Guo Y, Liu Y, Zhang X, Cheng G, Wang C, Zhang Y. Pharmaceutical salts/co-crystals of enoxacin with dicarboxylic acids: enhancing in vitro antibacterial activity of enoxacin by improving the solubility and permeability. Eur J Pharm Biopharm. 2020;154:62–73.
26. Mannava MC, Gunnam A, Lodagekar A, Shastri NR, Nangia AK, Solomon KA. Enhanced solubility, permeability, and tabletability of nicorandil by salt and co-crystal formation. CrystEngComm. 2021;23(1):227–37.
27. Zhao X, Li Q, Wang C, Hu S, He X, Sun CC. Simultaneous taste-masking and oral bioavailability enhancement of ligustrazine by forming sweet salts. Int J Pharm. 2020;577:119089.
28. Aitipamula S, Wong ABH, Kanaujia P. Evaluating suspension formulations of theophylline cocrystals with artificial sweeteners. J Pharm Sci. 2018;107(2):604–11.
29. Serrano DR, O'Connell P, Paluch KJ, Walsh D, Healy AM. Co-crystal habit engineering to improve drug dissolution and alter derived powder properties. J Pharm Pharmacol. 2016;68(5):665–77.
30. Yousef MA, Vangala VR. Pharmaceutical co-crystals: molecules, crystals, formulations, medicines. Cryst Growth Des. 2019;19(12):7420–38.
31. Duggirala NK, Perry ML, Almarsson Ö, Zaworotko MJ. Pharmaceutical co-crystals: along the path to improved medicines. Chem Commun. 2016;52(4):640–55.
32. Shinozaki T, Ono M, Higashi K, Moribe K. A novel drug-drug cocrystal of levofloxacin and Metacetamol: reduced hygroscopicity and improved photostability of levofloxacin. J Pharm Sci. 2019;108(7):2383–90.
33. Karimi-Jafari M, Padrela L, Walker GM, Croker DM. Creating co-crystals: a review of pharmaceutical co-crystal preparation routes and applications. Cryst Growth Des. 2018;18(10):6370–87.
34. Javoor MG, Mondal PK, Chopra D. Cocrystals: a review of recent trends in pharmaceutical and material science applications. Mater Sci Res India. 2017;14(1):09–18.
35. Fontana F, Figueiredo P, Zhang P, Hirvonen JT, Liu D, Santos HA. Production of pure drug nanocrystals and nano co-crystals by confinement methods. Adv Drug Deliv Rev. 2018;131:3–21.
36. Lusi M. Engineering crystal properties through solid solutions. Cryst Growth Des. 2018;18(6):3704–12.
37. Nia NY, Giordano F, Zendehdel M, Cinà L, Palma AL, Medaglia PG, Zakeeruddin SM, Grätzel M, Di Carlo A. Solution-based heteroepitaxial growth of stable mixed cation/anion hybrid perovskite thin film under ambient condition via a scalable crystal engineering approach. Nano Energy. 2020;69:104441.
38. Pawar N, Saha A, Nandan N, Parambil JV. Solution co-crystallization: a scalable approach for co-crystal production. Crystals. 2021;11(3):303.
39. Trask AV, Jones W. Crystal engineering of organic co-crystals by the solid state grinding approach. In: Organic solid state reactions; 2005. p. 41–70.
40. Gajda M, Nartowski KP, Pluta J, Karolewicz B. Continuous, one-step synthesis of pharmaceutical co-crystals via hot-melt extrusion from neat to matrix-assisted processing–state of the art. Int J Pharm. 2019;558:426–40.
41. Rehder S, Christensen NP, Rantanen J, Rades T, Leopold CS. High-shear granulation as a manufacturing method for co-crystal granules. Eur J Pharm Biopharm. 2013;85(3):1019–30.
42. Cuadra IA, Cabañas A, Cheda JA, Türk M, Pando C. Cocrystallization of the anticancer drug 5-fluorouracil and coformers urea, thiourea or pyrazinamide using supercritical CO2 as an antisolvent (SAS) and as a solvent (CSS). J Supercrit Fluids. 2020;160:104813.

43. Fang L, Gao Z, Wu S, Jia S, Wang J, Rohani S, Gong J. Ultrasound-assisted solution crystallization of fotagliptin benzoate: process intensification and crystal product optimization. Ultrason Sonochem. 2021;76:105634.
44. Hasa D, Schneider Rauber G, Voinovich D, Jones W. Cocrystal formation through mechanochemistry: from neat and liquid-assisted grinding to polymer-assisted grinding. Angew Chem. 2015;127(25):7479–83.
45. Friščić T, Trask AV, Jones W, Motherwell WS. Screening for inclusion compounds and systematic construction of three-component solids by liquid-assisted grinding. Angew Chem. 2006;118(45):7708–12.
46. Edwards AD, Shekunov BY, Kordikowski A, Forbes RT, York P. Crystallization of pure anhydrous polymorphs of carbamazepine by solution enhanced dispersion with supercritical fluids (SEDS™). J Pharm Sci. 2001;90(8):1115–24.
47. Walsh D, Serrano DR, Worku ZA, Norris BA, Healy AM. Production of co-crystals in an excipient matrix by spray drying. Int J Pharm. 2018;536(1):467–77.
48. Eddleston MD, Patel B, Day GM, Jones W. Cocrystallization by freeze-drying: preparation of novel multicomponent crystal forms. Cryst Growth Des. 2013;13(10):4599–606.
49. Chan HK, Kwok PC. Novel particle production technologies for inhalation products. Inhalation Drug Delivery: Techniques and Products. 2013;11:47–62.
50. Goyal S, Thorson MR, Zhang GG, Gong Y, Kenis PJ. Microfluidic approach to co-crystal screening of pharmaceutical parent compounds. Cryst Growth Des. 2012;12(12):6023–34.
51. Shi HH, Xiao Y, Ferguson S, Huang X, Wang N, Hao HX. Progress of crystallization in microfluidic devices. Lab Chip. 2017;17(13):2167–85.
52. Boksa K, Otte A, Pinal R. Matrix-assisted co-crystallization (MAC) simultaneous production and formulation of pharmaceutical co-crystals by hot-melt extrusion. J Pharm Sci. 2014;103(9):2904–10.
53. Titapiwatanakun V, Basit AW, Gaisford S. A new method for producing pharmaceutical co-crystals: laser irradiation of powder blends. Cryst Growth Des. 2016;16(6):3307–12.
54. Mehta PP, Pawar VS. Electrospun nanofiber scaffolds: technology and applications. In: Applications of nanocomposite materials in drug delivery. Woodhead Publishing; 2018. p. 509–73.
55. Urbanus J, Roelands CM, Mazurek J, Verdoes D, ter Horst JH. Electrochemically induced co-crystallization for product removal. CrystEngComm. 2011;13(8):2817–9.
56. am Ende DJ, Anderson SR, Salan JS. Development and scale-up of co-crystals using resonant acoustic mixing. Org Process Res Dev. 2014;18(2):331–41.
57. Mehta P. Dry powder inhalers: a focus on advancements in novel drug delivery systems. J Drug Deliv. 2016;2016:8290963.
58. Douroumis D, Ross SA, Nokhodchi A. Advanced methodologies for co-crystal synthesis. Adv Drug Deliv Rev. 2017;117:178–95.
59. Luque de Castro MD, Priego-Capote F. Ultrasound-assisted crystallization (sonocrystallization). Ultrason Sonochem. 2007;14(6):717–24.
60. Pasquali I, Bettini R, Giordano F. Supercritical fluid technologies: an innovative approach for manipulating the solid-state of pharmaceuticals. Adv Drug Deliv Rev. 2008;60(3):399–410.
61. Giry K, Péan JM, Giraud L, Marsas S, Rolland H, Wüthrich P. Drug/lactose co-micronization by jet milling to improve aerosolization properties of a powder for inhalation. Int J Pharm. 2006;321(1–2):162–6.
62. Mehta PP, Bothiraja C, Kadam SS, Pawar AP. Effect of USP induction ports, glass sampling apparatus, and inhaler device resistance on aerodynamic patterns of fluticasone propionate loaded engineered mannitol microparticles. AAPS PharmSciTech. 2019;20(5):197.
63. Young PM, Chan HK, Chiou H, Edge S, Tee TH, Traini D. The influence of mechanical processing of dry powder inhaler carriers on drug aerosolization performance. J Pharm Sci. 2007;96(5):1331–41.
64. Mehta PP, Dhapte-Pawar V. Role of surfactants in pulmonary drug delivery. In: Green sustainable process for chemical and environmental engineering and science. Academic Press; 2022. p. 559–77.

65. Mehta PP, Pawar AP, Mahadik KR, Kadam SS, Dhapte-Pawar V. Dry powder coating techniques and role of force controlling agents in aerosol. In: Polymer coatings: technology and applications; 2020. p. 41–74.
66. Mehta PP, Kadam SS, Pawar AP. Influence of modified induction port, modified DUSA assembly and device air-inlet geometry on the aerosolization pattern of a dry powder inhaler. J Drug Deliv Sci Technol. 2020;55:101416.
67. Lau M, Young PM, Traini D. A review of co-milling techniques for the production of high dose dry powder inhaler formulation. Drug Dev Ind Pharm. 2017;43(8):1229–38.
68. Chaurasiya B, Zhao YY. Dry powder for pulmonary delivery: a comprehensive review. Pharmaceutics. 2021;13(1):31.
69. Jong T, Li J, Morton DA, Zhou QT, Larson I. Investigation of the changes in aerosolization behavior between the jet-milled and spray-dried colistin powders through surface energy characterization. J Pharm Sci. 2016;105(3):1156–63.
70. Loh ZH, Samanta AK, Heng PW. Overview of milling techniques for improving the solubility of poorly water-soluble drugs. Asian J Pharmaceut Sci. 2015;10(4):255–74.
71. Babu NJ, Nangia A. Solubility advantage of amorphous drugs and pharmaceutical cocrystals. Cryst Growth Des. 2011;11(7):2662–79.
72. Ahmadi Tehrani A, Omranpoor MM, Vatanara A, Seyedabadi M, Ramezani V. Formation of nanosuspensions in bottom-up approach: theories and optimization. Daru. 2019;27(1):451–73.
73. Desai PP, Mapara SS, Patravale VB. Crystal engineering: upcoming paradigm for efficacious pulmonary drug delivery. Curr Pharm Des. 2018;24(21):2438–55.
74. Parmar PK, Wadhawan J, Bansal AK. Pharmaceutical nanocrystals: a promising approach for improved topical drug delivery. Drug Discov Today. 2021;26(10):2329–49.
75. Chen Z, Wu W, Lu Y. What is the future for nanocrystal-based drug-delivery systems? Ther Deliv. 2020;11(4):225–9.
76. Shete G, Bansal AK. NanoCrySP technology for generation of drug nanocrystals: translational aspects and business potential. Drug Deliv Transl Res. 2016;6(4):392–8.
77. Girdhar A, Thakur PS, Sheokand S, Bansal AK. Permeability behavior of nanocrystalline solid dispersion of dipyridamole generated using NanoCrySP technology. Pharmaceutics. 2018;10(3):160.
78. Shete G, Pawar YB, Thanki K, Jain S, Bansal AK. Oral bioavailability and pharmacodynamic activity of hesperetin nanocrystals generated using a novel bottom-up technology. Mol Pharm. 2015;12(4):1158–70.
79. Shete G, Modi SR, Bansal AK. Effect of mannitol on nucleation and crystal growth of amorphous flavonoids: implications on the formation of nanocrystalline solid dispersion. J Pharm Sci. 2015;104(11):3789–97.
80. Bhatt V, Shete G, Bansal AK. Mechanism of generation of drug nanocrystals in celecoxib: mannitol nanocrystalline solid dispersion. Int J Pharm. 2015;495(1):132–9.
81. Kaur A, Dubey G, Sharma N, Pant R, Bharatam PV, Tikoo K, Bansal AK. High dose nanocrystalline solid dispersion powder of voriconazole for inhalation. Int J Pharm. 2022;622:121827.
82. Kaur A, Bansal AK. Optimization of particle properties of nanocrystalline solid dispersion based dry powder for inhalation of voriconazole. J Pharm Sci. 2022;111(9):2592–605.
83. Demopoulos GP. Aqueous precipitation and crystallization for the production of particulate solids with desired properties. Hydrometallurgy. 2009;96(3):199–214.
84. MacFhionnghaile P, Svoboda V, McGinty J, Nordon A, Sefcik J. Crystallization diagram for antisolvent crystallization of lactose: using design of experiments to investigate continuous mixing-induced supersaturation. Cryst Growth Des. 2017;17(5):2611–21.
85. Thakur AK, Kumar R, Kumar VV, Kumar A, Gaurav GK, Gupta KN. A critical review on thermodynamic and hydrodynamic modeling and simulation of liquid antisolvent crystallization of pharmaceutical compounds. J Mol Liq. 2022;362:119663.
86. Ostergaard I, Szilagyi B, De Diego HL, Qu H, Nagy ZK. Polymorphic control and scale-up strategy for antisolvent crystallization using direct nucleation control. Cryst Growth Des. 2020;20(4):2683–97.

87. Öner M, Stocks SM, Abildskov J, Sin G. Scale-up modeling of a pharmaceutical antisolvent crystallization via a hybrid method of computational fluid dynamics and compartmental modeling. In: Computer Aided Chemical Engineering, vol. 46. Elsevier; 2019. p. 709–14.
88. Johnson MD, Burcham CL, May SA, Calvin JR, McClary Groh J, Myers SS, Webster LP, Roberts JC, Reddy VR, Luciani CV, Corrigan AP. API continuous cooling and antisolvent crystallization for kinetic impurity rejection in cGMP manufacturing. Org Process Res Dev. 2021;25(6):1284–351.
89. Simone E, Cenzato MV, Nagy ZK. A study on the effect of the polymeric additive HPMC on morphology and polymorphism of ortho-aminobenzoic acid crystals. J Cryst Growth. 2016;446:50–9.
90. Sadeghi F, Soleimanian Z, Hadizadeh F, Shirafkan A, Kamali H, Garekani HA. Anti-solvent crystallization of celecoxib in the presence of PVP for enhancing the dissolution rate: comparison of water and supercritical CO2 as two antisolvents. Chem Eng Res Des. 2022;177:741–50.
91. Li J, Yang R, Zeng T, Hu J, Tang W, Liu Z, Gong L. Preparation and growth mechanism of micro spherical ammonium dinitramide crystal based on ultrasound-assisted solvent-antisolvent method. Ultrason Sonochem. 2021;78:105716.
92. Kapourani A, Tzakri T, Valkanioti V, Kontogiannopoulos KN, Barmpalexis P. Drug crystal growth in ternary amorphous solid dispersions: effect of surfactants and polymeric matrix-carriers. Int J Pharmaceut X. 2021;3:100086.
93. Som S, Singh SK, Khatik GL, Kapoor B, Gulati M, Kuppusamy G, Anandhakrishnan NK, Kumar B, Yadav AK, Kumar R, Singh PK, Khursheed R, Kumar R, Pandey NK, Jyoti J, Mohanta S, Porwal O. Quality by design-based crystallization of curcumin using liquid anti-solvent precipitation: micromeritic, biopharmaceutical, and stability aspects. Assay Drug Dev Technol. 2020 Jan;18(1):11–33.
94. Wang X, Ponder CS, Kirwan DJ. Low molecular weight poly (ethylene glycol) as an environmentally benign solvent for pharmaceutical crystallization and precipitation. Cryst Growth Des. 2005;5(1):85–92.
95. Veiseh O, Gunn JW, Zhang M. Design and fabrication of magnetic nanoparticles for targeted drug delivery and imaging. Adv Drug Deliv Rev. 2010;62(3):284–304.
96. Murnane D, Marriott C, Martin GP. Comparison of salmeterol xinafoate microparticle production by conventional and novel antisolvent crystallization. Eur J Pharm Biopharm. 2008;69(1):94–105.
97. Murnane D, Marriott C, Martin GP. Developing an environmentally benign process for the production of microparticles: amphiphilic crystallization. Eur J Pharm Biopharm. 2008;69(1):72–82.
98. Murnane D, Marriott C, Martin GP. In situ and ex situ analysis of salmeterol xinafoate microcrystal formation from poly (ethylene glycol) 400– water co-solvent mixtures. Crystal Growth Design. 2008;8(6):1855–62.
99. Murnane D, Martin GP, Marriott C. Dry powder formulations for inhalation of fluticasone propionate and salmeterol xinafoate microcrystals. J Pharm Sci. 2009;98(2):503–15.
100. Tervonen T, Martinez FJ, Hanania NA, Heidenreich S, Eudicone JM, Gilbert I. Dual-combination maintenance inhaler preferences in asthma and chronic obstructive pulmonary disease: a patient-centered benefit-risk assessment. Respir Med. 2021;176:106278.
101. Mehta P, Bothiraja C, Kadam S, Pawar A. Potential of dry powder inhalers for tuberculosis therapy: facts, fidelity and future. Artific Cells Nanomed Biotechnol. 2018;46(sup3):S791–806.
102. Mehta PP, Dhapte-Pawar VS. Novel and evolving therapies for COVID-19 related pulmonary complications. Am J Med Sci. 2021;361(5):557–66.
103. Salem A, Khanfar E, Nagy S, Széchenyi A. Cocrystals of tuberculosis antibiotics: challenges and missed opportunities. Int J Pharm. 2022;623:121924.
104. Wang X, Du S, Zhang R, Jia X, Yang T, Zhang X. Drug-drug cocrystals: opportunities and challenges. Asian J Pharm Sci. 2021;16(3):307–17.
105. Banerjee M, Nimkar K, Naik S, Patravale V. Unlocking the potential of drug-drug cocrystals - a comprehensive review. J Control Release. 2022;348:456–69.

106. Darwish S, Zeglinski J, Krishna GR, Shaikh R, Khraisheh M, Walker GM, Croker DM. A new 1: 1 drug-drug cocrystal of theophylline and aspirin: discovery, characterization, and construction of ternary phase diagrams. Cryst Growth Des. 2018;18(12):7526–32.
107. Aitipamula S, Wong AB, Chow PS, Tan RB. Cocrystallization with flufenamic acid: comparison of physicochemical properties of two pharmaceutical cocrystals. CrystEngComm. 2014;16(26):5793–801.
108. Surov AO, Voronin AP, Manin AN, Manin NG, Kuzmina LG, Churakov AV, Perlovich GL. Pharmaceutical cocrystals of diflunisal and diclofenac with theophylline. Mol Pharm. 2014;11(10):3707–15.
109. Bommaka MK, Mannava MC, Suresh K, Gunnam A, Nangia A. Entacapone: improving aqueous solubility, diffusion permeability, and cocrystal stability with theophylline. Cryst Growth Des. 2018;18(10):6061–9.
110. Saikia B, Sultana N, Kaushik T, Sarma B. Engineering a remedy to improve phase stability of famotidine under physiological ph environments. Cryst Growth Des. 2019;19(11):6472–81.
111. Yeh KL, Lee T. Intensified crystallization processes for 1: 1 drug–drug cocrystals of sulfathiazole–theophylline, and sulfathiazole–sulfanilamide. Cryst Growth Des. 2018;18(3):1339–49.
112. Kakkar S, Bhattacharya B, Reddy CM, Ghosh S. Tuning mechanical behaviour by controlling the structure of a series of theophylline co-crystals. CrystEngComm. 2018;20(8):1101–9.
113. Grobelny P, Mukherjee A, Desiraju GR. Drug-drug co-crystals: temperature-dependent proton mobility in the molecular complex of isoniazid with 4-aminosalicylic acid. CrystEngComm. 2011;13(13):4358–64.
114. Liu F, Song Y, Liu YN, Li YT, Wu ZY, Yan CW. Drug-bridge-drug ternary cocrystallization strategy for antituberculosis drugs combination. Cryst Growth Des. 2018;18(3):1283–6.
115. Souza MS, Diniz LF, Vogt L, Carvalho PS Jr, D'vries RF, Ellena J. Mechanochemical synthesis of a multicomponent solid form: the case of 5-fluorocytosine isoniazid codrug. Cryst Growth Des. 2018;18(9):5202–9.
116. Banik M, Gopi SP, Ganguly S, Desiraju GR. Cocrystal and salt forms of furosemide: solubility and diffusion variations. Cryst Growth Des. 2016;16(9):5418–28.
117. Gopi SP, Banik M, Desiraju GR. New cocrystals of hydrochlorothiazide: optimizing solubility and membrane diffusivity. Cryst Growth Des. 2017;17(1):308–16.
118. Aitipamula S, Wong AB, Chow PS, Tan RB. Novel solid forms of the anti-tuberculosis drug, isoniazid: ternary and polymorphic cocrystals. CrystEngComm. 2013;15(29):5877–87.
119. Wang JR, Ye C, Mei X. Structural and physicochemical aspects of hydrochlorothiazide co-crystals. CrystEngComm. 2014;16(30):6996–7003.
120. Sanphui P, Babu NJ, Nangia A. Temozolomide cocrystals with carboxamide coformers. Cryst Growth Des. 2013;13(5):2208–19.
121. Sarcevica I, Orola L, Belyakov S, Veidis MV. Spontaneous cocrystal hydrate formation in the solid state: crystal structure aspects and kinetics. New J Chem. 2013;37(10):2978–82.
122. Drozd KV, Manin AN, Churakov AV, Perlovich GL. Novel drug–drug cocrystals of carbamazepine with Para-aminosalicylic acid: screening, crystal structures and comparative study of carbamazepine cocrystal formation thermodynamics. CrystEngComm. 2017;19(30):4273–86.
123. Bernasconi D, Bordignon S, Rossi F, Priola E, Nervi C, Gobetto R, Voinovich D, Hasa D, Duong NT, Nishiyama Y, Chierotti MR. Selective synthesis of a salt and a cocrystal of the ethionamide–salicylic acid system. Cryst Growth Des. 2019;20(2):906–15.
124. Agrawal U, Raju R, Udwadia ZF. Favipiravir: a new and emerging antiviral option in COVID-19. Med J Armed Forces India. 2020;76(4):370–6.
125. Tulbah AS, Lee WH. Physicochemical characteristics and in vitro toxicity/anti-SARS-CoV-2 activity of favipiravir solid lipid nanoparticles (SLNs). Pharmaceuticals (Basel). 2021;14(10):1059.
126. Wong SN, Weng J, Ip I, Chen R, Lakerveld R, Telford R, Blagden N, Scowen IJ, Chow SF. Rational development of a carrier-free dry powder inhalation formulation for respiratory viral infections via quality by design: a drug-drug cocrystal of favipiravir and theophylline. Pharmaceutics. 2022;14(2):300.

127. Ahn JM, Kassees K, Lee TK, Manandhar B, Yousif AM. Strategy and tactics for designing analogs: biochemical characterization of the large molecules. In: Paulo A, Castillo CC, Neidle III S, editors. Comprehensive medicinal chemistry III. Elsevier; 2017.
128. Garonzik SM, Li J, Thamlikitkul V, Paterson DL, Shoham S, Jacob J, Silveira FP, Forrest A, Nation RL. Population pharmacokinetics of colistin methanesulfonate and formed colistin in critically ill patients from a multicenter study provide dosing suggestions for various categories of patients. Antimicrob Agents Chemother. 2011;55(7):3284–94.
129. Zhou QT, Sun SP, Chan JG, et al. Novel inhaled combination powder containing amorphous colistin and crystalline rifapentine with enhanced antimicrobial activities against planktonic cells and biofilm of Pseudomonas aeruginosa for respiratory infections. Mol Pharm. 2015;12(8):2594–603.
130. Laws M, Jin P, Rahman KM. Efflux pumps in mycobacterium tuberculosis and their inhibition to tackle antimicrobial resistance. Trends Microbiol. 2022;30(1):57–68.
131. Gupta S, Tyagi S, Almeida DV, Maiga MC, Ammerman NC, Bishai WR. Acceleration of tuberculosis treatment by adjunctive therapy with verapamil as an efflux inhibitor. Am J Respir Crit Care Med. 2013;188(5):600–7.
132. Parumasivam T, Chan JG, Pang A, et al. In vitro evaluation of inhalable verapamil-rifapentine particles for tuberculosis therapy. Mol Pharm. 2016;13(3):979–89.
133. Kumon M, Kwok PC, Adi H, Heng D, Chan HK. Can low-dose combination products for inhalation be formulated in single crystalline particles? Eur J Pharm Sci. 2010;40(1):16–24.
134. Xu P, Chen H, Xu J, Wu M, Zhu X, Wang F, Chen S, Xu J. Moxifloxacin is an effective and safe candidate agent for tuberculosis treatment: a meta-analysis. Int J Infect Dis. 2017;60:35–41.
135. Eedara BB, Tucker IG, Zujovic ZD, Rades T, Price JR, Das SC. Crystalline adduct of moxifloxacin with trans-cinnamic acid to reduce the aqueous solubility and dissolution rate for improved residence time in the lungs. Eur J Pharm Sci. 2019;136:104961.
136. Dubuis E, Wortley MA, Grace MS, Maher SA, Adcock JJ, Birrell MA, Belvisi MG. Theophylline inhibits the cough reflex through a novel mechanism of action. J Allergy Clin Immunol. 2014;133(6):1588–98.
137. Wang L, Li S, Xu X, Xu X, Wang Q, Li D, Zhang H. Drug-drug cocrystals of theophylline with quercetin. J Drug Deliv Sci Technol. 2022;70:103228.
138. Li P, Chu Y, Wang L, Wenslow RM, Yu K, Zhang H, Deng Z. Structure determination of the theophylline–nicotinamide cocrystal: a combined powder XRD, 1D solid-state NMR, and theoretical calculation study. CrystEngComm. 2014;16(15):3141–7.
139. Zhang S, Chen H, Rasmuson ÅC. Thermodynamics and crystallization of a theophylline–salicylic acid cocrystal. CrystEngComm. 2015;17(22):4125–35.
140. Abourahma H, Urban JM, Morozowich N, Chan B. Examining the robustness of a theophylline cocrystal during grinding with additives. CrystEngComm. 2012;14(19):6163–9.
141. Goud NR, Khan RA, Nangia A. Modulating the solubility of sulfacetamide by means of cocrystals. CrystEngComm. 2014;16(26):5859–69.
142. Aitipamula S, Chow PS, Tan RB. Theophylline–gentisic acid (1/1). Acta Crystallogr Sect E: Struct Rep Online. 2009;65(9):o2126–7.
143. Zhu B, Zhang Q, Wang JR, Mei X. Cocrystals of baicalein with higher solubility and enhanced bioavailability. Cryst Growth Des. 2017;17(4):1893–901.
144. Tanaka R, Hattori Y, Otsuka M, Ashizawa K. Application of spray freeze drying to theophylline-oxalic acid co-crystal engineering for inhaled dry powder technology. Drug Dev Ind Pharm. 2020;46(2):179–87.
145. Alhalaweh A, Kaialy W, Buckton G, Gill H, Nokhodchi A, Velaga SP. Theophylline cocrystals prepared by spray drying: physicochemical properties and aerosolization performance. AAPS PharmSciTech. 2013;14(1):265–76.
146. Shadambikar G, Marathe S, Ji N, Almutairi M, Bandari S, Zhang F, Chougule M, Repka M. Formulation development of itraconazole PEGylated nano-lipid carriers for pulmonary aspergillosis using hot-melt extrusion technology. International Journal of Pharmaceutics: X. 2021;3:100074.

147. Pardeike J, Weber S, Zarfl HP, Pagitz M, Zimmer A. Itraconazole-loaded nanostructured lipid carriers (NLC) for pulmonary treatment of aspergillosis in falcons. Eur J Pharm Biopharm. 2016;108:269–76.
148. Pardeike J, Weber S, Haber T, Wagner J, Zarfl HP, Plank H, Zimmer A. Development of an itraconazole-loaded nanostructured lipid carrier (NLC) formulation for pulmonary application. Int J Pharm. 2011;419(1–2):329–38.
149. Lin L, Quan G, Peng T, et al. Development of fine solid-crystal suspension with enhanced solubility, stability, and aerosolization performance for dry powder inhalation. Int J Pharm. 2017;533(1):84–92.
150. Weng J, Wong SN, Xu X, Xuan B, Wang C, Chen R, Sun CC, Lakerveld R, Kwok PC, Chow SF. Cocrystal engineering of itraconazole with suberic acid via rotary evaporation and spray drying. Cryst Growth Des. 2019;19(5):2736–45.
151. Karashima M, Sano N, Yamamoto S, et al. Enhanced pulmonary absorption of poorly soluble itraconazole by micronized co-crystal dry powder formulations. Eur J Pharm Biopharm. 2017;115:65–72.
152. Soltani F, Kamali H, Akhgari A, Garekani HA, Nokhodchi A, Sadeghi F. Different trends for preparation of budesonide pellets with enhanced dissolution rate. Adv Powder Technol. 2022;33(8):103684.
153. Cimato A, Facorro G, Sarrasague MM. Budesonide associated with exogenous pulmonary surfactant in a novel formulation to improve the delivery to the lung. Respir Physiol Neurobiol. 2022;296:103825.
154. Ozturk OK, Turasan H. Applications of microfluidization in emulsion-based systems, nanoparticle formation, and beverages. Trends Food Sci Technol. 2021;116:609–25.
155. Ganesan P, Karthivashan G, Park SY, Kim J, Choi DK. Microfluidization trends in the development of nanodelivery systems and applications in chronic disease treatments. Int J Nanomedicine. 2018;13:6109–21.
156. Li Y, Deng L, Dai T, Li Y, Chen J, Liu W, Liu C. Microfluidization: a promising food processing technology and its challenges in industrial application. Food Control. 2021;137:108794.
157. Saboti D, Maver U, Chan HK, Planinsek O. Novel budesonide particles for dry powder inhalation prepared using a microfluidic reactor coupled with ultrasonic spray freeze drying. J Pharm Sci. 2017;106(7):1881–8.
158. Hu TT, Zhao H, Jiang LC, Le Y, Chen JF, Yun J. Engineering pharmaceutical fine particles of budesonide for dry powder inhalation (DPI). Ind Eng Chem Res. 2008;47(23):9623–7.
159. Besenhard MO, Neugebauer P, Scheibelhofer O, Khinast JG. Crystal engineering in continuous plug-flow crystallizers. Cryst Growth Des. 2017;17(12):6432–44.
160. Neugebauer P, Khinast JG. Continuous crystallization of proteins in a tubular plug-flow crystallizer. Cryst Growth Des. 2015;15(3):1089–95.
161. Jiang M, Braatz RD. Designs of continuous-flow pharmaceutical crystallizers: developments and practice. CrystEngComm. 2019;21(23):3534–51.
162. Hadiwinoto GD, Kwok PC, Tong HH, Wong SN, Chow SF, Lakerveld R. Integrated continuous plug-flow crystallization and spray drying of Pharmaceuticals for dry Powder Inhalation. Ind Eng Chem Res. 2019;58:16843–57.
163. Johnson BK, Prud'homme RK. Mechanism for rapid self assembly of block copolymer nanoparticles. Phys Rev Lett. 2003;91:118302.
164. Liu WJ, Ma CY, Liu JJ, Zhang Y, Wang XZ. Analytical technology aided optimization and scale-up of impinging jet mixer for reactive crystallization process. AICHE J. 2015;61:503–17.
165. Kiss N, Brenn G, Pucher H, Wieser J, Scheler S, Jennewein H, Suzzi D, Khinast J. Formation of O/W emulsions by static mixers for pharmaceutical applications. Chem Eng Sci. 2011;66(21):5084–94.
166. Dong Y, Ng WK, Hu J, Shen S, Tan RB. A continuous and highly effective static mixing process for antisolvent precipitation of nanoparticles of poorly water-soluble drugs. Int J Pharm. 2010;386(1–2):256–61.

167. Douroumis D, Scheler S, Fahr A. Using a modified shepards method for optimization of a nanoparticulate cyclosporine a formulation prepared by a static mixer technique. J Pharm Sci. 2008;97(2):919–30.
168. Douroumis D, Fahr A. Nano- and micro-particulate formulations of poorly water-soluble drugs by using a novel optimized technique. Eur J Pharm Biopharm. 2006;63(2):173–5.
169. Sheng F, Chow PS, Dong Y, Heng D, Lee SH, Tan RB. Continuous and scalable process for the production of hollow crystals of a poorly water-soluble active pharmaceutical ingredient for dissolution enhancement and inhaled delivery. Cryst Growth Des. 2019;19(6):3402–9.
170. Lenka M, Sarkar D. Combined cooling and antisolvent crystallization of L-asparagine monohydrate. Powder Technol. 2018;334:106–16.
171. Nagy ZK, Fujiwara M, Braatz RD. Modelling and control of combined cooling and antisolvent crystallization processes. J Process Control. 2008;18(9):856–64.
172. Ragab D, Rohani S, Samaha MW, El-Khawas FM, El-Maradny HA. Crystallization of progesterone for pulmonary drug delivery. J Pharm Sci. 2010;99(3):1123–37.
173. Yazdi AK, Smyth HD. Hollow crystalline straws of diclofenac for high-dose and carrier-free dry powder inhaler formulations. Int J Pharm. 2016;502(1–2):170–80.
174. Li W, Shi P, Jia L, Zhao Y, Sun B, Zhang M, Gong J, Tang W. Eutectics and salt of dapsone with hydroxybenzoic acids: binary phase diagrams, characterization and evaluation. J Pharm Sci. 2020;109(7):2224–36.
175. Chappa P, Maruthapillai A, Voguri R, Dey A, Ghosal S, Basha MA. Drug–polymer co-crystals of dapsone and polyethylene glycol: an emerging subset in pharmaceutical co-crystals. Cryst Growth Des. 2018;18(12):7590–8.
176. do Amaral LH, do Carmo FA, Amaro MI, et al. Development and characterization of dapsone cocrystal prepared by scalable production methods. AAPS PharmSciTech. 2018;19(6):2687–99.
177. Ma N, Liu Y, Ling G, Zhang P. Preparation of meloxicam-salicylic acid co-crystal and its application in the treatment of rheumatoid arthritis. J Drug Deliv Sci Technol. 2022;26:103542.
178. Chvatal A, Farkas Á, Balásházy I, Szabó-Révész P, Ambrus R. Aerodynamic properties and in silico deposition of meloxicam potassium incorporated in a carrier-free DPI pulmonary system. Int J Pharm. 2017;520(1–2):70–8.
179. Ikegami K, Kawashima Y, Takeuchi H, Yamamoto H, Isshiki N, Momose DI, Ouchi K. Improved inhalation behavior of steroid KSR-592 in vitro with Jethaler® by polymorphic transformation to needle-like crystals (β-form). Pharm Res. 2002;19(10):1439–45.
180. Rajalakshmi M, Shyju TS, Indirajith R, Gopalakrishnan R. Unidirectional growth of benzil crystal from solution by Sankaranarayanan-Ramasamy method and its characterization. Spectrochim Acta A Mol Biomol Spectrosc. 2012;86:27–32.
181. Chan HK, Gonda I. Respirable form of crystals of cromoglycic acid. J Pharm Sci. 1989;78(2):176–80.
182. McGinty J, Yazdanpanah N, Price C, ter Horst JH, Sefcik J. Chapter 1: Nucleation and Crystal Growth in Continuous Crystallization. In: The Handbook of Continuous Crystallization; 2020. p. 1–50. https://doi.org/10.1039/9781788013581-00001.
183. Pritula I, Sangwal K. Fundamentals of crystal growth from solutions. In: Handbook of crystal growth. Elsevier; 2015. p. 1185–227. https://doi.org/10.1016/B978-0-444-63303-3.00029-8.
184. Malamatari M, Ross SA, Douroumis D, Velaga SP. Experimental co-crystal screening and solution based scale-up co-crystallization methods. Adv Drug Deliv Rev. 2017;117:162–77.
185. Mehta P, Dhapte VV. Propulsive PAT paradigm: optimization of freeze drying process. Int J Pharm Sci Rev Res. 2014;28(2):240–6.
186. Nagy ZK, Braatz RD. Advances and new directions in crystallization control. Annu Rev Chem Biomol Eng. 2012;3:55–75. https://doi.org/10.1146/annurev-chembioeng-062011-081043.
187. Powell KA, Bartolini G, Wittering KE, Saleemi AN, Wilson CC, Rielly CD, Nagy ZK. Toward continuous crystallization of urea-barbituric acid: a polymorphic co-crystal system. Cryst Growth Des. 2015;15(10):4821–36.

188. Jadhav P, Patil P, Bhagwat D, Gaikwad V, Mehta PP. Recent advances in orthogonal analytical techniques for microstructural understanding of inhalable particles: present status and future perspective. J Drug Deliv Sci Technol. 2022;68:103089.
189. Mehta PP. Dry powder inhalers: a brief overview of the drug detachment techniques. Ther Deliv. 2020;11(3):139–43.
190. Murnane D, Marriott C, Martin GP. Crystallization and crystallinity of fluticasone propionate. Crystal Growth and Design. 2008;8(8):2753–64.
191. Chan HK. What is the role of particle morphology in pharmaceutical powder aerosols? Expert Opin Drug Deliv. 2008;5(8):909–14.

Thin-Film Freezing: A State-of-Art Technique for Pulmonary Drug Delivery

2

Piyush Pradeep Mehta and Vividha Dhapte-Pawar

Abstract

Thin-film freezing (TFF) is a freezing-induced precipitation system adopted from electrical and semiconductor industries to develop submicron particles. TFF is a "bottom-up" technique with excellent capability to fabricate high surface area nanoparticles and link them to shape nanostructured aggregates. TFF-processed low-density and high surface area, brittle matrix powders possess great potential in pulmonary drug delivery as they tend to form low-density respirable particles upon aerosolization. Thus, the fundamental purpose of the present chapter is to summarize the potential of TFF in pulmonary drug delivery. Primarily, the chapter starts by reviewing the necessary background of TFF, including technical aspects and applications in the pharmaceutical field. Later, the chapter deals with the aerosolization outcomes and biophysical properties of various TFF-processed respirable solids. Finally, the chapter provides an opinion on the present significance and identifies the existing bottlenecks for bench-to-bedside translation of TFF-processed respirable solids.

Keywords

Particle engineering · Thin-film freezing · "Bottom-up" technique · Nanostructured aggregates · Porous particles and Brittle matrix powder

P. P. Mehta (✉)
Global Respiratory R&D, Cipla R&D Center, Mumbai, Maharashtra, India
e-mail: piyu053@gmail.com

V. Dhapte-Pawar
Department of Pharmaceutics, Poona College of Pharmacy, Bharati Vidyapeeth (Deemed to be University), Pune, Maharashtra, India
e-mail: dhapte.vv@bharatividyapeeth.edu

P. P. Mehta, V. Dhapte-Pawar (eds.), *Pulmonary Drug Delivery Systems: Material and Technological Advances*,
https://doi.org/10.1007/978-981-99-1923-9_2

2.1 Introduction

Particle engineering is a theme that unites the elements of chemistry, solid-state physics, heat and mass transfer, colloidal and interfaces science, powder technology, and formulation science [1]. Particle engineering allows a good understanding of particle formation processes and offers a robust logical framework for designing and developing well-structured particles and powders [2]. Particle engineering is well explored for various dosage forms such as solid-oral [3], transdermal [4], and injectable formulations [5, 6] to improve the drug water solubility, drug release pattern, drug absorption, in vivo bioavailability and achieve controlled drug release [7, 8]. Similarly, particle engineering is used to generate drug microparticles in pure form or with carriers as combined materials of suitable size and morphology for effective pulmonary drug delivery [9, 10]. The critical objective of particle engineering in pulmonary drug delivery is to guarantee the reproducibility of the narrow particle size distribution (< 5 μm) [11], particle morphology (e.g., particle shape, internal microstructure, and surface attributes) [12, 13], powder dispersibility and eventually aerodynamic performance [14–16]. Several particle engineering methods formulate different pulmonary dosage forms such as dry powder inhalers (DPIs) [17–19], pressurized metered-dose inhalers [20], and nebulizers [21, 22]. Spray drying [23, 24], spray freeze drying [25], freeze drying [26], supercritical fluid technology [27, 28] are well-investigated particle engineering methods to design pulmonary dosage forms. Further, these methods can be used for porous microparticles to reduce the flow rate dependence performance and delivery efficiency of dry powders [29–31]. The most commonly used particle engineering techniques for the development of inhaled particles are summarized in Table 2.1. Apart from these techniques, a few novel platform technologies such as soft lithography-based molding, i.e., particle replication in nonwetting templates (PRINT) [32], "drop-on-demand" based inkjet printing (IJP) [33], PulmoSphere™ platform [34], fumaryl diketopiperazine-based Technosphere® technology [35], inorganic and organic salt-containing Inhaled small particles easily respirable and emittable (iSPERSE™) technology [19], hot-melt extrusion [36], dry powder coating [37], and surfactant therapy [38] are also well explored to develop engineered inhalable microparticles and nanoparticles. Likewise, TFF is also recently explored novel particle engineering technology to fabricate inhalable particles [39].

TFF is a "Bottom-up" particle engineering technique adopted from electrical and semiconductor industries. The application of TFF to formulate particles and powders is the logical progression of former freezing techniques. TFF is a freezing-induced precipitation technology in which liquefied droplets are impinged on a cold solid surface to form thin layers of frozen material. TFF includes the application of drug solution with a suitable stabilizer to a cryogenically cooled solid surface from a pharmaceutical viewpoint. It involves the drop-wise application of the solution to a rotating stainless steel drum from a height of roughly 10 cm [42]. As the solution droplet reaches the cryogenically cooled surface, a thin wafer is formed and speedily cooled until frozen. Afterward, a thin wafer is scraped off carefully before completing one full revolution of a rotating drum and gathered in a collecting pan,

Table 2.1 Particle engineering techniques used to develop inhaled particles and powders [39–41]

Techniques	Advantages	Limitations
Spray drying	• Easy to scale-up • Single-step process with simple set-up • Nozzle selection decides the droplet size formation • Fabricate uniform shape particles with consistent diameter • Fabricate both mesoporous and composite particles • **Powder characteristics:** Fine powder, hollow particles, spherical, and dimpled shape particles	• Low yield (50–70%) • Product is normally amorphous • Unsuitable for thermally sensitive compounds
Freeze drying	• Dose specific process • Reasonable control over moisture content • Suitable for heat-sensitive products • Offer appealing physical appearance • **Powder characteristics:** Form intact cake, impart high strength, good cake consistency with uniform color	• Multi-step process • Long batch processing time • High product exposure to ice–water interface
Spray freeze drying	• Offer high-speed cooling rate • Suitable for thermally-sensitive compounds • Apt to fabricate both micro or nanoparticles • Easy control of particle size by changing droplet size • Apt to fabricate porous and low-density powders with good flowability • **Powder characteristics:** Spherical shape with smooth surface, lightweight (low density) porous particle, high surface area particles	• High air–water interface • Two-step process with complex equipment • Particles experience stresses of freezing and dehydration (need apt cryoprotectant) • Product is generally amorphous and fragile due to high porosity (needs special care during processing and storage) • Tedious to scale up
Supercritical fluid drying	• Reasonable control over particle size distribution • Apt to fabricate single or multicomponent powders • Apt to develop porous and non-spherical particles • Assist in controlling polymorphic forms and crystallization conditions • **Powder characteristics:** Spherical particles, smooth surface particles	• Need anti-solvents to solubilize compounds • Selection of SCF is a tedious task • High possibility of clogging of dispersion nozzle owing to the fast formation of solids • High-pressure equipment and difficult to scaling-up

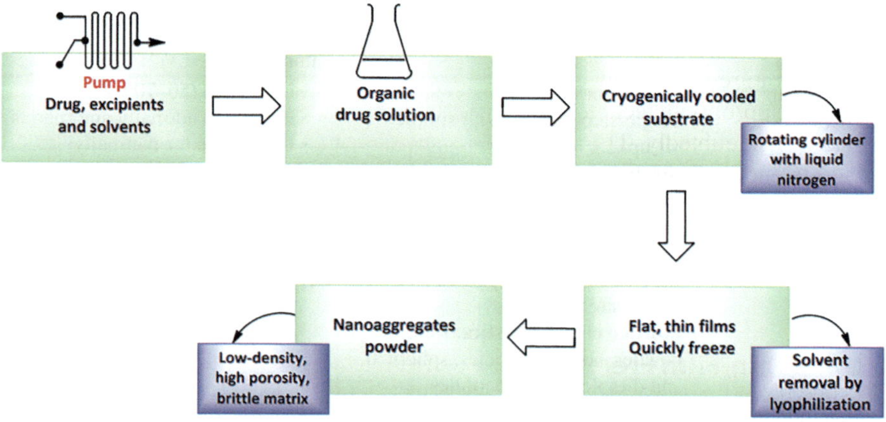

Fig. 2.1 Outline of the TFF particle formation process

cooled by cryogenic substances [43–45]. Subsequently, the collecting pan is subjected to a lyophilizer to eliminate solvent by sublimation (Fig. 2.1). Usually, the rotating drum is packed with cryogenic substances such as liquid nitrogen and dry ice. The cooling kinetics of the thin films can be regulated easily by varying the metal surface temperature. Additionally, the TFF assembly is located in a dry box with relative humidity <10% to control the generation of water vapor condensation on the stainless steel surface. Generally, the droplet has a mean diameter between 2 and 5 mm at room temperature, and it freezes within 50 and 150 milliseconds of contact with the cryogenically cooled solid surface [46]. TFF has the unique ability to shape a thin film with a thickness of <500 μm and a wide range of surface area to volume ratio, i.e., 25–500 cm^{-1} [46]. Furthermore, the TFF rotating drum presents scale-up benefits over other freezing-induced particle engineering methods due to the continuous freezing process and ease of humidity operations. Briefly, it is a cryogenically efficient freezing process. Based on processing, TFF can be classified with splat cooling [47], slat quenching solidification [48], ultra-rapid freezing [49], thermal spray coating [50], plasma or powder spray deposition [51], and spray forming method [52].

TFF has an excellent ability to produce respirable particles composed of a brittle matrix, nanostructured aggregates. It involves an ultra-rapid freezing rate (100–1000 K/s) which efficiently inhibits the crystallization and particle growth of the dissolved substance [53–56]. TFF produces nanoparticles of the drug and carrier and links them as nanostructured aggregates, while on the micron scale, these nanostructured aggregates are agglomerated as a brittle matrix powder (BMP). Briefly, TFF-based low-density and high surface area brittle matrix particles are suitable for deep lung deposition. Additionally, TFF produced more stable protein and peptides loaded respirable particles than other cryogenic techniques due to less denaturation potential and well-controlled air exposure. Furthermore, stable single-phase co-amorphous systems composed of two or more different substances (drug/drug or drug/carrier) can be formulated using TFF to improve the physical stability and dissolution kinetics of the co-amorphous system compared to amorphous drugs alone. Spray drying is also used to develop co-amorphous systems, yet elevated drying temperature lead to physical instability of sensitive drug substances. Thus, TFF is a more appropriate technique for the drug substances susceptible to thermal degradation. Undoubtedly, TFF is one of the potential alternatives for the design and development of respirable particles. By knowing the capability of TFF in the particle engineering segment, this chapter is primarily focused on the applications of TFF in pulmonary drug delivery. We have attempted to present the recent progress and loopholes relating to TFF technology in pulmonary drug delivery.

2.2 Thin-Film Freezing for Pulmonary Drug Delivery

In inhalation therapy, there is a growing demand for tailor-made respirable particles capable of deep lung delivery with the most optimal therapeutic results. A wide range of novel particle engineering technologies has materialized to match this growing need over the past years [9]. Particle engineering technologies have made a significant impact on particle morphology. The particle morphology can influence the aerosolization behavior at different phases, i.e., molecular, particulate, and aerosol [12]. TFF is a potential tool to control particle morphology as it can form different particle structures, i.e. (crystal habit and size) and surface attributes, i.e., surface area and energy. As discussed above, TFF has good potential to formulate nanostructured aggregates based on BMP. Longest and Hindle (2017) recently showed that nanoaggregates achieve superior airway absorption efficacy and better dose uniformity than traditional microparticles [57]. Additionally, brittle matrix morphology will ease the development of high-potency formulations with low amounts of excipient. Eventually, it decreases the powder burden for the patient while getting a total dose of the medicine and can reduce the deposited dose non-uniformity [58]. Therefore, from the pulmonary drug delivery viewpoint, TFF was thoroughly evaluated by various formulation scientists and reported in different exciting research articles, as summarized below.

2.2.1 TFF Processed Inhalable Tacrolimus Particles

Tacrolimus (TAR) is a macrolide, potent immunosuppressive agent obtained from *Streptomyces tsukubaensis*. It is a calcineurin inhibitor prescribed in prophylaxis against cellular rejection after organ transplantation [59]. During pulmonary transplantation, local immunosuppression therapy with inhaled TAR exhibited better clinical results with fewer adverse effects compared to systemic immunosuppressive therapy [60, 61]. Watts et al. (2012) prepared respirable low-density TAR BMP using three different carriers, i.e., lactose (TAR-LAC), mannitol (TAR-MAN), and raffinose (TAR-RAF) with acetonitrile: water co-solvent system by TFF method. TFF-processed TAR-LAC (49.10 μm), TAR-MAN (64.78 μm), and TAR-RAF (40.71 μm) dry powder showed tapped density and Hausner ratio of 0.0553, 0.0227,

0.0423 g/cm^3 and 1.58, 1.54, and 2.00, respectively. In aerosolization study using a high-resistance HandiHaler® inhaler at 51 L/min (4 s), TAR-LAC, TAR-MAN, and TAR-RAF dry powder displayed fine particle fraction (FPF) of 68.70%, 48.80%, and 69.20%, respectively. Further, TAR-LAC, TAR-MAN, and TAR-RAF dry powder showed mass median aerodynamic diameter (MMAD) of 2.71, 4.62, and 2.60 μm, respectively. Improved % FPF of TAR-LAC and TAR-RAF is mainly attributed to improved particle fracture during actuation compared to TAR-MAN particles. Besides, during storage at 50% relative humidity, TAR-MAN (4.10 μm) showed the most stable MMAD values as compared to TAR-LAC (7.59 μm) and TAR-RAF (8.17 μm) dry powder. Briefly, MAN-based TAR formulation is less influenced by humidity due to stabilizing effect of MAN on the amorphous form of the TAR. However, long-term stability studies are essential to understand the TFF-processed sugar-based powders [62].

Wang et al. (2014) fabricated a TFF-processed formulation for targeted and localized TAR delivery to the pulmonary airways with minimal systemic exposure. Mainly, co-solvent acetonitrile/water (60:40 v/v) containing TAR-MAN (1:1 w/w) was rapidly frozen on a cryogenically cooled rotating stainless steel surface (−50 °C) to obtain dry powder formulation. TFF-processed TAR-MAN formulation showed a highly porous nature with a specific surface area and mean particle size of 32.45 m^2/g and 17.84 μm, respectively, while fractured, irregularly-shaped micronized TAR-MAN formulation displayed a specific surface area and mean particle size of 2.87 m^2/g and 1.92 μm, respectively. Briefly, TFF-processed formulation exhibited acceptable particle size with low-density capable of attaining deep pulmonary deposition. During aerosolization study using Miat® monodose inhaler at 90 L/min (2.7 s), TFF-processed TAR-MAN formulation showed satisfactory FPF (83.3%) and lower MMAD (2.26 μm) values. Single-dose 24 h pulmonary pharmacokinetic assessment was performed in rats using nose-only dosing chamber equipped with a rotating brush generator. In this study, TFF-processed TAR-MAN formulation showed a marked improvement in pulmonary bioavailability (2.34-fold) and mean retention time (11.63-fold) in the pulmonary airways due to decreased clearance (2.33-fold) compared to the micronized TAR-MAN formulation. Moreover, TFF-processed TAR-MAN formulation caused a lower systemic TAR concentration following inhalation and minimized the adverse effects of the systemic circulation [63].

Recently, Sahakijpijarn et al. (2020) formulated inhalable TAR DPIs using LAC (bulking agent) and acetonitrile: water (60:40 v/v) solvent system by the TFF method. TFF-processed TAR-LAC (50:50) [TL-50/50-TFF] and TAR-LAC (95:5) [TL-95/5-TFF] powder exhibited brittle matrices of highly porous particles. Remarkably, the brittle matrix morphology of TFF-processed inhalable powder was mainly credited to the linking and agglomeration of primary particles. TFF-processed TL-50/50-TFF and TL-95/5-TFF powder exhibited a specific surface area of 143.13 and 73.58 m^2/g, respectively. In addition, LAC-containing TAR inhalable powder showed higher ($p < 0.05$) specific surface area than MAN and trehalose-containing TAR inhalable powders. In in vitro dissolution assessment using phosphate buffer containing polysorbate 80 (0.2%), TFF-processed powders displayed a higher rate of TAR dissolution (> 70% within 60 min) compared to crystalline TAR (25% within 60 min). In aerosolization study using RS01 Plastiape® inhaler at 60 L/min (4 s), TL-50/50-TFF powder showed emitted dose (ED) with FPF of 91.35 and 61.30%, respectively. However, TL-95/5-TFF powder showed ED and FPF of 95.01% and 69.31%, respectively. Additionally, in aerosolization study using high resistance (4 kpa and 60 L/min) and low resistance (4 kpa and 90 L/min), Plastiape® inhaler TL-95/5-TFF powder exhibited ED with FPF of 95.02 and 96.6 with 69.31 and 53.72%, respectively. Besides, the change in pressure drop from 4 kpa to 1 kpa showed only 5% lower FPF and indicated that TFF-processed powder can minimize the flow rate-dependent aerosolization outcome. Moreover, during a long-term 6-month stability study (40 °C/75% RH and 25 °C/60% RH), TFF-processed TAR powder retained the amorphous nature due to a lack of interaction between the drug and bulking agent. Furthermore, TL-95/5-TFF powder retained the aerosol functioning post 6 months of storage at 25 °C/60% RH. Concisely, TFF was found to be a promising option for the development of stable, inhalable dry powders [64].

2.2.2 TFF Engineered Inhalable Voriconazole Particles

Invasive fungal infections are progressive; life-threatening infections are triggered by yeasts or moulds. It is mainly caused by opportunistic fungi, for example, *Aspergillus* species. Primary infection with *Aspergillus* species frequently takes place in the pulmonary tract leading to invasive pulmonary aspergillosis [65]. Invasive pulmonary aspergillosis is a leading cause of morbidity and mortality in immunocompromised patients [65, 66]. The second-generation triazole antifungal agent, i.e., voriconazole (VOR) formulations are normally used to treat invasive pulmonary aspergillosis [67]. In most of the cases, VOR formulations showed improved therapeutic outcomes in immunocompromised patients with invasive pulmonary aspergillosis [68].

Beinborn et al. (2012) formulated respirable VOR formulations using the TFF system to enhance aerosolization performance and prolonged VOR concentrations in the airways for effective treatment of invasive pulmonary aspergillosis. Notably, respirable VOR formulations were fabricated using a 1,4-dioxane solvent system with and without (polyvinylpyrrolidone K12) stabilizing agent, i.e., VOR-PVP-K12-TFF and VOR-TFF. Microstructured crystalline VOR-TFF displayed bulk density and specific surface area of 0.034 g/cm^3 and 10.40 m^2/g, respectively, while nanostructured amorphous VOR-PVP-K12-TFF (1:2) exhibited bulk density and specific surface area of 0.015 g/cm^3 and 23.40 m^2/g, respectively. During aerodynamic analysis at 60 L/min (4 sec) using Handihaler® inhaler TFF-VOR displayed total ED, FPF, and MMAD of 80.60%, 43.10%, and 3.50 μm, respectively, whereas VOR-PVP-K12-TFF showed total ED, FPF, and MMAD of 96.50%, 42.40%, and 4.5 μm, respectively. Briefly, TFF with stabilizing agents and the TFF process parameters, such as solvent composition, dissolved solids percentage, and solution feed rate, have a crucial impact on the macroscopic powder properties and specific surface area of TFF-processed powders [69]. A similar research group also formulated respirable VOR formulations using a 1,4-dioxane solvent system with and without PVP K 25 stabilizing agent, i.e., VOR-PVP-K25-TFF and VOR-TFF. Microstructured crystalline VOR-TFF (0.45 μm) displayed bulk density and a specific surface area of 0.058 g/cm^3 and 9.38 m^2/g, respectively, whereas nanostructured amorphous VOR-PVP-K25-TFF (1:3; 0.05 μm) exhibited bulk density and specific surface area of 0.013 g/cm^3 and 43.35 m^2/g, respectively. In in vitro dissolution study using a next-generation impactor equipped with a dissolution cup with impaction insert (0.1 μm, pre-wetted polycarbonate filter membrane) at stage 4, VOR-PVP-K25-TFF showed 1.3-fold superior dissolution kinetics compared to VOR-TFF in phosphate buffer saline (pH 7.4). During aerodynamic analysis at 60 L/min (4 s) using Handihaler® inhaler, TFF-VOR displayed total ED and FPF of 81.20% and 37.80%, respectively, whereas VOR-PVP-K25-TFF showed total ED and FPF of 92.50% and 32.40%, respectively. In aerodynamic analysis, TFF-VOR and VOR-PVP-K25-TFF showed a marked 1.92 and 1.65-fold improvement in % FPF compared to InhaLac® 70 loaded micronized VOR (2% w/w), respectively. Moreover, in single-dose 24 h pulmonary pharmacokinetic analysis in male outbred ICR mice using DP-4 M Dry powder insufflator™, crystalline VOR-TFF showed 1.95-fold superior VOR concentration within the lung tissue as compared to amorphous VOR-PVP-K25-TFF at an equivalent dose (10 mg/kg of VOR). The quick dissolution kinetics of nanostructured amorphous VOR-PVP-K25-TFF formulation caused rapid transformation of VOR into the plasma and significantly reduced the retention and overall concentration of VOR in the lung tissue. Briefly, microstructured crystalline VOR-TFF formulation showed more favorable aerosolization properties with clinically relevant plasma concentrations than nanostructured amorphous VOR formulation. Therefore, microstructured crystalline VOR can potentially treat pulmonary infections like invasive pulmonary aspergillosis [70].

Recently, Moon et al. (2019) fabricated high-potency nanoaggregates containing crystalline VOR using different concentrations of surface texture-modifying agent, i.e., MAN and co-solvent water/acetonitrile 50:50 (v/v) by TFF method. Smooth micron-sized agglomerated VOR-TFF showed a specific surface area of 43.35 m^2/g. At the same time, a porous matrix of VOR-MAN-TFF (95:05) and VOR-MAN-TFF (50:50) displayed a specific surface area of 10.60 and 16.30 m^2/g, respectively. Mainly, a high-resolution 3D illustration of VOR-MAN-TFF (95:05) showed the rough surface texture with aggregates containing nanoparticles of <500 nm. During in vitro dissolution study using phosphate buffer saline (pH 7.4) by Franz cells method, VOR-MAN-TFF (95:05) exhibited significantly slower cumulative VOR release as compared to the amorphous VOR-PVP K25-TFF (25:75) formulation. At the end of 3 h, the cumulative VOR release from VOR-MAN-TFF (95:05), VOR-MAN-TFF (50:50), VOR-MAN-TFF (25:75), VOR-PVP K25-TFF (25:75) formulation was 22.80%, 35.30%, 46.30%, and 63.20%, respectively. In aerosolization study at 60 L/min (4 s) using highly resistant RS01 dry powder inhaler VOR-TFF, VOR-MAN-TFF (50:50), VOR-MAN-TFF (95:05), and VOR-MAN-TFF (97:03) showed % FPF of 24.60%, 28.50%, 38.10%, and 40.80%, respectively. Moreover, during the long-term 13 months (25 °C/ 60% RH) stability study, VOR-MAN-TFF (95:05) effectively maintained the aerosolization outcome. Principally, TFF-processed phase-separated submicron crystalline MAN assisted in reducing the

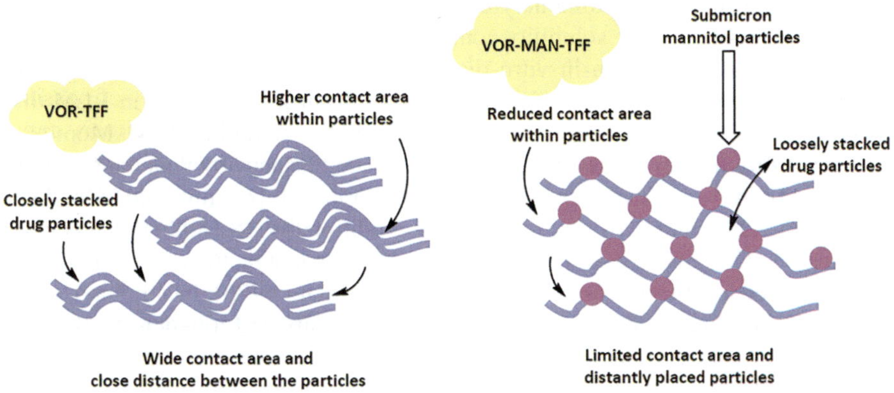

Fig. 2.2 Graphical illustration of contact area and distance between the TFF-processed particles with and without mannitol excipient

contact area and improving the distance between the VOR particles and thus enhanced the aerosolization performance (Fig. 2.2). In a few words, VOR nanoaggregates formulated using low concentrations of MAN (< 5%) showed good surface texture and a more feasible formulation option for local treatment of pulmonary infections such as invasive pulmonary aspergillosis [71].

Furthermore, a similar research cluster also judiciously evaluated several critical factors related to the process design space of TFF, such as processing temperature, solvent system composition, solid loading, and scale-up. To critically assess the process design space following conditions were used (i) VOR: MAN (95:05), (ii) acetonitrile: water mixture (30:70, 50:50, and 70:30 v/v), (iii) solid content (1–3% w/v), and (iv) cryogenically cooled stainless steel drum (−60 °C or − 150 °C) with a drop height of approximately 10 cm. The lower processing temperature (−150 °C) with acetonitrile: water mixture containing a higher concentration of water (70%) showed constructive aerosolization outcome. The higher solid content (3%) showed lower aerosolization fate, whereas powder conditioning (powder rolled for 30 min at 60 rpm and packed with a desiccant at room temperature) showed superior aerodynamic results. Moreover, in a comparative inhaler device resistance study, high-resistance Plastiape device showed consistent aerosolization results for crystalline VOR nanoaggregates at over a wide range of inhalation flow rates (30–90 L/min) and drug loading (10–20 mg) as compared to low resistance Plastiape device. Furthermore, 450-fold bigger manufacturing scale showed no remarkable variation in powder physicochemical properties and aerosolization outcome [72].

2.2.3 Respirable Remdesivir Particle Using TFF

Sahakijpijarn et al. (2020) fabricated respirable remdesivir (REM; 80% w/w) dry powder using Captisol® (modified cyclodextrin; 0.30% w/w) and acetonitrile: water (50:50 v/v) solvent system by a TFF method. Amorphous REM-Captisol® (REM-CAP) dry powder exhibited smaller nanostructure aggregates with mean particle size suitable for pulmonary delivery. Amorphous REM-CAP dry powder showed 20 times superior solubility compared to crystalline REM using simulated lung fluid. Additionally, an in vitro dissolution study using simulated lung fluid amorphous REM-CAP dry powder showed significant improvement in REM dissolution kinetics (> 80% within 90 min) compared to crystalline REM. Moreover, in aerosolization study using a high-resistance RS01 Plastiape® inhaler at 60 L/min (4 s) REM-CAP dry powder showed recovered dose, FPF (< 3 μm) and MMAD of 91.48%, 62.25%, and 2.16 μm, respectively. Additionally, REM-CAP dry powder demonstrated improved systemic uptake during pharmacokinetics assessment using rats compared to TFF-processed REM-leucine formulation [73]. In summary, ultra-rapid freezing kinetics of TFF-formulated high-porosity and low-density BMP of REM, which are easily aerosolizable by passive DPI device, and therefore permit high doses of REM to be delivered to the pulmonary airways [73]. A similar research group also studied the in vivo pharmacokinetic profile of REM DPI using Female Syrian hamsters. Mainly two TFF-based DPI formulations, i.e., REM-Captisol® (REM-CAP; 80/20 w/w) and REM-leucine (REM-LEU; 80/20 w/w), were intratracheally administered (10 mg/kg) using dry powder insufflator (DP-4, Penn-Century Inc., USA). REM-CAP exhibited eightfold higher maximum serum concentration (C_{max}) and 1.25-fold lower time to maximum plasma concentration (t_{max}) compared to REM-LEU. Moreover, REM-CAP showed 2.38-fold superior area under curve (AUC) for REM in the lungs compared to REM-LEU, demonstrating higher absorption of REM within the pulmonary airways. REM-LEU displayed 1.70-fold superior C_{max} and twofold lower t_{max} for GS-441524 (REM metabolite) compared to REM-CAP. GS-441524 showed a similar half-life using REM-CAP (7.12 h) and REM-LEU (7.38 h) formulations. Likewise, GS-441524 showed a biphasic elimination pattern using both dry powders. However, REM-CAP showed 1.70-fold higher AUC_{0-24} for GS-441524 in the lungs compared to REM-LEU. REM-CAP had a significant impact on the absorption kinetics of REM but not on the extent of absorption within the systemic circulation. Similarly, REM-LEU did not significantly impact absorption within the systemic circulation. However, regarding the efficacy of antiviral substances, pulmonary delivery of REM-LEU generates a superior C_{max} of GS-441524 within the short t_{max} and lower AUC of GS-441524, indicating that a lower total GS-441524 exposure is required to form high GS-441524 levels within the pulmonary airways for inhibiting virus replication. In summary, REM-LEU is also a constructive approach for further development. The plasma concentrations of both formulations were superior to reported maximal effective concentration (EC_{50}) values for REM and GS-441524. The detailed pharmacokinetic assessment suggested that REM twice-daily inhalation was appropriate for better therapeutic results [74].

2.2.4 Inhalable Monoclonal Antibodies Using TFF

Monoclonal antibodies (mAbs) constitute a significant class of therapeutic proteins in the biopharmaceutical discipline [75]. The mAbs are used in chemotherapy, radiation treatment, blocking target receptors, ligands, and flagging cancer cells to improve therapeutic outcomes in severe disease management [76]. The mAbs are stored as liquids or frozen liquids to ease their handling and manufacturing. However, in most of the cases, liquid storage is not favorable, as it causes aggregation, denaturation, and hydrolysis of mAbs and ultimately leads to poor storage stability. Consequently, formulating mAbs as dry powders help to address this concern of instability and dependence on cold chain storage. Hufnagel and co-workers (2022) formulated anti-programmed cell death protein (anti-PD-1-LL-PBS mAb) dry powders using LAC: leucine (60:40, w/w) in PBS by TFF and thoroughly studied for aerosol properties. TFF-processed anti-PD-1-LL-PBS mAb dry powder showed highly porous nanoaggregates with lower moisture content (1.5%). TFF-processed anti-PD-1-LL-PBS mAb dry powder showed glass transition temperature of 39 °C and was more stable compared to anti-PD-1-LL-PBS mAb liquid formulation. Furthermore, addition of PVP K40 (5%) in the formulation increased the glass transition temperature (152 °C) which further increase the storage stability of the anti-PD-1-LL-PBS mAb dry powder. In aerosolization assessment, using high-resistance RS01 Plastiape® inhaler at 60 L/min (4 s) anti-PD-1-LL-PBS mAb dry powder displayed a marked threefold higher % FPF and 2.3-fold lower MMAD values compared to conventional shelf freeze drying process. Briefly, TFF can be used to fabricate stable mAbs dry powders with suitable aerosol properties. The excipient concentration, type, properties, solvent as well as the mAbs composition and concentration showed impact on TFF-processed dry powder. Thus, in near future extra experimental efforts are needed to optimize the same [77].

2.2.5 TFF Routed Inhaled Powders

Cystic fibrosis (CF) is an inherited life-threatening condition that damages various key organs. It is a genetic, autosomal recessive disease, which particularly damage pulmonary and gastrointestinal tracts. CF is triggered by mutations of cystic fibrosis transmembrane conductance regulator (CFTR) gene. CF potentially influences the cells that supply mucus, sweat, and digestive fluids and drastically lessens life expectancy [78, 79]. CRISPR is a gene-editing system that emerged recently and shows its capacity to precisely restore CFTR mutations. With viral or non-viral cargos were used frequently to deliver CRISPR/Cas9 and surpass difficulties faced during gene therapy. However, in most of the cases CRISPR/Cas9 therapy showed insufficient therapeutic outcome due to its inability to offer targeted delivery to pulmonary airways for successful clinical application [80–82]. Recent in vitro and in silico investigations exhibited that PEGylated/chitosan nanocomplex-directed CRISPR/Cas9 delivery exhibited a dynamic improvement in mucus penetration, intracellular uptake and an offer effective cargo system for targeted delivery [83]. But the low stability of nanocomplex in an aqueous medium makes liquid atomization difficult during clinical application. Thus, Zhang et al. (2021) rationally fabricated the TFF-processed gene-editing delivery system, i.e., PEGylated chitosan/CRISPR-Cas9 DPI to surpass biopharmaceutical issues related to targeted pulmonary delivery. TFF-processed optimized nanocomplexes formulation contains cryoprotectant MAN (3%), PEGylated chitosan (0.875%) and DNA (50 ng/μL). Additionally, LEU (0.3%) as dispersion enhancer was added to final formulation. Lastly, fluorescein sodium salt (0.02%) was mixed with formulation to trace in vitro performance. MAN-nanocomplexes and LEU/MAN-nanocomplexes showed true densities of 1.68 and 1.55 g/cm^3 with specific surface areas of 3.13 and 3.80 m^2/g, respectively. Aggregated porous nanocomplexes microparticles displayed mean particle size (3.4 μm) suitable for effective pulmonary delivery. Aerodynamic analysis of DPI formulations were carried out using HandiHaler® and Plastiape Monodose® RS01 device. MAN-nanocomplexes using HandiHaler® device at 45 and 60 L/min flow rate displayed FPF of 18.4 and 43.3%, correspondingly while LEU/MAN-nanocomplexes using HandiHaler® device at 45 and 60 L/min flow rate showed FPF of 24 and 63%, individually. Similarly, MAN-nanocomplexes using Monodose® RS01 at 45 and 60 L/min flow rate displayed FPF of 23 and 60%, correspondingly while LEU/MAN-nanocomplexes using Monodose® RS01 at 45 and 60 L/min flow rate showed FPF of 39.2 and 62.3%, individually. The use of LEU as a force-controlling agent enhanced the aerodynamic outcome of DPI. This analysis confirmed that TFF-processed material is a hopeful way to formulate nucleic acid-containing polymer-based nanocomplex DPI [84].

Shingrix is a recombinant, adjuvanted vaccine developed by GSK Biologicals s.a. for prevention of herpes zoster and post-herpetic neuralgia in adults 50 years of age or older. A marketed pack of Shingrix vaccine contains 1 vial of powder and 1 vial of suspension. White-colored powder (antigen) composed of sucrose, dipotassium phosphate, polysorbate 80, and sodium dihydrogen phosphate dihydrate whereas an opalescent, colorless to pale brownish suspension (adjuvant) fabricated from dioleoyl phosphatidylcholine, cholesterol, potassium dihydrogen phosphate, disodium phosphate anhydrous, sodium chloride, and water for injections. For intramuscular injection both vials were reconstituted properly. After reconstitution, one dose of 0.5 mL vaccine contains varicella zoster virus-1 glycoprotein E antigen (50 μg) and adjuvanted with $AS01_B$ [3-O-desacyl-4′-monophosphoryl lipid A (MPL) from Salmonella Minnesota and plant extract Quillaja saponaria Molina, fraction 21 (QS-21)] liposomes (50 μg). To ease the manufacturing, handling, and administration of vaccine AboulFotouh et al. (2022) engineered TFF-processed dry powders of the $AS01_B$ liposomal adjuvant. TFF vaccine was developed using 4% w/v sucrose with sucrose to lipid ratio of 16:1 w/w. Shingrix vaccine and reconstituted TFF Shingrix vaccine showed similar mean particle size (105 nm) with negative zeta potential of 14 and 12 mV, individually. Electron microscopy analysis of reconstituted TFF Shingrix vaccine exhibited uniform small particles without any aggregation. Aerodynamic analysis was performed using a Plastiape Monodose RS01 device at 60 L/min TFF-processed DPI displayed a marked 12.50 and 11.19-fold higher delivered and recovered dose % FPF $_{(< 5\,\mu m)}$ compared to traditional shelf freeze drying processed DPI. TFF processed adjuvanted part when reconstituted with model ovalbumin antigen it retained the integrity of the liposomes with particle size and did not affect immunogenicity of the developed vaccine. Briefly, TFF scheme potentially offers a way to pack two different constituents of the Shingrix vaccine in a single ampoule as a dry powder. Dry powder dosage form is likely to be more stable than its liquid counterpart and do not need distinct storage conditions and therefore can potentially decrease vaccine wastage and handling issues. Moreover, single ampoule products can lessen the cost of the overall manufacturing campaign and increase the rate of production [85].

Niclosamide (NIC) is a poorly water-soluble, anthelmintic drug [86]. Recent preclinical investigations showed its therapeutic potential against severe acute respiratory syndrome coronavirus 2 (SARS-CoV-2 infection) [87]. Accordingly, Jara et al. (2021) developed an aerosol formulation to deliver the NIC to pulmonary airways and sustain niclosamide concentrations. Particularly, respirable NIC (1.66% w/w) dry powder is formulated using MAN (92.04% w/w), leucine (6.3% w/w), and tert-butanol: water (4:1 v/v) solvent system by a TFF method. The shift in melting peak of NIC to lower temperatures (230 °C–218 °C) and reduction of melting enthalpy validated the nanonization of NIC crystals by TFF method. During aerodynamic assessment using Plastiape® RS00 device at 58 L/min (4.1 s), NIC-TFF particles displayed FPF and MMAD of 76.6% and 1.1 μm, respectively. During a single-dose pharmacokinetic study using a female Syrian golden hamster model, NIC-TFF particles were administered intratracheally by DP-4 Penn-Century dry powder insufflator™ (Penn-Century Inc., USA) at two different concentrations of NIC, i.e., 145 and 290 μg/Kg. A non-compartmental analysis exhibited that NIC-TFF particles achieved concentrations above the concentration that will inhibit 90% of the virions (IC_{90}; 153.7 ng/mL) [88] of SARS-CoV-2 with satisfactory NIC concentrations (72.1 μg/Kg) within the lung tissue for 24 h. Moreover, the toxicology and histopathological studies of lung tissue showed that NIC-TFF powder has an adequate safety margin for dose modification, therefore acceptable for once-a-day inhalation therapy [89].

Carvalho et al. (2014) thoroughly compared the pharmacokinetic and aerosolization outcome of inhaled amorphous and crystalline rapamycin (RAP). Amorphous RAP was formulated using stabilizing agent LAC (1:1) by the TFF method (RAP-LAC-TFF). In contrast, the crystalline formulation was developed using RAP and LAC mixture by the wet ball milling method (RAP-LAC-WBM). Wet ball-milled RAP displayed a mean particle size (5.22 μm) suitable for pulmonary delivery with a surface area of 14.29 m^2/g, while RAP-LAC-TFF particles showed a specific surface area of 85.72 m^2/g. During aerosolization study using Handihaler™ at 54 L/min (4.4 s), amorphous RAP-LAC-TFF particles showed total ED, FPF, and MMAD of 97.14%, 72.11%, and 2.10 μm, respectively. At the same time, the crystalline RAP and LAC physical mixture exhibited ED, FPF, and MMAD of 78.92%, 36.79%, and 1.81 μm, respectively. In single-dose, 24 h pharmacokinetic studies using Sprague-Dawley rats after pulmonary administration using nose-only inhalation exposure, amorphous RAP-LAC-TFF particles showed 3.58-fold higher rapamycin concentration within the systemic circulation as compared to crystalline RAP formulation based on the five-compartmental modeling. In vivo pharmacokinetic analysis indicated similar pulmonary retention time for both amorphous and crystalline RAP powders; however, amorphous RAP-LAC-TFF particles exhibited superior systemic bioavailability because of their improved solubility and dissolution kinetics [90]. The improved solubility of amorphous RAP-LAC-TFF particles created a supersaturated milieu with a direct influence on lung absorption. Moreover, amorphous RAP-LAC-TFF particles stayed in the lung for the same time as the crystalline powder but with more surface area. Therefore, higher bioavailability was observed for amorphous RAP-LAC-TFF particles [90].

Co-administration of inhaled long-acting beta-agonist and corticosteroid for chronic respiratory disorders has decreased transience compared to either drug alone. Additionally, co-administration reduces exacerbations, emergency department visits, hospitalization, and health care costs. Therefore, by knowing this dual drug combination approach, Liu et al. (2015) developed BMP containing mometasone furoate (MF) and salmeterol xinafoate (SX) using the TFF process. During aerosolization study using Miat® monodose inhaler at 90 L/min (2.7 s), BMP-MFSX formulation showed improved aerodynamic properties compared to the micronized blends of the MFSX. Specifically, BMP-MFSX formulation showed a higher total ED for MF (93.91%) and SX (95.31%) as compared to micronized MF (71.18%) and SX (68.59%) blend. Aerodynamic analysis showed that BMP-MFSX formulation offered uniform drug deposition with consistent dose uniformity compared to micronized MFSX blend. Moreover, in long-term 6-month stability study, BMP-SXMF formulation delivered >50% of the loaded dose within the respirable range. In single-dose 24 h pharmacokinetic assessment in rats after pulmonary administration using the DP-4 M Dry insufflator™ device, BMP-SXMF formulation showed a marked improvement in pulmonary bioavailability [MF (5.85-fold) and SX (15.25-fold)] with mean retention time [MF (1.01-fold) and SX (1.13-fold)] within the lungs as compared to the crystalline micronized SXMF formulation. Briefly, fixed-dose combination BMP-SXMF formulation displayed a superior drug deposition profile within the pulmonary airways [91]. This novel drug combination approach can be used to tailor the pulmonary therapies. However, further animal studies are needed to balance the therapy and adverse effects because of the better absorption and higher concentration of drug substances [91].

2.2.6 TFF Engineered Powders for Metered Dose Inhaler

The pressurized metered-dose inhaler (pMDIs) are also one of the most frequently prescribed aerosol for treatment of asthma and chronic obstructive pulmonary disease [92, 93]. They are hand-held, pocket-size aerosols made up of propellants or a mixture of propellants, stabilizers, and co-solvents to deliver micronized drug particles to pulmonary airways. Mainly, pMDIs formulation is a hetero-dispersions delivered using a device made up of three key components, i.e., canister, metering valve, and an actuator. Recently, Engstrom and co-workers (2009) engineered high-aspect ratio particles using TFF to develop bovine serum albumin (BSA) pMDIs without the use of surfactants or co-solvents [94].

Particularly, BSA nanorods were fabricated using TFF and subsequently added to hydrofluoroalkane-227 (HFA-227) propellant to develop a respirable suspension. Particularly, BSA (5 mg/mL) was added in potassium phosphate buffer (10 mM, pH = 7.4) and passed through a 17 gauge stainless steel syringe needle (4 mL/min) over a hollow stainless steel drum filled with dry ice. The primary and secondary drying was performed at - 40 °C for 36 h and 25 °C for 24 h. BSA nanorods (< 100 nm) showed a web structure containing interconnected rods with a lower tapped density of 0.0064 g/cm^3. Developed formulation was packed in 17 mL glass vials equipped with metering valves containing 100 μL metering chambers. During dose delivery study, HFA-227 containing BSA nanorods showed superior delivered dose (915 μg) values as compared to milled (295 μg) and spray dried (308 μg) BSA particles. During cascade analysis using 15 cm spacer at 28.3 L/min, HFA-227 containing BSA nanorods displayed ED, FPF, and fine particle dose (FPD) of 695 μg, 47%, and 318 μg, respectively. Moreover, HFA-227 containing BSA nanorods demonstrated acceptable MMAD (3.2 and 3.1 μm) and geometric standard deviation (GSD; 1.6 and 1.9) values using aerodynamic particle sizer (5 L/min) and cascade impactor (28.3 L/min) assembly, respectively. Furthermore, BSA nanorods were found to be stable against settling without the use of any co-solvents and surfactants during long-term 1-year storage. Principally, quick and sticky attractive collisions of BSA nanorods ease the formation of low-density flocs, which stack upon every other floc to fill the whole solvent volume to avoid settling. In short, this formulation model can apply to several therapeutic agents without adding co-solvents or surfactants to stabilize the primary particles [94].

2.3 Clinical Overview

2.3.1 Voriconazole Inhalation Powder

TFF Pharmaceuticals, Inc. designed a phase 1 clinical trial (NCT04872231) to understand the safety and pharmacokinetic profile of voriconazole inhalation powder (VIP) in healthy adult subjects using single ascending dose (SAD) and multiple ascending doses (MAD) study. In a SAD study, VIP (10 mg) was administered using a Plastiape RS00 dry powder inhaler as one to eight 10 mg capsules, i.e., subjects in group 1 will receive 10 mg, group 2 (20 mg), group 3 (40 mg), and group 4 (80 mg). In a MAD study, VIP BID (10 mg) was administered using a Plastiape RS00 dry powder inhaler as one to eight 10 mg capsules, i.e., subjects in group 1 will receive 10 mg BID, group 2 (20 mg; BID), group 3 (40 mg; BID), and group 4 (80 mg; BID). In both studies, inhalations were conducted within a 10-minute time. TFF Pharmaceuticals, Inc. has completed the VIP phase 1 clinical trial and hoping for positive results [95]. Recently, TFF Pharmaceuticals, Inc. started recruitment for a Phase 1b (NCT04576325) randomized, double-blind, placebo-controlled clinical trial to assess VIP safety, tolerability, and pharmacokinetic profiles in adult subjects with well-controlled asthma [96].

2.3.2 Tacrolimus Inhalation Powder

Sahakijpijarn and team judiciously studied the safety and tolerability of TFF-processed TAR solution and dry powder in healthy volunteers. TAR respirable formulation was formulated using well-established protocol, i.e., TAR: LAC (50/50% w/w) was carefully dissolved in acetonitrile/water (60/40% v/v) solvent mixture. The prepared mixture was cautiously dropped on a cryogenic spinning cylinder and an ultra-rapid freezing kinetics fabricate the frozen discs. Frozen discs were transferred to a lyophilizer. The primary drying was performed at −40 °C for 20 h while secondary drying was performed at 25 °C for 20 h. Twenty human subjects (10 females and 10 males) between the ages of 18–55 years were involved in clinical research. Nebulized TFF TAR/LAC [TFF powder (3 mg TAR) dispersed in sterile water for injection and prepared colloidal dispersion was nebulized using Aeroneb Pro® nebulizer (Aerogen, Ireland)] and dry powder TAR/LAC [TFF powder (3 mg of TAR) delivered using HandiHaler® device] was administered to human subjects in the first and second phase of clinical study, correspondingly. Pharmacological assessment exhibited that orally inhaled colloidal dispersions and dry powder displayed low systemic absorption. Biochemical and spirometry data showed that both formulations were well-tolerated by human subjects without any major changes in complete blood count, lung, liver, and kidney functions. Few mild side effects, i.e., distaste, cough, and throat irritation were noted. In brief, pulmonary delivery of TFF-processed solid and liquid TAR formulation is a safe and hopeful treatment for lung transplant subjects [97].

2.4 Future Perspective

Several published articles showed that TFF has good potential and capability to formulate powders or particles suitable for deep lung delivery (Table 2.2). As reviewed above, many preliminary investigations used TFF to design nano/micron-size particles with understanding of particles' physicochemical properties and corroborate proof-of-concept. Afterward these can be logically compared to the same or other customary rapid freezing methods and milling techniques. Like well-established rapid freezing methods, TFF also efficiently formulated diverse particle morphologies and sizes with satisfactory reproducibility. TFF is an efficient scheme to transfer liquid droplets to inhalable dry powders without causing particle aggregation or loss of therapeutic potential.

The TFF processing parameters, solvent selection (viscosity, composition, and cryo-phase separation), and excipients concentration significantly impact TFF-processed powders. The higher viscosity co-solvent reduces the movement of molecules during the ultra-rapid freezing phase, which results in more uniform dispersal within the frozen state. In contrast, the low-viscosity co-solvent permits molecules to move more freely, enhancing the possibility of molecular aggregation and afterward decreasing the aerosol performance. The solvent composition

Table 2.2 Thin-film freezing driven inhaled formulations along with the key findings

Formulation	Particle properties	Outcome	References
TAR-LAC	Low-density brittle matrices (49.10 μm)	The formulation showed FPF and MMAD of 68.70% and 2.71 μm, respectively.	[62]
TAR-MAN	Low-density brittle matrices (64.78 μm)	The formulation showed FPF and MMAD of 48.80% and 4.62 μm, respectively, and during storage, at 50%, MAN matrices showed stable MMAD (4.10 μm) values.	[62]
TAR-RAF	Low-density brittle matrices (40.71 μm)	The formulation showed FPF and MMAD of 69.20% and 2.60 μm, respectively.	[62]
TAR-MAN (1:1)	Highly porous (17.84 μm)	The formulation showed satisfactory FPF (83.3%) and lower MMAD (2.26 μm) values using Miat® monodose inhaler at 90 L/min (2.7 s) and a marked improvement in pulmonary bioavailability (2.34-fold) and mean retention time (11.63-fold) compared to micronized TAR-MAN formulation.	[63]
TAR-LAC (95:5)	Highly porous brittle particles	The formulation showed ED with FPF of 95.02 and 96.6% using high-resistance Plastiape® inhaler (4 kpa and 60 L/min).	[64]
VOR-PVP K12 (1:2)	Nanostructured amorphous powder	The formulation showed ED, FPF, and MMAD of 96.50%, 42.40%, and 4.5 μm, respectively, using Handihaler® at 60 L/min (4 s).	[69]
VOR- PVP K25 (1:3)	Nanostructured amorphous powder (0.05 μm)	The formulation showed ED and FPF of 92.50% and 32.40%, respectively, using Handihaler® at 60 L/min (4 s).	[70]
VOR	Microstructured crystalline powder	The formulation showed ED and FPF of 81.20% and 37.80%, respectively, using Handihaler® at 60 L/min (4 s) and 1.95-fold superior VOR concentration within the lung tissue as compared to TFF-processed amorphous VOR powder at a similar drug dose (10 mg/kg).	[70]
VOR-MAN (95:05)	Rough surface nanoaggregates containing NPs of <500 nm	The formulation showed FPF of 28.50% and maintained aerosolization performance during long-term 13 months (25 °C/ 60% RH) storage. The formulation showed no remarkable variation in physicochemical and aerosolization properties at a 450-fold bigger manufacturing scale.	[71, 72]
VOR-MAN (50:50)	Porous matrix	The formulation showed FPF of 38.10% and cumulative VOR release of 35.30% at the end of 3 h using phosphate buffer (pH 7.4).	[71]

(continued)

Table 2.2 (continued)

Formulation	Particle properties	Outcome	References
REM-Captisol® (0.30%)	Amorphous nanostructure aggregates	The formulation showed FPF (< 3 μm) and MMAD of 62.25% and 2.16 μm, respectively, using a high-resistance RS01 Plastiape® inhaler at 60 L/min (4 s) and improved systemic uptake as compared to TFF-processed REM-leucine formulation.	[73]
Anti-PD-1-LL-PBS mAb) dry powder	Highly porous nanoaggregates	The formulation presented a marked threefold higher % FPF and 2.3-fold lower MMAD values compared to shelf freeze drying process product using a high-resistance RS01 Plastiape® inhaler at 60 L/min (4 s).	[77]
PEGylated chitosan/ CRISPR-Cas9 MAN dry powder	Highly organized nanocomplex	Nanocomplexes using HandiHaler at 45 and 60 L/min flow rate displayed FPF of 18.4 and 43.3%, correspondingly while using Monodose RS01 at 45 and 60 L/min flow rate displayed FPF of 23 and 60%, correspondingly.	[84]
PEGylated chitosan/ CRISPR-Cas9 MAN/LEU dry powder	Highly organized nanocomplex	Nanocomplexes using HandiHaler at 45 and 60 L/min flow rate showed FPF of 24 and 63%, individually while using Monodose RS01 at 45 and 60 L/min flow rate showed FPF of 39.2 and 62.3%, individually.	[84]
Shingrix vaccine	Uniform small particles without any aggregation (105 nm)	Aerodynamic analysis was performed using Plastiape Monodose RS01 (60 L/min) displayed a marked 12.50 and 11.19-fold higher delivered and recovered dose % FPF (< 5 μm) compared to traditional shelf freeze drying processed dry powders.	[85]
Niclosamide	Nano-sized crystalline powder	Aerodynamic assessment using Plastiape® RS00 device (58 L/min) displayed FPF and MMAD of 76.6% and 1.1 μm, respectively.	[89]
RAP-LAC (1:1)	Amorphous particles	The formulation showed twofold higher FPF and 3.58-fold higher RAP systemic concentration than crystalline RAP formulation.	[90]
MF and SX combination	Brittle matrix powder	The combination showed a higher emitted dose for MF (93.91%) as compared to micronized MF (71.18%) with consistent dose uniformity using Miat® monodose inhaler at 90 L/min (2.7 s) and enhanced pulmonary bioavailability (5.85-fold) and mean retention time (1.01-fold) as compared to the crystalline micronized MF formulation.	[91]

Table 2.2 (continued)

Formulation	Particle properties	Outcome	References
MF and SX combination	Brittle matrix powder	The combination showed a higher total emitted dose for SX (95.31%) as compared to micronized SX (68.59%) blend with consistent dose uniformity using Miat® monodose inhaler at 90 L/min (2.7 s) and enhanced pulmonary bioavailability of SX (15.25-fold)] with mean retention time (1.13-fold) as compared to the crystalline micronized SX formulation.	[91]
BSA	Nanorods (< 100 nm)	HFA-227 containing BSA nanorods showed a threefold higher delivered dose than milled BSA with FPF and MMAD of 47% and 3.2 μm, respectively.	[94]

of TFF mainly influences the droplet spreading across the cryogenically cooled surface and thus conclude the specific surface area and macroscopic powder properties. The polymeric stabilization at a drug to excipient ratio of at least 1 or 2 is necessary to fabricate nanostructured, amorphous solid dispersions by the TFF process [69, 72]. The droplet impingement and solidification onto a solid surface in freezing systems significantly impact particle morphologies. The cooling kinetics reached by the traditional freezing methods, i.e., spray freezing into liquid (10^3 K/s) and spray freeze drying (10^6 K/s), is superior as compared to TFF (10^2 K/s). However, TFF is quick to formulate stable powders or particles due to its unique air–liquid interface, droplet flow dynamics, and heat transfer processes [98]. Usually, TFF produces large droplets with an air–liquid interface of falling droplets and spreading films are lesser than the spray freeze-generated atomized droplets [98].

As reviewed in the above section, TFF-processed powders showed various advantages over conventional high-energy milling powders. TFF-processed amorphous powders showed suitable lung fluids solubility and increased in vivo systemic bioavailability compared to crystalline counterparts prepared using milling methods [90]. TFF-based powders retained good aerosol performance because the ultra-rapid freezing rate of TFF produces a low-density and high-porosity brittle matrix, which is effectively aerosolized by the low shear force of the inhaler. Additionally, TFF offers high drug loading as dry powder formulated by TFF needs only a small amount of excipient to stabilize its amorphous form [64, 73]. TAR [64], REM [73] powders, and fixed-dose combination, i.e., MF/SX powder [91] showed promising aerosol applications when prepared using the TFF process. TFF-processed dry powders exhibited better respirable fractions and higher drug deposition patterns in most of the studies. The powders manufactured using TFF are mainly composed of large interwoven matrices rather than discrete particles [62]. Apart from dry powder formulations, TFF can also formulate stable pMDIs. The uniquely designed nanorods-based flocs showed well-balanced van der Waals energies, presented effective floc formation, and broke up when suspended with the HFA [94]. Moreover, TFF-processed dry powders displayed prolonged drug retention within the airways with lower systemic concentration and decreased side effects because of minor dose variability compared to the micronized formulation [63].

Moreover, TFF-dried vaccine powders are not susceptible to freeze-thaw cycles. They can be stored at higher temperatures (40 °C) for months without losing the therapeutic potential. In contrast, TFF-dried powder can be effectively stored in a container with minimum humidity in longer-term storage. Furthermore, the TFF method was well verified and tested in the development of commercial human and veterinary vaccines [98]. Thus, the vaccination through respiratory airways can be performed successfully using TFF-dried powder for several immunization therapies. Briefly, such fresh formulation strategies and simple inhaler devices handling tactics offer ample room for developing new pulmonary products. Concisely, all such active formulations and clinical actions facilitate a firm therapeutic intervention for treating acute lung diseases.

2.5 Conclusion

This chapter has thoroughly summarized the applications of the TFF in pulmonary drug delivery with a specific focus on particle morphology and aerodynamic assessment. TFF showed satisfactory freezing and drying performance compared to well-established spray drying, spray freeze drying, and supercritical fluid processing techniques. TFF has excellent capability to fabricate low-density and high-porosity brittle powder matrix due to its unique process efficiency, i.e., lower freezing time (< 150 ms) and gas–liquid interface of fewer than 50 cm^{-1}. Moreover, the TFF showed superior performance than conventional micronization, dry/wet milling, and dry powder processing methods. TFF-derived brittle powder matrix displayed acceptable physicochemical, biophysical, pulmokinetics, and aerosolization results. Although there are numerous practical applications with TFF, additional research efforts are essential for understanding the correlation between co-solvent fraction, cooling substrate characteristics, drug: excipient ratio affecting powder morphology. Indisputably, the continued valuable efforts in the TFF field will stimulate the bench-to-bedside translation of pulmonary therapies.

References

1. Cun D, Zhang C, Bera H, Yang M. Particle engineering principles and technologies for pharmaceutical biologics. Adv Drug Deliv Rev. 2021;174:140–67.
2. Vehring R. Pharmaceutical particle engineering via spray drying. Pharm Res. 2008;25(5):999–1022.
3. Khadka P, Ro J, Kim H, Kim I, Kim JT, Kim H, Cho JM, Yun G, Lee J. Pharmaceutical particle technologies: an approach to improve drug solubility, dissolution and bioavailability. Asian J Pharmaceut Sci. 2014;9(6):304–16.

4. Alkilani AZ, McCrudden MT, Donnelly RF. Transdermal drug delivery: innovative pharmaceutical developments based on disruption of the barrier properties of the stratum corneum. Pharmaceutics. 2015;7(4):438–70.
5. Loftsson T. Cyclodextrins in parenteral formulations. J Pharm Sci. 2021;110(2):654–64.
6. Shi Y, Lu A, Wang X, Belhadj Z, Wang J, Zhang Q. A review of existing strategies for designing long-acting parenteral formulations: focus on underlying mechanisms, and future perspectives. Acta Pharm Sin B. 2021;11(8):2396–415.
7. Li Z, Lin X, Shen L, Hong Y, Feng Y. Composite particles based on particle engineering for direct compaction. Int J Pharm. 2017;519(1–2):272–86.
8. Shoyele SA, Cawthorne S. Particle engineering techniques for inhaled biopharmaceuticals. Adv Drug Deliv Rev. 2006;58(9–10):1009–29.
9. Chow AH, Tong HH, Chattopadhyay P, Shekunov BY. Particle engineering for pulmonary drug delivery. Pharm Res. 2007;24(3):411–37.
10. Kaialy W, Nokhodchi A. In: Nokhodchi A, editor. Particle engineering for improved pulmonary drug delivery through dry powder inhalers. Pulmonary drug delivery: advances and challenges. Martin, GP; 2015. p. 171–98.
11. Mehta PP, Kadam SS, Pawar AP. Influence of modified induction port, modified DUSA assembly and device air-inlet geometry on the aerosolization pattern of a dry powder inhaler. J Drug Deliv Sci Technol. 2020;55:101416.
12. Chan HK. What is the role of particle morphology in pharmaceutical powder aerosols? Expert Opin Drug Deliv. 2008;5(8):909–14.
13. Jadhav P, Patil P, Bhagwat D, Gaikwad V, Mehta PP. Recent advances in orthogonal analytical techniques for microstructural understanding of inhalable particles: present status and future perspective. J Drug Deliv Sci Technol. 2022:103089.
14. Nokhodchi A, Martin GP, editors. Pulmonary drug delivery: advances and challenges; 2015. p. 1–384.
15. Cun D, Wan F, Yang M. Formulation strategies and particle engineering technologies for pulmonary delivery of biopharmaceuticals. Curr Pharm Des. 2015;21(19):2599–610.
16. Mehta PP. Dry powder inhalers: a brief overview of the drug detachment techniques. Ther Deliv. 2020;11(3):139–43.
17. Mehta PP, Ghoshal D, Pawar AP, Kadam SS, Dhapte-Pawar VS. Recent advances in inhalable liposomes for treatment of pulmonary diseases: concept to clinical stance. J Drug Deliv Sci Technol. 2020;56:101509.
18. Mehta PP, Kadam SS, Pawar AP, Bothiraja C. Dendrimers for pulmonary delivery: current perspectives and future challenges. New J Chem. 2019;43:8396–409.
19. Mehta PP. Dry powder inhalers: upcoming platform technologies for formulation development. Ther Deliv. 2019;10(9):551–4.
20. Kumar R, Mehta P, Shankar KR, Rajora MA, Mishra YK, Mostafavi E, Kaushik A. Nanotechnology-assisted metered-dose inhalers (MDIs) for high-performance pulmonary drug delivery applications. Pharm Res. 2022;1-25
21. Mehta P. Dry powder inhalers: a focus on advancements in novel drug delivery systems. J Drug Deliv. 2016;2016:8290963.
22. Myrdal PB, Sheth P, Stein SW. Advances in metered dose inhaler technology: formulation development. AAPS PharmSciTech. 2014;15(2):434–55.
23. Alhajj N, O'Reilly NJ, Cathcart H. Designing enhanced spray dried particles for inhalation: a review of the impact of excipients and processing parameters on particle properties. Powder Technol. 2021;384:313–31.
24. Mehta PP, Bothiraja C, Kadam SS, Pawar AP. Effect of USP induction ports, glass sampling apparatus, and inhaler device resistance on aerodynamic patterns of fluticasone propionate loaded engineered mannitol microparticles. AAPS PharmSciTech. 2019;20(5):197.
25. Vishali DA, Monisha J, Sivakamasundari SK, Moses JA, Anandharamakrishnan C. Spray freeze drying: emerging applications in drug delivery. J Control Release. 2019;300:93–101.

26. Yu H, Teo J, Chew JW, Hadinoto K. Dry powder inhaler formulation of high-payload antibiotic nanoparticle complex intended for bronchiectasis therapy: spray drying versus spray freeze drying preparation. Int J Pharm. 2016;499(1–2):38–46.
27. Kankala RK, Xu PY, Chen BQ, Wang SB, Chen AZ. Supercritical fluid (SCF)-assisted fabrication of carrier-free drugs: an eco-friendly welcome to active pharmaceutical ingredients (APIs). Adv Drug Deliv Rev. 2021;176:113846.
28. Aguiar-Ricardo A. Building dry powder formulations using supercritical CO2 spray drying. Curr Opin Green Sustain Chem. 2017;5:12–6.
29. Cai Y, Chen Y, Hong X, Liu Z, Yuan W. Porous microsphere and its applications. Int J Nanomedicine. 2013;8:1111–20.
30. Shahin H, Vinjamuri BP, Mahmoud AA, Mansour SM, Chougule MB, Chablani L. Formulation and optimization of sildenafil citrate-loaded PLGA large porous microparticles using spray freeze-drying technique: a factorial design and in-vivo pharmacokinetic study. Int J Pharm. 2021;597:120320.
31. Zhang X, Qin L, Su J, Sun Y, Zhang L, Li J, Beck-Broichsitter M, Muenster U, Chen L, Mao S. Engineering large porous microparticles with tailored porosity and sustained drug release behavior for inhalation. Eur J Pharm Biopharm. 2020;155:139–46.
32. Garcia A, Mack P, Williams S, Fromen C, Shen T, Tully J, Pillai J, Kuehl P, Napier M, DeSimone JM, Maynor BW. Microfabricated engineered particle systems for respiratory drug delivery and other pharmaceutical applications. J Drug Deliv. 2012;2012
33. López-Iglesias C, Casielles AM, Altay A, Bettini R, Alvarez-Lorenzo C, García-González CA. From the printer to the lungs: inkjet-printed aerogel particles for pulmonary delivery. Chem Eng J. 2018;357:559–66.
34. Weers J, Tarara T. The PulmoSphere™ platform for pulmonary drug delivery. Ther Deliv. 2014;5(3):277–95.
35. Angelo R, Rousseau K, Grant M, Leone-Bay A, Richardson P. Technosphere insulin: defining the role of Technosphere particles at the cellular level. J Diabetes Sci Technol. 2009;3(3):545–54.
36. Shadambikar G, Marathe S, Ji N, Almutairi M, Bandari S, Zhang F, Chougule M, Repka M. Formulation development of itraconazole PEGylated nano-lipid carriers for pulmonary aspergillosis using hot-melt extrusion technology. Int J Pharm X. 2021;3:100074.
37. Mehta PP, Pawar AP, Mahadik KR, Kadam SS, Dhapte-Pawar V. Dry powder coating techniques and role of force controlling agents in aerosol. Polym Coat Technol Appl. 2020:41–74.
38. Mehta PP, Dhapte-Pawar V. Role of surfactants in pulmonary drug delivery. In: Green sustainable process for chemical and environmental engineering and science. Academic Press; 2022. p. 559–77.
39. Overhoff KA, Johnston KP, Tam J, Engstrom J, Williams RO III. Use of thin film freezing to enable drug delivery: a review. J Drug DelivSciTechnol. 2009;19(2):89–98.
40. Emami F, Vatanara A, Park EJ, Na DH. Drying Technologies for the Stability and Bioavailability of biopharmaceuticals. Pharmaceutics. 2018;10(3):131.
41. Gradon L, Sosnowski TR. Formation of particles for dry powder inhalers. Adv Powder Technol. 2014;25(1):43–55.
42. Engstrom JD, Lai ES, Ludher BS, Chen B, Milner TE, Williams RO, Kitto GB, Johnston KP. Formation of stable submicron protein particles by thin film freezing. Pharm Res. 2008;25:1334–46.
43. Overhoff KA, Engstrom JD, Chen B, Scherzer BD, Milner TE, Johnston KP, Williams RO. Novel ultra-rapid freezing particle engineering process for enhancement of dissolution rates of poorly water-soluble drugs. Eur J Pharm Biopharm. 2007;65:57–67.
44. Overhoff KA, McConville JT, Yang W, Johnston KP, Peters JI, Williams RO. Effect of stabilizer on the maximum degree and extent of supersaturation and oral absorption of tacrolimus made by ultra-rapid freezing. Pharm Res. 2008;25:167–75.
45. Overhoff KA, Moreno A, Miller DA, Johnston KP, Williams RO. Solid dispersions of itraconazole and enteric polymers made by ultra-rapid freezing. Int J Pharm. 2007;336:122–32.

46. Johnston KP, Engstrom J and Williams III RO. Formation of stable submicron peptide or protein particles by thin film freezing. US 2017 / 0360711 A1, 2017.
47. Kim-Ngan NT, Tkach I, Mašková S, Havela L, Warren A, Scott T. Cubic γ-phase U–Mo alloys synthesized by splat-cooling. Adv Nat Sci Nanosci Nanotechnol. 2013;4(3):035006.
48. Ghandehariun S, Wang Z, Naterer GF, Rosen MA. Experimental investigation of molten salt droplet quenching and solidification processes of heat recovery in thermochemical hydrogen production. Appl Energy. 2015;157:267–75.
49. Wang C, Wang Z, Song Y, Zhang X, Wang J, Liu P, Cao H, Xu B, Zhao M, Miao Y, Guo J. Facile synthesis of perovskite phosphors and nanocrystals using laundry detergent by ultrarapid freezing for light-emitting diodes application. J Lumin. 2021;233:117902.
50. Espallargas N, editor. Future development of thermal spray coatings: types, designs, manufacture and applications. Elsevier; 2015.
51. Yin J, Zhang X, Feng J, Zhang X, Song J, Liu M, Zeng D, Mao J, Deng C, Deng C, Deng Z. Effect of powder composition upon plasma spray-physical vapor deposition of 8YSZ columnar coating. Ceram Int. 2020;46(10):15867–75.
52. Liu Y, Yao C, Yin J. Test parameters optimization for constrained spray forming of aluminum alloy based on artificial neural network. Eng Res Exp. 2020;2(3):035025.
53. Gilkey JC, Staehelin LA. Advances in ultrarapid freezing for the preservation of cellular ultrastructure. J Electron Microsc Tech. 1986;3:177–210.
54. Zasadzinski JAN. A new heat transfer model to predict cooling rates for rapid freezing fixation. J Microsc. 1988;150:137–49.
55. Heuser JE, Reese TS, Landis DM. Preservation of synaptic structure by rapid freezing. Cold Spring Harb Symp Quant Biol. 1976;40:17–24.
56. Escaig J. New instruments which facilitate rapid freezing at 83 K and 6 K. J Microsc. 1982;126:221–30.
57. Longest PW, Hindle M. Small airway absorption and microdosimetry of inhaled corticosteroid particles after deposition. Pharm Res. 2017;34(10):2049–65. https://doi.org/10.1007/s11095-017-2210-7.
58. Brunaugh AD, Smyth HDC. Formulation techniques for high dose dry powders. Int J Pharm. 2018;547(1–2):489–98.
59. Malecic N, Young H. Tacrolimus for the management of psoriasis: clinical utility and place in therapy. Psoriasis (Auckl). 2016;6:153–63.
60. Ide N, Nagayasu T, Matsumoto K, Tagawa T, Tanaka K, Taguchi T, Sumida Y, Nakashima M. Efficacy and safety of inhaled tacrolimus in rat lung transplantation. J Thorac Cardiovasc Surg. 2007;133(2):548–53.
61. Schrepfer S, Deuse T, Reichenspurner H, Hoffmann J, Haddad M, Fink J, Fischbein MP, Robbins RC, Pelletier MP. Effect of inhaled tacrolimus on cellular and humoral rejection to prevent posttransplant obliterative airway disease. Am J Transplant. 2007;7(7):1733–42.
62. Watts AB, Wang YB, Johnston KP, Williams RO 3rd. Respirable low-density microparticles formed in situ from aerosolized brittle matrices. Pharm Res. 2013;30(3):813–25.
63. Wang YB, Watts AB, Peters JI, Liu S, Batra A, Williams RO 3rd. In vitro and in vivo performance of dry powder inhalation formulations: comparison of particles prepared by thin film freezing and micronization. AAPS PharmSciTech. 2014;15(4):981–93.
64. Sahakijpijarn S, Moon C, Ma X, Su Y, Koleng JJ, Dolocan A, Williams RO 3rd. Using thin film freezing to minimize excipients in inhalable tacrolimus dry powder formulations. Int J Pharm. 2020;586:119490.
65. Doffman SR, Agrawal SG, Brown JS. Invasive pulmonary aspergillosis. Expert Rev Anti-Infect Ther. 2005;3:613–27.
66. Xie Y, Zeng P, Wiedmann TS. Disease guided optimization ofthe respiratory delivery of microparticulate formulations. Expert Opin Drug Deliv. 2008;5:269–89.
67. Roffey SJ, Cole S, Comby P, Gibson D, Jezequel SG, Nedderman ANR, Smith DA, Walker DK, Wood N. The disposition of voriconazole in mouse, rat, rabbit, Guinea pig, dog, and human. Drug Metab Dispos. 2003;31:731–41.

68. Herbrecht R, Denning DW, Patterson TF, Bennett JE, Greene RE, Oestmann JW, Kern WV, Marr KA, Ribaud P, Lortholary O, Sylvester R, Rubin RH, Wingard JR, Stark P, Durand C, Caillot D, Thiel E, Chandrasekar PH, Hodges MR, Schlamm HT, Troke PF, de Pauw B. Voriconazole versus amphotericin B for primary therapy of invasive aspergillosis. N Engl J Med. 2002;347:408–15.
69. Beinborn NA, Lirola HL, Williams RO 3rd. Effect of process variables on morphology and aerodynamic properties of voriconazole formulations produced by thin film freezing. Int J Pharm. 2012;429(1–2):46–57.
70. Beinborn NA, Du J, Wiederhold NP, Smyth HD, Williams RO 3rd. Dry powder insufflation of crystalline and amorphous voriconazole formulations produced by thin film freezing to mice. Eur J Pharm Biopharm. 2012;81(3):600–8.
71. Moon C, Watts AB, Lu X, Su Y, Williams RO 3rd. Enhanced aerosolization of high potency nanoaggregates of voriconazole by dry powder inhalation. Mol Pharm. 2019;16(5):1799–812.
72. Moon C, Sahakijpijarn S, Koleng JJ, Williams RO III. Processing design space is critical for voriconazole nanoaggregates for dry powder inhalation produced by thin film freezing. J Drug Deliv Sci Technol. 2019;54:101295.
73. Sahakijpijarn S, Moon C, Koleng JJ, Christensen DJ, Williams Iii RO. Development of remdesivir as a dry powder for inhalation by thin film freezing. Pharmaceutics. 2020;12(11):1002.
74. Sahakijpijarn S, Moon C, Warnken ZN, Maier EY, DeVore JE, Christensen DJ, Koleng JJ, Williams RO III. In vivo pharmacokinetic study of remdesivir dry powder for inhalation in hamsters. Int J Pharmaceutics: X. 2021;3:100073.
75. Awwad S, Angkawinitwong U. Overview of antibody drug delivery. Pharmaceutics. 2018;10(3):83.
76. Lu RM, Hwang YC, Liu IJ, et al. Development of therapeutic antibodies for the treatment of diseases. J Biomed Sci. 2020;27:1.
77. Hufnagel S, Xu H, Sahakijpijarn S, Moon C, Chow LQ, Williams RO III, Cui Z. Dry powders for inhalation containing monoclonal antibodies made by thin-film freeze-drying. Int J Pharm. 2022;618:121637.
78. Burgel PR, Burnet E, Regard L, Martin C. The changing epidemiology of cystic fibrosis: the implications for adult care. Chest. 2022; https://doi.org/10.1016/j.chest.2022.07.004.
79. Despotes KA, Donaldson SH. Current state of CFTR modulators for treatment of cystic fibrosis. Curr Opin Pharmacol. 2022;65:102239.
80. Hodges CA, Conlon RA. Delivering on the promise of gene editing for cystic fibrosis. Genes Dis. 2019;6(2):97–108.
81. Boyd AC, Guo S, Huang L, Kerem B, Oren YS, Walker AJ, Hart SL. New approaches to genetic therapies for cystic fibrosis. J Cyst Fibros. 2020;19:S54–9.
82. Luthra R, Kaur S, Bhandari K. Applications of CRISPR as a potential therapeutic. Life Sci. 2021;284:119908.
83. Zhang H, Bahamondez-Canas TF, Zhang Y, Leal J, Smyth HDC. PEGylated chitosan for nonviral aerosol and mucosal delivery of the CRISPR/Cas9 system in vitro. Mol Pharm. 2018;15(11):4814–26.
84. Zhang H, Zhang Y, Williams RO 3rd, Smyth HDC. Development of PEGylated chitosan/CRISPR-Cas9 dry powders for pulmonary delivery via thin-film freeze-drying. Int J Pharm. 2021;605:120831.
85. AboulFotouh K, Xu H, Moon C, Williams RO III, Cui Z. Development of (inhalable) dry powder formulations of AS01B-containing vaccines using thin-film freeze-drying. Int J Pharm. 2022;622:121825.
86. Mehta PP, Dhapte-Pawar VS. Repurposing drug molecules for new pulmonary therapeutic interventions. Drug Deliv Transl Res. 2021;11(5):1829–48.
87. Xu J, Shi PY, Li H, Zhou J. Broad Spectrum antiviral agent Niclosamide and its therapeutic potential. ACS Infect Dis. 2020;6(5):909–15.
88. Arshad U, Pertinez H, Box H, Tatham L, Rajoli RKR, Curley P, Neary M, Sharp J, Liptrott NJ, Valentijn A, David C, Rannard SP, O'Neill PM, Aljayyoussi G, Pennington SH, Ward SA, Hill A, Back DJ, Khoo SH, Bray PG, Biagini GA, Owen A. Prioritization of anti-SARS-Cov-2

drug repurposing opportunities based on plasma and target site concentrations derived from their established human pharmacokinetics. Clin Pharmacol Ther. 2020;108(4):775–90.
89. Jara MO, Warnken ZN, Sahakijpijarn S, Moon C, Maier EY, Christensen DJ, Koleng JJ, Peters JI, Hackman Maier SD, Williams Iii RO. Niclosamide inhalation powder made by thin-film freezing: multi-dose tolerability and exposure in rats and pharmacokinetics in hamsters. Int J Pharm. 2021;603:120701.
90. Carvalho SR, Watts AB, Peters JI, Liu S, Hengsawas S, Escotet-Espinoza MS, Williams RO 3rd. Characterization and pharmacokinetic analysis of crystalline versus amorphous rapamycin dry powder via pulmonary administration in rats. Eur J Pharm Biopharm. 2014;88(1):136–47.
91. Liu S, Watts AB, Du J, Bui A, Hengsawas S, Peters JI, Williams RO 3rd. Formulation of a novel fixed dose combination of salmeterol xinafoate and mometasone furoate for inhaled drug delivery. Eur J Pharm Biopharm. 2015;96:132–42.
92. Mehta PP. Dry powder inhalers: a concise summary of the electronic monitoring devices. Ther Deliv. 2021;12(1):1–6.
93. Mehta P, Bothiraja C, Kadam S, Pawar A. Potential of dry powder inhalers for tuberculosis therapy: facts, fidelity and future. Artif Cells Nanomed Biotechnol. 2018;46(sup3):S791–806.
94. Engstrom JD, Tam JM, Miller MA, Williams RO 3rd, Johnston KP. Templated open flocs of nanorods for enhanced pulmonary delivery with pressurized metered dose inhalers. Pharm Res. 2009;26(1):101–17.
95. Single Ascending Dose and Multiple Ascending Dose Study of Voriconazole Inhalation Powder in Healthy Adult Subjects. NCT04872231. https://clinicaltrials.gov/ct2/show/NCT04872231?term=voriconazole+TFF+Pharmaceuticals&draw=2&rank=1
96. Pharmacokinetic profile of voriconazole inhalation powder in adult subjects with asthma. NCT04576325. https://clinicaltrials.gov/ct2/show/NCT04576325?term=voriconazole+TFF+Pharmaceuticals&draw=2&rank=2
97. Sahakijpijarn S, Beg M, Levine SM, Peters JI, Williams RO 3rd. A safety and tolerability study of thin film freeze-dried tacrolimus for local pulmonary drug delivery in human subjects. Pharmaceutics. 2021;13(5):717.
98. Blaine A. Pfeifer and Andrew Hill. Vaccine Delivery Technology. Springer Science+Business Media, LLC, part of Springer Nature. 2021. https://doi.org/10.1007/978-1-0716-0795-4.

Supercritical Fluid Technology as a Tool for Improved Drug Delivery to the Lungs

3

Alireza Ebrahimi, Hamed Hamishehkar, and Ali Nokhodchi

Abstract

To treat respiratory diseases, it is more efficient to deliver drugs through the respiratory system Drug delivery through the respiratory system assures efficient therapy in respiratory diseases. For this purpose, different types of inhaler formulations have been explored, where dry powder inhalers are more important than other inhalation formulations due to their ease of use, bioavailability, and high stability. The rate of transfer of the active pharmaceutical ingredients in the desired formulation depends on the size of the particles and other physicochemical parameters such as drug solubility, and other inhalation factors, e.g., fine particle dose (FPD) and delivered dose uniformity (DDU) fine particle fraction. Recently, the utilization of supercritical fluid (SCF) technology has attracted the attention of many researchers due to its non-flammability, non-toxic, non-reactive, economical, and non-polluting

A. Ebrahimi
Biotechnology Research Center, Tabriz University of Medical Sciences, Tabriz, Iran

Student Research Committee, and Department of Food Science and Technology, Faculty of Nutrition and Food Sciences, Tabriz University of Medical Sciences, Tabriz, Iran

H. Hamishehkar
Drug Applied Research Center, Tabriz University of Medical Sciences, Tabriz, Iran
e-mail: hamishehkarh@tbzmed.ac.ir

A. Nokhodchi (✉)
Lupin Inhalation Research Center, Coral Springs, FL, USA

School of Life Sciences, University of Sussex, Brighton, UK
e-mail: AliNokhodchi@lupin.com; a.nokhodchi@sussex.ac.uk

P. P. Mehta, V. Dhapte-Pawar (eds.), *Pulmonary Drug Delivery Systems: Material and Technological Advances*,
https://doi.org/10.1007/978-981-99-1923-9_3

properties. Furthermore, this technology is based on a green chemistry approach. Improving the solubility of poorly water-soluble pharmaceuticals by using organic solvents with little or no heat is one of the other characteristics of SCF. This chapter examines the recent applications and advances of SCF for the production of inhalation powders. The chapter also focuses on the effect of this technology on the aerosolization performance of actives from inhaler devices.

Keywords

Supercritical fluid · Inhalation drug delivery · Dry powder inhaler · Inhalable particles

3.1 Introduction

The poor physicochemical, pharmacodynamics, and pharmacokinetic properties of some inhalation drugs limit their therapeutic effects. Also, the low solubility of some of them causes frequent administrations, which can lead to serious adverse effects [1]. In recent decades, researchers have started employing nanotechnology approaches to enhance the bioavailability of drugs and also to overcome poor physicochemical properties such as poor solubility, [2, 3]. Furthermore, nanotechnology approaches are employed to design a drug delivery system that can release the drug more efficiently so that the therapeutic effect of the drug (as it is in nano-form) would be greater than its free form [4, 5]. In addition, it protects the drug from external factors such as oxygen, light, temperature, pH, enzymes, and others [6]. One of the methods to tailor the physicochemical properties of drugs for inhalation delivery is the supercritical fluid (SCF) technique which has several advantages over other drug production methods. These include (a) adjusting drug loading by changing operating conditions such as pressure, temperature, and time, as well as (b) changing the drug solubility.

Recently, SCF technology has attracted the attention of many researchers in the field to improve the properties of inhalable particles, and also to manufacture active pharmaceutical ingredients (API) on a large scale. The fluids used in SCF technology should have specific properties where they should become supercritical fluids at certain temperatures and pressures higher than their critical values [7]. The use of this technology has benefits such as being non-toxic, ineffective, cost-effective, environmentally friendly, and possessing high compressibility [8]. Of course, it should be borne in mind that the physical properties of SCFs are also controlled by temperature and pressure, which will be discussed below.

Carbon dioxide (CO_2) is the most commonly utilized SCF in the pharmaceutical industry, which is also used in the production of inhaled drugs as it is classified by the FDA as a safe solvent [9]. This gas is inert, colorless, odorless, non-toxic, non-flammable, economical, and recyclable. The critical temperature of CO_2 is 31.1 ° C and the critical pressure is 74 bar [10].

3.2 SCF-Based Manufacturing Technologies

Figure 3.1 presents the general scheme of the fabrication of API crystals without carriers (micronized and nanosized) using different strategies (solvent, anti-solvent and additive) and the effect of different processing and formulation aspects for various pharmaceutical applications.

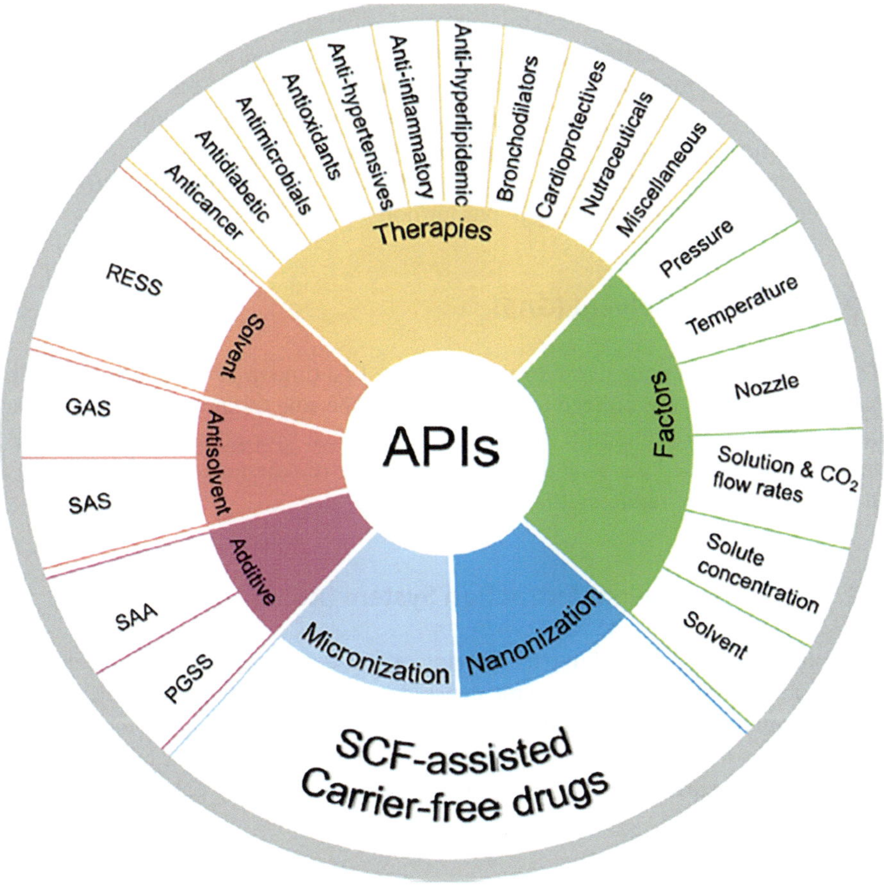

Fig. 3.1 The figure shows the overall view of the application of various SCF technology, the parameters involved, and the different types of therapeutic groups used in SCF technology. Reprinted with permission from [9]. Copyright 2022 Elsevier

3.2.1 Rapid Expansion of Supercritical Solutions (RESS) and Related Processes

The RESS technique, which has been reported for decades, is a process in which an SCF, such as CO_2, is employed as a solvent and then passed through a nozzle with a sudden drop in the pressure to precipitate powders in the system [11]. An abrupt decline in pressure results in a significant decrease in the solubility of the substance in the SCF causing the substance to precipitate. During this process, the temperature of the SCF is increased to get fine particles with a uniform size distribution which is due to an increase in the nucleation process and restricting the crystal growth. Bear in mind, it is possible to modulate the size of drug particles by adjusting the pressure and temperature [12–14].

This method of manufacturing or modification of particles for the production of dry powder inhalation has many advantages namely the production of powder without the presence of solvents, and this process is also environmentally friendly without causing any degradable effects on the environment [15]. Like any technique to modify particles, this method has a few disadvantages. One of the main disadvantages of powder production using SCF technology is the poor solubility of powder in water. Furthermore, considering that SCF-CO_2 is a nonpolar solvent, it is more used for nonpolar compounds. Generally, in order to overcome this restriction, a surfactant can be used, involving polyester/polyamine copolymer [16, 17].

3.2.2 Gas Anti-Solvent (GAS)

The use of this technique was introduced for the first time in 1989 [18]. In this technique, dense gases are utilized as anti-solvents. The role of dense gas is to expand the solution to precipitate the solute. Then, after the precipitation of the solutes, additional condensed gas is added to eliminate the remaining solvent and create a dry precipitate after reducing the pressure [19, 20].

3.2.3 Aerosol Solvent Extraction System (ASES)

The solution is first injected into the flowing or still-dense gas through a nozzle. Due to the solution being sprayed into the dense gas and the high degree of supersaturation, it leads to the production of sediment with fine particles. In this process, boosted mass transfer and hydrodynamic mixing are variables that may be modulated to control particle size, size distribution, and morphology [21, 22].

3.2.4 Solution-Enhanced Dispersion by Supercritical Fluids Process (SEDS)

The SEDS technique was established by York and Hanna from Bradford University in 1996 to improve the performance of the traditional SAS process [23]. The manufacturing of uniform-sized, fine particles in a single-phase equilibrium is the main objective of the invention of this technique while removing the organic solvent to deliver them in the dry form [24]. Typically, the powder is first solubilized in a suitable organic solvent and then rapidly sprayed together with the SCF through a coaxial nozzle into a high-pressure container. This mixture produces fine droplets due to breakage at the tip of the nozzle and subsequently creates fine particles. Hence, the nozzle as well as its characteristics (mainly internal diameters) play a significant role in the final particle size distribution [25, 26].

3.2.5 Precipitation with Compressed Anti-Solvent (PCA)

The PCA method is also known as a semi-continuous extraction system. In this system, an increase in mass transfer rate is required to minimize the particle accumulation that is often observed as well as to reduce or eliminate drying time. In most cases, this technique is used to produce biologically active powders such as gentamicin, naloxone, and naltrexone or polymeric microparticles. In the PCA method, a drug or polymer is dissolved in an organic solvent. Prior to this, solvent is continuously sprayed in small amounts into a container filled with SC-CO_2; the solvent dissolves in the SC-CO_2, but the solute does not. As a result, the particles precipitate in micronized form. Continuous process optimization, such as processing conditions and nozzle type and size, has resulted in controllable smaller particle formation [27].

3.2.6 Supercritical Anti-Solvent (SAS)

Due to the various applications of SCF in different sciences, the interest in its use by researchers has increased in order to produce pharmaceutical powders [28]. SAS is one of the techniques used to produce inhalation powders. SCFs are very dense and usually miscible with organic solvents. Usually, CO_2 anti-solvent gas is used in this technique. Anti-solvent has a high affinity with organic solvents, but it has little affinity with solutes [13]. In this technique, the solute (usually it is an API) is first solubilized in an organic solvent, and the solution is expanded by spraying in supercritical conditions. The solute undergoes nucleation and precipitation [29, 30]. In general, in this technique, by creating optimal operating conditions, such as choosing a suitable organic solvent, where the solvent and anti-solvent are fully or partially miscible with each other, it is possible to create integrated precipitation of solutes [31]. The main advantage of the SAS technique over the GAS method is the continuous injection of solution and SC-CO_2, which is a prerequisite for large-scale mass production of particles. Figure 3.2 shows a schematic depicting of the different particle formation processes using SC-CO_2. For better visualization of SAS and RESS techniques, these two techniques are compared as depicted in Fig. 3.3.

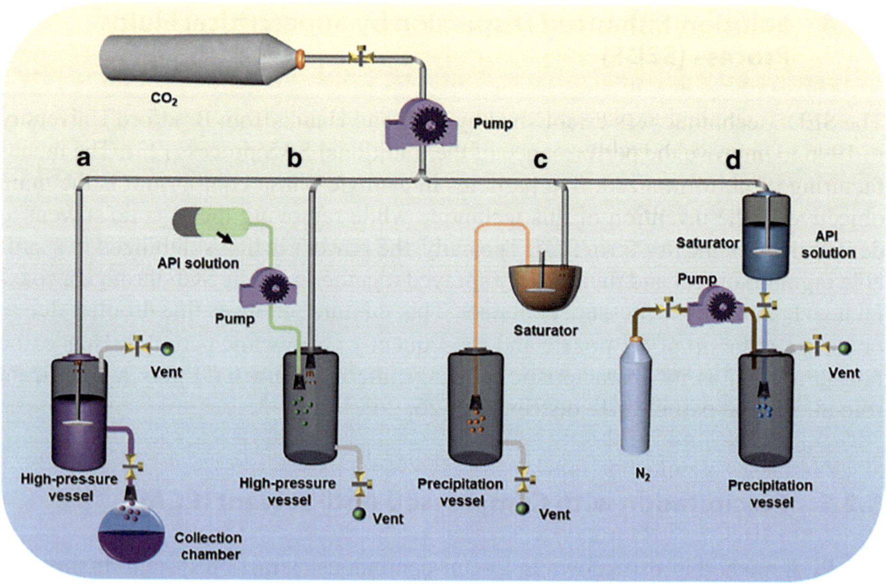

Fig. 3.2 Schematic shows a general of the different particle formation processes using SC-CO2 in different SCF technologies: (**a**) RESS, (**b**) SAS, (**c**) GAS, and (**d**) SAA. For more details refer to the main text for each SCF technology (Reprinted with permission from [9], Copyright 2022 Elsevier)

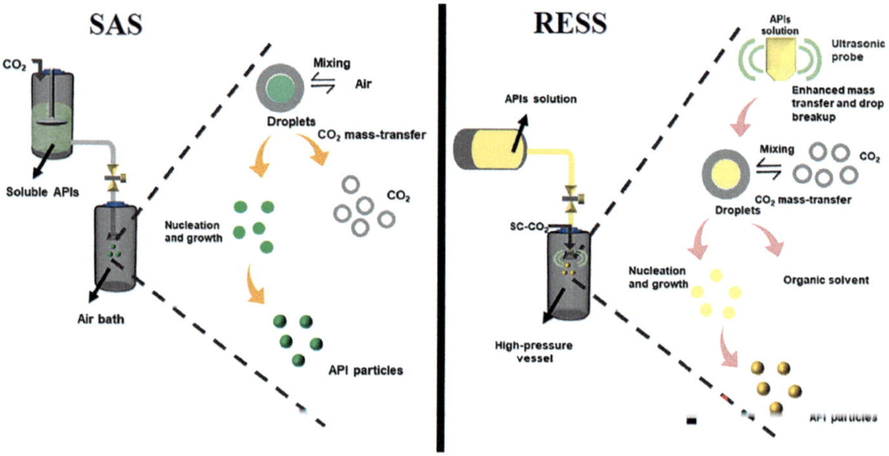

Fig. 3.3 General schematic of powder formation details using different processes (Reprinted with permission from [9]. Copyright 2022 Elsevier)

In this technique which was mentioned above, various key parameters affect particle size and morphology, including temperature, pressure, nature of the organic solvent, solute concentration, anti-solvent, solution flow rates, and nozzle type. Below is a detailed explanation of each parameter involved in the production of particles via the SAS technique.

3.2.6.1 Temperature

It has been proven that changes in the temperature of a supercritical system can modify the physical appearance of the particles. In other words, a rise in temperature causes a decrease in the size of the particles and vice versa [32]. The mechanism of particle size reduction with increasing temperature is attributed to a rise in mass transfer between the solvent and the SCF. This, in turn, can result in the formation of large nuclei with small particle sizes [33]. In this regard, [34] have produced simvastatin, a drug that is commonly utilized in the treatment of hypercholesterolemia but practically insoluble in water and not well absorbed through the gastrointestinal tract, using SCF technology. By elevating the temperature from 30 °C to 50 °C, the particle size was reduced from 87 nm to 54 nm, which was confirmed by the results of scanning electron microscopy (SEM), and dynamic light scattering.

3.2.6.2 Pressure

The effect of this factor is more on the SCF anti-solvent, which causes the solvent and anti-solvent to separate from each other [35]. Increasing the pressure leads to nucleation, which generates many nucleation sites, resulting in small-sized crystals [36]. Also, by decreasing the pressure, the solvent does not disappear completely and remains in the solution. This phenomenon affects the properties of the particles and the formation time. Thus, large-sized particles are produced that require a long formation time [37, 38].

3.2.6.3 Nature of Solvent

The selection of the solvent type is the most crucial part of the production of inhalation powders in the SAS method. The type of solvent and its compatibility with SCF is the most important part of the work. In the study performed by [39], they considered the solubility of SC-CO_2 in ethanol and acetone, and the results of the study exhibited that the solubility of CO_2 in acetone is much higher. SCFs are usually mixed with organic solvents, and when the process is performed at high pressures, smaller particles are precipitated into the solution droplets due to the increased anti-solvent diffusion. This is followed by an increase in solvent evaporation pressure [40, 41]. Various factors, such as vapor pressure, viscosity, and solubility in the anti-solvent can affect solvent evaporation, which affects particle size and crystal morphology [42].

3.2.6.4 Flow Rates and Nozzle Geometry

In critical conditions, the change in the flow rate of the solution and CO_2 results in a decline in its amount in the precipitant. The effect of a higher CO_2 flow rate leads to a reduction in the precipitating material and results in the creation of fine-size crystals [43]. In this regard, [44], examined the effect of nozzle diameter and CO_2 flow rate and concluded that a higher CO_2 flow rate and small nozzle diameter can lead to the production of spherical sediment particles. Also, different studies investigated the effect of different nozzles on particle size and concluded that the type of nozzle can affect the diameter, size distribution, and morphology of particles [45]. The image of different types of nozzles can be seen in Fig. 3.4.

Table 3.1 lists the name of APIs and the SCF method used to manufacture APIs for inhalation purposes. This table also shows the outcome of each technique employed to produce APIs.

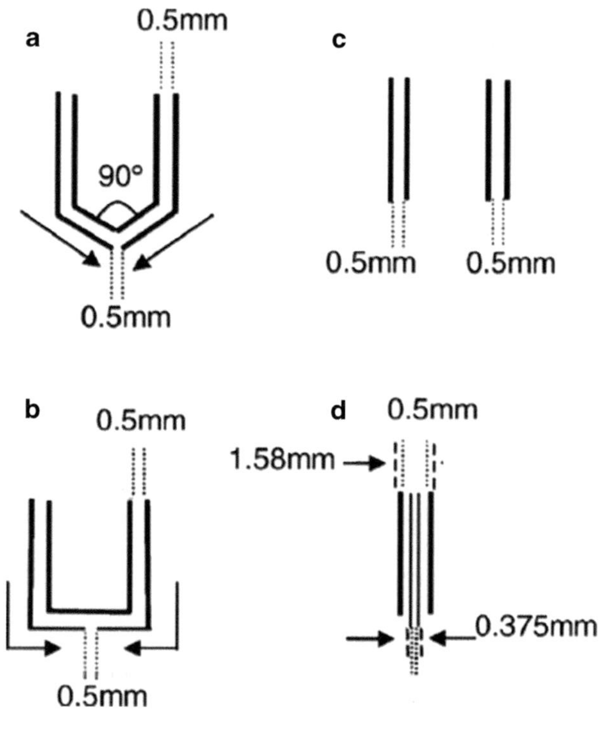

Fig. 3.4 A schematic presentation of common nozzles employed in the supercritical anti-solvent method. (**a**) V-type, (**b**) U-type, (**c**) parallel-type, and (**d**) two-channel coaxial nozzle

Table 3.1 Different SCF techniques in the production of different APIs

Type of material	SCF method	Outcome	Ref
Budesonide	SEDS	A remarkable enhancement in in vitro aerosolization performance of budesonide suspension (FPF 53.65% with MMAD of 3.53 μm) where the drug particles were engineered via supercritical fluid.	[46]
Puerarin	SEDS	Puerarin particles obtained via this SCF technique exhibited an average mean diameter of 190 nm with a spherical shape and a smooth surface. Puerain particles exhibited a small span value which is an indication of a narrow particle size distribution.	[47])
Insulin	PCA	Insulin-loaded poly-l-lactide porous microspheres obtained via PCA method showed rough and porous structures with a geometric mean diameter (dg) of 15.62 μm, but an acceptable aerodynamic diameter (Da) of 4.31 μm, where aerosolization performance is significantly improved (FPF of 65.57%).	[48]
Celecoxib	SC-CO_2	The obtained large porous particles of celecoxib containing poly lactic-co-glycolic acid (PLGA) showed 4.8-, 15.7-, and 2.1-fold enhancement in celecoxib levels in the lung, bronchoalveolar lavage fluid (BAL), and plasma respectively in comparison to conventional microparticles on day 21 after single intratracheal dosing of dry powders in A/J mice.	[49]
Chitosan oligomers	SAA	To obtain fine particles, hydrodynamic cavitation was employed to increase mass transfer and help the blending between SC-CO_2 and liquid solution. The method was able to produce individual spherical particles in respirable size with narrow distribution ranging from 0.5 to 3 μm.	[50]
A-tocopherol (vitamin E)	SFDD	This technique was capable of producing a narrow droplet size distribution of vitamin E between 0.3 and 3 μm. The decompressed aerosol stream was diluted with gaseous oxygen and directly given to rats for 10–30 min via inhalation. The results exhibited a remarkable increase in the concentration of vitamin E in the rats.	[51]

(continued)

Table 3.1 (continued)

Type of material	SCF method	Outcome	Ref
Terbutaline Sulfate (TS) and ipratropium bromide (IB)	ASES	In this study, a mixture of ethanol/dimethylformamide was used in ASES technique to obtain a micronized form of terbutaline sulfate (MMAD between 1.28 and 1.63). In the case of ipratropium bromide, pure DMF was used (MMAD between 2.2 and 3.26). To obtain respirable particle size different pressures and temperatures were used for both TS (100 bar and 50 °C) and for IB (200 bar and 40 °C).	[52]
Ono-2921	Combination of SC-CO_2 and RESAS	The combination method showed that the particles obtained exhibited a remarkable increase in the deposition of drug particles (24 times higher in stage 2 of twin-stage impinge compared to the plain Ono-12,921). The FPF value obtained for the engineered particles was 48.7% with an aerodynamic diameter of <4.7 μm).	[53]
Chitosan aerogel microparticles	SCF and FD	In this study, aerogels of chitosan were successfully made by SCF technology. The obtained aerogels showed an average particle size of 10 μm with very low tapped density (0.12 g/mL) with a high specific surface area (73–103 m^2/g). This was not the case for the particles obtained via freeze drying where the average particle size cryogels was 64 microns.	[54]
Terbutaline Sulfate (TS)	SCF	Researchers used a single-step production in SCF method to produce respirable size for terbutaline sulfate (TB) particles. Two different polymorphic forms of TB including a stoichiometric monohydrate and an amorphous form were obtained. Those particles showed low surface energy and exhibited better aerosolization performance compared to micronized particles (maximum FPF was 38.6% for processed materials).	[55]
Salmeterol (SX) Xinafoate, terbutaline (TBS) Sulfate and Fenoterol Hydrobromide (FHBr)	SC-CO_2	It was shown that all engineered particles obtained via SCF showed low surface energy (measured by IGC) led to an outstanding aerosolization performance in the dry powder inhalation formulation (the highest FPFs for SX, TBS, and FHBr were 57.8, 38.6, and 41.7%, respectively).	[56]

Table 3.1 (continued)

Type of material	SCF method	Outcome	Ref
Ipratropium bromide (IB)	ASES	The results presented in this study exhibited that the type of solvent used in the process of ASES can profoundly modify the particle size and morphology of IB particles. The best performance was obtained for the IB particles precipitated in the presence of DMF. The obtained particles presented ideal properties for inhalation formulations (not agglomerated with aerodynamic diameter between 0.6 and 3.0 μm).	[57]
Poly(D-L-Lactide-co Glycolide) (PLGA)	SASD	The modified particles obtained through SASD showed better aerosolization performance when it is tested using an Andersen cascade impactor (ACI). The value of FPF for the optimized formulation was 43.4% with MMAD values between 1.7 and 3.5 μm. The authors proved that the obtained particles have the requirements for potential pulmonary delivery in terms of cytotoxicity and pharmacokinetic properties.	[58]
Hydrocortisone (HC)	SEDS	The inhalation performance of the particles obtained through SEDS was assessed using multi-stage liquid impinge (MSLI). Different solvents produced particles with different morphology (acetone-produced needle shape and methanol-produced flakes or needles depending on the processing temperature and pressure). The results showed that HC particles obtained through SEDS exhibited better inhalation performance (98.9% delivered dose for engineered HC particles in acetone versus 66.1% for untreated HC particles when Easyhaler was used).	[59]
Glucocorticoids (budesonide and Flunisolide powders)	SEDS	This research showed that by employing SEDS methods it is possible to prepare budesonide and flunisolide particles with narrow distribution sizes where 90% of the particles were < 4 μm. Although both APIs showed particles in respirable size, the aerosolization performance of budesonide was much better than flunisolide form III. The MMAD for budesonide and flunisolide was around 3 μm and over 5 μm (due to aggregation of the particles in the case of flunisolide).	[60]

SEDS Solution-Enhanced Dispersion By Supercritical Fluids, *PCA* Precipitation of Compressed CO_2 Antisolvent, *SC-CO₂* Supercritical Carbon Dioxide, *SAA* Supercritical CO2-Assisted Aerosolization, *SFDD* Supercritical Fluid Drug Delivery, *ASES* Aerosol Solvent Extraction System, *RESAS* Rapid Expansion From Supercritical To Aqueous Solution, *SASD* Supercritical CO_2-Assisted Spray-Drying, *FD* Freeze Drying

3.3 Properties of Powders Produced by SCF Techniques

Among the methods based on high pressure, SCF is considered the most effective approach compared to the others. This causes an effective change in the polymorphism, size, and stability of the particles. Below is an explanation of each of them:

3.3.1 Polymorphism

The morphology and crystalline nature of inhalation powders play an important role in drug dissolution. In general, inhalation powders are based on morphologies such as polymorphic, pseudo-polymorphic, and amorphous [61]. Among all morphologies, crystalline forms have cohesive connections and low solubility, whereas amorphous forms are the opposite [61]. Considering that the SCF process can affect the morphology of the powder, in this regard, [62] used the RESS process to produce carbamazepine. The use of this method leads to an alteration in the crystal shape, reducing the particle size with improved dissolution. Also, [63] employed the RESS process to produce lonidamine (LND) and showed that in the conditions of LND concentartion.-0.5% (w/v), flow rate-1.0 mL/min, T-328.15 K, P-20.0 MPa, and CO_2 flow rate- 30.0 g/min, nano-powders were obtained which were converted from crystalline state to amorphous state with high solubility properties.

In this technique, temperature, pressure, and type of organic solvent can affect the process. In this regard, [64] investigated the morphology and size of acetaminophen using the SAS technique. The results showed that the solvents used such as ethanol and acetone as well as their combination led to the structural change of the powder using X-ray diffraction and morphological change using the SEM so that the ethanolic solution of acetaminophen led to the creation of monoclinic polymorphic form I while acetone led to orthorhombic form II, which improved solubility.

3.3.2 Particle Size

Particle size has a crucial effect on the performance of dry powder inhalers and also their dissolution in the lungs. Particle size can be related to the bioavailability of the powder by penetrating the smallest bronchi in the lung, and also the powder with smaller particles has a high dissolution rate [65, 66]. Using the SCF technique, medicinal nano-powders are produced, which are superior to other methods such as Spiral Jet Mill and Fluid bed jet mill. Of course, it should be noted that the environmental factors mentioned above can also affect the size of the particles. In this way, [67] micronized nabumetone, which is known as an effective nonsteroidal anti-inflammatory drug, using the RESS technique. The results showed that the lower temperature after expansion produced smaller particles. Naboomton dissolution was investigated before and after the RESS process, and the result revealed a 3.5-fold enhancement in its dissolution rate after the RESS process.

Furthermore, it is noted that the size of nanoparticles also has a positive effect on their dissolution, as Kakran et al. reported that quercetin nanoparticles obtained by the supercritical antisolvent with enhanced mass transfer (SAS-EM) approach with an average size of 120 nm showed 60% more dissolution than those particles that had an average size of 240 nm in 1.5 h [68].

In another study performed by [69], gambogic acid (GA) nanoparticles which are known as an anticancer drug with low aqueous solubility were produced using environmentally friendly SCF with the help of rapid expansion of a supercritical solution into a liquid solvent (known as RESOLV). In this technique, the equilibrium solubility of GA in SC-CO_2 was initially determined under modified supercritical conditions (P: 10.0 MPa–30.0 MPa, T: 308.15 K–328.15 K) followed by the production of GA nanoparticles using the above-mentioned process. Their results showed that GA nanoparticles led to an enhancement in the anti-proliferative effect in human breast cancer cells (MCF-7). The reason for this can be related to a rise in the solubility and also reducing the size of GA particles with the above technique so that the penetration power of drug nanoparticles is greater than that of the free drug inside tumor cells. The increase in the solubility of the drug particles produced by this method can also be related to an enhancement in the surface area of nanoparticles that can improve the dissolution of GA.

Considering that the main component for the successful administration of drugs and inhaled powders are the production of particles with suitable particle size (an aerodynamic diameter of 1–5 μm), which are optimal for lung deposition. Markedly, the particle size and aerodynamic diameter of the particles subsequently define the deposition of inhaled particles in the lung and the efficiency of the biological function. In this regard, [53] used SC-CO_2 synthesis techniques to prepare a poorly water-soluble drug (ONO-2921) as a dry inhalation powder. When the SC-CO_2-drug mixture was sprayed into the water using a RESAS device, a suspension containing sub-micron particles was acquired. The obtained formulation presented an aerodynamic diameter of <4.7 μm, which showed an inhalation performance of 24 times higher than the crude drug. In another case, an inhalable dry powder formulation of IB was produced based on the CO_2-based ASES process [57]. The above-made formula with a more regular shape and an aerodynamic diameter in the range of 2–3 μm has the efficiency of drug penetration into the small bronchi of the lungs, which is related to a significant reduction in the size of the obtained particles via CO_2-based ASES process [57].

3.3.3 Stability

The stability of pharmaceutical formulations and inhalation powders exhibits an important role in pharmaceutical and commercial capabilities, which may cause an alteration in their morphology, phase, structure, and dissolution rate. As in SCF processes, SC-CO_2 has non-reactive nature and also it is performed under mild conditions to prepare APIs for inhalation aims, therefore, it is highly unlikely the SCF methods and the processing involved in the production of APIs can cause any chemical instability issues [70]. This has also been proved for many APIs such as simvastatin [34], thermolabile compounds [71], progesterone [72], cromolyn sodium [73], ampicillin [74], lansoprazole [75] and citrus aurantifolia essential oil [76].

Among all the different stability characteristics, colloidal stability exhibits a crucial role in the formulation because it can cause an irregular dosage during the administration of the medicine., This, in turn, can result in serious side effects and fluctuations in the drug performance in patients [77–79].

It should be kept in mind that, although in the SCF, fine particles can be produced with a better dissolution rate for the inhaled particles with excellent bioavailability. These fine particles may undergo colloidal instability leading to aggregation or agglomeration problems. This phenomenon has been reported in several studies where aggregation of fine drug particles (micro- or nano-size particles) as a result of colloidal instability caused a reduction in bioavailability [80, 81]. Therefore, if the duration of drug storage is longer, colloidal stability should also be evaluated. In order to prevent this phenomenon, several approaches have been proposed. Classical approaches include ultrasound-based treatment and the addition of excipients to prevent particle aggregation [82].

Sanganwar et al. [83] investigated the fabrication and simultaneous mixing of nevirapine microparticles with excipients by the SAS method to enhance dissolution. They showed that the simultaneous mixing of nevirapine (a poorly water-soluble drug) with excipients (lactose and microcrystalline cellulose) during the SAS method can adsorb the drug particles on the particles' surface of the excipient, this, in turn, can reduce the aggregation of drug-drug particles to ensure faster dissolution. In comparison, nevirapine particles obtained by supercritical antisolvent-drug excipient mixing (SAS-DEM) showed more efficient deagglomeration or mixing than conventional SAS and physical mixtures of drugs-excipients processed with SAS. Remarkably, some environmental stress during the manufacture and storage of APIs may affect the final bioactivity due to degradation [84].

3.3.4 Commercial Application of SCF Technology

Although there are so many studies showing the success of SCF at the laboratory scale, moving from the laboratory scale to the industrial scale is in its infancy. This is due to a lack of fundamental investigations of this technology on large scale in the case of orally inhaled APIs.

One of the companies known as PIERRE FABRE CDMO Supercritical Fluids is taking steps for the commercialization of SCF with different pharmaceutical applications [85]. Based on the information provided by the company they have commercialized SCF technology for cleaning medical devices such as gaskets used in metered-dose inhalers. The cleaning of medical components by supercritical fluids is a green and efficient process. The other applications of their commercialized methods are in taste masking (Formulcoate®), improving the bioavailability of poorly water-soluble drugs (Formulplex®) and improving the functionality of excipients (Formuldisp®). To the best knowledge of the authors of this book chapter, there are no established marketed orally inhaled products manufactured by the SCF technology to date. It is believed that with further fundamental research on SCF and optimization, this technology can generate novel opportunities for the industrial processing of delivery systems in the future.

3.4 Conclusion

Formulation and manufacture of inhalable dry powders are very challenging but achievable. The advancement in SCF technology has now helped scientists to overcome some of the challenges associated with dry powder inhalation formulations in terms of performance and stability. Although SCF techniques have their own disadvantages, still it is important to employ SCF techniques to generate particles ideal for inhalation formulations. In this chapter, the use of SCF technology in the manufacturing of inhalable powders was discussed. SCT technology was used to develop inhalation formulations for various types of active ingredients from proteins (insulin) to herbs (Puerarin), hydrophobic medicines (Budesonide) to hydrophilic ones (Ipratropium Bromide), high dose medicines (Celecoxib) to low dose ones (Salmeterol), and even the development of hydrophilic (chitosan) and hydrophobic (PLGA) polymeric carriers. This implies the wide and massive capability of SCF technology in the development of respirable products. In general, the use of this technology in making inhalable powders has many advantages, and most importantly, this process is not harmful to the environment.

References

1. Savjani KT, Gajjar AK, Savjani JK. Drug solubility: importance and enhancement techniques. ISRN Pharmaceut. 2012;2012:1–10. https://doi.org/10.5402/2012/195727.
2. Chattopadhyay P, Huff R, Shekunov BY. Drug encapsulation using supercritical fluid extraction of emulsions. In: Journal of pharmaceutical sciences, vol. 95(3). John Wiley and Sons Inc.; 2006. p. 667–79. https://doi.org/10.1002/jps.20555.
3. Ye C, Chi H. A review of recent progress in drug and protein encapsulation: approaches, applications and challenges. Mater Sci Eng C. 2018;83:233–46. https://doi.org/10.1016/j.msec.2017.10.003.
4. Emilian Leucuta S. Nanotechnology for delivery of drugs and biomedical applications. Curr Clin Pharmacol. 2010;5(4):257–80. https://doi.org/10.2174/157488410793352003.
5. Saadh MJ, Jadullah RK. Nanotechnology in drug delivery. Pharmacologyonline. 2021;3:1129–35. https://doi.org/10.17148/ijarcce.2016.55255.
6. Movassaghian S, Merkel OM, Torchilin VP. Applications of polymer micelles for imaging and drug delivery. Wiley Interdiscip Rev Nanomed Nanobiotechnol. 2015;7(5):691–707. https://doi.org/10.1002/wnan.1332.
7. Yeo SD, Kiran E. Formation of polymer particles with supercritical fluids: a review. J Supercrit Fluids. 2005;34(3):287–308. https://doi.org/10.1016/j.supflu.2004.10.006.

8. Pavlova PL, Minakov AV, Platonov DV, Zhigarev VA, Guzei DV. Supercritical fluid application in the oil and gas industry: a comprehensive review. Mdpi: Com; 2022. https://doi.org/10.3390/su14020698.
9. Kankala RK, Xu PY, Chen BQ, Wang SB, Chen AZ. Supercritical fluid (SCF)-assisted fabrication of carrier-free drugs: an eco-friendly welcome to active pharmaceutical ingredients (APIs). Adv Drug Deliv Rev. 2021;176:113846. https://doi.org/10.1016/j.addr.2021.113846.
10. Hitchen SM, Dean JR. Properties of supercritical fluids. In: Applications of supercritical fluids in industrial analysis. Springer; 1993. p. 1–11. https://doi.org/10.1007/978-94-011-2146-0_1.
11. Sodeifian G, Sajadian SA, Saadati Ardestani N, Razmimanesh F. Production of Loratadine drug nanoparticles using ultrasonic-assisted rapid expansion of supercritical solution into aqueous solution (US-RESSAS). J Supercrit Fluids. 2019;147:241–53. https://doi.org/10.1016/j.supflu.2018.11.007.
12. Hezave AZ, Esmaeilzadeh F. Micronization of drug particles via RESS process. J Supercrit Fluids. 2010;52(1):84–98. https://doi.org/10.1016/j.supflu.2009.09.006.
13. Sauceau M, Rodier É. Particle generation for pharmaceutical applications using supercritical fluid technology. Powder Technol. 2004;141(3) https://doi.org/10.1016/j.powtec.2004.02.007ï.
14. Tom JW, Debenedetti PG. Particle formation with supercritical fluids-a review. J Aerosol Sci. 1991;22(5):555–84. https://doi.org/10.1016/0021-8502(91)90013-8.
15. Knez Ž, Pantić M, Cör D, Novak Z, Knez Hrnčič M. Are supercritical fluids solvents for the future? Chem Eng Process Process Intensificat. 2019;141:107532. https://doi.org/10.1016/j.cep.2019.107532.
16. Jagtap S, Magdum C, Jadge D, Jagtap R. Solubility enhancement technique: a review. J Pharmaceut Sci Res. 2018;10(9):2205–11. https://doi.org/10.31838/ijpr/2021.13.03.121.
17. Płotka-Wasylka J, Rutkowska M, Owczarek K, Tobiszewski M, Namieśnik J. Extraction with environmentally friendly solvents. TrAC - Trends Anal Chem. 2017;91:12–25. https://doi.org/10.1016/j.trac.2017.03.006.
18. Long B, Verma V, Ryan KM, Padrela L. Generation and physicochemical characterization of posaconazole cocrystals using GAS Antisolvent (GAS) and supercritical solvent (CSS) methods. J Supercrit Fluids. 2021;170:105134. https://doi.org/10.1016/j.supflu.2020.105134.
19. Beh CC, Mammucari R, Foster NR. Lipids-based drug carrier systems by dense gas technology: a review. Chem Eng J. 2012;188:1–14. https://doi.org/10.1016/j.cej.2012.01.129.
20. Tandya A, Mammucari R, Dehghani F, Foster NR. Dense gas processing of polymeric controlled release formulations. Int J Pharm. 2007;328(1 SPEC. ISS):1–11. https://doi.org/10.1016/j.ijpharm.2006.08.016.
21. Bleich J, Müller BW. Production of drug loaded microparticles by the use of supercritical gases with the aerosol solvent extraction system (ASES) process. J Microencapsul. 1996;13(2):131–9. https://doi.org/10.3109/02652049609052902.
22. Bleich J, Müller BW, Waßmus W. Aerosol solvent extraction system - a new microparticle production technique. Int J Pharm. 1993;97(1–3):111–7. https://doi.org/10.1016/0378-5173(93)90131-X.
23. Ram BG, Jae-Jin S. Particle design using supercritical fluids: literature and patent survey. Solubil Supercrit Carbon Dioxide. 2007;20:1–18. http://linkinghub.elsevier.com/retrieve/pii/S089684460100064X
24. Pasquali I, Bettini R, Giordano F. Supercritical fluid technologies: an innovative approach for manipulating the solid-state of pharmaceuticals. Adv Drug Deliv Rev. 2008;60(3):399–410. https://doi.org/10.1016/j.addr.2007.08.030.
25. Chan HK, Kwok PCL. Production methods for nanodrug particles using the bottom-up approach. Adv Drug Deliv Rev. 2011;63(6):406–16. https://doi.org/10.1016/j.addr.2011.03.011.
26. Jovanović N, Bouchard A, Hofland GW, Witkamp GJ, Crommelin DJA, Jiskoot W. Stabilization of proteins in dry powder formulations using supercritical fluid technology. Pharm Res. 2004;21(11):1955–69. https://doi.org/10.1023/B:PHAM.0000048185.09483.e7.
27. Ginty PJ, Whitaker MJ, Shakesheff KM, Howdle SM. Drug delivery goes supercritical. Mater Today. 2005;8(8 SUPPL):42–8. https://doi.org/10.1016/S1369-7021(05)71036-1.

28. Byrappa K, Ohara S, Adschiri T. Nanoparticles synthesis using supercritical fluid technology - towards biomedical applications. Adv Drug Deliv Rev. 2008;60(3):299–327. https://doi.org/10.1016/j.addr.2007.09.001.
29. Bristow S, Shekunov T, Shekunov BY, York P. Analysis of the supersaturation and precipitation process with supercritical CO2. J Supercrit Fluids. 2001;21(3):257–71. https://doi.org/10.1016/S0896-8446(01)00100-0.
30. Prosapio V, Reverchon E, De Marco I. Polymers' ultrafine particles for drug delivery systems precipitated by supercritical carbon dioxide + organic solvent mixtures. Powder Technol. 2016;292:140–8. https://doi.org/10.1016/j.powtec.2016.01.033.
31. Abuzar SM, Hyun SM, Kim JH, Park HJ, Kim MS, Park JS, Hwang SJ. Enhancing the solubility and bioavailability of poorly water-soluble drugs using supercritical antisolvent (SAS) process. Int J Pharm. 2018;538(1–2):1–13. https://doi.org/10.1016/j.ijpharm.2017.12.041.
32. Aymonier C, Loppinet-Serani A, Reverón H, Garrabos Y, Cansell F. Review of supercritical fluids in inorganic materials science. J Supercrit Fluids. 2006;38(2):242–51. https://doi.org/10.1016/j.supflu.2006.03.019.
33. Martín A, Cocero MJ. Numerical modeling of jet hydrodynamics, mass transfer, and crystallization kinetics in the supercritical antisolvent (SAS) process. J Supercrit Fluids. 2004;32(1–3):203–19. https://doi.org/10.1016/j.supflu.2004.02.009.
34. Fattahi A, Karimi-Sabet J, Keshavarz A, Golzary A, Rafiee-Tehrani M, Dorkoosh FA. Preparation and characterization of simvastatin nanoparticles using rapid expansion of supercritical solution (RESS) with trifluoromethane. J Supercrit Fluids. 2016;107:469–78. https://doi.org/10.1016/j.supflu.2015.05.013.
35. Rodrigues MA, Li J, Padrela L, Almeida A, Matos HA, de Azevedo EG. Anti-solvent effect in the production of lysozyme nanoparticles by supercritical fluid-assisted atomization processes. J Supercrit Fluids. 2009;48(3):253–60. https://doi.org/10.1016/j.supflu.2008.06.006.
36. Chen K, Zhang X, Pan J, Yin W. Recrystallization of andrographolide using the supercritical fluid antisolvent process. J Cryst Growth. 2005;274(1–2):226–32. https://doi.org/10.1016/j.jcrysgro.2004.09.089.
37. Banuti DT, Raju M, Ihme M. Between supercritical liquids and gases – reconciling dynamic and thermodynamic state transitions. J Supercrit Fluids. 2020;165:104895. https://doi.org/10.1016/j.supflu.2020.104895.
38. Supercritical fluid extraction. Chem Eng News. 1993;71(12):59–60.
39. Chang CJ, Day CY, Chen CY. Vapor-liquid equilibria and correlations of ethanol + CO2 and acetone + CO2 at elevated pressures. J Chin Inst Chem Eng. 1996;27(4)
40. Martín A, Cocero MJ. Micronization processes with supercritical fluids: fundamentals and mechanisms. Adv Drug Deliv Rev. 2008;60(3):339–50. https://doi.org/10.1016/j.addr.2007.06.019.
41. Shariati A, Peters CJ. Recent developments in particle design using supercritical fluids. Curr Opinion Solid State Mater Sci. 2003;7(4–5):371–83. https://doi.org/10.1016/j.cossms.2003.12.001.
42. Kalani M, Yunus R. Application of supercritical antisolvent method in drug encapsulation: a review. Int J Nanomedicine. 2011;6:1429–42. https://doi.org/10.2147/ijn.s19021.
43. Martin TM, Bandi N, Shulz R, Roberts CB, Kompella UB. Preparation of budesonide and budesonide-PLA microparticles using supercritical fluid precipitation technology. AAPS PharmSciTech. 2002;3(3):16–26. https://doi.org/10.1208/pt030318.
44. García-Casas I, Montes A, Pereyra C, Martínez de la Ossa EJ. Generation of quercetin/cellulose acetate phthalate systems for delivery by supercritical antisolvent process. Eur J Pharm Sci. 2017;100:79–86. https://doi.org/10.1016/j.ejps.2017.01.010.
45. Bouchard A, Jovanović N, de Boer AH, Martín Á, Jiskoot W, Crommelin DJA, Hofland GW, Witkamp GJ. Effect of the spraying conditions and nozzle design on the shape and size distribution of particles obtained with supercritical fluid drying. Eur J Pharm Biopharm. 2008;70(1):389–401. https://doi.org/10.1016/j.ejpb.2008.03.020.

46. Amani A, Chrystyn H, Clark BJ, Abdelrahim ME, York P. Evaluation of supercritical fluid engineered budesonide powder for respiratory delivery using nebulisers. J Pharm Pharmacol. 2010;61(12):1625–30. https://doi.org/10.1211/jpp.61.12.0006.
47. Chen AZ, Li Y, Chau FT, Lau TY, Hu JY, Zhao Z, Mok DK, wah. Application of organic nonsolvent in the process of solution-enhanced dispersion by supercritical CO2 to prepare puerarin fine particles. J Supercrit Fluids. 2009;49(3):394–402. https://doi.org/10.1016/j.supflu.2009.02.004.
48. Chen A-Z, Tang N, Wang S-B, Yong-Qiang Kang H-FS. Insulin-loaded poly-l-lactide porous microspheres prepared in supercritical CO2 for pulmonary drug delivery. J Supercrit Fluids. 2017;101:117–23.
49. Dhanda DS, Tyagi P, Mirvish SS, Kompella UB. Supercritical fluid technology based large porous celecoxib-PLGA microparticles do not induce pulmonary fibrosis and sustain drug delivery and efficacy for several weeks following a single dose. J Control Release. 2013;168(3):239–50. https://doi.org/10.1016/j.jconrel.2013.03.027.
50. Du Z, Shen YB, Tang C, Guan YX, Yao SJ, Zhu ZQ. Supercritical fluid assisted production of chitosan oligomers micrometric powders. Carbohydr Polym. 2014;102(1):400–8. https://doi.org/10.1016/j.carbpol.2013.11.029.
51. Hybertson BM, Repine JE, Beehler CJ, Rutledge KS, Lagalante AF, Sievers RE. Pulmonary drug delivery of Fine aerosol particles from supercritical fluids. Journal of Aerosol Medicine: Deposition, Clearance, and Effects in the Lung. 1993;6(4):275–86. https://doi.org/10.1089/jam.1993.6.275.
52. Kim YH, Sioutas C, Fine P, Shing KS. Effect of albumin on physical characteristics of drug particles produced by supercritical fluid technology. Powder Technol. 2008;182(3):354–63. https://doi.org/10.1016/j.powtec.2007.06.008.
53. Miyazaki Y, Sugihara H, Nishiura A, Kadota K, Tozuka Y, Takeuchi H. Application of combinational supercritical CO2 techniques to the preparation of inhalable particles. J Drug Deliv Sci Technol. 2016;36:1–9. https://doi.org/10.1016/j.jddst.2016.08.010.
54. Obaidat RM, Tashtoush BM, Bayan MF, Al Bustami T, R., & Alnaief, M. Drying using supercritical fluid technology as a potential method for preparation of chitosan aerogel microparticles. AAPS PharmSciTech. 2015;16(6):1235–44. https://doi.org/10.1208/s12249-015-0312-2.
55. Rehman M, Shekunov BY, York P, Lechuga-Ballesteros D, Miller DP, Tan T, Colthorpe P. Optimisation of powders for pulmonary delivery using supercritical fluid technology. Eur J Pharm Sci. 2004;22(1):1–17. https://doi.org/10.1016/j.ejps.2004.02.001.
56. Shekunov BY, Rehman M, Chow AHL, Tong HHY, York P, Group DD, Particle B. Production of crystalline powders for inhalation drug delivery using supercritical fluid technology. In: *6th Symposium on Supercritical Fluids*; 2003.
57. Kim YH, Shing KS. Supercritical fluid-micronized ipratropium bromide for pulmonary drug delivery. Powder Technol. 2008;182(1):25–32. https://doi.org/10.1016/j.powtec.2007.04.009.
58. Tavares M, Cabral RP, Costa C, Martins P, Fernandes AR, Casimiro T, Aguiar-Ricardo A. Development of PLGA dry powder microparticles by supercritical CO2-assisted spray-drying for potential vaccine delivery to the lungs. J Supercrit Fluids. 2017;128:235–43. https://doi.org/10.1016/j.supflu.2017.06.004.
59. Velaga SP, Ghaderi R, Carlfors J. Preparation and characterisation of hydrocortisone particles using a supercritical fluids extraction process. Int J Pharm. 2002;231(2):155–66. https://doi.org/10.1016/S0378-5173(01)00870-5.
60. Velaga SP, Bergh S, Carlfors J. Stability and aerodynamic behaviour of glucocorticoid particles prepared by a supercritical fluids process. Eur J Pharm Sci. 2004;21(4):501–9. https://doi.org/10.1016/j.ejps.2003.11.006.
61. Colombo P, Traini D, Buttini F. Inhalation drug delivery: techniques and products. In: Inhalation drug delivery: techniques and products; 2013. https://doi.org/10.1002/9781118397145.
62. Bolten D, Türk M. Micronisation of carbamazepine through rapid expansion of supercritical solution (RESS). J Supercrit Fluids. 2012;62:32–40. https://doi.org/10.1016/j.supflu.2011.06.014.

63. Chen BQ, Kankala RK, Wang SB, Chen AZ. Continuous nanonization of lonidamine by modified-rapid expansion of supercritical solution process. J Supercrit Fluids. 2018;133:486–93. https://doi.org/10.1016/j.supflu.2017.11.016.
64. Rossmann M, Braeuer A, Leipertz A, Schluecker E. Manipulating the size, the morphology and the polymorphism of acetaminophen using supercritical antisolvent (SAS) precipitation. J Supercrit Fluids. 2013;82:230–7. https://doi.org/10.1016/j.supflu.2013.07.015.
65. Alshora DH, Ibrahim MA, Alanazi FK. Nanotechnology from particle size reduction to enhancing aqueous solubility. Surf Chem Nanobiomater Appl Nanobiomater. 2016;3:163–91. https://doi.org/10.1016/B978-0-323-42861-3.00006-6.
66. Hofmann W, Sturm R. Stochastic model of particle clearance in human bronchial airways. J Aerosol Med Deposition Clearance Effects Lung. 2004;17(1):73–89. https://doi.org/10.1089/089426804322994488.
67. Su CS, Tang M, Chen YP. Micronization of nabumetone using the rapid expansion of supercritical solution (RESS) process. J Supercrit Fluids. 2009;50(1):69–76. https://doi.org/10.1016/j.supflu.2009.04.013.
68. Kakran M, Sahoo NG, Antipina MN, Li L. Modified supercritical antisolvent method with enhanced mass transfer to fabricate drug nanoparticles. Mater Sci Eng C. 2013;33(5):2864–70. https://doi.org/10.1016/j.msec.2013.03.002.
69. Xiang ST, Chen BQ, Kankala RK, Wang SB, Chen AZ. Solubility measurement and RESOLV-assisted nanonization of gambogic acid in supercritical carbon dioxide for cancer therapy. J Supercrit Fluids. 2019;150:147–55. https://doi.org/10.1016/j.supflu.2019.04.008.
70. Kankala RK, Zhang YS, Wang SB, Lee CH, Chen AZ. Supercritical fluid technology: an emphasis on drug delivery and related biomedical applications. In: Advanced healthcare materials, vol. 6, Issue 16. Wiley-VCH Verlag; 2017. https://doi.org/10.1002/adhm.201700433.
71. Adami R, Liparoti S, Reverchon E. A new supercritical assisted atomization configuration, for the micronization of thermolabile compounds. Chem Eng J. 2011;173:55–61.
72. Huang Z, Guo Y-H, Miao H, Teng L-J. Solubility of progesterone in supercritical carbon dioxide and its micronization through RESS. Powder Technol. 2014;258:66–77.
73. Reverchon E, Adami R, Caputo G. Production of cromolyn sodium microparticles for aerosol delivery by supercritical assisted atomization. AAPS PharmSciTech. 2007;8:E114.
74. Esfandiari N, Ghoreishi SM. Ampicillin nanoparticles production via supercritical CO2 gas antisolvent process. AAPS PharmSciTech. 2015;16:1263–9.
75. Ameri A, Sodeifian G, Sajadian SA. Lansoprazole loading of polymers by supercritical carbon dioxide impregnation: impacts of process parameters. J Supercrit Fluids. 2020;164:104892.
76. Akolade JO, Nasir-Naeem KO, Swanepoel A, Yusuf AA, Balogun M, Labuschagne P. CO2-assisted production of polyethylene glycol / lauric acid microparticles for extended release of Citrus aurantifolia essential oil. J CO2 Utilisation. 2020;38:375–84.
77. Healy AM, Amaro MI, Paluch KJ, Tajber L. Dry powders for oral inhalation free of lactose carrier particles. Adv Drug Deliv Rev. 2014;75:32–52. https://doi.org/10.1016/j.addr.2014.04.005.
78. Lechanteur A, Evrard B. Influence of composition and spray-drying process parameters on carrier-free DPI properties and behaviors in the lung: a review. Pharmaceutics. 2020;12(1) https://doi.org/10.3390/pharmaceutics12010055.
79. Pacheco LF, Carmona-Ribeiro AM. Effects of synthetic lipids on solubilization and colloid stability of hydrophobic drugs. J Colloid Interface Sci. 2003;258(1):146–54. https://doi.org/10.1016/S0021-9797(02)00103-0.
80. Kale K, Hapgood K, Stewart P. Drug agglomeration and dissolution - what is the influence of powder mixing? Eur J Pharm Biopharm. 2009;72(1):156–64. https://doi.org/10.1016/j.ejpb.2008.12.015.
81. Swanepoel E, Liebenberg W, de Villiers MM, Dekker TG. Dissolution properties of piroxicam powders and capsules as a function of particle size and the agglomeration of powders. Drug Develop Ind Pharm. 2000;26:1067–76.
82. Kompella UB, Amrite AC, Pacha Ravi R, Durazo SA. Nanomedicines for back of the eye drug delivery, gene delivery, and imaging. Prog Retin Eye Res. 2013;36:172–98. https://doi.org/10.1016/j.preteyeres.2013.04.001.

83. Sanganwar GP, Sathigari S, Babu RJ, Gupta RB. Simultaneous production and co-mixing of microparticles of nevirapine with excipients by supercritical antisolvent method for dissolution enhancement. Eur J Pharm Sci. 2010;39(1–3):164–74. https://doi.org/10.1016/j.ejps.2009.11.011.
84. Sellers SP, Clark GS, Sievers RE, Carpenter JF. Dry powders of stable protein formulations from aqueous solutions prepared using supercritical CO2-assisted aerosolization. J Pharm Sci. 2001;90(6):785–97. https://doi.org/10.1002/jps.1032.
85. pharmaceutical-technology.com. https://www.pharmaceutical-technology.com/contractors/contract-manufacturers/supercritical-fluids/#:~:text=Pierre%20Fabre%20CDMO%20Supercritical%20Fluids%20is%20a%20Supercritical%20CO2.proposes%20development%20and%20manufacturing%20services.

Nano-in-Microparticles for Pulmonary Drug Delivery

4

Nasser Alhajj, Niall J. O'Reilly, and Helen Cathcart

Abstract

Nanoparticles have been developed to overcome the limitation of free therapeutics and enable targeting of specific sites in a controlled manner; they have been proven to be effective as drug carriers and diagnostic tools. Pulmonary drug delivery is greatly preferred in the treatment of lung diseases and has its advantages over other drug delivery routes in the treatment of systemic diseases as well. However, delivering nanoparticles to the lung is hindered by their physical instability and poor lung deposition efficiency due to particle-particle interaction and low inertia, respectively. Developing inhalable nanoparticles as part of an inhalable solid state dry powder combines the advantages of pulmonary and nanoparticle drug delivery systems and offers unique advantages. Several particle engineering techniques have been utilized to combine nanoparticles within a form of micron-scale dry powder carrier with improved handling and aerosolization properties and the ability to release the nanoparticles upon deposition in the lungs. This chapter will describe the different classes of nanoparticles and the different inhalable nano-in-micro dry particle frameworks. The different particle

N. Alhajj (✉) · H. Cathcart
Pharmaceutical and Molecular Biotechnology Research Centre (PMBRC), Waterford Campus, South East Technological University, Waterford, Ireland
e-mail: nasser.al-hajj@postgrad.wit.ie; Helen.Fox@setu.ie

N. J. O'Reilly
Pharmaceutical and Molecular Biotechnology Research Centre (PMBRC), South East Technological University, Cork Road Waterford, Ireland

SSPC – The Science Foundation Ireland Research Centre for Pharmaceuticals, Limerick, Ireland
e-mail: Niall.OReilly@setu.ie

P. P. Mehta, V. Dhapte-Pawar (eds.), *Pulmonary Drug Delivery Systems: Material and Technological Advances*,
https://doi.org/10.1007/978-981-99-1923-9_4

engineering techniques used for developing these frameworks will also be discussed.

Keywords

Pulmonary drug delivery · Nanoparticles · Microparticles · Nano-in-microparticle engineering · Dry powder inhaler

4.1 Introduction

Respiratory tract diseases (RTDs) are among the top global causes of death with lower respiratory infections remaining the world's most deadly communicable disease [1]. RTDs caused more than 9 million deaths in 2016 which is about 15% of deaths worldwide [2]. RTDs are classified as chronic respiratory diseases (CRDs) and acute respiratory infections (ARIs). The conventional treatment of CRDs and ARIs (such as chronic obstructive pulmonary disease (COPD), asthma, cystic fibrosis, lung cancer, tuberculosis, and other bacterial and viral infections) faces a number of challenges. These challenges include, but are not limited to, late detection of the disease, unsuccessful therapeutic trials, adverse effects, multidrug-resistant strains, high treatment cost with poor compliance and adherence to long-term therapy which result in a significant decline in the patient's quality of life [3].

Nanotechnology has a promising role in overcoming many of the limitations associated with conventional therapies. A drug or diagnostic agent can be encapsulated within a matrix of nanocarriers or adsorbed/covalently bonded to their surfaces. Nanocarriers can also be loaded with a combination of drugs or drug/diagnostic agents for theragnostic applications. As a drug carrier, nanoparticles have been used for several purposes such as enhancing the solubility and permeability of active agents; boosting their stability against extracellular enzymes and clearance systems; enabling sustained drug release and active/positive targeting of a specific cell type or a disease site. A comprehensive description of the applications of nanotechnology in the different types of lung diseases can be found in a number of recent review articles such [4–11].

Delivering nanoparticles through inhalation is hindered by their submicron sizes. Nanoparticles have a high surface area to volume ratio and hence a high tendency to form large agglomerates (>5 μm) which deposit in the upper airways by impaction. Furthermore, any non-agglomerated nanoparticles that are inhaled would then be expelled during exhalation [12]. For nanoparticles to be successfully delivered to the lung via inhalation, they need to be transformed into some form of inhalable microparticles while maintaining the essential properties of the nanoparticles. This is a multistep formulation procedure that involves preparing the nanoparticles and then integrating them into microparticles. The focus of this chapter is to review the current approaches in fabricating nanoparticles-microparticles suitable for pulmonary delivery as a dry powder. The different formulation techniques and excipients used for this purpose along with the different fabricated framework of inhalable nanoparticles-microparticles systems will be discussed. The synthesis and properties of nanoparticles are not the main focus of this chapter, but it is essential to understand the different types of nanoparticles as building units. Hence, the different classes of nanoparticles are introduced and key differences between these classes are highlighted.

4.2 Types of Nanoparticle

Nanoparticles are defined as particles with at least one dimension within the range of 1–100 nm [13]. However, pharmaceutical nanoparticles refer to particles with at least one dimension below 1000 nm. The size of pharmaceutical nanoparticles commonly varies in the range of 100–500 nm depending on their applications, e.g., improving targeted delivery, cellular uptake, solubility, or sustained release [14]. Nanoparticle drug delivery systems can be divided into three main groups: lipid-, polymeric-, and inorganic-based nanoparticles.

4.2.1 Lipid-Based Nanoparticles

Lipid-based nanoparticles are the most common FDA-approved nanomedicines [15]. As a drug carrier, lipid-based nanoparticles offer a number of advantages such as biocompatibility, high bioavailability, self-assembly, formulation simplicity, and suitability for controlled release applications. Lipid-based nanoparticles have gained a special interest in pulmonary delivery due to the similarity between their chemical structures and lung surfactants [16]. Figure 4.1 represents the structures of the different types of lipid nanoparticles.

4.2.1.1 Lipid Nano-Emulsions

Lipid nano-emulsions (LNEs) are a dispersion of oil in aqueous phase stabilized by one or more surfactants. Due to the thermodynamical instability of LNEs [17, 18], emulsifiers such as phospholipids, proteins, polysaccharides, and polyvinyl alcohol (PVA) are added to reduce the interfacial tension and prevent collision as well as coalescence between the droplets to increase the kinetic stability of the nano-emulsions [19]. LNEs can be prepared by high-energy or low-energy approaches. The later approach is more suitable for thermolabile molecules and commonly used to prepare peptide-loaded LNEs [18].

4.2.1.2 Liposomes

Liposomes are spherical vesicles composed of amphiphilic molecules, i.e., natural or synthetic phospholipids. Cholesterol is commonly added to the phospholipids to boost the in vitro and in vivo stability of the liposomes. Upon dispersion in an aqueous medium, phospholipid molecules spontaneously assemble into spherical vesicles comprised of at least one lipid bilayer, with the polar heads facing the aqueous medium and the hydrophobic tails facing each other within the layers. As a result,

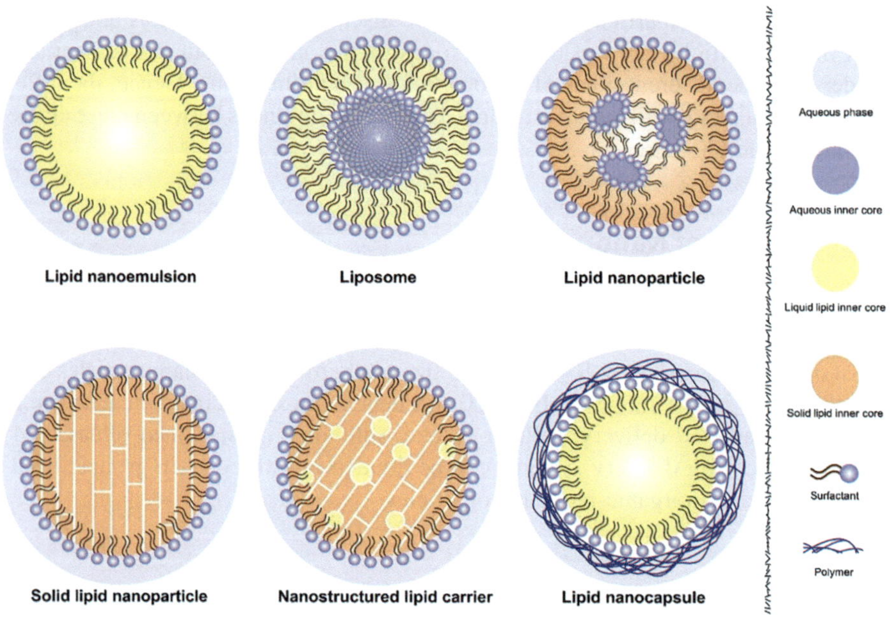

Fig. 4.1 Schematic representation of lipid nanoparticles

one (unilamellar) or more (multilamellar) lipid bilayer rings surrounding an aqueous internal cavity are formed [20]. Thus, liposomes can carry drugs of different solubilities. Liposomes can entrap hydrophilic or lipophilic compounds in their aqueous core and lipid bilayer, respectively. Furthermore, both types of drugs, hydrophilic and lipophilic, can be entrapped simultaneously in the same liposome system [21].

Liposomes can be prepared by a number of different methods. Some methods use solvents, such as the reversed-phase evaporation technique, stable plurilamellar vesicle procedure, and ethanol injection technique; others are solvent free such as dry-film or thin-film hydration, freeze–thaw liposomes, freeze-dehydration liposomes and osmotically dried vesicles [22].

4.2.1.3 Lipid Nanoparticles

Lipid nanoparticles (LNs) are liposome-like structured particles that exist in a solid state at room and body temperatures. LNs are composed of solid lipids or a mixture of solid and liquid lipids. Some LNs have a similar structure to liposomes; they exist as a series of concentric vesicles each with a continuous lipid bilayer surrounding an aqueous core. Other LNs form micellar structures within their core and their outer lipid layer is not a continuous bilayer. LNs are commonly used for the delivery of nucleic acids. LNs consist of cationic or ionizable lipids which form a complex with the charged genetic material along with phospholipids and cholesterol. PEGylated lipids are usually added to improve the stability and circulation [16].

4.2.1.4 Solid Lipid Nanoparticles

Solid lipid nanoparticles (SLNs) are colloidal carriers composed of long-chain triglycerides or partial glycerides, fatty acids, phospholipids, and waxes. Surfactants such as polysorbate 80, poloxamer 188, and lecithin are usually added at concentrations between 1 and 5% to improve the stability of the SLNs [20]. SLNs consist of a solid hydrophobic core surrounded by a monolayer phospholipid; the hydrophobic tail of the phospholipid is embedded in the solid core. Compared to LNEs, drugs are loaded into SLNs by dissolving or dispersing them in a solid fatty matrix instead of a liquid lipid core which reduces the mobility and enhances the stability and controlled release of the loaded drug. However, SLNs have the disadvantages of low drug-loading capacity and the possibility of drug leaking or matrix explosion during storage. This happens when a low-ordered lipid modification of the matrix transforms into a more stable high-ordered modification during storage. A high ordered modified lipid matrix has a perfect crystal lattice with few imperfections and hence, there is a little room for drug accommodation [23]. In order to overcome polymorph transformation-related problems in SLNs, nanostructured lipid carriers (NLC) were developed as a second-generation lipid nanoparticles.

4.2.1.5 Nanostructured Lipid Carriers

Nanostructured lipid carriers (NLCs) were developed to increase the loading capacity and enhance the stability of SLNs. The core matrix of NLCs consists of a mixture of solid lipids and liquid lipids (at room and body temperature) at ratios ranging from 70:30 to 99.9:0.1, respectively [20]. The inclusion of liquid lipids within the core matrix reduces the degree of crystallinity of the solid lipids and hence increases the NLCs' drug-loading capacity and long-term physical stability [24]. Surfactant are also commonly used at content ranges between 1.5 and 5% to improve NLCs stability [20]. A large number of approaches have been reported for the preparation of SLNs and NLCs. These approaches can be categorized into three main groups: approaches that use organic solvents, high- and low-energy approaches; extensive details of these methods have been reported by Ganesan and Narayanasamy [25].

4.2.1.6 Lipid Nanocapsules

Lipid nanocapsules (LNCs) have a hybrid structure between that of liposomes and polymeric nanoparticles. An LNC's structure consists of a hydrophobic oily core composed of medium chain triglycerides which are surrounded by a hydrophilic shell of a mixture of PEGylated surfactants and optionally hydrophobic lecithin or other co-surfactants. LNCs are usually prepared by a phase inversion temperature which is a solvent-free low-energy method. In this method, the components are mixed together to form an initial emulsion which is then subjected to heating and cooling cycles between 60 °C and 90 °C to obtain reversible emulsion phase inversions [26].

4.2.2 Polymeric Nanoparticles

Polymeric nanoparticles (PNs) can be prepared from natural polymers, synthetic polymers, or preformed polymers which allows a wide variety of possible structures and characteristics with variable drug delivery and co-delivery capabilities. Polymeric nanoparticles can be used for the delivery of small molecules, biological macromolecules, proteins, and vaccines. Furthermore, their surface can be easily modified for active targeting. However, polymeric nanoparticles have the disadvantages of particle aggregation and toxicity [16]. Only a small number of polymeric nanoparticles have been approved by FDA but a large number of polymeric nanoparticle formulations are under clinical trials [15]. Figure 4.2 presents the structures of the different types of polymeric nanoparticles.

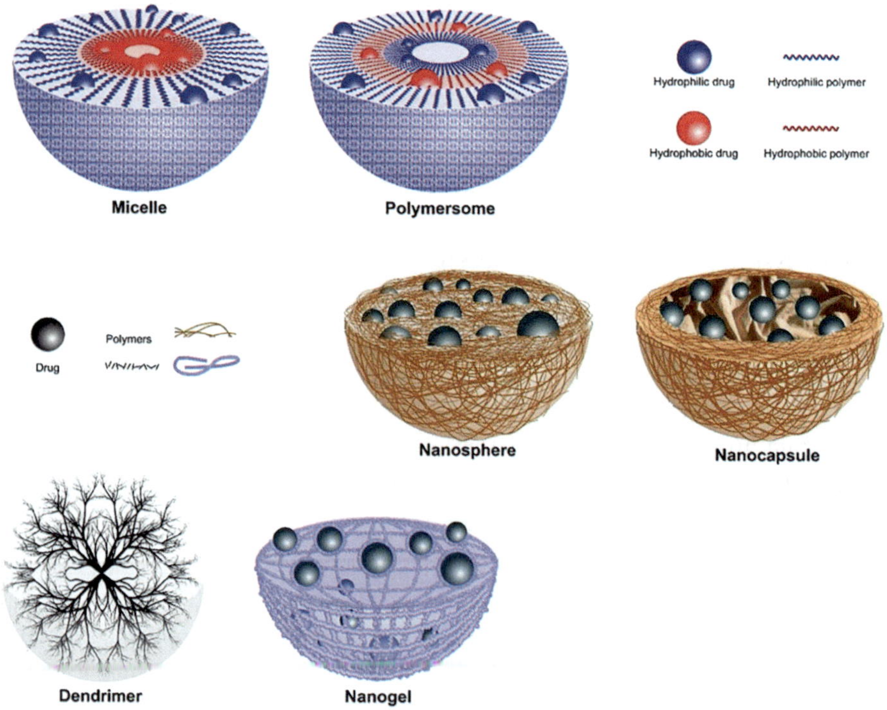

Fig. 4.2 Schematic representation of diverse structures of polymeric nanoparticles

4.2.2.1 Micelles

Micelles are prepared by the self-assembly of amphiphilic copolymers. Depending on the type of copolymer and the interchain interaction, micelles can be designed to have a hydrophobic core and a hydrophilic corona or a hydrophilic core and a hydrophobic corona (reverse micelle). Thus, micelles can be designed for the delivery of hydrophilic or hydrophobic drugs. Furthermore, both hydrophilic and hydrophilic drugs can be loaded into a single system utilizing the core and corona of the micelles. Compared to surfactant-based micelles, polymeric micelles have a lower critical micelle concentration. Hence, polymeric micelles are more stable and require less material (polymer) during preparation.

4.2.2.2 Polymersomes

Polymersomes (nanovesicles) are composed of a hydrophilic core and a hydrophilic shell separated by a hydrophobic layer. Polymersomes can carry either hydrophilic or hydrophobic drugs or both simultaneously [27]. Polymersomes are structurally similar to liposomes, but they are composed of amphiphilic copolymers instead of lipids. Copolymers have a larger molecular weight (1–10 kDa) compared to lipids (0.1–1 kDa) [28] and hence polymersomes demonstrate better physical and chemical stability [27, 29].

4.2.2.3 Nanocapsules

Nanocapsules are hollow structured nanoparticles composed of a hollow core surrounded by a cross-linked polymeric shell. The inner space usually contains an oily liquid. Drugs are accommodated within the core and protected by the shell. Thus, nanocapsules prevent drug degradation, reduce drug toxicity on normal cells, mask unpleasant tastes, and provide controlled release. Nanocapsules are usually prepared by nanoprecipitation, nanoemulsion template methods, or layer-by-layer methods [30].

4.2.2.4 Nanospheres

Nanospheres are composed of a polymeric solid matrix. Drugs can be entrapped, dissolved, adsorbed or chemically bound to the matrix. Nanospheres usually exhibit higher polydispersity compared to micelles. Nanospheres are mostly prepared by solvent displacement (nanoprecipitation) techniques. Other approaches such as solvent evaporation, emulsification/solvent diffusion, and emulsification/reverse salting-out methods have been also used to prepare nanospheres [31].

4.2.2.5 Dendrimers

Dendrimers are highly branched polymers with complex 3D architectures and a well-controlled mass, size, shape, and surface chemistry. Charged polymers such as poly(ethylenimine) (PEI) and poly(amidoamine) (PAMAM) are commonly used in preparing dendrimers. Different dendrimers exhibit different architectures depending on the molecular chirality, composition, and branching of the initial generation (i.e., the molecular used at the core). Dendrimers have been widely investigated for the delivery of nucleic acids and small molecules. Drugs can be loaded into the interior part while biomolecules or contrast agents can be conjugated to the active functional groups present at the surface [32].

4.2.2.6 Nanogels

Polymeric nanogels are nanosized, swellable hydrogels prepared via physical or chemical crosslinking of hydrophilic or amphiphilic polymer chains [33]. Nanogels have been developed with different structural designs such as simple nanogel spheres, core-shell, hollow, functionalized, hairy, and multilayer [34]. Nanogels have gained a lot of interest in recent years due to their high loading capacity and stability. Also, nanogels are responsive to different stimuli such as temperature, pH, light, magnetic field, and biomolecule recognition and hence nanogels can be prepared as stimuli-responsive nanocarriers [35].

4.2.3 Nanocrystals

Nanocrystals are suspensions of nanosized crystalline drug particles; they are also known as nanosuspensions. Unlike other nanoparticles such as lipidic and polymeric nanoparticles, nanocrystals are carrier-free nanoparticles which distinguish them in terms of superior loading capacity and stability. Nanocrystals were initially developed to improve the solubility and bioavailability of poorly water-soluble drugs. This is achieved due to their large specific surface area and the curvature of the nanosized drug particles which results in improved solubility and dissolution rates [36]. Nanocrystals can be produced via top-down approaches, such as high-pressure homogenization and media milling, or bottom-up approaches, such as precipitation or spray drying. Bottom-up approaches are less energy-consuming and produce particles with super crystallinity and narrow particle size distribution. However, stabilizers such as polymers and surfactants are usually added to prevent crystal growth (Ostwald ripening) and particle aggregation by coating the growing interfaces. Stabilizers may reduce drug loading and increase potential toxicity when used at high concentrations [37].

4.2.4 Inorganic Nanoparticles

Inorganic nanoparticles such as gold, iron oxide, and silica have attracted great attention as drug delivery systems due to their unique plasmonic and magnetic features. The magnetic, radioactive, and plasmonic properties distinguish inorganic nanoparticles as unique nanoparticles with applications such as diagnostics, imaging, and photothermal therapies. For instance, magnetic nanomaterials have been developed as a promising theranostics tool in cancer therapy due to their intrinsic magnetic properties and multifunctional design. For instance, tumor ablation can be achieved by applying an interchangeable magnetic field to superparamagnetic iron oxide nanoparticles (SPIONs) which results in magnetic hyperthermia-generating heat [38]. Gold nanoparticles have been extensively investigated for the delivery of nucleic acids. This is due to the cationic metal ions of gold nanoparticles which preferentially bind to the positively charged nucleic acids [39]. Mesoporous silica nanoparticles have a large surface area and pore volume and hence hydrophobic drugs can be loaded into the pores and entrapped via electrostatic or covalent interaction with silanol groups. Also, the silanol groups on the exterior surface can be conjugated with organic molecules to achieve specific targeting [40]. However, in spite of the unique features and potential application of inorganic nanoparticles, they have the disadvantages of poor solubility and toxicity, especially where the nanoparticles are synthesized using heavy metals [16].

Inorganic nanoparticles can be prepared by top-down or bottom-up approach. In top-down approach, a large piece of material is broken down into nanosized particles mostly via attrition or milling. The main disadvantage of this approach is the heterogeneity in the size and shape of the produced particles. Also, the surface of particles can suffer defects and crystallographic damage. In addition, significant amounts of impurities from the milling medium may exist in the product. By contrast, bottom-up approach usually produce particles with narrower size and shape distributions, less defects, and less impurities. In this approach, single atoms and molecules are assembled in their inorganic salts into nanostructures via crystallization process [41].

4.3 Inhalable Nanoparticles Frameworks

As mentioned earlier in this chapter, nanoparticles are not suitable for pulmonary delivery due to their submicron sizes and thus a nanoparticle system needs to be transformed into another system suitable for lung delivery. One, alternative, approach to achieve this is by producing a nanoparticle/microparticle powder system suitable for lung delivery. However, the most common approach is to transform the nanoparticles into some form of inhalable microparticles, i.e., Trojan particles. This approach can be achieved with or without the addition of a matrix-forming agent(s). It is noteworthy that inhalable Trojan particles have been formed in many different frameworks. Some frameworks have been presented in the literature with more than one name, and some names have been used interchangeably for different frameworks. In order to remove confusion, throughout this chapter a descriptive term is selected to name each framework, and other common names used in the literature are highlighted as and when necessary. Figure 4.3 presents the structures of different types of inhalable nanoparticle frameworks.

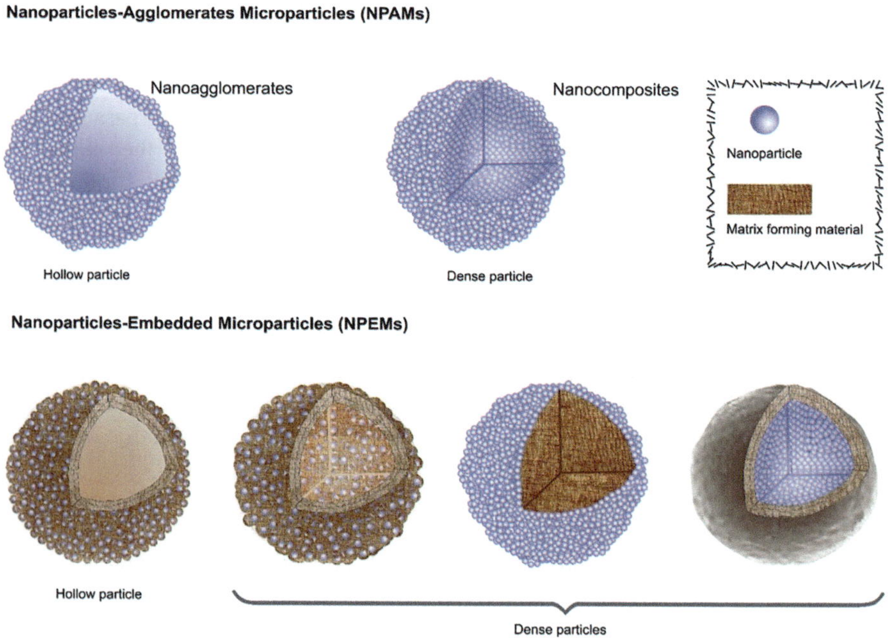

Fig. 4.3 Schematic representation of inhalable nanoparticles frameworks

4.3.1 Nanoparticle-Microparticle Powder Systems (NPMPs)

This approach depends on designing a dry powder system composed of particles within the nano and micro size range. All particles are <5 μm with a narrow and unimodal particle size distributions, suitable for pulmonary deposition in the mid- and deep lung regions. This system is mostly produced by spray drying the drug(s) with [42] or without [43] the addition of an excipient(s). Another approach for producing such a system is to separately prepare nanoparticle and microparticle systems and then physically mix the two systems at the required mass ratio [44–46]. In this system, the cohesive nature of the nanoparticles is mitigated by their distribution on the surface of the microparticles while, simultaneously, the nanoparticles act as spacers to prevent the microparticles from forming large aggregates [44].

4.3.2 Nanoparticle-Agglomerate Microparticles (NPAMs)

Without the addition of a matrix-forming agent, nanoparticles can be brought together in a finely controlled manner to form micrometer-sized particles with inhalable properties. The nanoparticles can be arranged together to form either nanoagglomerate or nanocomposite microparticles. Nanoagglomerate microparticles are also known as porous nanoparticle-aggregate particles (PNAPs) [47], large porous nanoparticle (LPNP) aggregates [48], and large hollow nanoparticulate aggregates (LHNPA) [49, 50]. Nanoagglomerate microparticles are hollow microparticles with a wall composed of one or more layers of nanoparticles that are held together by controlling their inter-particle forces. Such microparticles have a low density which reduces their MMAD and hence improve their aerosolization [51]. Unlike nanoagglomerate microparticles, nanocomposite microparticles are dense microparticles (not hollow) composed of a cluster of nanoparticles. In spite of their dense structure, nanocomposite microparticles exhibit good aerosolization performance due to their porous nature.

The advantage of this approach is that the dried powder contains drug-loaded nanoparticles which increases the amount of the drug and reduces the amount of powder to be inhaled compared to excipient-containing particles. However, NPAMs face challenges in maintaining the integrity of the initial nanoparticle during the process of forming the microparticles. Also, the redispersion of the nanoparticles from NPAMs might be hindered by strong interactions between the nanoparticles. To help solve these challenges, another approach involving the addition of a matrix-forming agent has been developed.

4.3.3 Nanoparticle-Embedded Microparticles

Nanoparticle-embedded microparticles (NPEMs) are also known as nano-in-microparticles (NIMs) [52], nanocomposite microparticles (NCMPs) [53], and nanoparticle-containing microparticles. In this approach, nanoparticles are embedded in a microparticle matrix composed of an inert matrix-forming agent, mainly a carbohydrate sugar such as mannitol, with or without the addition of other excipients. The matrix agent helps protect the nanoparticles' essential properties during the process of transforming them into microparticles and helps assure the redispersion of the nanoparticles from the microparticles upon application. The spatial distribution of the nanoparticles and the matrix agent(s) in the microparticles can occur in different manners as reported in the literature for different NPEMs structures. Some NPEMs have a hollow structure composed of an empty core and a shell that contains the nanoparticles homogenously distributed within the matrix material. Other NPEMs are dense with a homogenous distribution of the nanoparticles throughout the matrix. NPEMs may also exist as nanoparticles encapsulated by a shell formed from the matrix material or as a shell made of the nanoparticles surrounding the matrix material. Table 4.1 presents a summary of dry powder formulations of different inhalable nanoparticle frameworks.

Table 4.1 The summary of nano-in-micro particles systems used for pulmonary drug delivery

Drug	Nanoparticle system	Microparticles framework	Matrix agent	NPs		MPs				Ref.
				Size (nm)	Z-potential (mV)	D_{V50} (μm)	MMAD (μm)	ED (%)	FPF (%)	
Tobramycin	NA	NPMPs	NA	NA	NA	0.766	5.58	92.92	27.85	[43]
Azithromycin	NA	NPMPs	NA	NA	NA	0.676	2.16	88.93	53.45	[43]
Tobramycin	NA	NPMPs	Mannitol	NA	NA	0.620	4.55	96.89	28.70	[42]
Azithromycin	NA	NPMPs	Mannitol	NA	NA	0.59	2.71	92.03	38.26	[42]
NA	Polymeric NPs (chitosan)	NPMPs (mixture)	Lactose	347.27	+20.17	5.43	3.85	78.5	32.42	[44]
NA	Polymeric NPs (chitosan)	NPMPs (mixture)	Lactose	684.00	+27.86	10.44	3.16	79.66	43.47	[45]
Montelukast	Polymeric NPs (chitosan)	NPMPs (mixture)	Lactose	382.88	+11.40	5.60	1.34	100	83.22	[46]
Prothionamide	Polymeric NPs (chitosan)	NPMPs (mixture)	Lactose	314.37	+80.65	–	1.76	82.37	81.19	[54]
Bedaquiline	Polymeric NPs (chitosan)	NPMPs (mixture)	Lactose	109.7	+36	–	3.98	–	15.38	[55]
Tobramycin	Nanosuspensions	NPAMs	NA	400	–	0.7	–	–	63	[56]
Clarithromycin	Nanosuspensions	NPAMs	NA	400	–	0.7	–	–	62	[56]
Indomethacin	Nanosuspensions	NPAMs	NA	263	–	3.37	4.1	–	43.2	[57]
Ovalbumin	Polymeric NPs (PLGA)	NPAMs	NA	270	–	2.9	3.3	–	51	[58]
Curcumin	Polymeric NPs (ac-DEX)	NPAMs	NA	199.3	−14.5	1.52	1.61	> 95	> 95	[59]
Ethambutol HCl	SLNs (Compritol)	NPAMs	NA	57.65	–	1.153	5.63	91.76	23.98	[60]
Erlotinib	SLNs (Compritol)	NPAMs	NA	82.32	–	< 4	5.53	87.16	24.25	[61]
Alendronate	SLNs (Compritol)	NPAMs	NA	< 100	–	5	6.39	96.34	11.87	[62]
Calcium phosphate	Nanocrystal	NPEMs	Mannitol	80	−25	3.1	–	≈ 90	74.3	[63]
Ciprofloxacin	Nanoplex	NPEMs	Mannitol	306	−23	69	2.5	79	31	[64]
Isoniazid	Polymeric NPs (chitosan)	NPEMs	Mannitol	249.72	+25.92	–	3.37	93.28	43.95	[65]
Pyrazinamide	Polymeric NPs (chitosan)	NPEMs	Mannitol	249.72	+25.92	–	3.28	95.03	41.03	[65]
Insulin	Polymeric NPs (chitosan)	NPEMs	Mannitol	218	+33	–	2.80	–	–	[66]
Bedaquiline	Polymeric NPs (chitosan)	NPEMs	Lactose	109.7	+36	–	3.38	–	28.27	[55]
NA	Polymeric NPs (chitosan, CM-β-CD)	NPEMs	Mannitol	282	+29	–	2.2	–	–	[67]

CRISPR-Cas9	Polymeric NPs (PEGylated chitosan)	NPEMs	Mannitol	184.1	≈ + 15	3.4	4.6	71.5	62.3	[68]
PspA	Polymeric NPs (PGA-co-PDL)	NPEMs	Leucine	322.83	−13.23	1.99	1.70	–	74.31	[53]
BSA	Polymeric NPs (PGA-co-PDL)	NPEMs	Leucine	348.36	−1.44	2.09	2.80	–	70.67	[69]
Methotrexate	Gelatin NPs	NPEMs	Leucine	148.45	−22.56	–	2.59	96.08	49.53	[70]
HNP120	Hydrogel NPs	NPEMs	Mannitol	82.6	–	–	4.82	78.5	31.7	[71]
siRNA	Dendriplexes	NPEMs	Mannitol	363	+40	2.4	4.8	50.4	38.5	[72]
NA	MINPs	NPEMs	Mannitol	159.0	–	0.85	4.5	92.0	34.3	[73]
Ciprofloxacin	Liposomes	NPEMs	Mannitol	196	−10	58	3.3	75	5	[64]
Ciprofloxacin	Liposomes	NPEMs	Lactose	101.5	−9.93	< 2	1.75	84.3	76.2	[74]
Ciprofloxacin	Liposomes	NPEMs	Sucrose	101.5	−9.93	< 2	2.48	85.0	61.7	[74]
Ciprofloxacin	Liposomes	NPEMs	Sucrose	122.3	–	1.14	2.61	89.9	66.9	[75]
Ethambutol HCl	SLNs (Comritol)	NPEMs	Mannitol	57.65	–	1.12	4.15	94.70	30.91	[60]
Erlotinib	SLNs (Comritol)	NPEMs	Mannitol	82.32	–	< 4	3.93	94.91	30.98	[61]
Alendronate	SLNs (Comritol)	NPEMs	Mannitol	< 100	–	6	6.72	90.06	9.030	[62]
Gefitinib	SLNs (lecithin, cholesterol)	NPEMs	Mannitol	187.23	−15.53	–	4.48	98.35	44.41	[76]
Gefitinib	SLNs (lecithin, cholesterol)	NPEMs	Lactose	187.23	−15.53	–	5.70	96.51	26.45	[76]
siRNA	SLNs (lecithin, cholesterol, lipid-PEG)	NPEMs	Mannitol	136.3	−37.5	–	3.96	–	37.01	[77]
NA	LNCs	NPEMs	Lactose	59	–	2.87	5.4	76.0	29.0	[78]

n/a Not applicable, *Ac-DEX* acetalated dextran, *BSA* Bovine serum albumin, *CM-β-CD* carboxymethyl-β-cyclodextrin, Dendriplexes TPP-conjugated dendrimers, *MINPs* magnetic iron oxide nanoparticles, *PEG* polyethylene glycol, *PGA-co-PDL* poly(glycerol adipate-co-ω-pentadecalactone, *PLGA* poly(lactic-co-glycolic acid, *PspA* Pneumococcal surface protein A

4.4 Approaches for Producing Inhalable Nanoparticles

Majority of the above-mentioned inhalable nanoparticle frameworks published in literature are produced by spray drying. Other techniques such as spray freeze drying, supercritical CO_2-assisted spray drying, and thin-film freeze drying have been used as well but to a lesser extent. Figure 4.4 represents the different techniques used for the preparation of inhalable nanoparticulate frameworks.

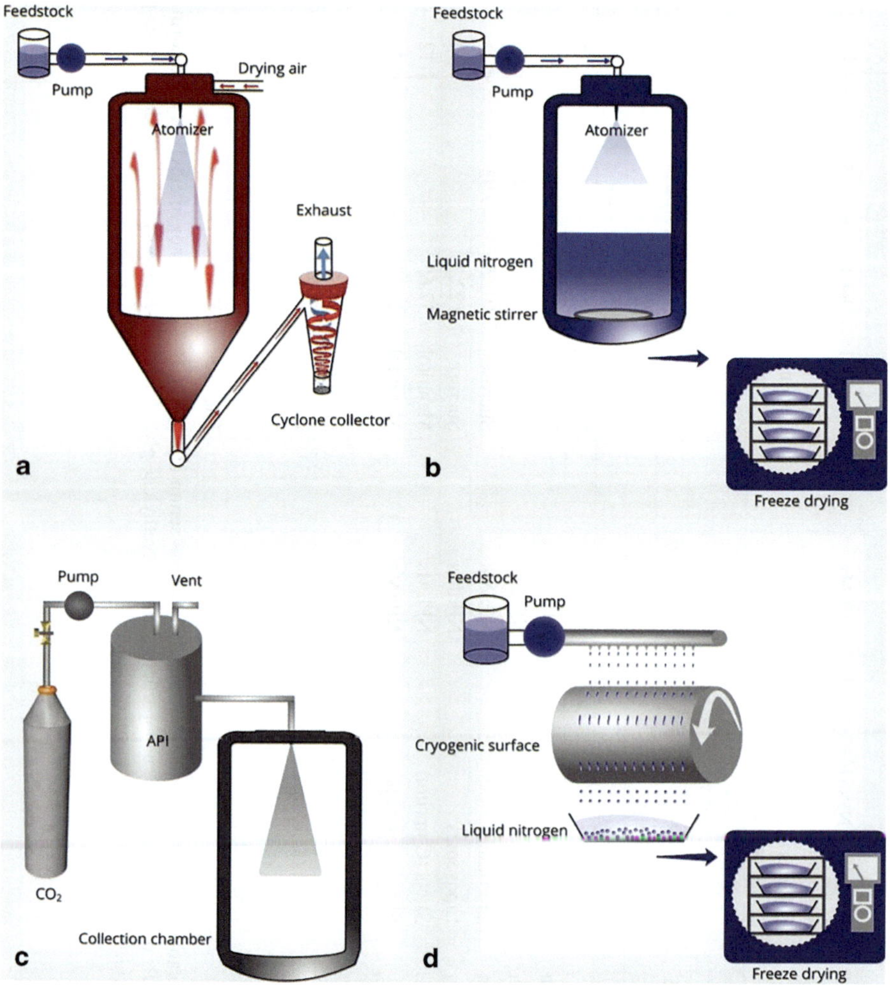

Fig. 4.4 Schematic illustrating of various techniques used in the preparation of inhalable nanoparticulate frameworks: (**a**) spray drying (concurrent mode); (**b**) spray-freezing drying (into vapor over liquid); (**c**) Supercritical CO_2-assisted spray drying (rapid expansion of supercritical solution—RESS); (**d**) thin film freeze drying (TFFD)

4.4.1 Spray Drying

Spray drying is a one-step technique that transforms liquids (e.g., solution, emulsion, and suspension) into dry powders. The properties of the dried particles such as size, shape, morphology, components distribution, and solid state are formed under the influence of a wide variety of parameters including the physicochemical properties of the starting materials, i.e., solvents and solutes (molecules and/or nanoparticles) and the spray drying processing parameters. The formation of particles during spray drying is often not fully understood because of the multifactor effect and the complexity of the thermodynamic phenomena related to the two-phase flow during drying [79]. However, the formation process, morphology, and density of the dried particles can be anticipated by the Péclet number (Pe) which is a dimensionless mass transport number that characterizes the relative importance of the diffusion and convection during the drying process [80]. Pe is the ratio of two critical times for the spray drying process: the diffusion time (τ_{diff}) which is the time required for a substance to move from the external interface to the center of a drying droplet and the drying time (τ_{dry}) of a droplet.

$$Pe = \frac{\tau_{\text{diff}}}{\tau_{\text{dry}}} = \frac{R^2}{\tau_{\text{dry}} D_i} = \frac{\kappa}{8D_i}$$

where R is the radius of a droplet, Di is the diffusion coefficient of solutes (molecules or nanoparticles), and κ is the evaporation rate.

Spray dried particles formed at large Pe, Pe > 1, are usually large hollow particles. A large Pe is the result of one or more of the following conditions: firstly, a large droplet size (R^2) and thus a long diffusion time (τ_{diff}); secondly, a high evaporation rate (κ) and hence a low drying time (τ_{dry}); or thirdly, a low diffusion coefficient (D_i), and hence a long diffusion time (τ_{diff}). All these conditions hinder the movement of a solute from the external face to the center of a drying droplet. By contrast, a small Pe is the result of the reverse of these conditions which facilitates the movement of the solute to the center of a drying droplet and leads to the formation of small dense particles thereafter [79].

In the context of spray drying nanoparticles, the surface of a drying droplet contains more and more nanoparticles as the droplet evaporation proceeds [81]. When nanoparticles are spray dried without a matrix agent, the Pe of the nanoparticles dictates the morphology of the dried particles. Particles formed at Pe > 1 are usually large hollow particles composed of an empty core and a shell made of a dense layer/s of nanoparticles. For a drying droplet, increasing the distance between the surface and the center of the droplet (R^2) increases the required diffusion time (τ_{diff}) for nanoparticles to move from the droplet surface to the center. Also, if the droplet dried fast (high κ) and/or the diffusion coefficient (D_i) of nanoparticles was low (which is usually true for nanoparticles), the diffusion time (τ_{dry}) will not be long enough for the nanoparticles to move from the surface to the center of the droplet. Thus, the nanoparticles will form a dense shell at the surface surrounding an empty core large hollow particles (nano-agglomerates microparticles) will be formed. By contrast, dense small particles are formed at Pe <1. In this case, the low evaporation rate (κ) and/or the small size of a droplet (R^2) enable the nanoparticles to move from the surface to the center of a drying droplet and small dense microparticles (nanocomposites microparticles) formed thereafter [81].

The above-mentioned scenarios of particle formation get more complicated when a matrix-forming agent is co-spray dried with nanoparticles (e.g., forming nanoparticle-embedded microparticles (NPEMs)). In this case, the formation of the microparticles is dictated by both the Pe of the nanoparticles and the Pe of the matrix agent(s). Also, the presence of a matrix agent at a higher mass ratio can lead to some outcomes which seem to contradict the concept of Pe. For example, some spray dried NPEMs displayed an external shell composed of small molecule matrix agents which encapsulate the nanoparticles. Usually, small molecule agents have a lower Pe compared to nanoparticles and hence the nanoparticles are expected to form the external shell. However, the high mass ratio of the matrix agent can result in supersaturation and fast crystallization leading to the formation of the observed external shell [82, 83].

4.4.1.1 Nanoparticle-Microparticle Powder Systems

Li et al. (2014) used spray drying to produce Nanoparticle-Microparticle Powder Systems (NPMPs) of tobramycin and azithromycin antibiotics. Different powder systems were prepared by spray drying 0.1% w/v methanol solution of each drug at three different pump rates. All formulations produced particles with narrow and unimodal size distributions and sizes within the microparticulate/nanoparticulate range. For example, the size range of the tobramycin and the azithromycin particles were 0.51–1.61 μm and 0.16–1.78 μm, respectively (at the highest pump speed, i.e., 100%). In comparison with the azithromycin (hydrophobic) particles, the tobramycin (hydrophilic) particles absorbed moisture and formed large aggregates which reduced their aerosol performance. The mass median aerodynamic diameter (MMAD), emitted dose (ED), and fine particle fraction (FPF) of tobramycin were 5.58 μm, 88.93%, and 27.85% while azithromycin MMAD, ED, and FPF were 2.16 μm, 92.92%, and 53.45%, respectively [43]. In another study by Muralidharan et al. (2020), mannitol was co-spray dried with fluticasone propionate and salmeterol xinafoate (drug molar ratio 5:1, respectively) at 1:3 drugs to mannitol molar ratio. The solutes were dissolved in methanol and spray dried at three different pump speeds under an inlet temperature of 130 °C. Among the prepared powder systems, the one prepared at the highest pump speed (i.e., 100%) had the best aerosolization parameters with ED, FPF, and MMAD of 97.53%, 38.67%, and 4.65 μm, respectively [84].

4.4.1.2 Nanoparticle-Agglomerate Microparticles (NPAMs)

Nanosuspensions

Nanosuspensions comprising of a mixture of tobramycin and clarithromycin at mass ratio 10:1 were prepared with sodium glycocholate stabilizer (surfactant) at mass ratios of 1%, 2%, and 5% relative to the tobramycin mass. Nanosuspensions were prepared in isopropanol by high-pressure homogenization. The nanosuspensions were spray dried resulting in the formation of porous microparticles build of amorphous clarithromycin accommodating tobramycin nanoparticles size of about 400 nm. The de-agglomeration tendency of the particles increased upon increasing the percentage of the surfactant used in the initial nanosuspensions. The aerosol performance of the spray dried formulations with 2% surfactant was superior to a physical mixture of the two drugs at the same mass ratio. The FPF increased from 35% and 31% for the physical blend of tobramycin and clarithromycin, respectively, to 63% and 62% for the spray dried formulation [56]. In a study by Malamatari et al. (2015), indomethacin nanosuspension was prepared by wet bead milling using a planetary micro mill and stabilized with Pluronic® F68. Nanosuspension was transformed into microparticles composed of nano-agglomerates with and without matrix-forming agents (mannitol and leucine). The microparticles formed with matrix agents had higher FPF and lower MMAD values in comparison with microparticles formed without matrix agents indicating better aerosolization performance. Particles prepared with matrix agent had FPF and MMAD of 61.9% and 3.4 μm while particles without matrix agent had FPF and MMAD of 43.2% and 4.1 μm. However, the drug content of the microparticles prepared without matrix agent was about 91% which is more than two times higher than the drug content of particles prepared with mannitol, i.e., about 41%. Thus, a larger amount of powder would be needed for inhalation in the case of using mannitol [57].

Polymeric Nanoparticles

Polymeric nanoparticles composed of acetalated dextran polymer and loaded with curcumin were prepared via an oil/water emulsion solvent evaporation with sizes of 200 nm. An aqueous suspension of the nanoparticles was spray dried at a concentration of 0.5% w/v. During the spray drying process, the nanoparticles agglomerated and formed a shell surrounding the droplet of water which then evaporated leaving an empty center. As a result, hollow particles were initially formed which then shrank as the droplets dried leaving nanocomposite microparticles with a raisin-like morphology. The microparticles dissociated into 200 nm nanoparticles upon reconstitution in water. All nanocomposite microparticle systems exhibited desirable in vitro aerodynamic performance with MMAD between 1.61 and 2.41 μm, and FPF above 70% [59]. Ovalbumin loaded-poly-lactic-co-glycolic acid (PLGA) nanoparticles were prepared by double emulsion with a size of around 270 nm then spray dried from their suspension without the addition of matrix-forming excipients. Nanocomposite microparticles with desirable inhalable parameters (MMAD of 3.3 μm and FPF of 51%).were produced. The microparticles were easily redispersed in water and nanoparticle size was unchanged [58].

Lipid Nanoparticles

Nemati et al. (2019) fabricated ethambutol hydrochloride-loaded SLN and subjected it to spray drying with and without mannitol to form inhalable microparticles. Ethambutol hydrochloride-loaded SLN showed a mean particle size of 57.65 nm while the microparticles produced with and without mannitol showed mean particle sizes of 1120 and 1153 nm, respectively. Scanning electron microscope (SEM) analysis did not show any agglomerates or clusters of nanoparticles. Additionally, nanoparticles were not re-dispersed from the microparticles which make it tricky to realize the fate of the nanoparticles during spray drying process. During the aerodynamic assessment of microparticles by Next Generation Impactor (NGI) (60 L/min) microparticles fabricated with and without mannitol displayed FPF of 30.91 and 23.98%, respectively. The use of mannitol assists to develop spherical shaped, less adhesive particles and thus showed higher FPF values when compared to microparticles fabricated without mannitol [60].

4.4.1.3 Nanoparticle-Embedded Microparticles (NPEMs)

In this approach, nanoparticles are dispersed in a solution of a matrix-forming agent. Mannitol is the most commonly used agent. Other sugars, such as lactose, sucrose, trehalose, raffinose, along with polymers, such as chitosan and cyclodextrin, and amino acids, such as leucine, have also been investigated, though to a lesser extent, for this purpose.

Nanosuspensions

Quarta et al. (2021) fabricated respirable microparticles by spray drying calcium phosphate nanoparticles disperse in an aqueous mannitol solution. The various ratios of calcium phosphate nanoparticles to aqueous mannitol solution (0:1, 1:4, 1:1, and 14:1) were studied and process was optimized by factorial design. The morphology of the spray dried particles was studied using an SEM equipped with Gallium focused ion beam (Fig. 4.5). Cross-section imaging of spray dried mannitol particles alone showed spherical-shaped particles with smooth surfaces. Spherical particles were hollow with a central cavity of about 1 μm. The addition of nanoparticles at 1:4 mass ratio resulted in spherical particles with slightly rough surfaces and a central hollow surrounded by small pores. Compared to mannitol-only particles, the presence of these pores reduced the true density of the particles which reduces their aerodynamic diameter. Particles formed from 14:1 mass ratios showed larger particle size (D_{v50} 2.1 μm) compared to 1:4 particles (D_{v50} 1.6 μm). The composition with higher nanoparticle concentration showed doughnut-shaped structures. During spray drying, the nanoparticles became concentrated at the surface of the droplets due to their low diffusion coefficient. The liquid at the droplet's core evaporated; leaving a central hollow and hence the surface shell collapsed giving the doughnut like shape. Particles produced from 1:1 mass ratio exhibited a rough surface and dimpled shape. These particles showed an incomplete doughnut formation due to their higher mannitol content which made the surface shell stronger and more resistance to collapse during drying. The aerodynamic performance of spray dried powders was analyzed using a fast screening impactor (60 L/min) equipped with RS01® medium resistance monodose inhaler. All spray dried particles exhibited ED above 75%. Optimized spray dried particles showed satisfactory FPF (70%) due to surface properties such as surface roughness and shape [63].

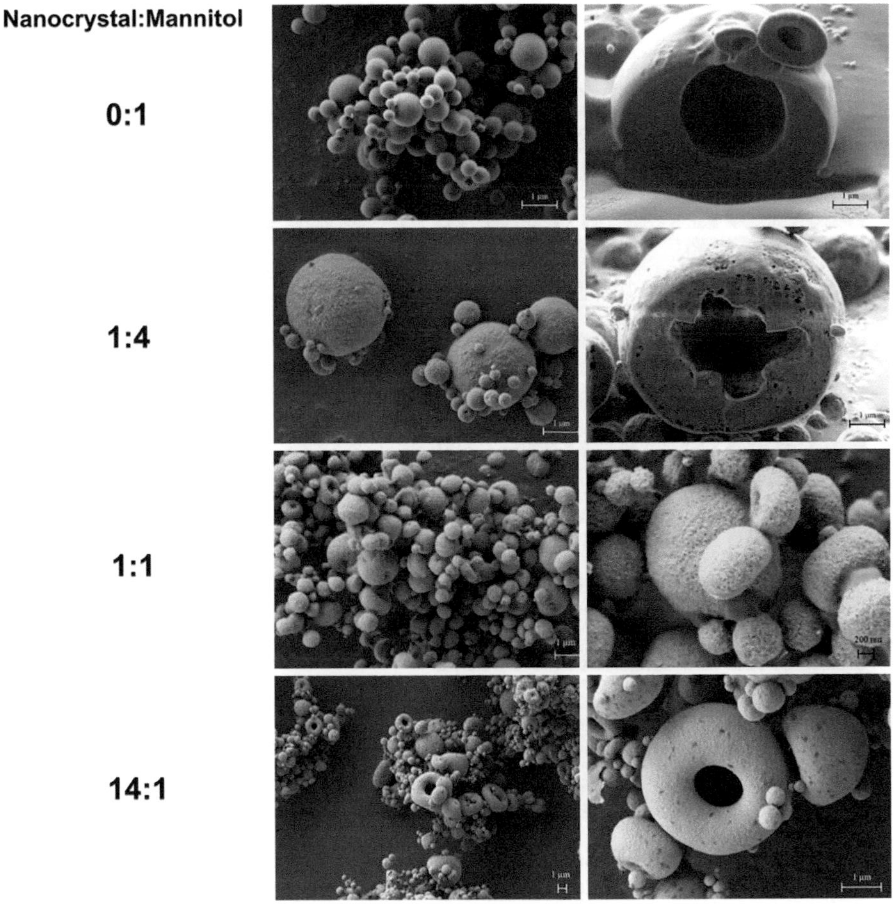

Fig. 4.5 SEM images of formulations with nanocrystal to mannitol mass ratios of 0:1, 1:4, 1:1, and 14:1. Reprinted from Quarta E, Sonvico F, Bettini R, De Luca C, Dotti A, Catalucci D, et al. Inhalable Microparticles Embedding Calcium Phosphate Nanoparticles for Heart Targeting: The Formulation Experimental Design. Vol. 13, Pharmaceutics. 2021

Polymeric Nanoparticles

Torge et al. (2017) investigated the influence of mannitol's mass ratio on the properties of the spray dried nanoparticle-loaded microparticles. PLGA nanoparticles (0%, 20%, 33%, 50%, and 100%) were suspended in a mannitol aqueous solution and subjected to spray dryer (inlet temperature of 80 °C) to obtain NPEMs. The component distribution and particle morphology of NPEMs were studied by confocal laser scanning microscopy (CLSM) and SEM. The NPEMs containing 20, 33, and 50% nanoparticles showed hollow morphology with a shell composed of mannitol matrix and nanoparticles which was well probed by the co-localization of the two fluorescent dyes during CLSM analysis (Fig. 4.6I). SEM analysis of NPEMs containing 0% nanoparticles showed solid and spherical shape particles whereas the

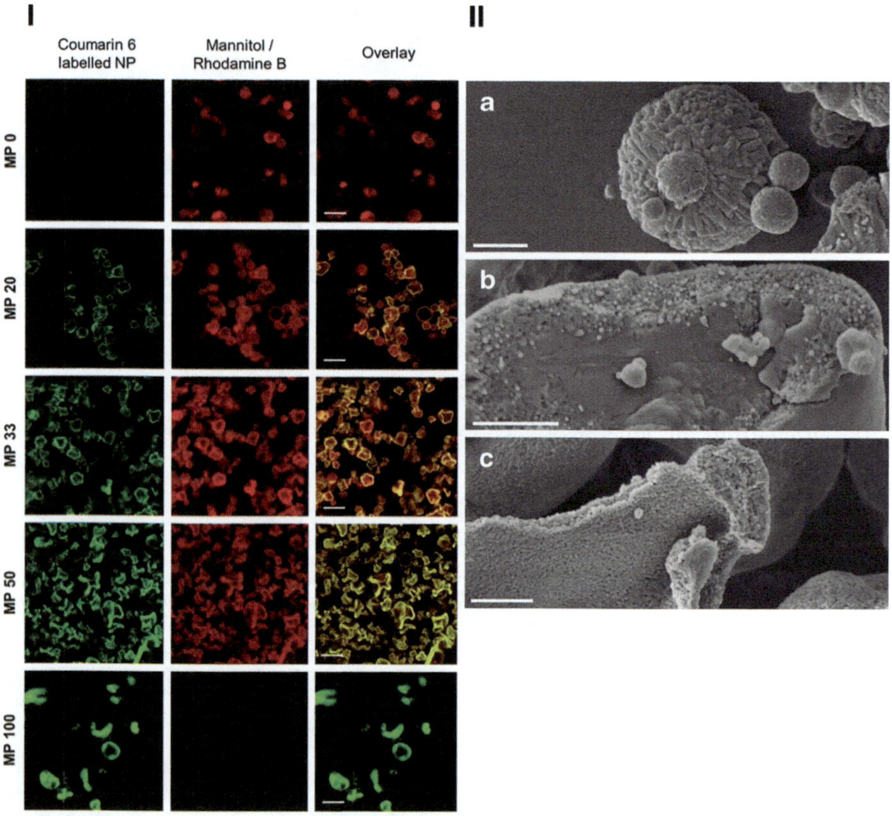

Fig. 4.6 I CLSM, the left row shows the fluorescence of the coumarin 6 (nanoparticles), the second row displays rhodamine B fluorescence (matrix) and in the right row the overlay of both is shown; Scale bars: 10 µm; II SEM, (**a**) spray dried mannitol particles without nanoparticles (MP 0); (**b**) Nanoparticles are embedded in the mannitol matrix (MP 50); (**c**) spray dried nanoparticles without mannitol (MP 100); Scale bars: 2 µm. Reprinted from Torge A, Grützmacher P, Mücklich F, Schneider M. The influence of mannitol on morphology and disintegration of spray dried nano-embedded microparticles. Eur J Pharm Sci. 2017;104:171–9, Copyright 2017, with permission from Elsevier

addition of nanoparticles resulted in the formation of raisin-shaped hollow particles. This was attributed to the difference in the Péclet numbers of the nanoparticles and mannitol. The Péclet number of the nanoparticles was >200 times higher compared to mannitol. Assuming a fixed evaporation rate, the nanoparticles had a much lower diffusion coefficient compared to mannitol. Thus, spray drying nanoparticles without mannitol resulted in the formation of hollow particles with a shell composed of densely packed nanoparticles while spray drying pure mannitol resulted in dense spherical particles (Fig. 4.6IIA) (Fig. 4.6IIC). Spray drying nanoparticles with mannitol resulted in hollow particles with a shell composed of nanoparticles embedded in mannitol matrix (Fig. 4.6IIB). The presence of mannitol at the shell was explained by two steps, i.e., (1) the crystallization of mannitol upon supersaturation at the surface of the drying droplets and (2) the formation of hydrogen bonds between mannitol and PVA (a stabilizer located at the nanoparticle surface). Increasing the amount of mannitol in NPEMs increased their disintegration rate while microparticles prepared without mannitol failed to disintegrate and release nanoparticles [83]. During aerodynamic analysis using high resistance HandiHaler® device spray dried powder (20 mg/ hard gelatin capsule) showed FPF in the range of 19.9–28.6%. Particularly, NPEMs containing 33 and 100% nanoparticles showed lowest and highest FPF values, respectively. Briefly, the NPEMs composed of mannitol and PLGA nanoparticles denote a promising delivery system due to desired deposition pattern of nanoparticles in the airways [83].

In another study by Al-Qadi et al. (2014), hybrid polymeric nanoparticles (composed of chitosan, carboxy methyl-β-cyclodextrin, and a gelling counterion) were spray dried from an aqueous mannitol solution at three nanoparticles:mannitol mass ratios, i.e., 1:9, 2:8, and 3:7 and inlet temperature of 170 °C. All dried particles were spherical in shape. Increasing the mass ratio of the nanoparticles reduced the agglomeration tendency of the dried microparticles. The MMAD of NPEMs produced at mass ratio of 1:9, 2:8, and 3:7 were 2.2, 2.6, and 2.3 µm, respectively. The MMAD was measured using a TSI Aerosizer® LD, equipped with an Aerodisperser® [67]. Fluorescent imaging by CLSM (Fig. 4.7) showed a different distribution of nanoparticles within the mannitol matrix in the microparticles compared to the matrix presented by Torge et al. (2017) in the above-mentioned study [83]. The nanoparticles were homogeneously distributed in the core of the microparticles and almost completely encapsulated in a pure mannitol shell [67]. The author attributed the presence of mannitol at the surface to the fact that it was labelled with a hydrophobic fluorophore (Bodiby®) which resulted in the mannitol localizing at the air/liquid interface on the surface of drying droplets.

In another study, chitosan/tripolyphosphate nanoparticles with a size of 382 nm and zeta potential of +33.5 mV were co-spray dried with mannitol at a mass ratio of 20:80, respectively, and an inlet temperature of 160 °C. CLSM images showed a

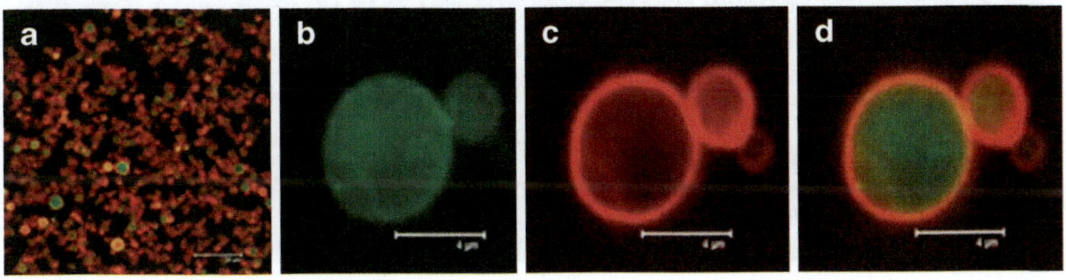

Fig. 4.7 (**a**) A section of mannitol microsphere loaded with hybrid nanoparticles, nanoparticles/ mannitol 3:7, Scale bar: 10 µm; (**b**) a green channel representing the fluorescently labelled nanoparticles, (**c**) a red channel representing mannitol labelled with Bodipy®, (**d**) both channels overlapping. Reprinted from Al-Qadi S, Remuñán-López C. A micro- and nanostructured drug carrier based on biocompatible, hybrid polymeric nanoparticles for potential application in dry powder inhalation therapy. Polymer (Guildf). 2014;55 (16):4012–21, Copyright 2014, with permission from Elsevier

different type of nanoparticle-embedded microparticle matrix compared to the matrix reported by the two above-mentioned studies [67, 83]. In this study, the formed particles were dense, the mannitol was distributed across the particle and nanoparticles were homogenously mixed with the mannitol. Both mannitol and nanoparticles were detected at the surface of particles, but mannitol was detected at a higher level which is due to its high mass ratio, 80% w/w (Fig. 4.8a). Image series of three different sections of microparticle, obtained by moving the detector from the middle to the top of the microparticles (Fig. 4.8b), reveal that the mannitol is distributed throughout the whole particle, forming a kind of continuous matrix, where the nanoparticles are homogeneously dispersed (Fig. 4.8b). The aerodynamic performance of the microparticles was tested using a TSI Aerosizer® LD equipped with an Aerodisperser®; the microparticle exhibited favorable aerodynamic properties with MMAD of 2.71 μm [82].

Lipid Nanoparticles

Mannitol was also used to transform rifabutin-loaded SLNs, prepared by hot high-shear homogenization, into microparticles. Mannitol was individually co-spray dried with glyceryl dibehenate and glyceryl tristearate SLNs at mannitol to SLNs

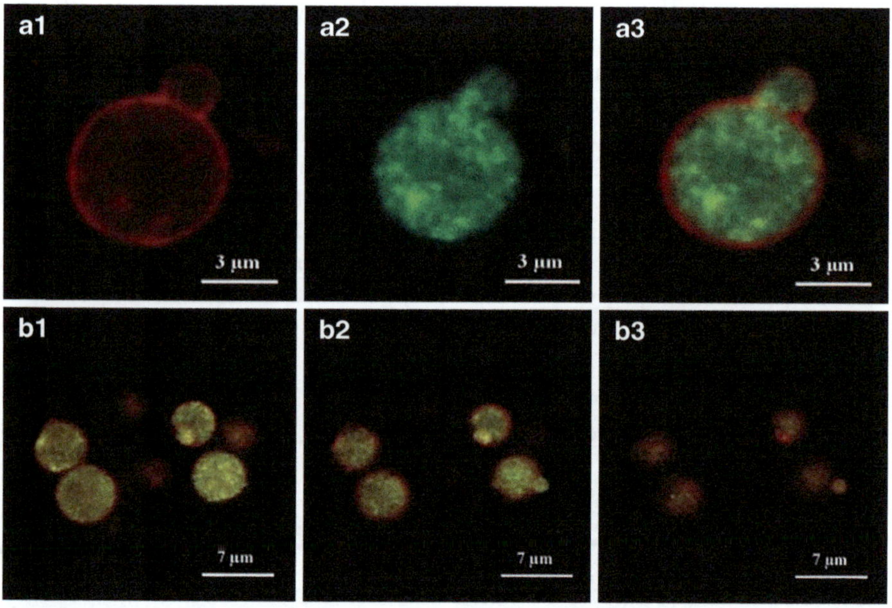

Fig. 4.8 CLSM images of microparticles with mannitol labeled with Bodipy (red) and nanoparticles labeled with FITC-BSA (green): A1: red channel; A2: green channel; A3: overlapping of both channels. B1, B2, and B3 correspond to the overlapping of both channels in different sections of the same series, separated in the *z*-axis by 0.8 μm. Reprinted (adapted) with permission from Grenha A, Seijo B, Serra C, Remuñan-López C. Chitosan nanoparticle-loaded mannitol microspheres: structure and surface characterization. Biomacromolecules. 2007 Jul; 8 (7):2072–9. Copyright 2007 American Chemical Society

mass ratio of 3:1 and inlet temperature of 103 °C. CLSM images showed a homogenous distribution of glyceryl dibehenate and glyceryl tristearate SLNs with mannitol throughout the microparticle's matrix (Fig. 4.9) [85]. This matrix structure differs from the matrix discussed earlier due to the disappearance of the surface shell of mannitol or nanoparticles. The particle size of rifabutin loaded in glyceryl tristearate SLN and glyceryl dibehenate SLN were 4.87 μm and 5.54 μm, respectively. In vitro lung deposition study using a Twin Stage Liquid Impinger (60 L/min) showed respirable fractions of 51.6 and 44.1% for rifabutin loaded in glyceryl tristearate SLN and glyceryl dibehenate SLN, respectively [85].

Inorganic Nanoparticles

Mannitol was used to transform iron oxide magnetic nanoparticles (MNPs) with a size of 159.0 nm into inhalable microparticles. MNPs were dispersed in mannitol solution at two concentrations: 0.20 mg/ml and 0.05 mg/ml (i.e., MF20 and MF5). Mannitol was also spray dried without MNPs. The transmission electron microscopy images (Fig. 4.10) showed a homogenous distribution of MNPs throughout mannitol matrix. The amount of the MNPs in the microparticle composite was in agreement with the initial concentration used in the feedstock indicating the possibility of controlling the amount of MNPs loaded into the composites which is

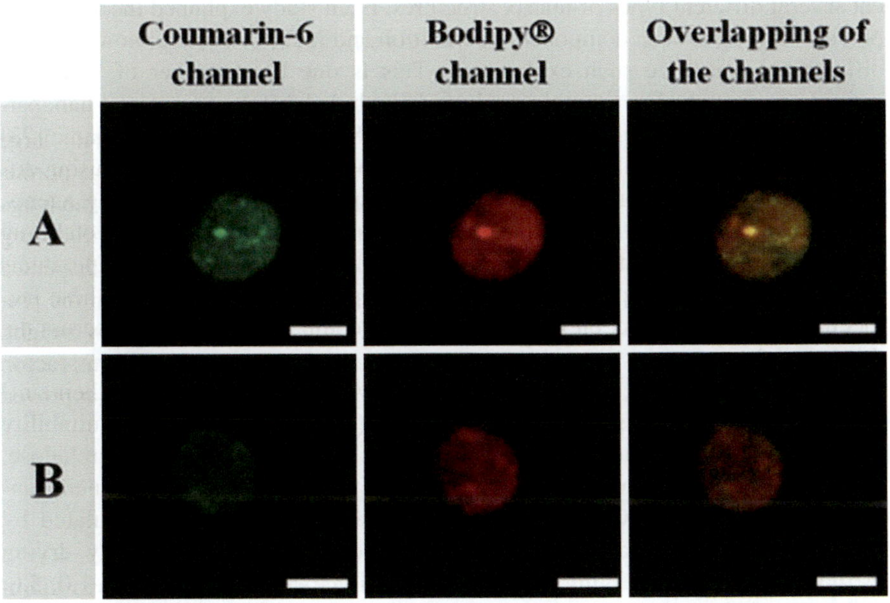

Fig. 4.9 CLSM images of mannitol microspheres containing: (**a**) glyceryl dibehenate SLN and (**b**) glyceryl tristearate SLN. SLN were labeled with coumarin-6 (green channel) and mannitol was stained with Bodipy (red channel) (scale bar: 5 μm). Reprinted (adapted) with permission from Gaspar DP, Gaspar MM, Eleutério C V., Grenha A, Blanco M, Gonçalves LMD, et al. Microencapsulated Solid Lipid Nanoparticles as a Hybrid Platform for Pulmonary Antibiotic Delivery. Mol Pharm. 2017 Sep;14 (9):2977–90. Copyright 2017 American Chemical Society

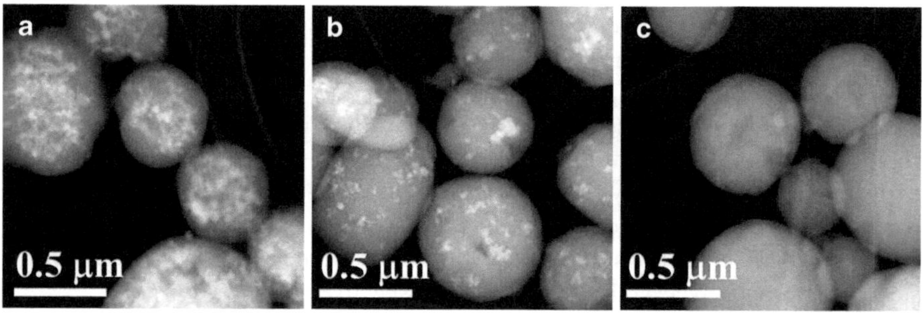

Fig. 4.10 Transmission electron microscopy images of (**a**) MF20, (**b**) MF5, and (**c**) MF0 dry powders. Reprinted from Stocke NA, Meenach SA, Arnold SM, Mansour HM, Hilt JZ. Formulation and characterization of inhalable magnetic nanocomposite microparticles (MnMs) for targeted pulmonary delivery via spray drying. Int J Pharm. 2015 Feb 20;479 (2):320–8, Copyright 2014, with permission from Elsevier

necessary for the development of these composites for specific medicinal applications. The aerosol performance of microparticles was tested using NGI at (60 L/min). MF20 and MF5 had MMADs of 4.5 μm and 4.5 μm, and FPFs of 34.3% and 31.2%, respectively [73].

Although mannitol was used in all the above-mentioned studies, the studies present several different types of matrix structures. Each study explained the formation process leading to such component distribution and matrix structure; however, it is difficult to generalize such explanation. This is due to a number of variations between each study. For example, each study included different classes of nanoparticles such as nanosuspensions [63], polymeric [83], lipid [85], and inorganic [73] nanoparticles. Additionally, the same class nanoparticles were fabricated using different types of ingredients such as polymers, lipids, and matrix-forming agents. Furthermore, spray drying processing parameters, i.e., solute mass ratio, total drug concentrations, the feedstock viscosity, feed rate, and inlet/outlet temperatures showed a significant impact on final particle properties. Particularly, all these factors have a critical impact on Péclet number and component distribution across the dried particles.

Apart from mannitol, several other sugars have been investigated for their suitability as matrix-forming agent. Umerska et al. (2020) investigated the suitability of different carbohydrate-based microparticles, i.e., lactose, mannitol, trehalose, sucrose, and raffinose as a matrix-forming agent for entrapping lipid nanocapsules (LNCs). The LNCs (59 nm) based Trojan particles were formulated by entrapping LNCs in carbohydrate-based microparticles through spray drying method. During spray drying, sugar concentrations were changed (0.5, 1.0, 5.0, and 10% w/v) while LNCs: sugar mass ratio (1:25) was kept constant. The morphology of the dried particles was studied using SEM (Fig. 4.11). The key physicochemical and aerodynamic features of LNCs-based Trojan particles are shown in Table 4.2. The use of LNCs showed a substantial impact on the mean particle size and surface morphology of the spray dried particles. The use of the LNCs

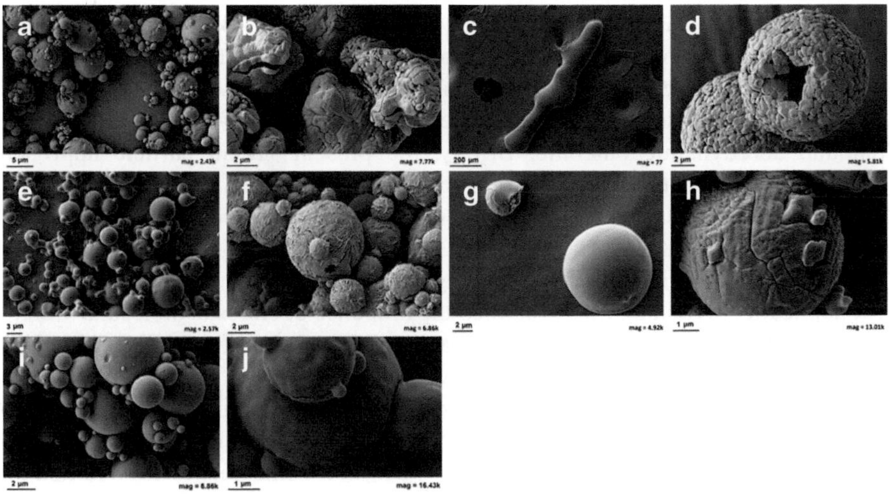

Fig. 4.11 (**a**) Mannitol particles spray dried from 1% solution; (**b**) LNC-loaded mannitol particles; (**c**) sucrose spray dried from 5% solution (**d**) LNC-loaded sucrose particles; (**e**) lactose particles spray dried from 1% solution; (**f**) LNC-loaded lactose particles; (**g**) trehalose particles spray dried from 1% solution; (**h**) LNC-loaded trehalose particles; (**i**) raffinose particles spray dried from 1% solution; (**j**) LNC-loaded raffinose particles. Reprinted from Umerska A, Mugheirbi NA, Kasprzak A, Saulnier P, Tajber L. Carbohydrate-based Trojan microparticles as carriers for pulmonary delivery of lipid nanocapsules using dry powder inhalation. Powder Technol. 2020;364:507–21., Copyright 2020, with permission from Elsevier

Table 4.2 The key physicochemical and aerodynamic parameters of LNCs-based Trojan particles

Parameters	Lactose/LNCs	Trehalose/ LNCs	Sucrose/ LNCs	Raffinose/ LNCs	Mannitol/ LNCs[a]
Particle size (µm)	3.71	6.11	9.62	5.00	11.09 to 12.57
Particle shape	Spherical shape, hollow particles with rough surfaces. Particles contain holes, cracks while shell was composed of plate-like elements	Spherical with small dimples and the surface was rough with few cracks	Spherical, hollow with shells composed of small crystals	Spherical with small dimples and smooth surface	Fused aggregates
MMAD (µm)	5.40	6.10	NA	6.2	–
GSD	2.30	3.00	NA	2.5	–
ED (%)	76.00	55.00	91.00	75.00	–
FPF (%)	29.00	16.70	15.90	19.40	–

[a]Due to the presence of agglomerates and particle morphology deviating from the spherical shape, aerodynamic analysis was not performed for mannitol/LNCs

altered the solid state properties of the processed sugars. But, the spray drying process did not show a significant impact on the re-dispersibility potential of LNCs and effectively preserves their mean particle size. Solid state analysis showed that raffinose is the most suitable to preserve LNC re-dispersibility potential and aerodynamic outcome. Briefly, the present Trojan particle approach appears to be robust delivery option for pulmonary delivery of drugs. However, further detail studies are required to understand the biological consequences of delivery systems [78].

Chitosan was utilized for encapsulating nanoparticles for pulmonary drug delivery applications. Ni et al. (2017) studied the influence of chitosan molecular weight and concentration on encapsulating cinaciguat nanocrystals. Initially, cinaciguat nanosuspension (424–473 nm) was prepared by high-pressure homogenization and suspended in different feedstock solutions of chitosan. Chitosan 50 kDa was used to prepare chitosan solutions of 0.3, 0.4, 0.5, and 0.6% w/v while chitosan 100 kDa was used to prepare chitosan solutions of 0.5% w/v. SEM images showed that all dried particles had wrinkled shapes. This shape was attributed to the high Peclet number of the nanocrystals and the formation of a surface shell which then deformed afterward due to water evaporation. In this study, chitosan concentration and molecular weight had no significant influence on the particle's morphology or aerosol performance while drug release was significantly decreased by increasing chitosan concentration or molecular weight. Chitosan 50 kDa and chitosan 100 kDa based microparticles had MMADs of 3.91 μm and 4.48 μm and FPFs of 44.90% and 39.25%, respectively (tested using NGI at 100 L/min). In vitro release study was performed under sink condition using phosphate buffer ($pH = 7.4$) containing 2.0% (w/v) Tween-80. When comparing microparticles containing 50 kDa chitosan and 100 kDa chitosan, the burst release decreased from $\approx$ 40% (50 kDa) to 15% (100 kDa) [86].

Leucine is one of the most widely used surface modification agent in respirable dry powders of small molecules and nanoparticles [87]. In a few works, leucine has been utilized instead of carbohydrate as a matrix-forming agent to encapsulate nanoparticles. The produced nanocomposite microparticles had rough surfaces with a wrinkled texture which resulted in good inhalation performance. For example, pneumococcal surface protein A (PspA)-adsorbed PGA-co-PDL nanoparticles were co-spray dried with leucine at mass ratios of 1:1.5, nanoparticle: leucine. Microparticle aerosol performance was tested using NGI at a flow rate of 60 L/min, and microparticles had MMAD of 1.70 μm and FPF of 74.31% [53]. Similarly, bovine serum albumin (BSA)-adsorbed PGA-co-PDL nanoparticles were co-spray dried with leucine at mass ratios of 1:1.5, nanoparticle: leucine. Microparticles had MMAD of 2.80 μm and FPF of 70.67%, tested using NGI at 60 L/min [69]. Methotrexate-loaded gelatin nanoparticles were prepared via a desolvation method with a mean particle size ranging from 180.31 nm to 469.56 nm. The nanoparticles were transformed into nanocomposite microparticles via co-spray drying with leucine at a ratio of 1:1.5, respectively. The nanocomposite microparticles had doughnut shapes with rough surfaces. Microparticles had MMAD of 2.59 μm and a respirable fraction of 49.53%, tested using NGI at 60 L/min [70].

Spray Freeze Drying

Spray freeze drying (SFD) is a two-step technique which has gained attention in formulating dry powder for inhalation for its ability to produce porous particles with appealing properties for pulmonary delivery [88]. In conventional SFD, a feedstock is atomized and small droplets are released into a freezing medium cooled by a cryogen (typically liquid nitrogen, −196 °C) which turns the fine droplets into frozen droplets. The cryogen liquid is usually stirred during the process to prevent particle aggregation. In the next step, the solvent is sublimated from the frozen droplet (lyophilization) and, as a result, porous particles with low density are formed. In other modified SFD, the atomized droplets can be injected directly beneath the surface of the cryogenic medium (i.e., SFL: spray freezing into liquid) or into a counter-current stream of air refrigerated to −60 °C (i.e., SFG: spray freezing into gas) [89]. However, due to the time consuming and high cost of the SFD technique, its use is usually limited to formulating thermolabile substances such as proteins and antibodies [90]. SFD has been used to transform different types of nanoparticles such as nanosuspension, polymeric- and lipid nanoparticles into inhalable dry powders. Usually, a lyoprotectant/matrix-forming agent is co-spray dried with the nanoparticles in order to protect the integrity of the nanoparticles and facilitate their redispersion from the microparticles. Yu et al. (2021) prepared ivacaftor-colistin nanosuspensions by anti-solvent method and stabilized them using N-(methylpolyoxyethylene oxycarbonyl)-1,2-distearoyl-sn-glycero-3--phosphoethanolamine sodium salt (DSPG-PEG-OMe). The mass ratio of ivacaftor, colistin, and DSPG-PEG-OMe was 1:1:1. The nanosuspension was transformed into microparticles using SFD without the addition of cryoprotectants. The SEM images showed microparticles with rough surfaces with the presence of some nanoparticles on the surface. The ED of ivacaftor and colistin were 97.8% and 97.6% with FPFs of 61.4% and 63.3%, respectively [91].

Ivacaftor-encapsulated bovine serum albumin (BSA) nanoparticles were spray freeze dried from an aqueous solution of antibiotic colistin, which was used as a matrix-forming agent as well. Leucine was added to the feedstock to improve the aerosolization of the dry powder. Six formulations were prepared with different solid concentrations, i.e., 1%–10% w/w. With increasing the initial solid content, the morphology of the particles changed from irregular to spherical (Fig. 4.12). This change was attributed to the formation of a fragile matrix, which collapsed during freeze drying, or the formation of a discontinuous matrix at low solid content. However, the spherical particles, prepared at 3–10% w/w, exhibited lower aerosol performance compared with the irregular particles which were prepared at 1% and 2% w/w. The MMADs of formulations prepared with 1% and 2% w/w were 1.4 μm and 2.0 μm while formulations prepared at 3%–10% w/w had MMADs ranging from 6.6 μm to 14.9 μm. This was due to their large size and high density which increased their MMADs and reduced their FPFs [92].

Ali and Lamprecht (2014) used SFD and spray drying to transform several polymeric nanoparticles, i.e., (poly (DL-lactide-co-glycolide) (PLGA), poly(meth)acrylate (EDRL), and ethyl cellulose (EC)) and lipid nanoparticles, i.e., SLN and LNC into inhalable microparticles. Feedstocks for SFD and spray drying were prepared

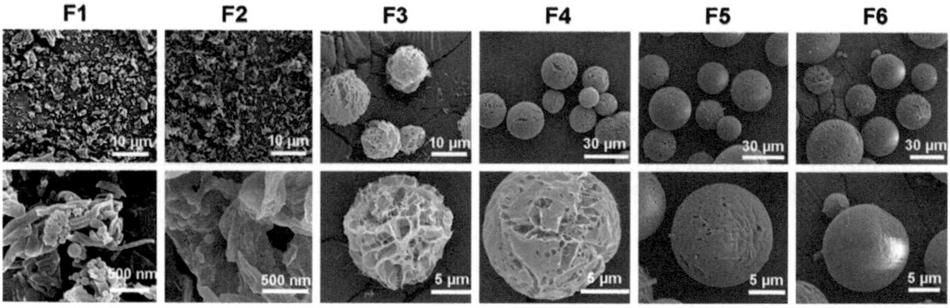

Fig. 4.12 SEM images of formulations F1-F6 at two different magnifications. Reprinted with permission from Zhu C, Chen J, Yu S, Que. C, Taylor LS, Tan W, et al. Inhalable nanocomposite microparticles with enhanced dissolution and superior aerosol performance. Mol Pharm. 2020;17 (9):3270–80. Copyright 2020 American Chemical Society

by dispersing polymeric nanoparticles at 1% w/v in a 5% w/v aqueous solution of maltodextrin and lipid nanoparticles at 1% w/v in a 5% w/v aqueous solution of trehalose. Polyvinylpyrrolidone was add to all feedstocks at 5% w/v. Polymeric NPEMs were successfully produced by SFD and spray drying whereas lipid nanoparticle-embedded microparticles were formed by SFD only. The high temperatures in spray drying (inlet and outlet temperatures were 110 °C and 80 °C, respectively) melted the lipid core and the droplets cohered into waxy material. By comparison, SFD produced microparticles with a lower density (0.02 g/cm^3) and larger specific surface areas (67–77 m^2/g) than the corresponding microparticles produced by spray drying (0.24–0.27 g/cm^3 and 1.8–2.4 m^2/g). The FPFs were between 43% and 46% (for SFD microparticles) and between 35% and 40% (for spray dried microparticles). However, SFD was superior to spray drying in maintaining the size of all investigated nanoparticles after reconstitution (size after reconstitution/initial size ratio for SFD ≈ 1 versus >1.5 for spray drying) [93].

SFD was used to produce two types of inhalable microparticles of poly(caprolactone) (PCL) nanoparticles using mannitol and PVA as adjuvants. For both adjuvants, the nanoparticles aggregated to form large, hollow, spherical microparticles (Fig. 4.13). The PCL nanoparticles were physically dispersed in a porous mannitol matrix, whereas the PVA was coated onto the nanoparticle's surface. Compared to the mannitol formulation, PVA led to a more stable morphology with a better aqueous percentage re-dispersibility (≈ 90% vs. 70%), a lower S_f/S_i ratio (≈1 vs. 1.5; S_f: size of re-dispersed nanoparticles and S_I: initial size of nanoparticles), and a higher production yield (≈ 30–80% vs. 15–40%) [94]. Mannitol/PCL particles were effectively aerosolized using the entrainment tube with ED of ≈ 94%, however, the majority of particles were deposited in the pre-separator resulted in low FPF of ≈ 10% [94].

Colistin and ciprofloxacin co-loaded liposomes were prepared using a composition of hydrogenated soybean phosphatidylcholine (HSPC), 1,2-distearoyl-sn-glycero-3-phosphoglycerol sodium salt (DSPG), PEG and cholesterol in a 3:2:0.5:1.7 mass ratio via thin film evaporation and sonication method. Colistin and

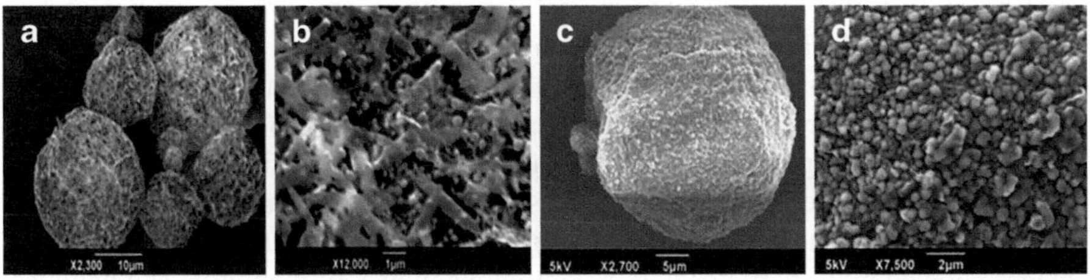

Fig. 4.13 (**a**) PCL–mannitol nano-aggregates; (**b**) PCL nanoparticles are physically dispersed in porous mannitol matrix; (**c**) PCL–PVA nano-aggregates; (**d**) closely packed PCL nanoparticles. Reprinted from Cheow WS, Ng MLL, Kho K, Hadinoto K. Spray-freeze-drying production of thermally sensitive polymeric nanoparticle aggregates for inhaled drug delivery: effect of freeze-drying adjuvants. Int J Pharm. 2011;404 (1, 2):289–300, Copyright 2010, with permission from Elsevier

ciprofloxacin were added at a concentration of 4 mg/ml. Nanoparticles had a diameter of 102.1 nm and zeta potential of −10.4 mV. Nanoparticles were transformed into microparticles via ultrasonic spray freeze drying. The dried liposomes were hydrated by adding mannitol (5% w/v) solution containing 0.5 mol/L ammonium sulfate as an internal lyoprotectant. Sucrose (5%, w/v), mannitol (10%, w/v), and leucine (5%, w/w) were used as external lyoprotectants/aerosolization enhancers. The formed particles were hollow, porous, and spherical with EDs of 97% and 95% and FPFs of 45.8% and 43.6% for ciprofloxacin and colistin, respectively. In vitro antimicrobial activities test against *P. aeruginosa* H133880624 showed no killing observation following colistin solution (4 and 8 mg/L) and ciprofloxacin solution (4 and 8 mg/L) monotherapies. However, combination solutions of colistin and ciprofloxacin resulted in a total eradication of bacteria with the combination solution of colistin (8 mg/L) and ciprofloxacin (8 mg/L) [95].

Supercritical CO_2-Assisted Spray Drying (SASD)

A supercritical fluid (SCF) is any fluid compressed at a pressure and temperature above its critical point, where distinct liquid and gas phases do not exist. Due to its unique physical properties, an SCF can penetrate substances as a gas and dissolve materials as a liquid. The most commonly used SCF in pharmaceuticals is supercritical carbon dioxide (sc-CO_2) because of its easily achievable critical temperature, i.e., 31.15 °C and pressure, i.e., 7.4 MPa in addition to nontoxicity and low-cost recyclability [96]. The use of SCF in particle engineering includes several different processes such as precipitation from gas saturated solutions (PGSS), precipitation from saturated solutions using supercritical fluid as an anti-solvent (SAS), atomized rapid injection solvent extraction process (ARISE), and rapid expansion of supercritical solution (RESS) [97]. The sc-CO_2 assisted spray drying, also known as supercritical CO_2-assisted atomization is an RESS process where sc-CO_2 is pumped with the feedstock into a static mixture to achieve a near-equilibrium mixture which is sprayed through the nozzle into the precipitator at atmospheric pressure. The rapid flow of the heated compressed gas causes rapid evaporation of liquids from droplets and the formation of particles [98].

Peng et al. (2021) investigated the influence of several processing parameters such as CO_2/liquid mass flow ratio, total mass concentration, and nanoparticle/mannitol ratio on the size and morphologies of chitosan nanoparticles embedded in mannitol microparticles. CO_2/liquid mass flow ratios of 1.8, 1.5, and 1.2 were achieved by changing the mass flow of the feedstock with a fixed CO_2 mass flow. The microparticles generated at a 1.8 mass ratio had a well-defined spherical shape and an average geometric diameter of 1.46 μm with narrow size distribution. SEM images showed microparticles with a continuous matrix of mannitol. The nanoparticles appeared on the surface and encapsulated inside the microparticles as shown by TEM and a cross-section SEM images. However, reducing CO_2/liquid mass flow ratio (i.e., increasing the liquid mass flow) led to the formation of a coalescence of irregular microparticles with wide size distribution. This was attributed to two reasons: firstly, increasing the liquid content and hence reducing the evaporation efficiency; secondly, the hygroscopic nature of mannitol which caused a deformation and coalescence between microparticles deposited on the collector. The effect of total solid concentration was investigated at three concentrations, i.e., 20 g/L, 30 g/L, and 40 g/L. Increasing the total solid insignificantly increased the microparticles' average geometric diameters from 1.33 to 1.76 μm. The insignificant increment is expected due to the incorporation of sc-CO_2 which mitigated the impact of the solid concentration on the viscosity and surface tension of the droplets. Similarly, the impact of nanoparticle to mannitol ratios on the liquid viscosity was attenuated by the sc-CO_2. At nanoparticle to mannitol mass ratios of 10:90, 20:80, and 30:70, the microparticles' average geometric diameters were 1.46, 1.33, and 1.33 μm, respectively. However, SEM images showed an increased number of nanoparticles on the surface of microparticles with increasing the nanoparticles mass ratio [99]. In comparison to other techniques, the applications of SASD in producing dry powders for inhalation or transforming nanoparticles into inhalable microparticles are relatively new. However, the latter application has become more frequent in the recent few years. For instance, SASD has been utilized to encapsulate siRNA-loaded nanoparticles [100], gold-coated magnetite nanocomposites [101], POxylated gold nanoparticles [102], and paclitaxel/doxorubicin-loaded POXylated polyurea dendrimer nanoparticles [103] in chitosan microparticles. Also, sildenafil-loaded POXylated polyurea dendrimer nanoparticles were encapsulated into poly(D,L-lactide coglycolide)-cholesterol microparticles using SASD [104].

Shelf Freeze Drying (FD) and Thin Film Freeze Drying (TFFD)

Shelf freeze drying (lyophilization) is a common approach in processing thermolabile drug substances, biologics, etc. In this approach, a liquid formulation is frozen in a lyophilizer by gradually reducing the shelf temperature (cooling rate around 0.017 K/s) to between −40 °C and − 55 °C. The freezing phase is followed by a primary drying phase in which a pre-set heat input (shelf temperature) and vacuum are applied to sublimate the frozen liquid, forming a porous solid cake or compact with high moisture content ~7–8%. In a secondary drying phase, the drying process continues at a higher temperature to remove the remaining bound water and reduce the moisture content to ~1–2% [105]. The application of lyophilization in formulating dry powders for inhalation requires a further dispersion step to produce submicron particles from the lyophilized cake [106, 107].

In thin film freeze drying, droplets of a liquid formulation (diameter ~ 2–4 mm) are applied onto the chilled surface of a rotating steel drum, filled with cryogens such as liquid nitrogen. The droplets spread out to form a frozen film at a freezing rate of 102–103 K/s [108]. The frozen film is scraped off the drum, collected in cryogen-cooled receptacle under the drum and transferred to a lyophilizer. TFFD has several advantages over other techniques in producing dry powder for inhalation. Firstly, the super rapid freezing rate minimizes the crystal growth and enables the production of powders with large specific areas and particle sizes in the range of 1–5 μm. Secondly, TFFD does not involve atomization which reduces unwanted shear stress and allows the process of formulation of high concentration and viscosity. Thirdly, TFFD does not involve heating which eliminates the risk of substance degradation. Finally, the TFFD processing parameters are less complicated compared to spray drying and spray freeze drying [68, 107].

Wang et al. (2021) compared the feasibility of three techniques: spray drying, FD, and TFFD in engineering inhalable dry powder of SLN. Mannitol was used as a matrix/cryoprotective agent. The ratio of mannitol to SLN (30:1) was kept constant in every technique. The size of the reconstituted SLN from the dried powders showed no significant difference for spray dried (178.1 nm) and thin film freeze-dried powder (165.3 nm) while these values were three times higher for shelf freeze-dried powder (370.0 nm) compared to SLN initial size (136.3 nm). In comparison to TFFD, the low freezing rate in shelf freeze drying resulted in the formation of larger ice crystals and generates more shears to the particles. During spray drying (inlet and outlet temperatures, 115 °C and 65 °C, respectively), the SLNs remained intact and no melting or fusing of the lipids was observed. SEM images of powder prepared by spray drying showed bead-like microstructures whereas that prepared by TFFD was fluffy and brittle with a porous texture. The specific surface area of the TFFD powders was 20 times higher than that of the spray dried powder, 19.34 vs. 0.92 m^2/g. Microparticle aerosolization performance was tested using NGI at 60 L/min. The TFFD powder demonstrated better aerosol performance than the spray drying powder. TFFD powder showed good MMADs and FPF of 3.96 μm compared to spray drying powder, i.e., MMAD of 5.97 μm and FPF of 37.01% [77].

Zhang et al. (2021) used TFFD to transform PEGylated chitosan/clustered regularly interspaced short palindromic repeats (CRISPR-Cas9) nanoparticles to inhalable microparticles utilizing mannitol, sucrose, or trehalose at several concentrations between 0.1% and 10% w/v with and without the addition of 0.3% leucine. After lyophilization, the dry powder had a disc-like shape with different degrees of shrinkage. Shrinkage decreased upon increasing the concentration of the cryoprotective agents. Also, disc-like powders prepared with leucine showed less shrinkage compared to their counterparts without leucine. The size, PDI, and zeta potential of the nanoparticles after dispersion from the powders increased upon decreasing the concentrations of cryoprotective agents which indicated that a higher concentration of cryoprotective agents minimized the aggregation of the nanoparticles after redispersion. Based on the properties of the reconstituted nanoparticles, powders prepared at 3% cryoprotective agents with and without leucine were selected for pulmonary delivery. SEM images showed that mannitol particles exhibited a higher degree of porosity while sucrose and trehalose particles were smooth and solid. The geometric particle size of sucrose and trehalose particles were significantly larger (D_{50} between 8.1 μm and 42.5 μm) which indicate they are not suitable for lung delivery. The in vitro aerosolization test confirmed that only the mannitol particles (without and with leucine) were suitable for inhalation with MMADs of 4.8 μm and 4.6 μm, and FPFs of 44.5%, and 44.2%, respectively [68].

4.5 Conclusion

Transforming nanoparticle drug carriers into inhalable microparticles is a formulation strategy which results in complex and advantageous drug delivery systems. Nano-in-micro inhalable particles can be produced using a variety of techniques with different conditions allowing the formulation of a wide variety of nanoparticle classes and therapeutic substances. However, special attention needs to be paid to the number and amount of excipients used in designing nano-in-micro inhalable particles. The mass ratio of a drug in a final powder can be as low as <1% which necessitates the inhalation of a large amount of powder. Further work is required to achieve a better understanding of all factors that dictate the formation and properties of the nano-in-micro inhalable particles which can help in developing these formulations with less use of excipients.

Acknowledgment Funding from the South East Technological University PhD scholarship programme (Nasser Alhajj) is gratefully acknowledged.

References

1. World Health Organization. The top 10 causes of death [Internet]. 2020 [cited 2022 Jul 19]. Available from: https://www.who.int/news-room/fact-sheets/detail/the-top-10-causes-of-death
2. de Menezes BRC, Rodrigues KF, Schatkoski VM, Pereira RM, Ribas RG, do Amaral Montanheiro TL, et al. Current advances in drug delivery of nanoparticles for respiratory disease treatment. J Mater Chem B. 2021;9(7):1745–61.
3. Cappa V, Marcon A, Di Gennaro G, Chamitava L, Cazzoletti L, Bombieri C, et al. Health-related quality of life varies in different respiratory disorders: a multi-case control population based study. BMC Pulm Med [Internet]. 2019;19(1):32. https://doi.org/10.1186/s12890-019-0796-8.
4. Alhajj N, Chee CF, Wong TW, Rahman NA, Abu Kasim NH, Colombo P. Lung cancer: active therapeutic targeting and inhalational nanoproduct design [Internet]. Vol. 15, Expert opinion on drug delivery. Taylor & Francis; 2018 [cited 2019 Mar 19]. p. 1223–1247. Available from: https://www.tandfonline.com/doi/full/10.1080/17425247.2018.1547280
5. Huang Z, Kłodzińska SN, Wan F, Nielsen HM. Nanoparticle-mediated pulmonary drug delivery: state of the art towards efficient treatment of recalcitrant respiratory tract bacterial infections. Drug Deliv Transl Res [Internet]. 2021;11(4):1634–54. https://doi.org/10.1007/s13346-021-00954-1.

6. Virmani T, Kumar G, Virmani R, Sharma A, Pathak K. Nanocarrier-based approaches to combat chronic obstructive pulmonary disease. Nanomedicine [Internet]. 2022;17:1833. https://doi.org/10.2217/nnm-2021-0403.
7. Liu D, Long M, Gao L, Chen Y, Li F, Shi Y, et al. Nanomedicines targeting respiratory injuries for pulmonary disease management. Adv Funct Mater [Internet]. 2022;32(22):2112258. https://doi.org/10.1002/adfm.202112258.
8. Zhong W, Zhang X, Zeng Y, Lin D, Wu J. Recent applications and strategies in nanotechnology for lung diseases. Nano Res. 2021;14(7):2067–89.
9. Sabuj MZR, Islam N. Inhaled antibiotic-loaded polymeric nanoparticles for the management of lower respiratory tract infections. Nanoscale Adv. 2021;3(14):4005–18.
10. Wang J, Zhou T, Liu Y, Chen S, Yu Z. Application of nanoparticles in the treatment of lung cancer with emphasis on receptors. Front Pharmacol. 2022;12:3808.
11. Carrasco-Esteban E, Domínguez-Rullán JA, Barrionuevo-Castillo P, Pelari-Mici L, Leaman O, Sastre-Gallego S, et al. Current role of nanoparticles in the treatment of lung cancer. J Clin Transl Res. 2021;7(2):140.
12. Forest V, Pourchez J. Nano-delivery to the lung - by inhalation or other routes and why nano when micro is largely sufficient? Adv Drug Deliv Rev [Internet]. 2022;183:114173. Available from: https://www.sciencedirect.com/science/article/pii/S0169409X22000631
13. Commission EU. Recommendation on the definition of nanomaterial (2011/696/EU). Off J Eur Com. 2011;275:38–40.
14. Rizvi SAA, Saleh AM. Applications of nanoparticle systems in drug delivery technology. Saudi Pharm J. 2018;26(1):64–70.
15. Anselmo AC, Mitragotri S. Nanoparticles in the clinic: an update. Bioeng Transl Med. 2019;4(3):e10143.
16. Mitchell MJ, Billingsley MM, Haley RM, Wechsler ME, Peppas NA, Langer R. Engineering precision nanoparticles for drug delivery. Nat Rev Drug Discov. 2021;20(2):101–24.
17. Hörmann K, Zimmer A. Drug delivery and drug targeting with parenteral lipid nanoemulsions — a review. J Control Release [Internet]. 2016;223:85–98. Available from: https://www.sciencedirect.com/science/article/pii/S0168365915302686
18. Plaza-Oliver M, Santander-Ortega MJ, Lozano MV. Current approaches in lipid-based nanocarriers for oral drug delivery. Drug Deliv Transl Res [Internet]. 2021;11(2):471–97. Available from: http://europepmc.org/abstract/MED/33528830
19. Aswathanarayan JB, Vittal RR. Nanoemulsions and their potential applications in food industry [Internet]. Front Sustain Food Syst. 2019;3:95. Available from: https://www.frontiersin.org/articles/10.3389/fsufs.2019.00095
20. Buya AB, Witika BA, Bapolisi AM, Mwila C, Mukubwa GK, Memvanga PB, et al. Application of lipid-based nanocarriers for antitubercular drug delivery: a review. Pharmaceutics. 2021;13
21. Andra VVSNL, Pammi SVN, Bhatraju LVKP, Ruddaraju LK. A comprehensive review on novel liposomal methodologies, commercial Formulations, clinical trials and patents. Bionanoscience [Internet] 2022;12(1):274–291. Available from: https://doi.org/10.1007/s12668-022-00941-x.
22. Swaminathan J, Ehrhardt C. Liposomes for pulmonary drug delivery. In: Controlled pulmonary drug delivery. Springer; 2011. p. 313–34.
23. Weber S, Zimmer A, Pardeike J. Solid lipid nanoparticles (SLN) and nanostructured lipid carriers (NLC) for pulmonary application: a review of the state of the art. Eur J Pharm Biopharm [Internet]. 2014;86(1):7–22. Available from: https://www.sciencedirect.com/science/article/pii/S0939641113002920
24. Scioli Montoto S, Muraca G, Ruiz ME. Solid lipid nanoparticles for drug delivery: pharmacological and biopharmaceutical aspects. Front Mol Biosci. 2020;7:587997.
25. Ganesan P, Narayanasamy D. Lipid nanoparticles: different preparation techniques, characterization, hurdles, and strategies for the production of solid lipid nanoparticles and nanostructured lipid carriers for oral drug delivery. Sustain Chem Pharm [Internet]. 2017;6:37–56. Available from: https://www.sciencedirect.com/science/article/pii/S2352554117300347

26. Matougui N, Boge L, Groo A-C, Umerska A, Ringstad L, Bysell H, et al. Lipid-based nanoformulations for peptide delivery. Int J Pharm [Internet]. 2016;502(1):80–97. Available from: https://www.sciencedirect.com/science/article/pii/S0378517316301120
27. Meerovich I, Dash AK. Polymersomes for drug delivery and other biomedical applications. In: Materials for Biomedical Engineering Elsevier; 2019. p. 269–309.
28. Rideau E, Dimova R, Schwille P, Wurm FR, Landfester K. Liposomes and polymersomes: a comparative review towards cell mimicking. Chem Soc Rev. 2018;47(23):8572–610.
29. Lee JS, Feijen J. Polymersomes for drug delivery: design, formation and characterization. J Control Release [Internet]. 2012;161(2):473–83. Available from: https://www.sciencedirect.com/science/article/pii/S0168365911009576
30. Deng S, Gigliobianco MR, Censi R, Di Martino P. Polymeric nanocapsules as nanotechnological alternative for drug delivery system: current status, challenges and opportunities. Nano. 2020;10(5):847.
31. Zielińska A, Carreiró F, Oliveira AM, Neves A, Pires B, Venkatesh DN, et al. Polymeric nanoparticles: production, characterization, toxicology and ecotoxicology. Molecules. 2020;25(16):3731.
32. Palmerston Mendes L, Pan J, Torchilin VP. Dendrimers as Nanocarriers for nucleic acid and drug delivery in cancer therapy. Molecules. 2017;22:1401.
33. Suhail M, Rosenholm JM, Minhas MU, Badshah SF, Naeem A, Khan KU, et al. Nanogels as drug-delivery systems: a comprehensive overview. Ther Deliv. 2019;10(11):697–717.
34. Sabir F, Asad MI, Qindeel M, Afzal I, Dar MJ, Shah KU, et al. Polymeric Nanogels as versatile Nanoplatforms for biomedical applications. J Nanomater [Internet]. 2019;2019:1526186. https://doi.org/10.1155/2019/1526186.
35. Preman NK, Barki RR, Vijayan A, Sanjeeva SG, Johnson RP. Recent developments in stimuli-responsive polymer nanogels for drug delivery and diagnostics: a review. Eur J Pharm Biopharm [Internet]. 2020;157:121–53. Available from: https://www.sciencedirect.com/science/article/pii/S093964112030309X
36. Chen Z, Wu W, Lu Y. What is the future for nanocrystal-based drug-delivery systems? Therapeut Deliv. 2020;11:225–9.
37. Tuomela A, Hirvonen J, Peltonen L. Stabilizing agents for drug nanocrystals: effect on bioavailability. Pharmaceutics. 2016;8(2):16.
38. Mukherjee S, Liang L, Veiseh O. Recent advancements of magnetic nanomaterials in cancer. Therapy. 2020;12:Pharmaceutics.
39. Mendes BB, Conniot J, Avital A, Yao D, Jiang X, Zhou X, et al. Nanodelivery of nucleic acids. Nat Rev Methods Prim [Internet]. 2022;2(1):24. https://doi.org/10.1038/s43586-022-00104-y.
40. Liang R, Wei M, Evans DG, Duan X. Inorganic nanomaterials for bioimaging, targeted drug delivery and therapeutics. Chem Commun [Internet]. 2014;50(91):14071–81. https://doi.org/10.1039/C4CC03118K.
41. Salas G, Costo R, del Morales MP. Chapter 2 - synthesis of inorganic nanoparticles. In: de la Fuente JM, Grazu VBT-F of N, editors. Nanobiotechnology [Internet]. Elsevier; 2012. p. 35–79. Available from: https://www.sciencedirect.com/science/article/pii/B9780124157699000029.
42. Li X, Vogt FG, Hayes D, Mansour HM. Design, characterization, and aerosol dispersion performance modeling of advanced co-spray dried antibiotics with mannitol as respirable microparticles/nanoparticles for targeted Pulmonary delivery as dry powder inhalers. J Pharm Sci [Internet]. 2014;103(9):2937–49. Available from: https://www.sciencedirect.com/science/article/pii/S0022354915304287
43. Li X, Vogt FG, Hayes D, Mansour HM. Physicochemical characterization and aerosol dispersion performance of organic solution advanced spray-dried microparticulate/nanoparticulate antibiotic dry powders of tobramycin and azithromycin for pulmonary inhalation aerosol delivery. Eur J Pharm Sci. 2014 Feb;52(1):191–205.
44. Alhajj N, Zakaria Z, Naharudin I, Ahsan F, Li W, Wong TW. Critical physicochemical attributes of chitosan nanoparticles admixed lactose-PEG 3000 microparticles in pulmonary

inhalation. Asian J Pharm Sci [Internet]. 2020;15(3):374–84. Available from: http://www.sciencedirect.com/science/article/pii/S1818087618313187
45. Alhajj N, Naharudin I, Colombo P, Quarta E, Wong TW. Probing critical physical properties of lactose-polyethylene glycol microparticles in pulmonary delivery of chitosan nanoparticles. Pharmaceutics [Internet]. 2021;13(10) https://doi.org/10.3390/pharmaceutics13101581.
46. Ullah F, Shah KU, Shah SU, Nawaz A, Nawaz T, Khan KA, et al. Synthesis, characterization and in vitro evaluation of chitosan nanoparticles physically admixed with lactose microspheres for Pulmonary delivery of Montelukast. Polymers (Basel). 2022;14(17):3564.
47. Yang L, Luo J, Shi S, Zhang Q, Sun X, Zhang Z, et al. Development of a pulmonary peptide delivery system using porous nanoparticle-aggregate particles for systemic application. Int J Pharm [Internet]. 2013;451(1):104–11. Available from: https://www.sciencedirect.com/science/article/pii/S0378517313003888
48. Tsapis N, Bennett D, Jackson B, Weitz DA, Edwards DA. Trojan particles: large porous carriers of nanoparticles for drug delivery. Proc Natl Acad Sci [Internet]. 2002;99(19):12001–5. Available from: http://www.pnas.org/content/99/19/12001.abstract
49. Hadinoto K, Zhu K, Tan RBH. Drug release study of large hollow nanoparticulate aggregates carrier particles for pulmonary delivery. Int J Pharm. 2007;341(1–2):195–206.
50. Gosens I, Post JA, de la Fonteyne LJJ, Jansen EHJM, Geus JW, Cassee FR, et al. Impact of agglomeration state of nano-and submicron sized gold particles on pulmonary inflammation. Part Fibre Toxicol. 2010;7(1):1–11.
51. Sung JC, Pulliam BL, Edwards DA. Nanoparticles for drug delivery to the lungs. Trends Biotechnol. 2007;25(12):563–70.
52. Porsio B, Cusimano MG, Schillaci D, Craparo EF, Giammona G, Cavallaro G. Nano into micro Formulations of tobramycin for the treatment of Pseudomonas aeruginosa infections in cystic fibrosis. Biomacromolecules [Internet]. 2017;18(12):3924–35. https://doi.org/10.1021/acs.biomac.7b00945.
53. Kunda NK, Alfagih IM, Miyaji EN, Figueiredo DB, Gonçalves VM, Ferreira DM, et al. Pulmonary dry powder vaccine of pneumococcal antigen loaded nanoparticles. Int J Pharm. 2015;495(2):903–12.
54. Debnath SK, Saisivam S, Debanth M, Omri A. Development and evaluation of chitosan nanoparticles based dry powder inhalation formulations of Prothionamide. PLoS One. 2018;13(1):e0190976.
55. Rawal T, Patel S, Butani S. Chitosan nanoparticles as a promising approach for pulmonary delivery of bedaquiline. Eur J Pharm Sci [Internet]. 2018;124:273–87. Available from: https://www.sciencedirect.com/science/article/pii/S0928098718304032
56. Pilcer G, Rosière R, Traina K, Sebti T, Vanderbist F, Amighi K. New co-spray-dried tobramycin nanoparticles-clarithromycin inhaled powder systems for lung infection therapy in cystic fibrosis patients. J Pharm Sci. 2013 Jun;102(6):1836–46.
57. Malamatari M, Somavarapu S, Bloxham M, Buckton G. Nanoparticle agglomerates of indomethacin: the role of poloxamers and matrix former on their dissolution and aerosolisation efficiency. Int J Pharm [Internet]. 2015;495(1):516–26. Available from: https://www.sciencedirect.com/science/article/pii/S0378517315302118
58. Scherließ R, Janke J. Preparation of poly-lactic-co-glycolic acid nanoparticles in a dry powder formulation for pulmonary antigen delivery. Pharmaceutics. 2021;13:1196.
59. Wang Z, Gupta SK, Meenach SA. Development and physicochemical characterization of acetalated dextran aerosol particle systems for deep lung delivery. Int J Pharm [Internet]. 2017;525(1):264–74. Available from: https://www.sciencedirect.com/science/article/pii/S0378517317303629
60. Nemati E, Mokhtarzadeh A, Panahi-Azar V, Mohammadi A, Hamishehkar H, Mesgari-Abbasi M, et al. Ethambutol-loaded solid lipid nanoparticles as dry powder inhalable formulation for tuberculosis therapy. AAPS PharmSciTech. 2019;20(3):120.
61. Bakhtiary Z, Barar J, Aghanejad A, Saei AA, Nemati E, Ezzati Nazhad Dolatabadi J, et al. Microparticles containing erlotinib-loaded solid lipid nanoparticles for treatment of non-

small cell lung cancer. Drug Dev Ind Pharm [Internet]. 2017;43(8):1244–53. https://doi.org/10.1080/03639045.2017.1310223.
62. Ezzati Nazhad Dolatabadi J, Hamishehkar H, Valizadeh H. Development of dry powder inhaler formulation loaded with alendronate solid lipid nanoparticles: solid-state characterization and aerosol dispersion performance. Drug Dev Ind Pharm [Internet]. 2015;41(9):1431–7. https://doi.org/10.3109/03639045.2014.956111.
63. Quarta E, Sonvico F, Bettini R, De Luca C, Dotti A, Catalucci D, et al. Inhalable microparticles embedding calcium phosphate nanoparticles for heart targeting: The Formulation Experimental Design Vol. 13, Pharmaceutics . 2021.
64. Tran TT, Yu H, Vidaillac C, Lim AYH, Abisheganaden JA, Chotirmall SH, et al. An evaluation of inhaled antibiotic liposome versus antibiotic nanoplex in controlling infection in bronchiectasis. Int J Pharm. 2019;559:382–92.
65. Changsan N, Sinsuebpol C. Dry powder inhalation formulation of chitosan nanoparticles for co-administration of isoniazid and pyrazinamide. Pharm Dev Technol [Internet]. 2021;26(2):181–92. https://doi.org/10.1080/10837450.2020.1852570.
66. Al-Qadi S, Taboada P, Remuñán-López C. Micro/nanostructured inhalable formulation based on polysaccharides: effect of a thermoprotectant on powder properties and protein integrity. Int J Pharm [Internet]. 2018;551(1):23–33. Available from: https://www.sciencedirect.com/science/article/pii/S0378517318306276
67. Al-Qadi S, Remuñán-López C. A micro- and nano-structured drug carrier based on biocompatible, hybrid polymeric nanoparticles for potential application in dry powder inhalation therapy. Polymer (Guildf) [Internet]. 2014;55(16):4012–21. Available from: https://www.sciencedirect.com/science/article/pii/S0032386114005400
68. Zhang H, Zhang Y, Williams RO, Smyth HDC. Development of PEGylated chitosan/CRISPR-Cas9 dry powders for pulmonary delivery via thin-film freeze-drying. Int J Pharm [Internet]. 2021;605:120831. Available from: https://www.sciencedirect.com/science/article/pii/S0378517321006360
69. Kunda NK, Alfagih IM, Dennison SR, Somavarapu S, Merchant Z, Hutcheon GA, et al. Dry powder pulmonary delivery of cationic PGA-co-PDL nanoparticles with surface adsorbed model protein. Int J Pharm [Internet]. 2015;492(1):213–22. Available from: https://www.sciencedirect.com/science/article/pii/S037851731530034X
70. Abdelrady H, Hathout RM, Osman R, Saleem I, Mortada ND. Exploiting gelatin nanocarriers in the pulmonary delivery of methotrexate for lung cancer therapy. Eur J Pharm Sci. 2019;133:115–26.
71. Stocke NA, Arnold SM, Hilt JZ. Responsive hydrogel nanoparticles for pulmonary delivery. J Drug Deliv Sci Technol. 2015;29:143–51.
72. Bielski E, Zhong Q, Mirza H, Brown M, Molla A, Carvajal T, et al. TPP-dendrimer nanocarriers for siRNA delivery to the pulmonary epithelium and their dry powder and metered-dose inhaler formulations. Int J Pharm [Internet]. 2017;527(1):171–83. Available from: https://www.sciencedirect.com/science/article/pii/S0378517317304581
73. Stocke NA, Meenach SA, Arnold SM, Mansour HM, Hilt JZ. Formulation and characterization of inhalable magnetic nanocomposite microparticles (MnMs) for targeted pulmonary delivery via spray drying. Int J Pharm [Internet]. 2015;479(2):320–8. Available from: http://www.sciencedirect.com/science/article/pii/S0378517314009478
74. Khatib I, Ke W-R, Cipolla D, Chan H-K. Storage stability of inhalable, controlled-release powder formulations of ciprofloxacin nanocrystal-containing liposomes. Int J Pharm [Internet]. 2021;605:120809. Available from: https://www.sciencedirect.com/science/article/pii/S0378517321006141
75. Khatib I, Khanal D, Ruan J, Cipolla D, Dayton F, Blanchard JD, et al. Ciprofloxacin nanocrystals liposomal powders for controlled drug release via inhalation. Int J Pharm. 2019 Jul;566:641–51.
76. Satari N, Taymouri S, Varshosaz J, Rostami M, Mirian M. Preparation and evaluation of inhalable dry powder containing glucosamine-conjugated gefitinib SLNs for lung cancer therapy.

Drug Dev Ind Pharm [Internet]. 2020;46(8):1265–77. https://doi.org/10.1080/03639045.2020.1788063.

77. Wang J-L, Hanafy MS, Xu H, Leal J, Zhai Y, Ghosh D, et al. Aerosolizable siRNA-encapsulated solid lipid nanoparticles prepared by thin-film freeze-drying for potential pulmonary delivery. Int J Pharm [Internet]. 2021;596:120215. Available from: https://www.sciencedirect.com/science/article/pii/S0378517321000193
78. Umerska A, Mugheirbi NA, Kasprzak A, Saulnier P, Tajber L. Carbohydrate-based Trojan microparticles as carriers for pulmonary delivery of lipid nanocapsules using dry powder inhalation. Powder Technol [Internet]. 2020;364:507–21. Available from: http://www.sciencedirect.com/science/article/pii/S0032591020301339
79. Alhajj N, O'Reilly NJ, Cathcart H. Designing enhanced spray dried particles for inhalation: a review of the impact of excipients and processing parameters on particle properties [Internet], vol. 384. Powder Technol; 2021. p. 313–31. Available from: https://www.sciencedirect.com/science/article/pii/S0032591021001340
80. Anton N, Jakhmola A, Vandamme TF. Trojan microparticles for drug delivery. Pharmaceutics [Internet]. 2012 Jan 6 [cited 2015 Nov 19];4(1):1–25. Available from: http://www.mdpi.com/1999-4923/4/1/1/htm
81. Wei Y, Deng W, Chen R-H. Effects of insoluble nano-particles on nanofluid droplet evaporation. Int J Heat Mass Transf [Internet]. 2016;97:725–34. Available from: https://www.sciencedirect.com/science/article/pii/S0017931015313764
82. Grenha A, Seijo B, Serra C, Remuñan-López C. Chitosan nanoparticle-loaded mannitol microspheres: structure and surface characterization. Biomacromolecules [Internet]. 2007 Jul [cited 2015 May 18];8(7):2072–2079. Available from: http://www.ncbi.nlm.nih.gov/pubmed/17585804
83. Torge A, Grützmacher P, Mücklich F, Schneider M. The influence of mannitol on morphology and disintegration of spray-dried nano-embedded microparticles. Eur J Pharm Sci [Internet]. 2017;104:171–9. Available from: http://www.sciencedirect.com/science/article/pii/S0928098717301975
84. Muralidharan P, Mallory EK, Malapit M, Phan H, Ledford JG, Hayes D, et al. Advanced design and development of nanoparticle/microparticle dual-drug combination lactose carrier-free dry powder inhalation aerosols. RSC Adv. 2020;10(68):41846–56.
85. Gaspar DP, Gaspar MM, Eleutério CV, Grenha A, Blanco M, Gonçalves LMD, et al. Microencapsulated solid lipid nanoparticles as a hybrid platform for Pulmonary antibiotic delivery. Mol Pharm. 2017 Sep;14(9):2977–90.
86. Ni R, Zhao J, Liu Q, Liang Z, Muenster U, Mao S. Nanocrystals embedded in chitosan-based respirable swellable microparticles as dry powder for sustained pulmonary drug delivery. Eur J Pharm Sci [Internet]. 2017;99:137–46. Available from: https://www.sciencedirect.com/science/article/pii/S0928098716305528
87. Alhajj N, O'Reilly NJ, Cathcart H. Leucine as an excipient in spray dried powder for inhalation. Drug Discov Today [Internet]. 2021; Available from: https://www.sciencedirect.com/science/article/pii/S1359644621001975
88. Rostamnezhad M, Jafari H, Moradikhah F, Bahrainian S, Faghihi H, Khalvati R, et al. Spray freeze-drying for inhalation application: process and formulation variables. Pharm Dev Technol. 2022;27(3):251–67.
89. Gradon L, Sosnowski TR. Formation of particles for dry powder inhalers. Adv Powder Technol [Internet]. 2014;25(1):43–55. Available from: https://www.sciencedirect.com/science/article/pii/S0921883113001921
90. Cun D, Zhang C, Bera H, Yang M. Particle engineering principles and technologies for pharmaceutical biologics. Adv Drug Deliv Rev [Internet]. 2021;174:140–67. Available from: https://www.sciencedirect.com/science/article/pii/S0169409X2100106X
91. Yu S, Pu X, Ahmed MU, Yu HH, Mutukuri TT, Li J, et al. Spray-freeze-dried inhalable composite microparticles containing nanoparticles of combinational drugs for potential treatment of lung infections caused by Pseudomonas aeruginosa. Int J Pharm [Internet].

2021;610:121160. Available from: https://www.sciencedirect.com/science/article/pii/S0378517321009662
92. Zhu C, Chen J, Yu S, Que C, Taylor LS, Tan W, et al. Inhalable nanocomposite microparticles with enhanced dissolution and superior aerosol performance. Mol Pharm. 2020;17(9):3270–80.
93. Ali ME, Lamprecht A. Spray freeze drying for dry powder inhalation of nanoparticles. Eur J Pharm Biopharm [Internet]. 2014;87(3):510–7. Available from: http://www.sciencedirect.com/science/article/pii/S0939641114000861
94. Cheow WS, Ng MLL, Kho K, Hadinoto K. Spray-freeze-drying production of thermally sensitive polymeric nanoparticle aggregates for inhaled drug delivery: effect of freeze-drying adjuvants. Int J Pharm [Internet]. 2011 Feb 14 [cited 2015 Aug 1];404(1–2):289–300. Available from: http://www.sciencedirect.com/science/article/pii/S0378517310008707
95. Yu S, Wang S, Zou P, Chai G, Lin Y-W, Velkov T, et al. Inhalable liposomal powder formulations for co-delivery of synergistic ciprofloxacin and colistin against multi-drug resistant gram-negative lung infections. Int J Pharm [Internet]. 2020;575:118915. Available from: https://www.sciencedirect.com/science/article/pii/S0378517319309603
96. Aguiar-Ricardo A. Building dry powder formulations using supercritical CO2 spray drying. Curr Opin Green Sustain Chem [Internet]. 2017;5:12–6. Available from: https://www.sciencedirect.com/science/article/pii/S2452223616300748
97. Kankala RK, Xu P-Y, Chen B-Q, Wang S-B, Chen A-Z. Supercritical fluid (SCF)-assisted fabrication of carrier-free drugs: an eco-friendly welcome to active pharmaceutical ingredients (APIs). Adv Drug Deliv Rev [Internet]. 2021;176:113846. Available from: https://www.sciencedirect.com/science/article/pii/S0169409X21002386
98. Reverchon E, Adami R, Campardelli R, Della Porta G, De Marco I, Scognamiglio M. Supercritical fluids based techniques to process pharmaceutical products difficult to micronize: Palmitoylethanolamide. J Supercrit Fluids [Internet]. 2015;102:24–31. Available from: https://www.sciencedirect.com/science/article/pii/S0896844615001436
99. Peng H-H, Wang Z-D, Guan Y-X, Yao S-J. Supercritical CO2 assisted preparation of chitosan-based nano-in-microparticles with potential for efficient pulmonary drug delivery. J CO2 Util [Internet]. 2021;46:101486. Available from: https://www.sciencedirect.com/science/article/pii/S2212982021000536
100. Silva AS, Shopsowitz KE, Correa S, Morton SW, Dreaden EC, Casimiro T, et al. Rational design of multistage drug delivery vehicles for pulmonary RNA interference therapy. Int J Pharm [Internet]. 2020;591:119989. Available from: https://www.sciencedirect.com/science/article/pii/S0378517320309741
101. Silva MC, Silva AS, Fernandez-Lodeiro J, Casimiro T, Lodeiro C, Aguiar-Ricardo A. Supercritical CO2-assisted spray drying of strawberry-like gold-coated magnetite nanocomposites in chitosan powders for inhalation. Materials. 2017;10:74.
102. Silva AS, Sousa AM, Cabral RP, Silva MC, Costa C, Miguel SP, et al. Aerosolizable gold nano-in-micro dry powder formulations for theragnosis and lung delivery. Int J Pharm. 2017;519(1–2):240–9.
103. Restani RB, Pires RF, Tolmatcheva A, Cabral R, Baptista PV, Fernandes AR, et al. POxylated dendrimer-based Nano-in-micro dry powder Formulations for inhalation chemotherapy. ChemistryOpen. 2018;7(10):772–9.
104. Restani RB, Pires RF, Baptista PV, Fernandes AR, Casimiro T, Bonifácio VDB, et al. Nano-in-micro sildenafil dry powder formulations for the treatment of pulmonary arterial hypertension disorders: the synergic effect of poxylated polyurea dendrimers, plga, and cholesterol. Part Part Syst Charact [Internet]. 2020;37(6):1900447. https://doi.org/10.1002/ppsc.201900447.
105. Siow CRS, Wan Sia Heng P, Chan LW. Application of freeze-drying in the development of oral drug delivery systems. Expert Opin Drug Deliv [Internet]. 2016;13(11):1595–608. https://doi.org/10.1080/17425247.2016.1198767.

106. Chen L, Okuda T, Lu XY, Chan HK. Amorphous powders for inhalation drug delivery. Adv Drug Deliv Rev. 2016;100:102–15.
107. Chang RYK, Chow MYT, Khanal D, Chen D, Chan H-K. Dry powder pharmaceutical biologics for inhalation therapy. Adv Drug Deliv Rev [Internet]. 2021;172:64–79. Available from: https://www.sciencedirect.com/science/article/pii/S0169409X21000636
108. Hufnagel S, Sahakijpijarn S, Moon C, Cui Z, Williams RO III. The development of thin-film freezing and its application to improve delivery of biologics as dry powder aerosols. KONA Powder Part J. 2022;39:2022010.

Porous Particle Technology: Novel Approaches to Deep Lung Delivery

5

Somchai Sawatdee, Narumon Changsan, Teerapol Srichana, and Basavaraj Nanjwade

Abstract

Porous particle technologies are widely used to deliver small and large molecular drugs to the lung for local and systemic treatment. Porous particles are able to deliver high drug payload via dry powder inhalers. The greatest advantages of porous particles are their simple dispersion and reduced aerodynamic diameter from their low density. When compared to non-porous particles, inhaled porous particles showed deep lung deposition. Carrier particle morphology has significant impacts on aerodynamic distribution. Additionally, large particle size also restricts immune response, which is beneficial in treating a number of lung diseases and makes them suitable carriers for antibiotics and anticancer molecules. The design of porous materials for controlled drug delivery is experimentally successful. The commercial application of porous particles for the pulmonary delivery of drugs with systemic or controlled release would be feasible when clinical outcomes are successful. Several technologies have been successfully

S. Sawatdee
Drug and Cosmetics Excellence Center and School of Pharmacy, Walailak University, Nakhon Si Thammarat, Thailand
e-mail: somchai.sa@wu.ac.th

N. Changsan
College of Pharmacy, Rangsit University, Pathum Thani, Thailand
e-mail: narumon.c@rsu.ac.th

T. Srichana (✉)
Drug Delivery System Excellence Center and Department of Pharmaceutical Technology, Faculty of Pharmaceutical Sciences, Prince of Songkla University, Songkhla, Thailand
e-mail: teerapol.s@psu.ac.th

B. Nanjwade
Surya College of Pharmacy and MRMH Pharma Pvt Ltd., Bangalore, Karnataka, India
e-mail: drbasavaraj@mrhmpharma.com

P. P. Mehta, V. Dhapte-Pawar (eds.), *Pulmonary Drug Delivery Systems: Material and Technological Advances*,
https://doi.org/10.1007/978-981-99-1923-9_5

employed in the production of porous particles, i.e., double-emulsion solvent evaporation technology, spray drying technology, supercritical fluid-assisted anti-solvent technology, spray freeze drying technology, and co-suspension delivery technology. Commercial porous particle technology includes PulmoSphere™ and Aerogel. In this chapter, the mentioned technologies are discussed along with examples of pertinent research and their applications for pulmonary delivery.

Keywords

Porous particles · PulmoSphere™ · Pulmonary drug delivery · Dry powder inhaler · Metered dose inhaler and aerodynamic distribution

5.1 Introduction

Pulmonary drug delivery is an important route to deliver drug molecules or macromolecules since the delivered substance potentially have either a local or systemic effect. Due to its non-invasive nature and especially high bioavailability, the inhalation route is an alternative to the parenteral route [1]. There are four different types of aerosolized mechanisms including nebulizer, pressurized metered dose inhaler (pMDI), dry powder inhaler (DPI) and the latest is soft mist inhaler (SMI). Nebulizers turn a solution or suspension-based composition into an inhalable aerosol using compressed air. Patients commonly inhale the medicine mist through the mouthpiece adaptor wherein minimum coordination is required. Nebulizer has the drawbacks of being cumbersome, non-portable, and time consuming to use [1, 2]. The pMDIs delivery pressurized propellant containing the medication through the valve fitted with an actuator. Even though pMDIs are portable and user-friendly, the usage of propellants like hydrofluoroalkanes (HFA) is problematic in certain nations. The pMDIs cannot deliver high doses and require coordination of inhalation and actuation for effective therapy, which is difficult for a specific patient class such as pediatric patients [3]. The Respimat® is the only SMI available today. The SMI device works by pushing a metered dose of drug solution through a unique and precisely engineered nozzle (the uniblock), producing two fine jets of liquid that converge at a pre-set angle. The delicate mist is created by the collision of these two jets [4]. The Respimat® SMI was designed to reduce patient inspiratory effort, avoid the usage of propellants, increase patient usability, and enhance drug delivery. In most of the cases, Respimat® SMI showed a high lung deposition compared to oropharyngeal drug deposition [5].

DPIs are among the most frequently used techniques due to their portability, ease of usage, and more stable formulation properties. DPI mainly contains drug/carrier powder mixture and creates aerosol cloud using patient inspiratory force. Therefore, a rapid rate of inhalation is necessary to produce the requisite energy for aerosolization, which is tricky for some patients. Effective particle deagglomeration is crucial for the effective use of DPIs [6, 7]. Breath-activated DPIs are common as the aerosol formation starts after the inhalation begins. However, in the "passive" DPIs powder diffuses with the airstream and powder aggregates disperse into individual particles with the help of patients breathing. As a result, the patient's capability to inhale a sufficient volume of air at a suitable inspiratory flow rate affects the outcome of therapy [8].

The effective particle size and size distribution of inhaled medications for deep lung delivery should be 1–5 μm, since these size ranges provide the greatest therapeutic outcomes [9–11]. Particles suitable for deep lung deposition are the most common challenge in pulmonary drug delivery. Powder production techniques, such as milling or spray drying, freeze drying, and spray freeze drying can help in obtaining powders with respirable particle sizes. It is very challenging to deliver de-aggregated dry powders to the lungs due to numerous interrelated issues, i.e., surface properties, interparticle adhesion (electrostatic and van der Waals forces), cohesive and adhesive balance (CAB) and high surface energy [8]. Thus, normally micronized drug particles were combined with large carrier particles (e.g., lactose or mannitol) to enhance powder flow, dispersion, and handling. Inhaler device also plays a key role in delivery of dry powder. In most cases, the inhaler device's internal arrangement offers mechanical energy to break up interparticle interactions and successfully dispense the particles for inhalation [8]. Interparticle interactions and CAB can be adjusted using elongated particles, such as fibers. Nevertheless, it is still very challenging to accomplish effective dispersion for such particles.

Large porous particles are one of the important classes of particles used for pulmonary drug delivery. Large porous particles disperse more readily compared to typical non-porous particles of similar aerodynamic size. Large porous particles offer excellent dispersion without the need for carrier particles due to their decreased aggregation [8, 12]. Large porous particles containing DPIs showed longer therapeutic effects and higher systemic bioavailability since they are cleared gradually from the pulmonary airways [12]. By having these backgrounds, the present chapter provides a summary of porous particle applications in pulmonary drug delivery. This chapter contains manufacturing technologies, formulation delivery methods, and other key aspects, i.e., local treatment, systemic drug delivery, and controlled release of porous particles.

5.2 Factors Affecting Inhaled Porous Particles Deposition

There are three basic mechanisms for particle deposition in the respiratory system: inertial impaction, sedimentation, and Brownian motion. These mechanisms are primarily regulated by the aerodynamic diameter (d_{ae}) [13]. Stokes' Law is employed to characterize aerodynamic diameters based on particle density. The terminal setting velocity of porous particle deposition in the lungs is able to underline the design of therapeutic aerosols. To understand the challenges and parameters involved with the design of porous particles, it is important to know the concept of aerodynamic diameter and its relationship to the respiratory deposition of inhaled particles. Airborne particles are subjected to gravity's downward force and the opposing drag force of air resistance. When the velocity-dependent drag force quickly balances the gravitational attraction, the particles have attained their terminal settling velocity. For example, the terminal settling velocity of spherical unit density particles of 1 and 10 μm reached 10^{-5} and 10^{-3} s, respectively [11]. The aerodynamic drag force is the major driving force underlying the velocity of airborne particles. The Stokes terminal settling velocity is calculated by equating the gravitational force and the aerodynamic drag force as in Eq. (5.1).

$$mg = \frac{3\pi\eta dV}{C_c} \tag{5.1}$$

where m is the mass, g is the gravity, and $3\pi\eta dV$ is the drag force where η is the air viscosity, d is the particle diameter, the particle velocity is V, and the Cunningham slip correction factor is C_C. By transforming the particle mass in terms of its particle density (ρ_p) and volume, the terminal settling velocity is expressed as Eq. (5.2).

$$V_{TS} = \frac{gC_c\rho_p d^2}{18\eta} \tag{5.2}$$

In aerosol physics, particle density is a crucial variable. Particularly intriguing is its application in estimating the aerodynamic diameter of a particle (d_{ae}). d_{ae} is the diameter of a unit density sphere with the same settling velocity. The d_{ae} is computed using Eq. 5.3.

$$d_{ae} = \sqrt{\rho_p}\, d_g \tag{5.3}$$

Aerodynamic diameter, particle density, and geometric dimensions are interrelated, as demonstrated by Eq. (5.3). A geometric particle size with varying densities will always have a varied aerodynamic diameter. As a result, if the density is low enough, a particle with a large geometric diameter can be considered as an aerodynamic diameter of less than 3 μm and be acceptable for inhalation. This characteristic facilitates the production of large porous particles that can penetrate deeply into the lungs. The particles have a lower density than the unit density due to their porous architecture, which traps air volume inside the particle. Despite having a geometric particle size of more than 10 μm, these particles acquire aerodynamic diameters within the respirable size range [14].

A shape factor is important and accounted for non-spherical particles, in Eq. (5.2). The shape factor (X) ranging from 1.1 to 1.88 multiplies the Eq. (5.2) as illustrated in Eq. (5.4).

$$V_{TS} = \frac{gC_c\rho_p d^2}{18\eta X} \tag{5.4}$$

As demonstrated in Eq. (5.4), the simplified Stokes law represents the relationship between the aerodynamic size and the particle size, morphology, shape, and density of inhaled particles and rewrite into Eq. (5.5).

$$d_{ae} = d_g \sqrt{\frac{\rho_b}{X}} \tag{5.5}$$

where ρ_b is the bulk density, and X is described as the deviation from sphericity of inhaled particles based on their shape, surface roughness, and surface area. According to the equation above, the adequate aerodynamic diameters influence by decreasing the ρ_b or raising the X as illustrated in Eq. (5.5). Although needle-shaped particles have a higher X value ($X = 1.7$), industrial manufacturing of needle-shaped inhaled particles is challenging. Spherical particles, on the other hand, have a dynamic shape factor of one and are easily produced at large production scales. The mass median aerodynamic diameter (MMAD), one of the most widely, employed diameters for analyzing the deposition efficacy of an inhaled aerosol, is the particle size at 50% cumulative mass. Due to the variability of particle sizes, the total deposited mass of an inhaled polydisperse aerosol cannot be anticipated by the MMAD alone. Aerosol size distributions are described by a log-normal distribution with geometric standard deviation (GSD). A larger particle size tail in the distribution is inferred by a larger GSD. In contrast, despite having a higher GSD, aerosols with lower MMADs (1–3 μm) are more respirable [11]. A geometric particle size greater than 10 μm may provide d_{ae} within the respirable size limit for particles with a density of less than 1 g/cm^3. Engineering large porous particles having low density and large geometric dimensions has attracted interest as a potential solution for the efficient transfer of therapeutic drugs to the alveolar region. By successfully providing space or pores between the particles, they are able to reduce van der Waals force, which contributes to their improved dispersibility. During effective dose administration to patients, it is important to maximize the density and powder volume contained in the dose package at the expense of each particle's total mass. Numerous techniques are available to fabricate low-density porous or hollow particle's [14].

5.3 Preparation of Porous Particles in Lung Delivery

The porous aerosol particles do have low mass density and a large particle size, although their average d_{ae} falls within the range of 1–3 μm. Large size and low mass density of porous aerosol particles have two key benefits for aerosolization. Firstly, larger particles have less of a tendency to aggregate, which increases the effectiveness of aerosolization. Secondly, once particle size exceeds 2–3 μm, macrophage phagocytosis of the particles decreases. As a result, very large particles may not be cleared by alveolar macrophages when they deposit in the lungs [15]. According to pore diameter, porous particles are classified into 3 categories: macroporous (pore diameter >50 nm), mesoporous (2–30 nm), and microporous (<2 nm). Table 5.1 provides an overview of the various powder methods to generate porous particles for pulmonary drug delivery.

Table 5.1 Porous particles and their properties used for pulmonary drug delivery systems

Porous particles and drug	Excipients	Physical properties	Aerosol properties	Key outcomes	References
Aerogel and salbutamol sulfate	Alginic acid sodium salt and calcium chloride	Bulk density of 0.013 g/cm^3	d_{ae} 4.0 μm and FPF of 49.7%	Aerogel microporous spheres were formulated by inkjet printing and followed by supercritical drying. This method showed good potential and reproducibility to fabricate highly porous alginate-based microspheres with nanoporous range. It is useful for sustained release of the drug.	[10]
PulmoSphere™ and tobramycin	DSPC, calcium chloride and PFOB	Tapped density of 0.05 g/cm^3	d_{ae} 3 μm and total lung drug deposition of 50–80%	Delivery system nature retains the crystalline form of drug and improves physicochemical stability of dry powder. It improved treatment compliance (reduced delivery time) and clinical outcomes (FEV_1) in chronic *Pseudomonas aeruginosa*-infected cystic fibrosis subjects.	[16–19]
PulmoSphere™ for MDI and Salbutamol sulfate	DSPC, calcium chloride dihydrate, PFOB and HFA-134a	Tapped density of 0.12 g/cm^3	MMAD = 3.61 μm GSD and FPF are 1.79 and 27%	Formulation showed improved physical stability, good content uniformity, and better aerosolization efficiency.	[20, 21]
Trojan or hollow particles	DPPC, DMPE and lactose monohydrate	Density of <0.4 g/cm^3	d_{ae} of 1–3 μm	Large porous nanoparticles prepared by spray drying techniques showed hollow pores and sphere shaped particles.	[22, 23]
Nanoporous Particles and Octreotide acetate	Sodium deoxycholate, Soybean phosphatidyl choline, Glycol tristearate, Glyceryl monostearate, Oleic acid	MVD and tapped density of 5.68 μm and 0.27 g/cm^3	d_{ae} of 2.94 μm	Spray freeze drying was used to prepare porous nanoparticle-aggregates. During biological assessment using rats, pulmonary delivery showed significantly higher bioavailability than subcutaneous injection.	[24]

Porous particles and drug	Excipients	Physical properties	Aerosol properties	Key outcomes	References
Spray dried porous microspheres and cyclosporine A	HTCC, HP-β-CD, and ammonium bicarbonate	MVD of >4 μm	MMAD of 3.53 μm, GSD of 1.96 and FPF of 55–88%.	Spray drying method used to formulate porous microspheres of cyclosporine A during in vivo study using an asthmatic rat model, microspheres effectively suppress the inflammation.	[25]
Porous microparticles and cyclosporine A	Mannitol, *tert*-butyl alcohol, and water (solvent)	Particle size (D_{50}) between 2.20 to 7.89 μm and SSA of 111.1 m^2/g	FPF and MMAD of 31% and 9.48 μm.	Spray freeze drying method was used to formulate porous microparticles of cyclosporine A during dissolution study, using a low-volume, simulated respiratory fluid pH 6.6 porous microparticles showed 2–8 fold higher dissolution rate compared to non-porous particles.	[26]
Large porous and isoniazid, Rifampicin, IDR-1018 Peptide (1535.93 Da short peptide; Sequence RLIVAVRIWRR-NH_2), Bovine serum albumin (BSA)	Poly-L-lactide (PLLA), Polyvinyl alcohol (PVA), Hydrogen peroxide, Sodium bicarbonate, Sodium chloride, Polyethylene glycol, Sucrose, Ammonium bicarbonate, Phosphate buffer saline	Mass densities = 0.21 g/cm^3 (microporous), 0.17 g/cm^3 (mesoporous), 0.10 g/cm^3 (macroporous) Surface area = 377.45 m^2/g (microporous), 109.77 m^2/g (mesoporous), 10.02 m^2/g (macroporous)	For microporous particle: MMAD = 4.1 μm GSD = 1.5 FPF = 37.2% For mesoporous particle: MMAD = 3.7 μm GSD = 1.7 FPF = 36.0% For macroporous particle: MMAD = 3.1 μm GSD = 1.8 FPF = 52.6%	Porous poly –L-lactide (PLLA) microspheres were prepared by double-emulsion solvent evaporation method with either open or closed porosity. Microporous, mesoporous, and macroporous particles gave burst release and controlled release patterns. Size and porosity influence the phagocytosis of macrophages.	[27]

(continued)

Table 5.1 (continued)

Porous particles and drug	Excipients	Physical properties	Aerosol properties	Key outcomes	References
Insulin porous microspheres	Poly –L-lactide (PLLA), Poloxamer-127, Ammonium bicarbonate and Dichloromethane	Geometric mean diameter of 15.62 μm	MMAD of 4.31 μm	The necessary aerodynamic deposition and particle size distribution were attained due to the solvent-free residues. Satisfactory in vivo hypoglycemic performance was obtained from extended-release microspheres compared to subcutaneous injection.	[6]
Dexamethasone mesoporous particles	TEOS and APTES	Hydrodynamic size of 230 nm–1.5 μm and SSA of 900 m^2/g	NA	During in vivo study using lipopolysaccharide-induced inflammatory rats, mesoporous silica loaded with dexamethasone showed a significant reduction in inflammatory responses.	[28]
Gentamicin sulfate PLGA large porous particles	PLGA, ammonium bicarbonate, PVA and mannitol	Bulk and tapped density of 0.075 and 0.104 g/cm^3	MMAD, GSD and FPF of 4.98 μm, 2.92 and 39.07%	PLGA large porous particles produced by the double emulsification method showed highly porous particles and burst release in 30 minutes in Franz diffusion cell containing phosphate buffer saline pH 7.4.	[29]
Lysozyme, Doxorubicin PLGA large Porous particles	PLGA, Ammonium bicarbonate, PVA	Particle size (D_{50}) of 2.7 to 18.2 μm	MMAD = 4.6 μm and FPF = 29.1–33.8%)	Double-emulsion processed PLGA large porous microparticles avoided phagocytosis by macrophages and showed active drug release over 4 days. Thus, they can be used as sustained local pulmonary delivery.	[30]

Porous particles and drug	Excipients	Physical properties	Aerosol properties	Key outcomes	References
Meloxicam large porous particles	L-leucine, ammonium bicarbonate, sodium hyaluronate	Bulk and tapped density of 0.10 and 0.19 g/cm^3	MMAD =2.41–2.32 μm and FPF = 65.8%	During aerodynamic assessment (30 L/min) carrier-free meloxicam large porous particles and non-porous meloxicam particles showed similar MMAD. Meloxicam large porous particles showed significantly higher FPF (1.8 folds) compared to non-porous meloxicam particles.	[31]
Dexamethasone palmitate large porous particles (LPPs)	DPPC and HA	MVD 13.7 μm. Tapped density = 0.053 g/cm^3	FPF and ED of 40% and 98%	Drug loaded LPPs showed a sustained release pattern up to 24 h in phosphate buffer saline. The systemic exposure of large porous particles is far less than the local effect.	[32]
Voriconazole porous particle	Mannitol	No data	MMAD = 3.41 μm, GSD =2.38 and FPF = 40%	Porous particles with high concentrations of voriconazole and tert-butyl alcohol achieved the optimal FPF. The dissolution rate of voriconazole porous particles was 100% within 5 min compared to 2 h of non-porous particles.	[33]
Budesonide-loaded nanoporous mannitol carrier particles	Mannitol, ammonium bicarbonate	D_{50} = 5.05 μm Bulk density = 0.09 g/cm^3 Tapped density = 0.38 g/cm^3 Specific surface area = 20.33m^2/g pore diameter = 5.90 nm.	FPF = 38%.	Micronized budesonide and nanoporous mannitol carrier (1:30) mixture showed a 2- and 3.2-fold higher bioavailability compared to non-porous mannitol carrier and commercial Pulmicort®, respectively.	[34, 35]

(continued)

Table 5.1 (continued)

Porous particles and drug	Excipients	Physical properties	Aerosol properties	Key outcomes	References
Rifampicin porous particles	Phytoglycogen	D_{50} = 7.21 μm	MMAD = 5 μm FPF = 38.9%	Porous particles containing phytoglycogen enhanced phagocytosis by alveolar macrophages.	[36]
Salbutamol sulfate micro-balloon shaped (hollow particle)	Ethyl acetate	Bulk density and porosity of 0.19 g/cm^3 and 74.32%.	MMAD, FPD, and ED of 3.76 μm, 37.40%, and 95.07%	With the aid of ultrasound and a stirring process, inhalable balloon-like hollow, spherical salbutamol sulfate microparticles, carrier-free DPI were created. Mass drug diffusion from the interior to the external surface of the droplet produced throughout the process creates a hollow structure.	[37]
Pyrazinamide LPPs	DPPC, leucine, HA, ammonium bicarbonate	Tap density and D_{50} of 0.09 g/cm^3 and 5.8 μm	MMAD of 1.75 μm	The LPPs of pyrazinamide are stable for more than 4 weeks.	[38]

1,2-dimyristoyl-sn- glycero-3-phosphoethanolamine (DMPE), 1,2-Dipalmitoyl-sn-glycero-3-phosphocholine (DPPC), 1,1,1,2-Tetrafluoroethane (HFA-134a) Aerodynamic diameter (d_{ae}), Aminopropyl triethoxysilane (APTES), Bovine serum albumin (BSA), Distearoylphosphatidylcholine (DSPC), Emitted dose (ED), Fine particle dose (FPD), Fine particles fraction (FPF), Forced expiratory volume in 1 second (FEV_1), Geometric standard deviation (GSD), Hyaluronic Acid (HA), Hydroxypropyl-β-cyclodextrin (HP-β-CD), Large porous particle, Median diameter (D_{50}), Metered dose inhaler (MDI), Mass median aerodynamic diameter (MMAD), Median volume geometric diameter (MVD), N-[(2-Hydroxyl)-propyl-3-trimethyl ammonium] chitosan chloride (HTCC), Not applicable (NA), Perfluorooctyl bromide (PFOB), Poly (lactic-co-glycolic acid) (PLGA), Poly-L-lactide (PLLA), Polyvinyl alcohol (PVA), Specific surface area (SSA), tetraethyl orthosilicate (TEOS)

5.3.1 Method of Preparations

Porous particles are generated by adding and removing certain substances known as porogens or templating agents, or by utilizing a specific procedure, in order to form pores inside particles. Figure 5.1 depicted the creation of porous particles utilizing porogens or template agents. First, porogens or template agents are introduced into particles, and then they are removed to form pores. In practice, porogens and templates employ different methods to generate pores. Porogens create gas or volatilization to form pores. Hydrogen peroxide, for example, is a porogen to generate pores within PLGA microspheres. As hydrogen peroxide reacts with catalase enzymes, oxygen gas bubbles form within microspheres, resulting in the formation of internal pores. Ammonium bicarbonate (NH_4HCO_3) is also a porogen to produce porous mannitol prepared by spray drying technique since ammonium bicarbonate sublimes at a temperature higher than 50 °C and creates voids in particles during spray drying process due to the escape of water, ammonia (NH_3), and carbon dioxide (CO_2). Camphor, menthol, and thymol were porogens to make porous granules, which depend on their sublimation under vacuum. In contrast to porogens, templating agents need to be removed via heat treatment or solvent washing in the second step. Sulfuric acid, in some cases, is used to remove a copolymer template. Many substances can be templates such as (1) inorganic salts, (2) polystyrene and other polymers, and (3) food-grade acids and ethanol-soluble sugars. Additionally, emulsion comprising two-phase systems made up of a continuous phase and a dispersed phase may serve as a unique templating agent. Pore formation results from the

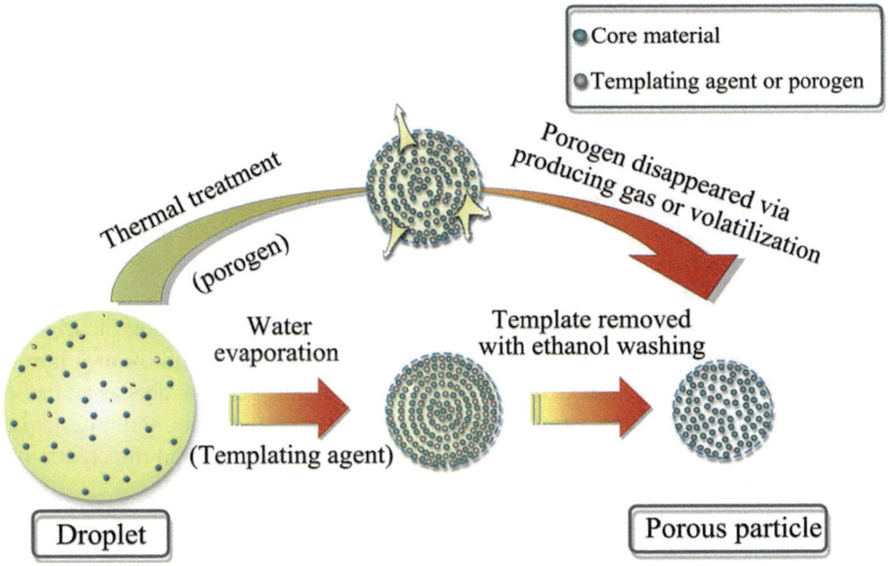

Fig. 5.1 The preparation of porous particles by using porogen or templating agent (Reprint from [39] with permission by Royal Society of Chemistry by Creative Commons Attribution)

continuous phase of the emulsion being polymerized and then the dispersed phase being removed with ethanol [39].

5.3.2 Double Emulsion Solvent Evaporation Production Method

Large porous particles were first formulated by double-emulsion solvent evaporation. In this method, the active ingredient dissolves in primary, internal aqueous phase, which may contain cyclodextrin, solubilizing agents, polymeric additives, or pH modifiers. The external organic phase is composed of a polymeric material such as PLGA or polyethylene glycol (PEG) dissolved in dichloromethane or methylene chloride with the presence of an emulsifier (e.g., phosphatidylcholine). The secondary aqueous phase contains different concentrations of polymers such as PVA and other necessary ingredients, for example, sodium chloride. The primary aqueous internal phase and organic exterior phase are homogenized to form a primary emulsion, which is then mixed by injecting it with a secondary aqueous phase. The double emulsion is stirred further so that the organic solvent can evaporate, enabling large porous particles to form. The large porous particles are subsequently separated by centrifugation, washing, and freeze drying (Fig. 5.2).

5.3.3 Spray Drying Technology

Spray drying is a widely used technique in the pharmaceutical industry. This technique involves atomizing a solution into liquid droplets and exposing them to a drying gas stream. In the spray drying chamber, the solvent evaporates from the liquid droplets to generate solid particles. Majority of the dried powder is gathered via cyclonic collection, and any remaining fine powder from the exhaust gas stream is removed from filter bags. By modifying spray drying variables such as inlet temperature, feed pressure, solution feed rate, airflow rate, and nozzle type; it is possible to produce particles with diverse physicochemical properties such as surface quality, shape, and size. Using a modified spray drying process, large porous particles with a high porosity and a low tapped density are generated and used in commercial pharmaceutical formulations [40]. Dimensions between 5 and 30 μm best describe large porous particles. Typically, a suitable porogen (often ammonium bicarbonate) is required to produce large porous particles. Ammonium bicarbonate releases ammonia and carbon dioxide rapidly, leading to the creation of a porous structure (Fig. 5.3). Cyclodextrin is another common porogen that acts as an osmogen, generating an osmotic pressure differential between the inner and outer aqueous phases, resulting in water infiltration into the organic phase and the creation of pores in the matrix [13].

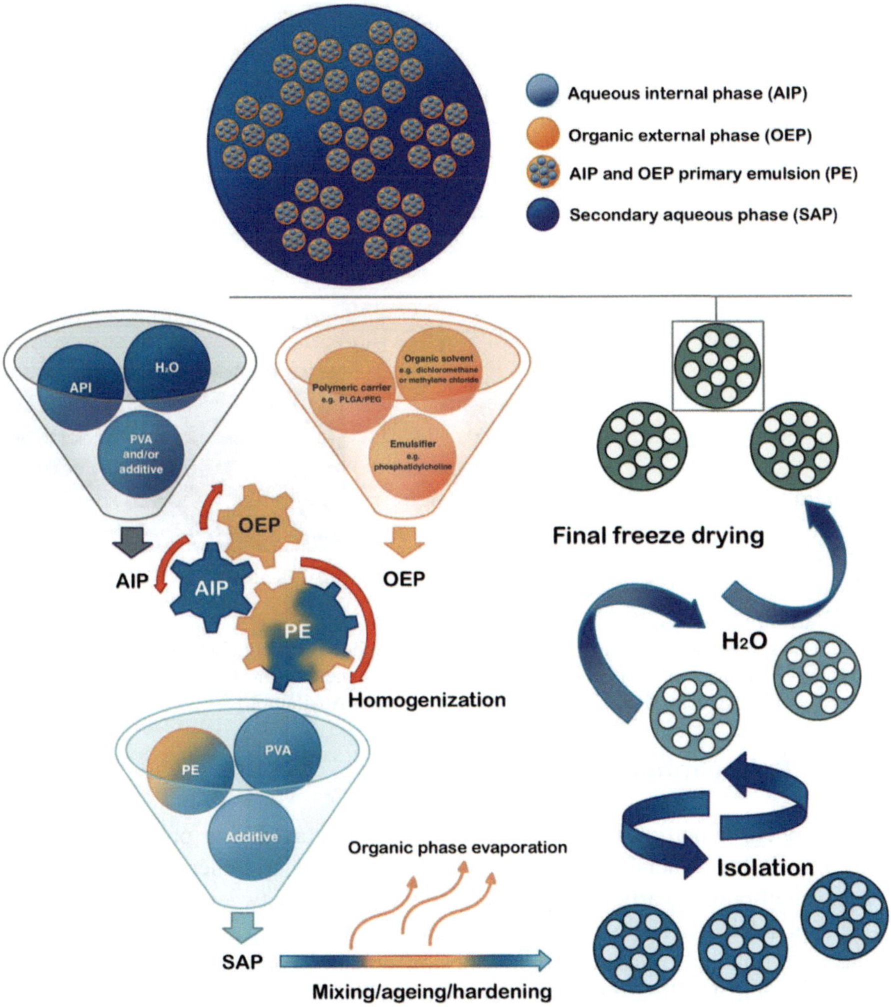

Fig. 5.2 Schematic diagram of large porous particles production by using double-emulsion solvent evaporation production method. *API* active pharmaceutical ingredient, *PVA* polyvinyl alcohol, *AIP* aqueous internal phase, *OEP* organic external phase, *PE* primary emulsion, *SAP* secondary aqueous phase (Redrawn from [40] with permission by Elsevier B.V.)

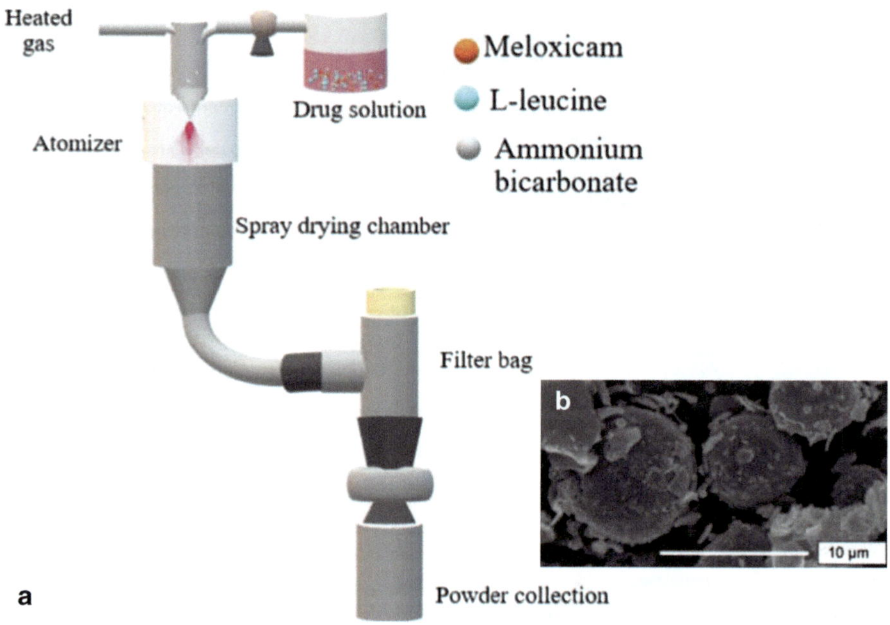

Fig. 5.3 A schematic illustration of the spray drying method for producing large porous particles (**a**), and a SEM image of large porous particles containing meloxicam (**b**) (Reprint from [13] with permission by Frontiers Media S.A. by Creative Commons Attribution)

5.3.4 PulmoSphere™

PulmoSphere™ particles were developed using a spray drying process (Figs. 5.4 and 5.5). PulmoSphere™ technique uses spray drying to generate phospholipid-based small porous particles with geometric diameters between 1 and 5 μm with a low tapped density. The porous nature of particles reduces particle–particle interactions, resulting in decreased particle cohesion, enhanced particle flow, and improved aerosolization. The liquid feed for PulmoSphere™ formulations is a submicron oil-in-water emulsion, with the oil droplets acting as pore formers to generate the required sponge-like particle shape. The liquid feed is an emulsion of PFOB in water. DSPC, a key component of endogenous pulmonary surfactant, is employed in a mole ratio of 2:1 with calcium chloride to stabilize the submicron PFOB emulsion (Fig. 5.5a). The emulsion comprises several tiny droplets. Heat energy from spray drying causes the PFOB to evaporate from the emulsion, leaving pores in the particle's structure [19].

PulmoSphere™ particles production methods include oil-in-water (o/w) emulsion containing the fluorocarbons perfluorodecalin or PFOB (Perflubron™) whereas phosphatidylcholine or DSPC is used as an emulsifier in the aqueous continuous phase. Before spray drying, the aqueous solutions containing the API and other components are mixed with the homogenized o/w emulsion. During spray drying,

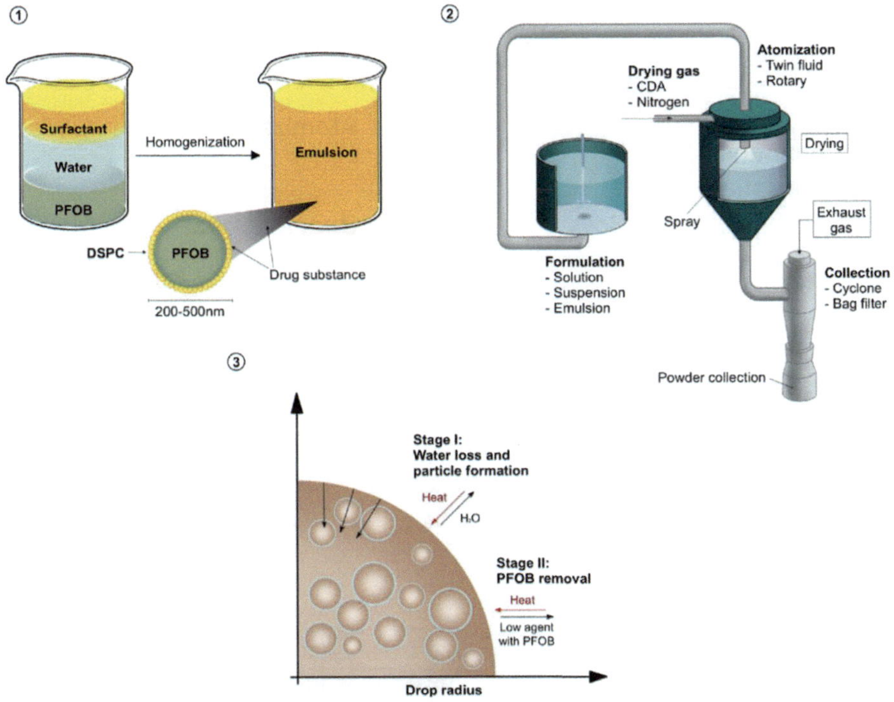

Fig. 5.4 PulmoSphere™ particles by an emulsion-based spray drying method. (1) High-pressure homogenization produces submicron emulsion droplets. (2) The spray drying process. (3) PulmoSphere™ formulations. *CDA* Clean dry air, *DSPC* Distearoylphosphatidylcholine, *PFOB* Perfluorooctyl bromide (Modified from [18] with permission by Future Science Ltd.)

water evaporates first, followed by fluorocarbon, which works as a pore-former, resulting in the pore formation within the particle [40]. Based on the characteristics of the desirable solid state, three distinct types of PulmoSphere™ were used to incorporate APIs within porous particles [13].

(1) Solution-based PulmoSphere™ —The APIs are simply dissolved in the emulsion's continuous phase before being spray dried to produce porous particles containing amorphous drugs. TOBI Podhaler™ (tobramycin) is one example of a PulmoSphere™ solution-based formulation (Fig. 5.5b). **(2) Suspension-based PulmoSphere™**—Micronized API particles are dispersed to the emulsion and the resulting suspension is spray dried to generate an amorphous or crystalline drug coated with a porous surface (Fig. 5.5c). **(3) Carrier-based PulmoSphere™**—Small, porous PulmoSphere™ particles are employed as carriers for micronized APIs. Small porous placebo particles and micronized drug particles are dispersed in a non-solvent-fluorinated liquid feed. Spray drying leads to the production of agglomerates of the drug and small porous carrier particles upon evaporation of the liquid feed (Fig. 5.5d).

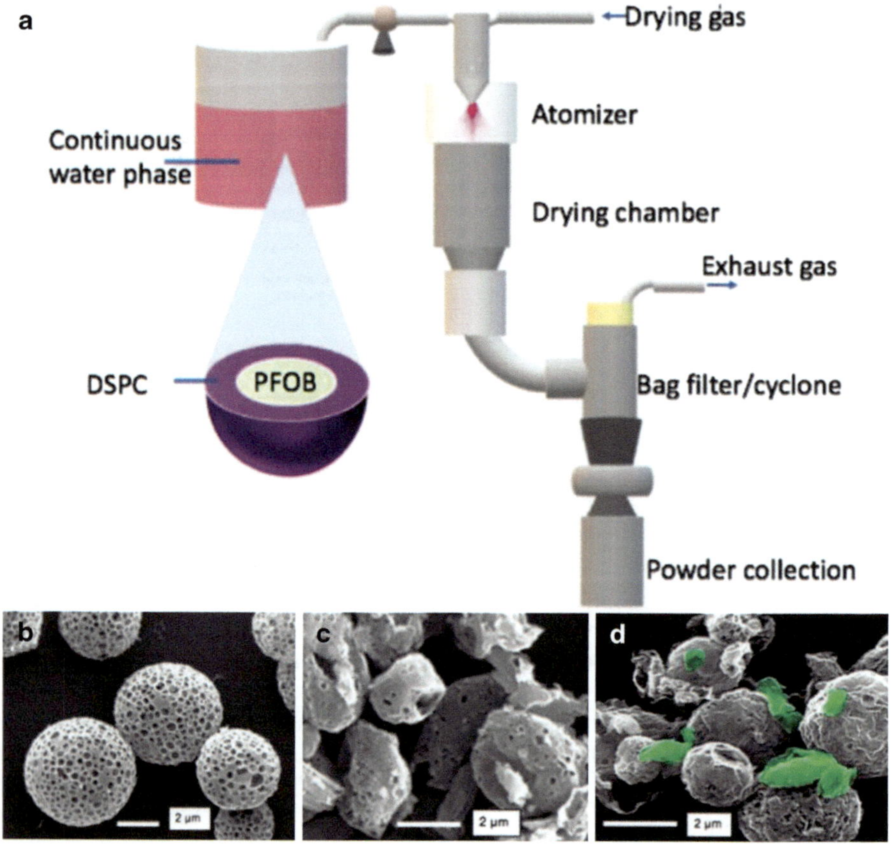

Fig. 5.5 Schematic diagram of PulmoSphere™ production technology (**a**), SEM image of solution-based tobramycin PulmoSphere™ (**b**), SEM image of suspension-based ciprofloxacin PulmoSphere™ (**c**), and SEM image of carrier-based PulmoSphere™ (**d**) (Reprint from [13] with permission by Frontiers Media S.A. by Creative Commons Attribution)

5.3.5 Supercritical Fluid-Anti-Solvent Technology

In DPI formulation, a supercritical fluid (SCF) is a green and non-toxic method to formulate dry powders. By using less amount of organic solvent, the dry powder particles can be successfully prepared with desired particle size. Supercritical CO_2 ($scCO_2$) is a United States Food and Drug Administration-approved solvent and the most extensively used media in SCF technology due to its safety, and non-flammability. In addition, it is inexpensive and made from recycled materials. Furthermore, due to its low viscosity, high diffusivity, and zero surface tension, $scCO_2$ may infuse porous materials easily under mild conditions. Particle technology makes use of the affinity and strong solvation power of $scCO_2$ to various organic solvents (such as acetone, ethanol, and dichloromethane) via anti-solvent procedures. The

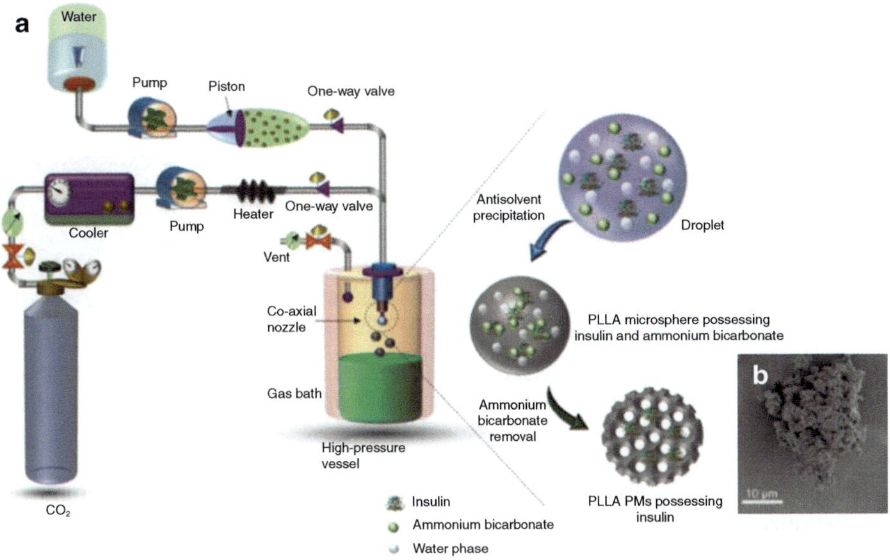

Fig. 5.6 Porous particles from supercritical fluid-anti-solvent technology (**a**), SEM of insulin-loaded porous microspheres (**b**) (Reprint from with permission by Frontiers Media S.A. by Creative Commons Attribution)

precipitation of compressed CO_2 anti-solvent (PCA) and supercritical fluid anti-solvent process (SAS) are particularly beneficial to develop particles with desired aerodynamic size. The PCA approach was used to produce insulin-loaded PLLA porous microspheres with ammonium bicarbonate as the porogen (Fig. 5.6). A Pluronic F-127 oil phase containing PLLA in dichloromethane was brought into contact with an aqueous phase consisting of insulin solution and ammonium bicarbonate. The resulting water-in-oil emulsion was treated with compressed CO_2 in the PCA method. Simultaneously, vacuum drying was used to remove the ammonium bicarbonate porogen. During in vivo hypoglycemic study in a rat model, pulmonary delivery of insulin porous microspheres showed prolonged release of insulin similar to subcutaneous injection. Additionally, SCF-treated porous microspheres showed low inflammatory reactions due to organic solvent-free processing [6].

5.3.6 Spray Freeze Drying

The manufacturing of large porous particles for pulmonary drug delivery using spray freeze drying (SFD) is an innovative process with high production yields that is particularly suited for compounds that are thermally sensitive. There are three related sub-processes, i.e. **(1) atomization**—the drug solution is atomized with the aid of an atomization gas into a refrigerant medium; **(2) freezing**—this process is

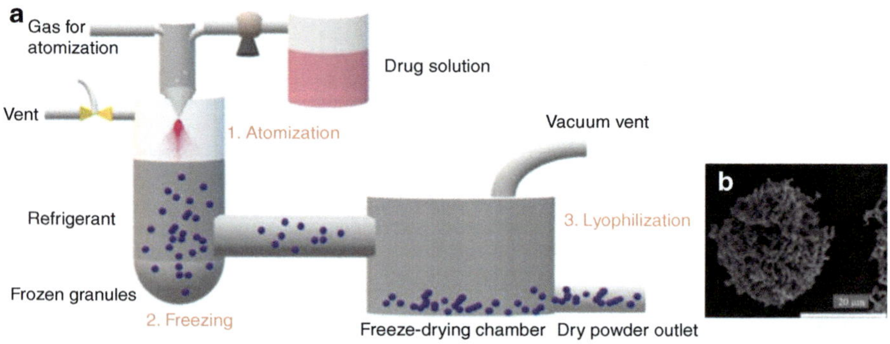

Fig. 5.7 Production of porous particles by spray freeze drying (**a**), and SEM of inhaled voriconazole by spray freeze drying (**b**) (Reprint from with permission by Frontiers Media S.A. by Creative Commons Attribution)

carried within the chamber to create frozen granules, and liquid nitrogen is employed as the refrigerant to provide rapid cooling and **(3) lyophilization**—the solvent is sublimated under high vacuum venting to create porous particles (Fig. 5.7) [13]. In most of the cases, SFD created porous particles meet the requirements for aerodynamic deposition. Voriconazole-loaded large porous particles were manufactured by SFD for the treatment of pulmonary aspergillosis. Voriconazole large porous particles showed complete dissolution within 5 min while untreated voriconazole showed complete release after 2 h [33].

5.3.7 Aerogel

Aerogels are extremely light porous materials used in drug delivery systems. Aerogels' large surface area, accessible pores, high aerodynamic characteristics, and physicochemical stability make them attractive for various delivery routes. In general, inorganic aerogels, such as silica aerogels, have larger surface area than organic aerogels, which improves drug loading efficiency. Biopolymer or polysaccharide aerogels, such as gelatin, agar, cellulose, alginate, chitin, and pectin are biodegradable and biocompatible, making them ideal for use in biomedicine [41]. Aerogel preparation proceeds as follows: sol-gel, solvent exchange, and wet-gel solvent removal via drying. Drugs can be introduced to aerogels by one of the four methods, which determine the drug release mechanisms: (1) before gelling, (2) during solvent exchange, (3) during drying, or (4) with prepared aerogels impregnated with SCF. The loading approach is determined by the drug's solubility and stability in the chosen solvent. The two common techniques for producing biopolymer-based aerogel particles are external gelation (dripping) and internal gelation (emulsification) as shown in Fig. 5.8 [13]. Traditional dripping procedures employ pipettes, vibrating nozzles, electrovalves, or syringes. Large aerogel beads with droplet sizes of a few millimeters are formed because of gravity forcing a droplet of polymer

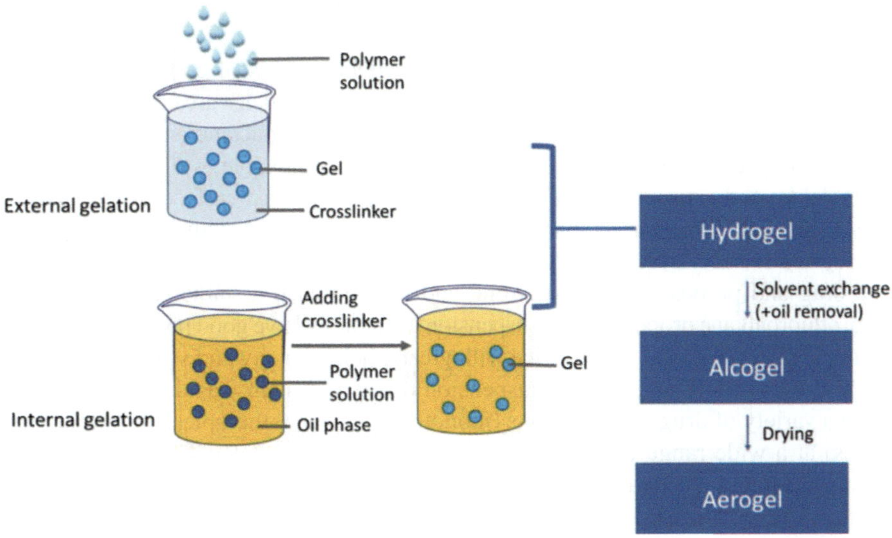

Fig. 5.8 Schematic diagram of aerogel particles by external and internal gelation (Adapted with permission from Frontiers Media S.A. by Creative Commons Attribution)

solution to fall into the gelation bath; nevertheless, these aerogel beads do not meet the requirements of inhaled formulations. Recently, a modified dripping approach involving thermal inkjet printing was developed to produce aerogel microspheres 10–20 μm for DPI formulations [13]. In the solvent-emulsification procedure, gel microparticles formed via internal gelation. The system is emulsified using an emulsifier with a hydrophilic–lipophilic balance (HLB) of 3–6. Using polymer solution as the aqueous phase distributed in the oil phase at a ratio of 1:2–1:10, the emulsion is produced. The water-to-oil phase's viscosity ratio must remain below 1 in order to form an emulsion.

In order to create aerogel particles, the gel precursor must be dried. The three most widely used techniques for drying wet gels are fluid-assisted supercritical drying, ambient pressure drying, and freeze drying. Xerogels are dried at room temperature and pressure. By this approach, the delicate porous structure of wet gels is damaged because of the high capillary pressure that occurs when the solvent evaporates. Cryogels are porous solids created using the freeze drying method, which involves solvent sublimation. Crystals that develop in the pores during freezing might create tension and damage the polymeric structure of the gels. The most reliable approach to make aerogels is SCF-assisted drying. This is attributed to low surface tension and high diffusivity of CO_2. As a result, SCF is the most effective method for maintaining the physiochemical characteristics of aerogels and preventing pores from collapsing. During nitrogen adsorption-desorption studies, xerogels did not show much porosity compared to aerogels. However, in cryogels, partial network collapse causes a significant decrease in the porosity [13].

5.3.8 Co-suspension Delivery Technology

The "co-suspension delivery technology" for MDIs consists of co-suspending drug crystals with porous phospholipid particles. The porous particles are composed of two substances that exist in the lungs: phospholipids, a lung surfactant, and calcium chloride. Therefore, porous phospholipid particles are biocompatible, well tolerated, readily soluble, and consequently release drugs into the mucosal surface of the airways when inhaled. The porous particles have a nanostructure with an amphiphilic-corrugated surface, which is advantageous for aerosolization. The association between the drug and porous particles is strong and they produce a stable suspension with hydrofluoroalkane propellant. Co-suspension technology have good potential to deliver multiple drugs from the same device with greater homogeneity and stability [42].

The Aerosphere inhaler's co-suspension delivery system was designed to work with a variety of drug combinations (monotherapies, dual therapies, and triple therapies) at a wide range of doses for inhaled therapy (Fig. 5.9). This approach has been developed for dual and triple fixed-dose combination (FDC) inhalers with comparable FPFs, MMADs, and particle size distributions for each component [42]. A triple combination of inhaled corticosteroid (ICS)/ long-acting muscarinic antagonist (LAMA)/long-acting β2-agonist (LABA), budesonide/glycopyrrolate/formoterol fumarate (BGF), and a dual combination of LAMA and LABA, glycopyrrolate/formoterol fumarate (GFF) MDI (Bevespi Aerosphere®), both have been developed using co-suspension delivery technology [42].

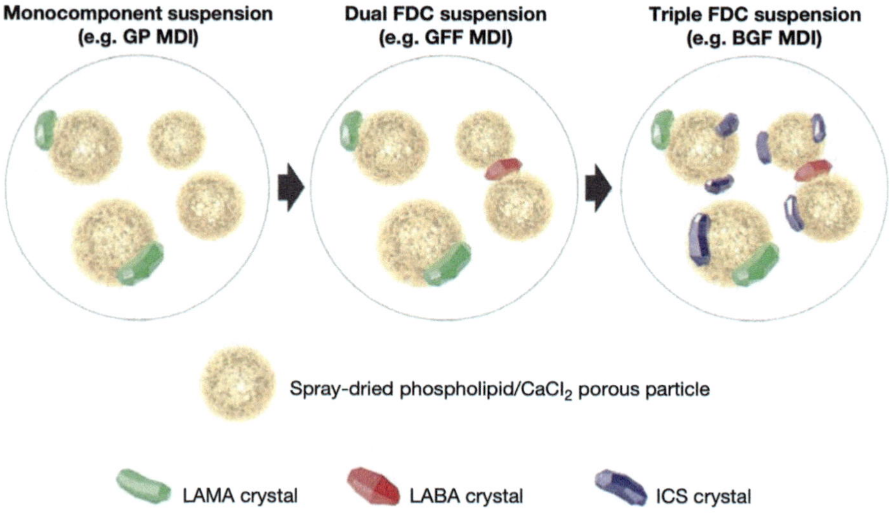

Fig. 5.9 Aerosphere inhaler's co-suspension delivery system Abbreviations: *BGF* budesonide/glycopyrrolate/formoterol fumarate, *$CaCl_2$* calcium chloride, *FDC* fixed-dose combination, *GFF* glycopyrrolate/formoterol fumarate, *GP* glycopyrrolate, *ICS* inhaled corticosteroid, *LABA* long-acting β2-agonist, *LAMA* long-acting muscarinic antagonist, *MDI* metered dose inhaler (Reprinted from [43] with permission by Taylor & Francis Ltd.)

5.4 Porous Particles in Metered Dose Inhaler

The porous particles are also used in the MDI. The PulmoSphere™ technique was used to produce the hollow porous particles containing cromolyn sodium, albuterol sulfate, or formoterol fumarate particles. Hand-filled aluminum or glass canisters were used to hold the tailored drug particles along with HFA-134a. Porosity, sponge-like appearance, and average tap density of 0.06–0.19 g cm^{-1} are all characteristics of PulmoSphere™ powder. All active medications in the PulmoSphere™ distributes in HFA-134a with remarkably little sedimentation over period of time (Fig. 5.10) [21]. Additionally, there were no variations whether the MDI was activated immediately following shaking or 30 seconds later. After 3 months of storage under accelerated circumstances (40 °C 75%RH), PulmoSphere™ dose uniformity was within 20% of the label claim and continued to meet specifications. With MMAD of 3.4–3.6 μm and geometric standard deviation (GSD) of 1.7–2.0, the FPF of the cromolyn, albuterol, and formoterol PulmoSphere™ MDIs were 68.1, 66.9, and 68.7%, respectively. Hollow particles dispersed in HFA propellant effectively. The propellant penetration inside hollow porous particles enhances suspension stability in two ways. First, the low difference between the density of continuous phase and fluid-filled particles helps to reduce the probability of sedimentation. Second, van der Waals' attractive forces are reduced due to spheronization of particles and accumulation of fluid within the pores. Particle aggregation is reduced by the decrease in interparticle attractive forces [21]. During in vitro analysis using cascade impactor PulmoSphere™ MDI showed FPF and MMAD of 27% and 3.61 μm, respectively. During in vivo study using radiolabeling method (^{99m}Tc-pertechnetate) PulmoSphere™ MDI presented twofold higher drug deposition compared to micronized drugs containing MDI (28.5 vs 14.5%). This study showed that porous particles loaded MDIs are more efficient than using a traditional MDI formulation [20].

Rennard et al. (2014) evaluated the pharmacokinetics and safety of hollow, porous particles containing glycopyrronium formulated as HFA-MDI. Glycopyrronium is a muscarinic medication that has been tested for efficacy and safety in chronic obstructive pulmonary disease (COPD) patients. The mean changes in FEV_1, FEV_1 area under the curve (AUC) for 0–12 h (FEV_1 AUC_{0-12}), FEV_1 AUC_{0-24}, and FEV_1 AUC_{12-24} were seen in glycopyrronium porous particle MDI (Fig. 5.11). In research participants with mild to moderate COPD, glycopyrronium MDI showed clinically significant greater bronchodilation compared to placebo MDI [44].

The micro and nanoporous particles are physically and chemically stable in the MDIs over a long period. When the discharged particles from MDIs are exposed to the warmth and humidity of the airways, the porous particles absorb moisture from lungs. As a result, the particles completely disintegrate and merge with the mucosal airways. The technical difficulties are incorporating multiple drugs into a single MDI while maintaining homogeneous suspension, physicochemical stability, and uniform drug delivery with low drug-drug interactions. The challenges to overcome co-suspension delivery are summarized in Table 5.2 [45].

Fig. 5.10 Time-lapse photos of albuterol sulfate MDI suspensions of Proventil HFA formulation and hollow porous particles in HFA-134a (Reprint from [21] with permission by Springer Nature)

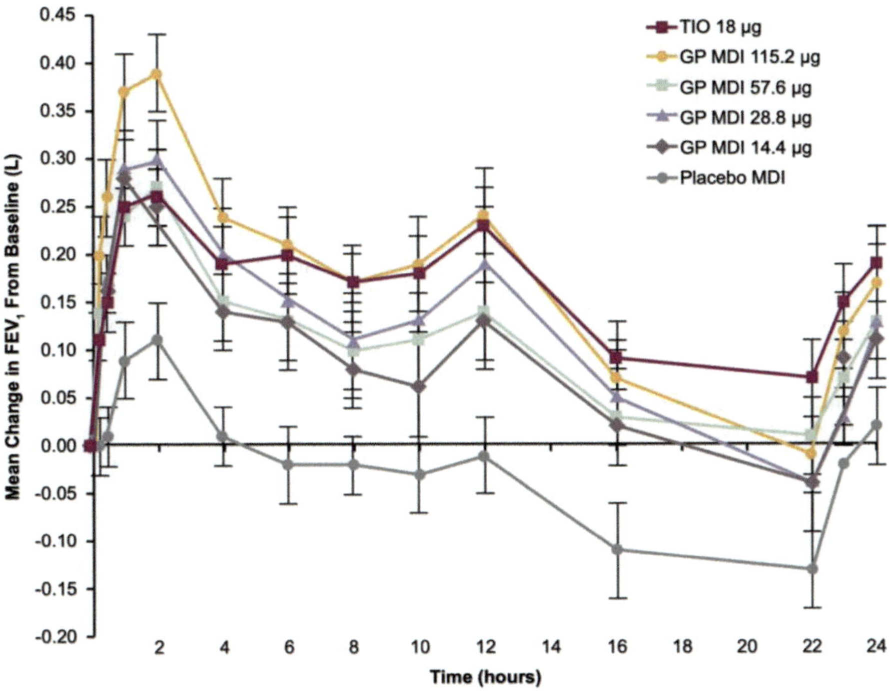

FEV$_1$=forced expiratory volume in 1 second; GP MDI=glycopyrronium metered-dose inhaler; Tio=tiotropium.

Fig. 5.11 Mean change from baseline in FEV$_1$ over 24 h during treatment (Reprint from [44] with permission by Springer Nature)

Vehring et al. (2012) created co-suspensions of drug microcrystals and customized microparticles for the uniform administration of therapeutically active compounds (glycopyrrolate, formoterol fumarate dihydrate, and mometasone furoate). Firstly, air-jet milling was employed to generate drug microcrystals, whereas spray drying was used to generate phospholipid/calcium chloride porous microparticles. Mometasone furoate, glycopyrrolate, formoterol fumarate dihydrate, and spray-dried porous microparticles all had appropriate mean particle sizes for pulmonary drug delivery (1.0–2.3 µm). Additionally, the surfaces of the designed microparticles had corrugations, and the surfaces of the amphiphilic particles produced a layer with a thickness of around 100 nm. Surface corrugation and the amphiphilic characteristic of surfaces revealed the reduced cohesiveness between microparticles, which greatly lowered the propellant's rate of aggregation. Porous microparticles and drug microcrystals were suspended in HFA-134a and packed into 14 mL aluminum canisters capped with fluorinated ethylene polymer and 50 µL valves. The corrugated surface of the microparticles and the shape-fitting contact feature of the drug microcrystals resulted in improved adhesion. When subjected to aerodynamic testing, co-suspensions showed outstanding colloidal stability and homogeneity, and agglomerates remained unbroken after activation. Achievable aerodynamic

Table 5.2 Challenges of MDI inhalers overcome by co-suspension delivery technology

Challenges	Results with co-suspension delivery technology
Combining multiple drugs into a single inhaler without altering the delivery of each drug	• The drug crystals are associated with the porous particles to ensure uniform drug delivery. • Avoids drug–drug interaction within a single inhaler device, retaining drug delivery characteristics.
Poor suspension stability	• Creates a uniform suspension of multiple drugs. • Porous particle–drug mixtures remain floating. Gentle shaking is often sufficient to redisperse the co-suspension into a uniform suspension. • Irreversible binding of drug to porous particles creates stable colloids, where the drug crystal attached to the particles remains in suspension, rather than the need for individual components to go into suspension.
Formation of large drug–crystals agglomerates	• The engineered phospholipid particles prevent drug-drug agglomeration by keeping them apart from each other. • Ensures consistent dose delivery and retains the physicochemical properties of each drug. • Gentle shaking redisperses co-suspension flocculates.
Inconsistent and irreproducible drug delivery over time	• Reduced drug–drug interaction in co-suspension leads to consistent dose delivery and aerosolization over time, regardless of nature and number of drug types, and quantity of each drug. • Provides broad applicability.
High spray velocity causes high oropharyngeal deposition	• Co-suspension utilizes the low spray-velocity HFA propellant. • Low-density porous particles tend to form a more uniform cloud, with particle size suitable for lung delivery.

(Reprint from [45] with permission by Elsevier B.V.)

results in terms of dosage uniformity and FPF primarily reflect the properties of the produced porous microparticles [46].

Taylor et al. (2018) prepared a FDC of LAMA and LABA (glycopyrronium/formoterol dihydrate) using co-suspension delivery technology and evaluated the pulmonary, extrathoracic, and regional lung deposition patterns using gamma scintigraphy. Co-suspension delivery technique uses a strong non-specific interaction between micronized drug crystals and phospholipid-based porous particles suspended in the MDI propellant HFA 134a to generate uniform suspensions and reliable dose delivery. In a randomized, placebo-controlled Phase I study involving 10 healthy male subjects, 38.4% of the radiolabeled glycopyrronium/formoterol dihydrate emitted dose was deposited in the lungs, whereas 61.4% was deposited in the oropharynx and stomach (Fig. 5.12). The exhalation filter identified not more than 0.25% of the discharged dosage. Standardized central/peripheral ratios of 1.85 and normalized outer/inner ratios of 0.57 revealed that the formulations of the developed co-suspension delivery method were dispersed throughout the airways [47].

Co-suspension technology, which combines drug crystals with porous phospholipid particles to create stable, homogenous solutions that dissolve when they enter

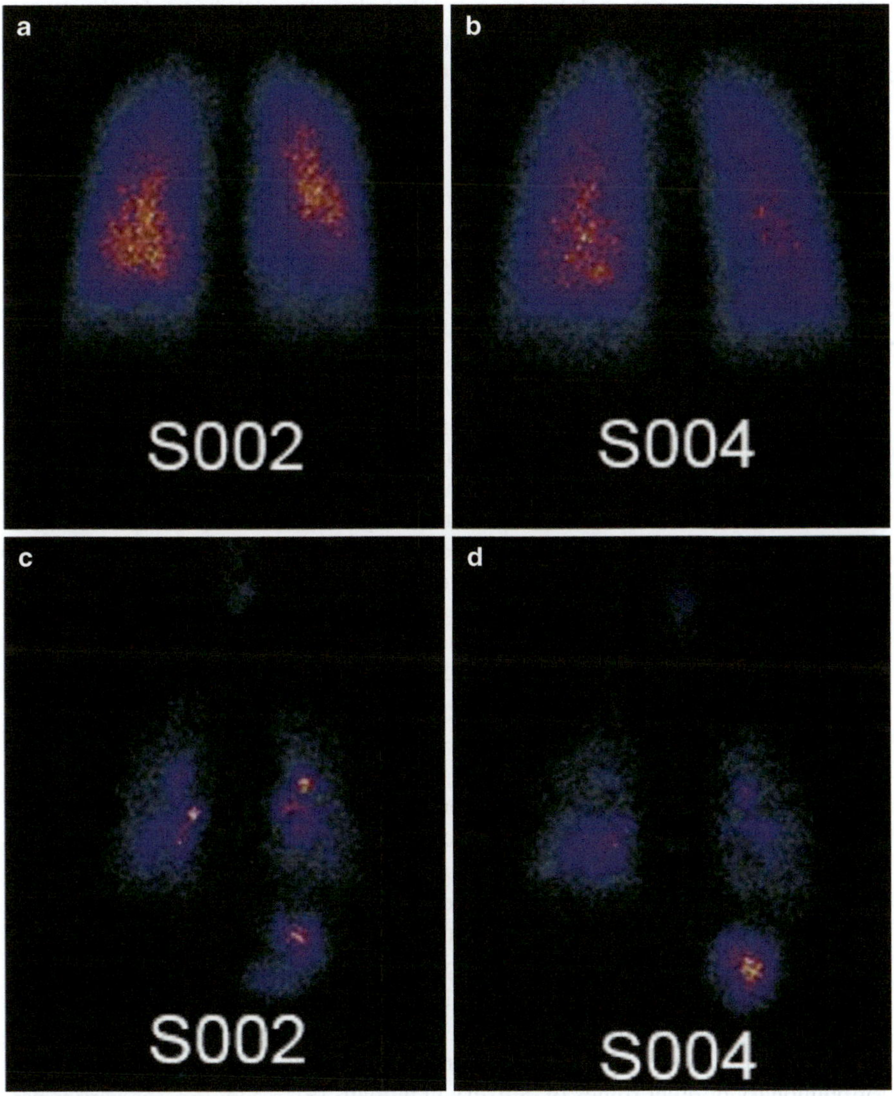

Fig. 5.12 Gamma scintigraphy of lung deposition of glycopyrronium/formoterol dihydrate pMDI and placebo pMDI (**a**, **b**) ^{81m}Kr ventilation, (**c**) placebo pMDI and (**d**) glycopyrronium/formoterol dihydrate pMDI (Reprint from [47] with permission by Elsevier B.V.)

the airways, was used by Aerosphere™ technology to develop glycopyrrolate/formoterol fumarate pMDI and budesonide/glycopyrrolate/formoterol fumarate MDI. Even with simulated patient handling errors, the Aerosphere™ inhaler consistently delivers dosages. Combination drugs provided with this method demonstrated a consistent FPF and the appropriate particle size distribution for delivery through the central and peripheral airways. Using gamma scintigraphy and functional respiratory imaging, research has revealed that glycopyrrolate/formoterol fumarate pMDI is successfully deposited in the central and peripheral airways and exhibited clinically significant changes in airway volume and resistance throughout the lung. Studies suggested that the co-suspension delivery system used in formulation of the Aerosphere™ inhaler may have advantages over conventional formulations, including ideal particle size, FPF, and effective delivery to the central and peripheral airways. It may also deliver multiple components consistently under a variety of patient handling conditions. The clinical advantages of these technological advancements in MDI drug administration may be further supported by research conducted in the future [42].

In a randomized study, 34 patients with moderate-to-severe chronic obstructive pulmonary disease (COPD) were administered formoterol fumarate in a porous particle pMDI by Quinn et al. (2014). This work represents the clinical investigation of formoterol fumarate pMDI prepared using a novel lipid-based porous particle engineering approach. It is predicted that this pMDI will lead to better lung function FEV_1 during a 12 h period when compared to a placebo. The in vivo testing has proven that the active medication contained as pMDIs porous particles are efficacious, and aerosol characteristics of the porous particles are unaffected by either the release of medication from the actuator or the evaporation of propellant from the porous particles [48].

5.5 Application of Porous Particles in Pulmonary Drug Delivery

Large porous particles are beneficial for rapid administration of relatively large doses of drugs into the lungs or systemic circulation [15]. Additionally, large porous particles can serve as excellent carriers for the prolonged drug delivery to the lungs and solid form is more stable at room temperature. To achieve the required drug dose, drug stability, or drug release kinetics, the drug and excipients are dispersed in the porous particle matrix in specific proportions. The drug fractions in the prepared formulations range from less than 1% by weight to close to 100% by weight. The excipient used in porous particles for pulmonary delivery is an important factor. The selected excipients must be either compatible with the human lung or approved for pulmonary or other routes of delivery in humans [9].

5.5.1 Local Treatment of Respiratory Diseases

Inhaled porous microspheres with low mass density (<0.4 g/cm ^{3}) and large diameters result in an acceptable aerodynamic diameter (1–3 μm) for efficient drug delivery into deep lungs with high respiratory fraction of inhaled dose. Besides, large porous particles enhance deep lung deposition and can bypass macrophage uptake. The deposited particles have the potential to release glycopyrrolate/formoterol fumarate into the pulmonary airways [49]. To generate the large porous microparticles, osmogens or porogens such as pluronic, cyclodextrins, hydrogen peroxide, or catalase (gas bubbles) and gas foaming (ammonium bicarbonate) are utilized. Both osmogens and porogens form pores inside the polymeric matrix [50]. Localized chemotherapy refers to the administration of anticancer medications directly to the damaged organs, which can result in increased tumor concentrations compared to non-target areas [51–57]. Inhaled chemotherapy has proven effective against lung malignancies. A significant concentration of inhaled medication is frequently observed in lymph nodes through lymphatic circulation. Consequently, inhaled chemotherapy may also be effective for treating lung cancer that has spread to the lymph nodes. In addition, medications that are absorbed into the lymphatic circulation can redistribute in peripheral airways, giving access to parts of the lungs that are otherwise inaccessible [51].

Electrospraying is a technique that can be used to produce fine particles that are widely used for biomedical applications, particularly drug encapsulation. The size and morphology of electrosprayed particles can be controlled to a certain extent by the factors related to polymer solutions [e.g., concentration, shear viscosity, polymer molecular weight, and solvent] and the electrospraying process [e.g., flow rate, electric potential difference, distance between the tip of the needle and the collector] [58]. Electrospray porous microparticles (EPMs) loaded with oridonin utilizing PLGA showed high in vitro lung deposition. The FPF of the EPMs loaded with oridonin was 19.1%, which was greater than the FPF (12%) of non-porous solid PLGA microspheres [49]. When oridonin-loaded EPMs, oridonin powder, and gemcitabine were used to treat lung cancer in rat models, the oridonin-loaded EPMs treated group had virtually no nodes, whereas the oridonin powder or gemcitabine-treated group had fewer tumor nodes than the control group (Fig. 5.13). As a result, the therapeutic efficacy of oridonin-loaded EPMs may be superior to gemcitabine, the first-line anticancer drug in clinical practice. Oridonin-loaded electrospray porous microparticles (EPMs) have a high anti-lung cancer effect due to their high lung deposition, prompt release of oridonin, and avoidance of macrophage ingestion. During pharmacodynamic study using rat cancer model, the pulmonary delivery of oridonin-loaded PLGA EPMs exhibited a strong anticancer effect as verified by computed tomography method [49].

Spray drying was employed to produce microporous spheres containing N-[(2-hydroxyl)-propyl-3-trimethyl ammonium] chitosan chloride (HTCC) and HP-β-CD for the solubilization of the hydrophobic cyclosporine A. Following pulmonary administration, the efficiency of the microporous spheres was evaluated in vivo in an asthmatic rat model. These microspheres lowered the amount of total protein, lactate dehydrogenase (LDH), inflammatory cells, interleukin (IL)-4, and tumor necrosis factor (TNF)-α cytokines as compared to chicken ovalbumin (OVA)-induced control group. Microporous spheres reduced OVA-induced inflammatory infiltration and mucus hypersecretion in an asthmatic rat, according to lung histological tests (Fig. 5.14). Based on these findings, HTCC microporous spheres are suitable drug carriers for pulmonary drug delivery [25].

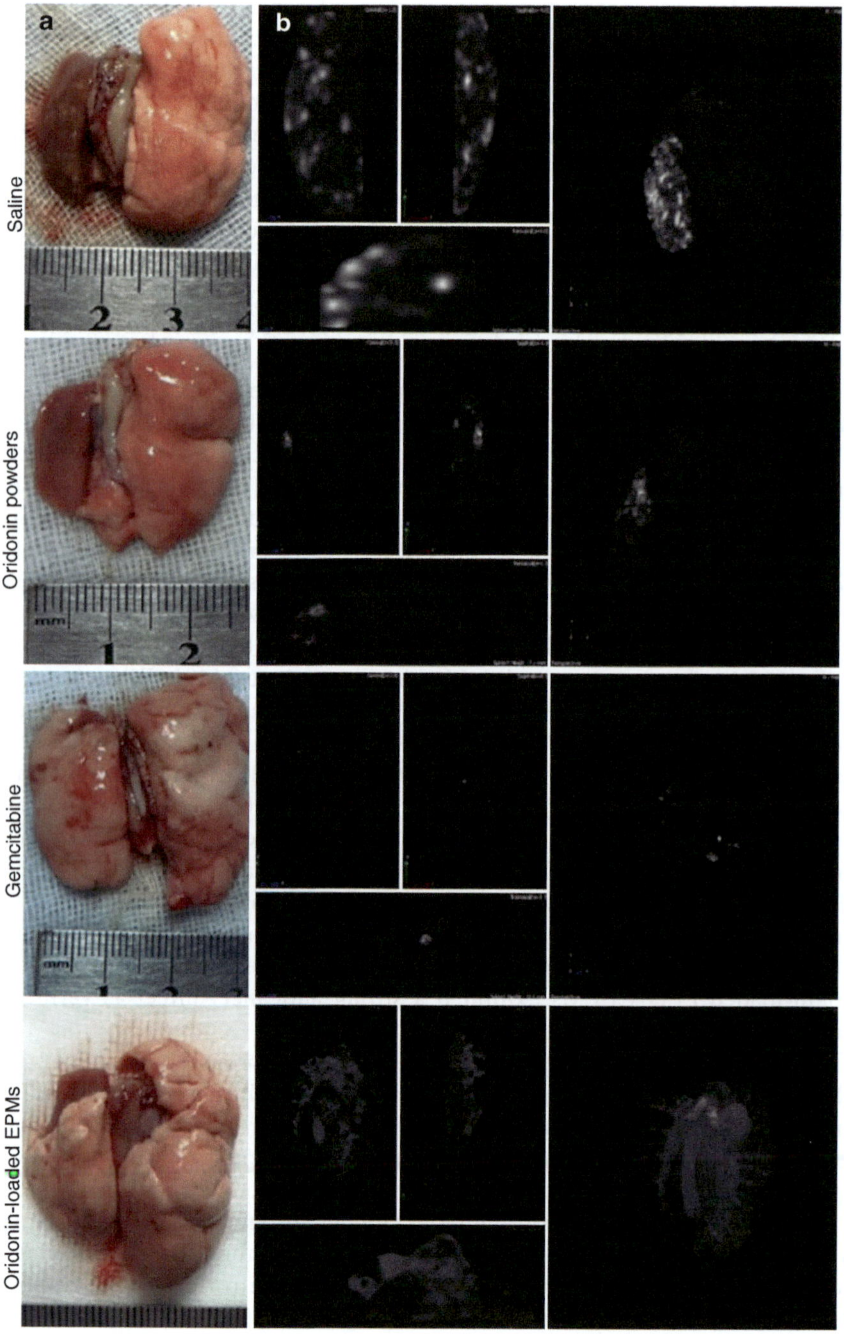

Fig. 5.13 Lung appearances (**a**) and computed tomography (CT) images (**b**) of the primary lung cancer rats treated with saline, oridonin, gemcitabine, and oridonin-loaded EPMs via pulmonary delivery. The light points in the CT images indicate tumor nodes. (Reprint from [49] with permission by American Chemical Society by Creative Commons Attribution)

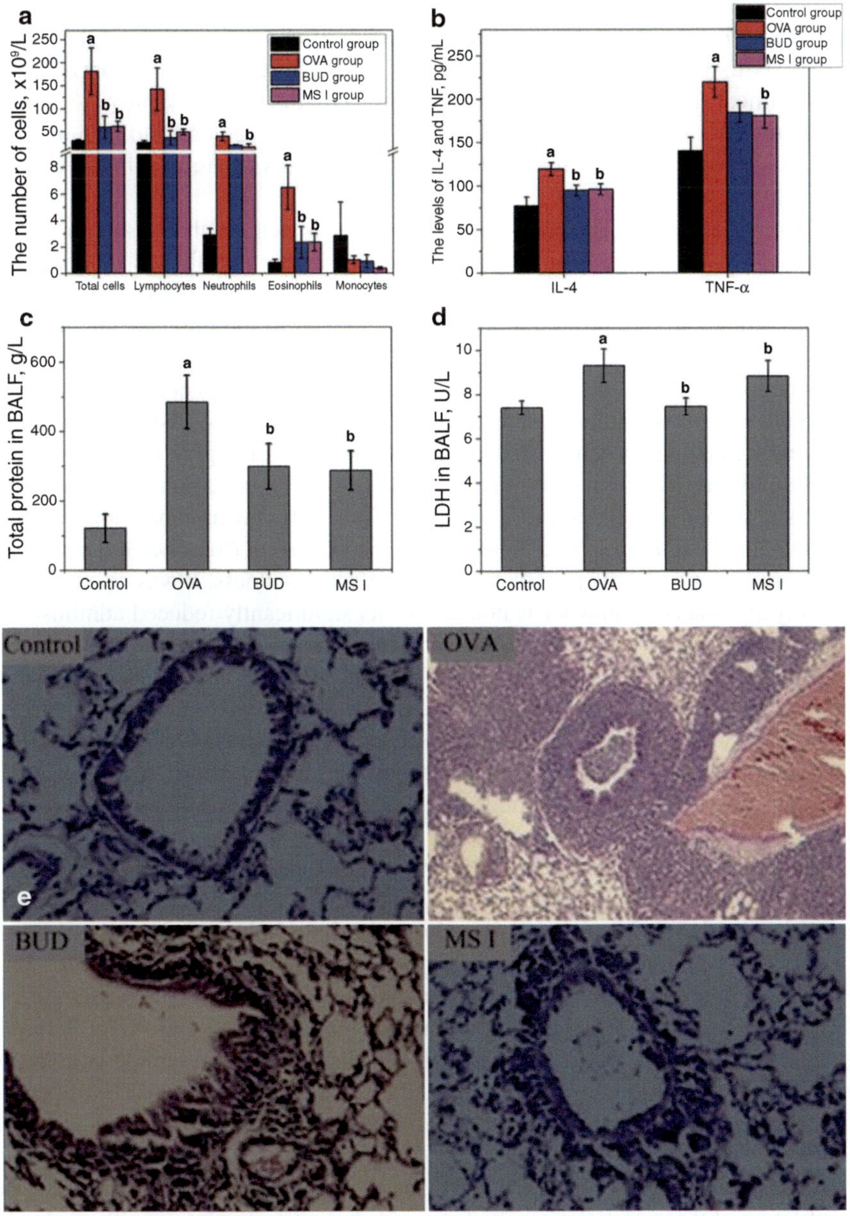

Fig. 5.14 In vivo efficacy in the asthmatic rat model includes effects of microsphere I (MS I) on OVA-induced total and differential leukocytes in bronchoalveolar lavage fluid (BALF) (**a**), effects of MS I on OVA-induced levels of IL-4 and TNF-α (**b**), effects of MS I on OVA-induced levels of total protein (**c**) and LDH (**d**) in BALF, and effects of lung histopathological changes (**e**). Bars represent standard errors of five replications. "a" compared with the control group ($p < 0.05$) was observed according to a two-tailed paired t-test. "b" compared with the OVA-treated group ($p < 0.05$) was observed according to a two-tailed paired t-test (Reprint from [25] with permission by Taylor & Francis Ltd.)

Patel et al. (2014) examined the use of large porous PEG-PLGA particles for the treatment of asthma in rat asthma models. Low molecular weight heparin (LMWH)-loaded large porous PEG-PLGA particles decreased LDH levels 2.8–3 times. The thickness of the airway wall decreased from 47.37 μm to 21.35 μm after treatment with LMWH PEG-PLGA particles. Additionally, goblet cell hyperplasia was diminished in asthmatic rats treated with LMWH particles [59].

Various pathogenic bacteria cause pulmonary infections that result in variable degrees of lung damage. The primary defense reaction to pathogen-induced tissue damage is inflammation. The general process of inflammation consists of metamorphosis, exudation, and proliferation, which correlate to the onset of inflammation, tissue injury, and repair, and involve many cellular pathway alterations and signal transduction processes [60]. Among numerous delivery systems or tactics, pulmonary drug delivery offers distinct benefits, such as no first-pass effect, high bioavailability, allowing treatments to deliver directly to lung lesions. Emerging research indicates that various antiviral pulmonary inhalations are currently undergoing clinical investigations in response to the ongoing pandemic [60]. Tobramycin inhalation powder as porous particles (PulmoSphere™) is used to treat persistent *Pseudomonas aeruginosa* infection in cystic fibrosis patients with lung infectious illness. It was discovered that tobramycin inhalation powder is notable for its significantly reduced administration time, mobility, and compliance. Tobramycin inhalation powder has been developed as a therapeutic alternative to tobramycin inhalation solution. The results demonstrated that tobramycin inhalation powder is non-inferior with regard to the primary endpoint of pulmonary function improvement and secondary endpoints, which include pulmonary exacerbations. Tobramycin inhalation powder has been shown to be delivered in about a fourth of the time of tobramycin inhalation solution, resulting in a significant reduction in administration time [16, 17].

As an alternative technique for achieving higher drug concentrations adjacent to granulomatous lesions, pulmonary administration of antitubercular medicines has been proposed. Targeting drugs directly to the lung help to reduce the dose and probability of drug resistance. The pyrazinamide large porous particles were developed for pulmonary delivery. The optimized powder possesses a median size of 5.1 μm, ED of 99.3%, FPF about 30% and tap density of 0.09 g/cm^3. The estimated d_{ac} is around 1.75 μm. After intratrachcal insufflation of pyrazinamide large porous particles (4.22 mg/kg) into rats, plasma and Epithelial lining fluid (ELF) of pyrazinamide were measured and compared to intravenous administration of a pyrazinamide solution (5.82 mg/kg). The in vivo pharmacokinetic study of pyrazinamide-loaded large porous particles in rats demonstrated that intratracheal insufflation of these particles resulted in rapid absorption in plasma with an absolute bioavailability of 66%. This demonstrates that pyrazinamide readily passes the pulmonary barrier to reach the systemic circulation and pyrazinamide large porous particles disintegrate rapidly upon deposition. After intratracheal delivery, pyrazinamide concentrations in ELF were 1.28 times higher than intravenous treatment, and the ratio of ELF concentrations to plasma concentrations was two times higher [61]. A mouse model was used to develop mesoporous silica particles as carriers for anti-inflammatory medicines (dexamethasone) in the treatment of airway inflammation. Large mesoporous particles (1 μm) and small mesoporous particles (200 nm) were 100% weight-loaded with dexamethasone. The results from mice exposed to melphalan showed that mesoporous silica particles are feasible drug carriers for the pulmonary administration of corticosteroids in the treatment of chemically induced lung injury [28].

Large porous particles formulations demonstrated highly efficient delivery to the lung deposition of 40–70% of the nominal dose while dramatically reducing dose variability to 10–20% [18, 62]. The total lung deposition and interpatient variability of commercially available inhaled corticosteroid formulations were compared to PulmoSphere™ formulation [18]. The commercial formulation including Diskus® (GlaxoSmithKline) consists of lactose-blend formulation of fluticasone propionate while Turbuhaler® (AstraZeneca) is spheronized particles of budesonide formulation was compared with the PulmoSphere™. The PulmoSphere™ formulation is made up of spray dried budesonide particles that are given by the capsule-based Eclipse® DPI. The PulmoSphere™ formulation produced a higher lung deposition (58% w/w) when compared to spheronized particles (Turbuhaler®) (36% w/w) and lactose mix (Diskus®) (12% w/w). Furthermore, interpatient variance is reduced from around 40% to 12% with the PulmoSphere™ formulation [18].

Metal-organic frameworks (MOFs) are a new class of porous materials composed of metal ions and organic linkers [62, 63]. The morphology of MOF was cubic structure containing large spherical pores of 17 Å in diameter. The pores of MOFs are self-assembled by coordination bonding between metal ions and organic links without using any porogens [64]. Cyclodextrin-based metal-organic framework (CD-MOF) composed of potassium ions and γ-cyclodextrin was biocompatible. The homogenous nanoporous structure of CD-MOF showed excellent aerosolization performance (MMAD = 1.82 μm and FPF of 57.99%) as compared to commercial inhalation carrier (lactose, InhaLac® 230). Fluorescence imaging showed a high deposition rate of CD-MOF in the lungs [64].

Carrier can facilitate the porous particles in improved drug delivery. Nanoporous mannitol was prepared by spray drying with ammonium carbonate. This technique produced nanoporous mannitol with a specific surface area of 42.29 m^2/g and pore diameter of 5.90 nm. The concentration of ammonium carbonate affected the pore size of the particles. After blending nanoporous mannitol with micronized budesonide (M_4), the FPF is 38% with 2.16-fold higher than non-porous mannitol (M_0). The in vivo pharmacokinetics studies in Sprague-Dawley rats in comparison with Pulmicort™ are shown in Fig. 5.15. The peak drug concentration (C_{max}) of lung budesonide concentration for M_4 was twofold compared with M_0 and 1.6-fold compared with Pulmicort™. The highest $AUC_{0\text{-}8h}$ was found in M_4, which showed a twofold higher than the Pulmicort™ and a 3.2-fold higher than M_0. The highest drug delivery efficiency was obtained by using nanoporous particles as carriers. A comparatively tiny contact area between the non-porous, smooth-surfaced particles and the drug led to low drug-interparticulate force as demonstrated in Fig. 5.15c [34].

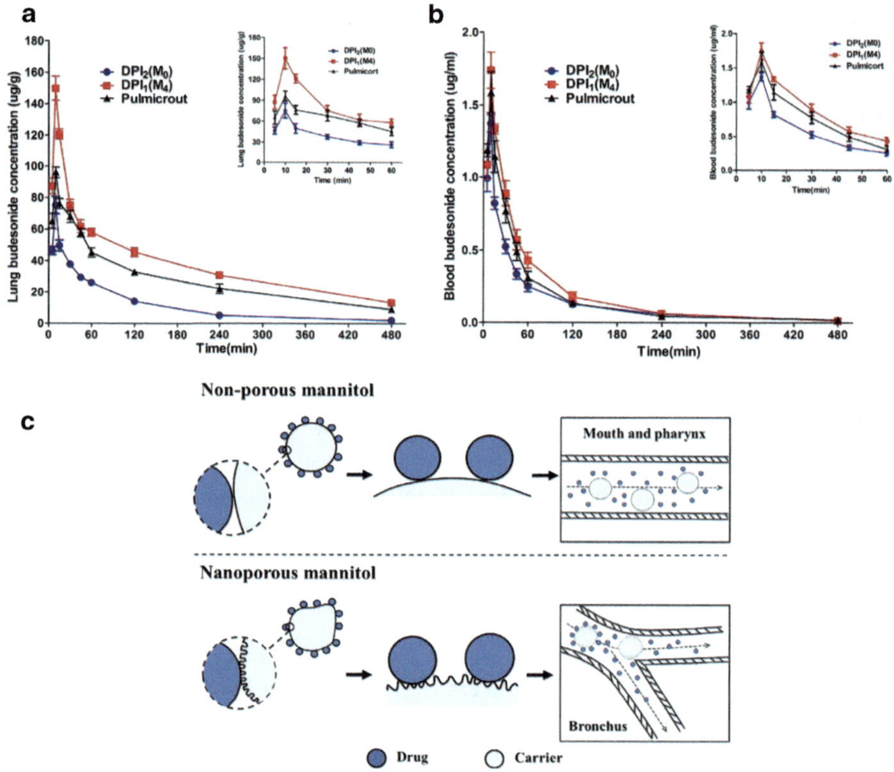

Fig. 5.15 Budesonide concentration-time curve in the lung (**a**) and plasma (**b**) after administration of different DPI formulations and the schematic diagram to illustrate the influence of the nanoporous and non-porous carrier on the DPI aerosolization performance (**c**). Abbreviations: DPI_1 (M_4) is the budesonide DPI formulation with nanoporous mannitol carrier produced by spray drying with the outlet temperature maintained at 75 °C and the ratio of mannitol and ammonium carbonate set at 3:1; DPI_2 (M_0) is the budesonide DPI control formulation with mannitol carrier alone without ammonium carbonate produced by spray drying with the outlet temperature maintained at 75 °C (Reprint from [34] with permission by Springer Nature by Creative Commons Attribution)

5.5.2 Systemic Treatment Via Pulmonary Delivery

The use of the lungs as a route for drugs to enter the systemic circulation has advanced significantly over the past decade, and drug administration via inhalation has emerged as a crucial component in the management of respiratory and systemic diseases. Currently, there are numerous examples in which large porous particles are used to treat systemic illnesses. A few interesting case studies are discussed below.

5.5.2.1 Thrombosis

Thrombosis can be divided into arterial thrombosis and venous thrombosis which occurs when a blood clot forms within a blood vessel and obstructs blood flow via the vasculature [23, 65, 66]. Immunosuppressive medicines, such as corticosteroids, are used to diminish the immune response to blood vessel inflammation, while anticoagulants, such as heparin and warfarin, are used to dissolve clots. After an initial dose of LMWH or unfractionated heparin (UFH), the vitamin K antagonist warfarin was administered. Because of its polydispersity and polyanionic nature, heparin is frequently administered intravenously or subcutaneously to avoid interpatient variance. However, due to local pain at the injection site and ulceration associated following the intravenous and subcutaneous heparin treatment, patient adherence is low. Oral administration of warfarin prevents a number of these side effects; however, the dose-response is inconsistent. Due to these limitations, an alternate, noninvasive method of delivering these medicines into the systemic circulation is required.

The pharmacokinetic and pharmacodynamic effects of pulmonary administration of heparin-loaded porous particles were investigated using both in vitro and in vivo models. The d_{ae} of LMWH large porous particles is less than 5 μm with a mass density of 0.45 g/cm^3. The LMWH large porous particles showed faster onset of heparin (time to peak drug concentration (t_{max} of 40 min vs. 2–4 h) for subcutaneous administration), and a comparable C_{max} of 0.9 g/mL vs. 1.1 g/mL. Repeated inhaled heparin treatment resulted in stable pharmacokinetics in terms of t_{max} and C_{max} with no noticeable adverse effects. Despite the fact that the results are favorable, the safety of using drugs that affect tight junctions is an ongoing research, because epithelial cells' tight junction alteration has been linked to progressive lung disease [23, 67]. Patel et al. (2014) studied the use of heparin-loaded large porous PEG-PLGA particles for enhanced pulmonary heparin administration. The particles exhibited a significant drug uptake, enhanced drug release, and decreased absorption by alveolar macrophages. The large porous particles had a 19-h elimination half-life compared to 4 h for the subcutaneous administration, allowing for less frequent heparin administration over a longer period. Overall, these investigations showed that large porous particles could transport drugs through the alveolar membrane into the systemic circulation and attain pharmacokinetic parameters that are comparable to or better than those of conventional delivery systems [59].

5.5.2.2 Tuberculosis

Tuberculosis (TB) is an infectious disease caused by *Mycobacterium tuberculosis* that affects the lungs, abdomen, kidneys, bones, and joints [23]. The major objective of the antitubercular therapy is to eradicate the microbes entirely. Oral treatment will take 6–9 months with a combination of antibacterial drugs. Numerous individuals do not adhere to chemotherapy due to lengthy duration, high cost, and severe side effects. Repetitive administration and long-term drug course usually lead to resistance. Since the majority of first-line anti-TB drugs are ineffective against *M. tuberculosis* strains, second-line anti-TB drugs with limited safety, efficacy, and significant toxicity are introduced [68]. Antimicrobials administered via the pulmonary route may be preferable for treating TB due to high concentrations at the target site. Direct pulmonary delivery has the potential to reduce the dose and improve the onset of action, resulting in lower costs, shorter treatment times, and leading to increased patient compliance. Moreover, water-soluble drugs used to treat tuberculosis are easily dissolved in the alveolar fluid facilitating the killing of *M. tuberculosis* strains outside the alveolar macrophage. In guinea pigs, the pharmacokinetic profiles of inhaled rifampicin are large porous particles compared with intravenous and oral administration. Rifampicin generated from large porous particles demonstrated higher plasma concentration than that oral rifampicin and intravenous administration [23, 69]. The local and systemic concentrations of para-aminosalicylic acid (PAS) increased following administration of lower dosages of PAS-large porous particles compared to oral administration of PAS. Although the pulmonary dose was lower, experiments in guinea pigs utilizing the second-line medication capreomycin-loaded large porous particles and capreomycin delivered intravenous and intramuscular routes indicated comparable plasma concentrations 2–6 h after treatment for both formulations. The pulmonary route of capreomycin elimination was slower than the intravenous and intramuscular methods. Antituberculosis medications have been found to maintain appropriate local and systemic concentrations to combat the infection in both pulmonary and extrapulmonary organs when administered via pulmonary route [23].

Pham et al. (2015) developed pyrazinamide (PZA) loaded large porous particles (PZA-LPPs) and evaluated plasma and ELF in rats after intratracheal insufflation. In vivo pharmacokinetic evaluation of PZA-LPPs in rats demonstrated that intratracheal insufflation of PZA-LPPs results in rapid absorption in plasma with a bioavailability of 66%. This reveals that PZA-LPPs degrade rapidly upon deposition, and that PZA readily crosses the lung barrier into the systemic circulation. After intratracheal treatment, PZA concentrations in ELF were 1.28-fold higher than after intravenous administration, and the ratio of ELF to plasma concentrations was twofold higher. Although these improvements are small, lung delivery of PZA appears to be an attractive alternative to oral delivery of the drug, and its efficacy against *M. tuberculosis* should be examined in an infected animal model [61]. In addition, the formulation of PZA-LPPs has advantage over crystal growth by using suitable excipients such as hyaluronic acid in combination with DPPC [38].

Rifampicin and new inhaled carrier phytoglycogen (a hydrophilic polymeric polysaccharide) produced porous particles by spray drying techniques. The morphology of particles was wrinkled structures with D_{50} of 58 nm and loading capacity of Rifampicin >95%. This new carrier demonstrated safety on A549, Calu-3, and RAW264.7 cells without cytotoxic effects in rats. This approach demonstrated an enhancement of phagocytic activity of macrophage-like cells capable of reaching the alveoli [36].

5.5.2.3 Human Growth Hormone Deficiency and Luteinizing Hormone-Releasing Hormone

The pituitary glands release a 191-amino acid protein known as human growth hormone (GH). It is used to treat Turner syndrome, adult growth hormone deficit, pediatric pituitary dwarfism, and chronic renal insufficiency. GH subcutaneous or intramuscular injections are currently used as treatments for GH deficiency [23, 70].

The short GH half-life, renal toxicity, low patient compliance due to frequent, uncomfortable injections, and high healthcare expenses are some of the disadvantages of this therapy [23, 70]. As a result, an extended-release, noninvasive formulation could improve the efficacy of GH therapy. Pre-clinical studies and clinical trials have been conducted on aerosol formulations of GH, although none has yet been commercialized, and only a small number of formulations with sustained release have been tried. Inhaled large porous particles may serve as depots for the steady release of GH across the alveolar membrane and into the bloodstream. Numerous studies have shown GH-loaded porous, biodegradable polymeric microspheres allow for a sustained release of GH over a 1-month period [23, 71]. The efficacy of GH dry powder with porous particles in pulmonary delivery was examined, as well as the impact of excipients in promoting absorption through the alveolar membrane. The tap density was between 0.05 and 0.08 g/cm^3, d_{ae} was between 1.0 and 1.3 μm indicating the powder was suitable for deposition in the deep lung. In vivo pharmacokinetic experiments on rats revealed that dry porous particles containing DPPC as an excipient absorbed GH faster (t_{max} = 23 mins) than subcutaneous injection (t_{max} = 64 mins) and spray-instillation of GH solution (t_{max} = 51 mins). Additionally, larger C_{max} values (234 ng/mL) were produced by GH-loaded porous particles than by subcutaneous injection (55 ng/mL) or spray-instilled GH (46 ng/mL). A phase I clinical trial has been completed, with outcomes including concentration-time profiles of inhaled GH comparable to those of subcutaneous administration and the ability to administer patients' higher doses without causing negative effects. These results are encouraging and give reason to believe that the therapeutic development of large porous hollow particles for systemic administration is feasible [23, 72]. Koushik et al. (2004) [73] developed systemic administration of deslorelin, a synthetic luteinizing hormone-releasing hormone (LHRH) agonist. This study was performed in Sprague-Dawley rats by intratracheal deslorelin solution or powder or deslorelin-HP-β-CD solution vs. deslorelin LPPs. Plasma deslorelin concentrations were 120- and 2.5-fold higher after 7 days in the large porous deslorelin-PLGA particle group compared to deslorelin powder and conventional deslorelin-PLGA particles, respectively. Co-administration of HP-β-CD increased plasma deslorelin concentrations by a factor of 2, 3, and 3 compared to deslorelin solution at 3 h, 1 day, and 3 days, respectively. At the end of 7 days following intratracheal instillation of different deslorelin formulations, the concentrations of deslorelin in BALF and plasma were as follows: deslorelin LPPs > deslorelin small conventional particles > deslorelin-HP-β-CD complexes solution > deslorelin powder > deslorelin solution [73].

5.5.2.4 Cancer

Spray freeze drying is used to make porous nanoparticle-aggregate particles (PNAPs) out of octreotide acetate. The in vivo data showed that PNAP intrapulmonary delivery method for octreotide acetate had a much larger AUC and plasma aspartate aminotransferase higher levels than the subcutaneous injection method [24].

Porous scaffolds frequently have an extracellular matrix-like structure, which promotes cell adhesion and growth. The porosity of particles can be improved or modified depending on the porogen introduced into them. The double-emulsion process is used to create porous PLGA microparticles from gelatin, sodium bicarbonate, or poly (N-isopropylacrylamide)-based particles (PMPs). Compared to lung cancer cells (A549 lung adenocarcinoma cells) grown on porous particles exhibit higher drug resistance, which suggests that porous microparticle substrates are potential for in vitro tumor or tissue development [74].

5.5.2.5 Diabetes

Lin et al. (2018) [6] developed insulin-loaded porous microspheres for the treatment of diabetes via inhalation route. The MMAD of these microparticles is smaller than 5 μm, the d_g was 15.62 μm indicating that the density was less than 1 g/cm^3. The PLLA porous microspheres are less toxic to lung-derived cells compared to the negative control. This means that cell viability and levels of inflammatory mediators like IL-6, IL-8, and TNF-α do not change in a significant way. Insulin-loaded PLLA porous microspheres (insulin doses of 6 IU/kg) are easily deposited in the trachea and bronchi of the upper lobes of the lungs, where they have effects on lowering blood sugar in diabetic rats. The results showed that insulin-loaded PLLA porous microspheres could be alternative to treat diabetes through inhalation [6].

5.5.3 Controlled Release

There is a significant absorptive surface area in the lungs due to the numerous alveoli and the thin alveolar-vascular permeable membrane, which facilitates the ability to deliver drugs to the systemic circulation. Furthermore, direct access to the site of action for lung disorders is provided through pulmonary administration, which enables a rapid response to systemic conditions [75, 76]. In comparison to intravenous or oral delivery, this would enable a lower dose. However, the rapid removal of inhaled medications from lung tissue because of the lungs' rapid absorption to them could limit their local therapeutic efficacy. Therefore, repeated drug doses might be required to maintain their local concentration in the lung tissue within the therapeutic window, which would lead to poor patient adherence to the treatment regimen [76, 77]. The initial pattern of particle deposition in the lungs, which is influenced by the physical and chemical characteristics of the aerosol, such as particle interactions and aerodynamic particle size distribution, as well as physiological factors, such as airflow and breathing patterns, determines the clearance of inhaled particles from the lung [78, 79]. Regulated pulmonary drug delivery in terms of lung deposition, dissolution, and absorption is highly desirable for the treatment of chronic diseases due to the possibility to reduce dosage frequency and improve powder loading [80]. When used locally, drugs can be actively or passively administered into capillary blood vessels, and then withdrawn from lung tissue once they have entered the systemic circulation. A further challenge to the development of controlled pulmonary drug administration is clearance by metabolic breakdown. Despite the fact that pulmonary drug metabolism is significantly lower than that of the liver, the lungs express cytochrome P450 (CYP) families that act as a barrier against xenobiotics. Five of the most prevalent CYP enzymes are CYP1B1, CYP2B6, CYP2E1, and CYP2J2 [81]. Although conventional microparticles are appropriate for inhalation, alveolar macrophages destroy them, resulting in a limited half-life. Using a pore-forming technique, particles with large geometrical dimensions lower alveolar phagocytosis and the potential for particle aggregation, while their low bulk density reduces the aerodynamic diameter and promotes aerosolization effectiveness. The w/o/w emulsion technique with a porosity-inducing porogen is the most frequently used technique to produce large porous particles. Ammonium bicarbonate or cyclodextrins frequently work well as porogens for entrapping water-soluble substances like peptides. During emulsification, ammonium bicarbonate breaks down into ammonia and carbon dioxide, which rapidly escape to create a porous matrix. The single emulsion, high-voltage electrostatic anti-solvent, and supercritical fluid technology are the techniques to fabricate large porous particles with sustained drug release properties [81]. Cyclodextrins are porogens in the production of large porous particles. The purpose of osmotically active additives is to create a pressure difference between the internal and external aqueous phases, which causes water to move into the organic phase and create porosity. Due to its superior safety profile and capacity to enhance macromolecule absorption via the respiratory epithelium, HP-β-CD is the most frequently use than other types of cyclodextrins to produce large porous particles [82]. Gupta and coworker (2011) formulated PLGA microparticles for hydrophobic drugs. Prostaglandin E1 and HP-β-CD were dissolved, and the freeze drying method was used to process the inclusion complexes. Prostaglandin E1 and HP-β-CD encapsulated in LPPs enhanced drug release in simulated interstitial lung fluid and extended the biological half-life in rats. Drug loss may occur during particle hardening leading to inadequate drug entrapment. Additional porogens include Pluronic F68, F127, sodium oleate, and oils. Large porous particles are useful for administering controlled pulmonary medications, however, the technology is still in its immaturity and has many limitations [82].

There is no standardized manufacturing platform with the needed drug encapsulation effectiveness and predictable release patterns for a wide range of medications. In addition, the porosity enhanced the particles' friability and exacerbated the production that adds to the challenges [81]. Little research has shown porous nanoparticle-aggregate particles (PNAP) to be a practical drug delivery system with in vivo testing. Sung et al. (2009) developed rifampicin with PNAP technology by encapsulating rifampicin in PLGA nanoparticles using solvent evaporation method and then spray dried into PNAPs. The FPF was 44.7% with rifampicin loading at 10%. The in vitro release of rifampicin-loaded PNAPs resulted in an initial burst followed by a persistent release for more than 8 h. Similar to this, systemic levels of rifampicin persisted for 6–8 h following the insufflation of PNAPs into guinea pigs. In addition, rifampicin had a longer lasting effect on lung tissue and bronchoalveolar lavage than it did on particles containing free rifampicin. Additional studies also showed that spray freeze drying was a potentially viable strategy for manufacturing the solid dosage form of thermolabile nanoparticles since the created PNAPs had considerably more water redispersibility than those produced by spray drying. According to recent research, intrapulmonary distribution of PNAPs created by spray freeze drying of nanostructured lipid carriers significantly increased the bioavailability of octreotide acetate compared to subcutaneous injection [81, 83]. The biological effectiveness of insulin-loaded PLGA-based large porous particles and non-porous particles were investigated in rats after intratracheal administration. Only the large porous particles produced higher insulin blood levels after 4 h, but insulin secretion remained rather steady over 96 h. Compared to subcutaneous injection, the bioavailability of insulin was 87.5% for large porous particles and 12% for non-porous particles. The presence of exogenous insulin in the systemic blood circulation several days after inhalation demonstrated that large porous particles are extended and non-phagocytosed in the deep lungs. After inhalation immediately, the rat's lung was lavaged and then again 48 h later. There were 30.3% of phagocytic cells retained non-porous particles, while 39.5% did so after 48 h compared to 8.2% and 12.5% for large porous particles after inhalation immediately and after 48 h, respectively. The results supported the theory that large porous particles have a reduced phagocytic clearance [40].

N'Guessan et al. (2018) formulated large porous particles of dexamethasone palmitate (DXP) for lung delivery. During in vitro release study using PBS, DXP large porous particles showed controlled release for up to 21 days. Briefly, DXP large porous particles showed low systematic exposure compared to dexamethasone alone. Moreover, the lipophilic prodrug-loaded large porous particles delivery system can be used with other molecules to improve local action and control unwanted systematic exposure (Fig. 5.16).

Stimuli-responsive polymers were used to control drug diffusion through porous pulmonary carriers. Drug delivery and retention in the lungs are essential for appropriate biological outcomes. Various biodegradable polymers were used to produce porous pulmonary carriers including PLGA, poly (lactic acid), chitosan, gelatin, and alginate. Tumor necrosis factor (TNF)-related apoptosis-inducing ligand (TRAIL) is a safe anticancer agent with little or no toxic effect on normal cells because it specifically binds to death receptors. The anticancer effect of TRAIL is enhanced by co-treating with certain chemotherapeutics. Inhalable porous PLGA microparticles containing doxorubicin and surface-attached with TRAIL were fabricated using a w/o/w double emulsification method using ammonium bicarbonate as a gas-foaming agent. Strong ionic interactions between TRAIL and poly (ethylene-*alt*-maleic anhydride) (PEMA) altered the surface of microparticles. Over the course of 7 days, the inhalable microparticles showed

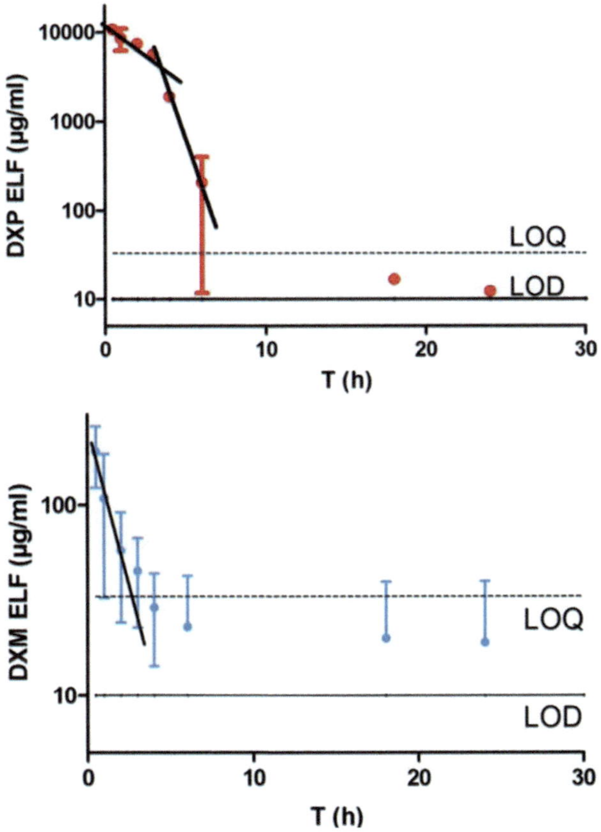

Fig. 5.16 Concentration of DXP (top) and DXM (bottom) in plasma after intratracheal insufflation of 3 mg of DXP (DXP dose = 150 μg). Data are presented as mean ± SEM, *n* = 3 (Reprint from [32] with permission by Elsevier B.V.)

effective drug delivery to the lung by releasing doxorubicin and leaching TRAIL from their surrounding ionic bonds [84, 85]. The schematic of the synergistic antitumor effects of TRAIL/doxorubicin PLGA microparticles administered by inhalation in lung cancer is shown in Fig. 5.17. The release of doxorubicin from TRAIL/doxorubicin PLGA microparticles was slower than the release of surface-coated TRAIL molecules. The results showed H226 cells influenced by the cytotoxic effect of TRAIL plus doxorubicin whereas individual treatments of TRAIL and doxorubicin had negligible cytotoxic effects. Compared to mice treated with TRAIL (22 μg) or doxorubicin PLGA microparticles alone (10 μg), tumors in H226-implanted mice treated with the 10 μg doxorubicin and 22 μg TRAIL were noticeably smaller and fewer in number [84]. Large porous microparticles were accumulated in the mice's lungs throughout a 7-day period as a result of the sequential release of TRAIL and doxorubicin.

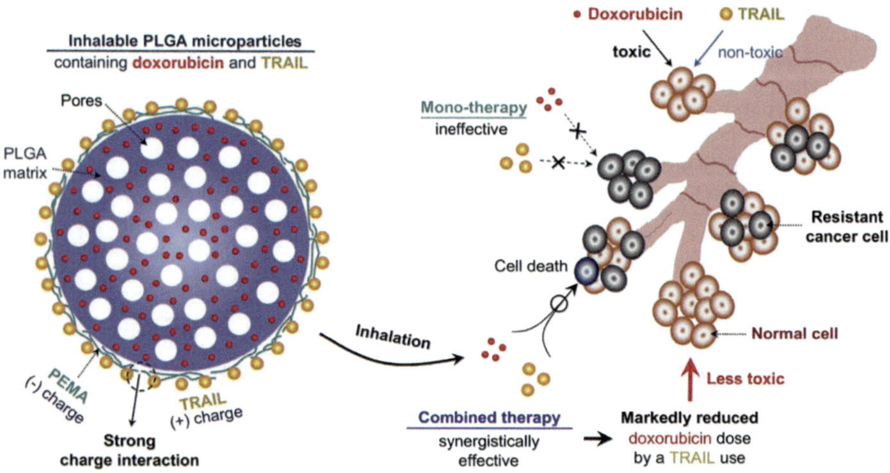

Fig. 5.17 Schematic demonstration of inhalable TRAIL/doxorubicin-loaded PLGA porous microspheres for anti-lung tumor treatment (Reprint from [84] Kim et al., 2013 with permission by Elsevier B.V.)

5.6 Summary

The obstacle in delivering drugs via the pulmonary route, whether for a local or systemic effect can be surpassed by creating particles suitable for deep lung deposition. Large porous particles are designed to alleviate the difficulty of ordinary non-porous particles. As shown in the above sections porous aerosol particles with large sizes and low mass density have a lower tendency to agglomerate, increasing the efficiency of aerosolization. Furthermore, particle sizes larger than 2–3 μm reduced macrophage phagocytosis and had longer therapeutic effects since they were slowly eliminated from the lungs after inhalation. For pulmonary drug delivery applications, porous particles can be manufactured by introducing porogens or templating agents, or by utilizing a specific pores-generating procedure. Porous particles production includes double-emulsion solvent evaporation, spray drying, PulmoSphere™, supercritical fluid-anti-solvent technology, spray freeze drying, Aerogel, and co-suspension delivery technology. The porous particles for pulmonary drug delivery route, especially PulmoSphere™ is available in the market today with pre-clinical safety and clinical data. Other porous particle-based pulmonary drug delivery platforms are still under safety and clinical efficacy studies. The risks associated with developing the new technology and the perceived regulatory hazards will diminish as these items are commercialized. Porous particles platform will be applied to produce new pulmonary drug delivery systems for idiopathic pulmonary fibrosis, pulmonary arterial hypertension, asthma/COPD, and cystic fibrosis.

References

1. Boddu S, Nesamony J. Nanocarrier systems for lung drug delivery. In: Pathak Y, Islam N, editors. Handbook of lung targeted drug delivery systems: recent trends and clinical evidences. Boca Raton: CRC Press; 2022.
2. Katiyar SK, Gaur SN, Solanki RN, et al. Indian guideline on nebulization therapy. Indian J Tuberc. 2022;69:S1–S191.
3. Lavorini F. The challenge of delivering therapeutic aerosols to asthma patients. Allergy. 2013:102418.
4. Dalby R, Spallek M, Voshaar T. A review of the development of Respimat® soft mist™ inhaler. Int J Pharm. 2004;283(1–2):1–9.
5. Iwanaga T, Tohda Y, Nakamura S, Suga Y. The Respimat® soft mist inhaler: implications of drug delivery characteristics for patients. Clin Drug Investig. 2019;39:1021–30.
6. Lin X, Kankala RK, Tang N, Xu P, Hao L, Yang D, Wang S, Zhang YS, Chen A. Supercritical fluid-assisted porous microspheres for efficient delivery of insulin and inhalation therapy of diabetes. Adv Healthc Mater. 2018:1800910.
7. Mahato RI, Narang A. Inserts, implants, and devices. In Mahato, RI, Narang AS, editors. Pharmaceutical dosage forms and drug delivery. Boca Raton: CRC Press; 2018. p563–581.
8. Batycky R, Deaver D, Dwivedi S, Hrkach J, Johnston L, Olsen T, Wright JE. Chapter 75: the development of large porous particles for inhalation drug delivery. In: Rathbone MJ, Hadgraft J, Roberts MS, editors. Modified-release drug delivery technology. Marcel Dekker, Inc.; 2003. p. 891–902.
9. Vanbever R, Mintzes JD, Wang J, Nice J, Chen D, Batycky R, Langer R, Edwards DA. Formulation and physical characterization of large porous particles for inhalation. Pharm Res. 1999;16:1735–42.
10. López-Iglesias C, Casielles AM, Altay A, Bettini R, Alvarez-Lorenzo C. García-González CA from the printer to the lungs: inkjet-printed aerogel particles for pulmonary delivery. Chem Eng J. 2019;357:559–66.
11. Musante CJ, Schroeter JD, Rosati JA, Crowder TM, Hickey AJ, Martonen TB. Factor affecting the deposition of inhaled porous drug delivery. J Pharm Sci. 2002;91:1590–600.
12. Batycky R, Nice J, Chen D, Sung J, Lipp M, Mintzes J, Dunbar C, Niven R, Edwards D. Production and characterization of large porous particles for pulmonary drug delivery. Mat Res Soc Symp Proc. 1999;550:95–100.
13. Duong T, López-Iglesias C, Szewczyk PK, Stachewicz U, Barros J, Alvarez-Lorenzo C, Alnaief M, García-González CA. A pathway from porous particle technology toward tailoring aerogels for pulmonary drug administration. Front Bioeng Biotechnol. 2021;9:671381.
14. Crowder TM, Rosati JA, Schroeter JD, Hickey AJ, Martonen TB. Fundamental effects of particles morphology on lung delivery: predictions of stokes' law and the particular relevance to dry powder inhaler formulation and development. Pharm Res. 2002;19:239–45.
15. Edwards DA, Ben-Jebria A, Langer R. Recent advances in pulmonary drug delivery using large, porous inhaled particles. J Appl Physiol. 1998;85:379–85.
16. Lam J, Vaughan S, Parkins MD. Tobramycin inhalation powder (TIP): an efficient treatment strategy for the management of chronic Pseudomonas aeruginosa infection in cystic fibrosis. Clin Med Insights: Circ Respir Pulm Med. 2013;7:61–77.
17. Konstan MW, Geller DE, Minic P, Brockhaus F, Zhang J, Angyalosi G. Tobramycin inhalation powder for P. aeruginosa infection in cystic fibrosis: the EVOLVE trial. Pediatr Pulmonol. 2011;46:230–8.
18. Weers J, Tarara T. The PulmoSphere™ platform for pulmonary drug delivery. Ther Deliv. 2014;5:277–95.
19. Weers JG, Miller DP, Tarara TE. Spray-dried PulmoSphere™ formulations for inhalation comprising crystalline drug particles. AAPS PharmSciTech. 2019;20:103.

20. Hirst PH, Pitcairn GR, Wees JG, Tarara TE, Clark AR, Dellamary LA, Hall G, Shorr J, Newman SP. In vivo lung deposition of hollow porous particles from a pressurized metered dose inhaler. Pharm Res. 2002;19:258–64.
21. Dellamary LA, Tarara TE, Smith DJ, Woelk CH, Adractas A, Costello ML, Gill H, Weers JG. Hollow porous particles in metered dose inhalers. Pharm Res. 2000;27:168–74.
22. Tsapis N, Bennett D, Jackson B, Weitz DA, Edwards DA. Trojan particles: large porous carriers of nanoparticles for drug delivery. Proc Natl Acad Sci U S A. 2002;99:12001–5.
23. Gharse S, Fiegel J. Large porous particles: lightweight champions of pulmonary drug delivery. Curr Pharm Des. 2016;22:2463–9.
24. Yang L, Luo J, Shi S, Zhang Q, Sun X, Zhang Z, Gong T. Development of a pulmonary peptide delivery system using porous nanoparticle-aggregate particles for systemic application. Int J Pharm. 2013;451:104–11.
25. Yang T, Wen B, Liu K, Qin M, Gao Y, Ding D, Li W, Zhang Y, Zhang W. Cyclosporine a/ porous quaternized chitosan microspheres as a novel pulmonary drug delivery system. Artif Cells Nanomed Biotechnol. 2018;46(sup2):552–64.
26. Niwa T, Mizutani D, Danjo K. Spray freeze-dried porous microparticles of a poorly water-soluble drug for respiratory delivery. Chem Pharm Bull. 2012;60:870–6.
27. Sharma A, Vaghasiya K, Verma RK. Inhalable microspheres with hierarchical pore size for tuning the release of biotherapeutics in lungs. Micropor Mesopor Mat. 2016;235:195–203.
28. Gulin-Sarfraz T, Jonasson S, Wigenstam E, Haartman EV, Bucht A, Rosenholm JM. Feasibility study of mesoporous silica particles for pulmonary drug delivery: therapeutic treatment with dexamethasone in a mouse model of airway inflammation. Pharmaceutics. 2019;11:149.
29. Shiehzadeh F, Tafaghodi M, Laal-Dehghani M, Mashhoori F, Sedigheh B, Bazzaz F, Imenshahidi M. Preparation and characterization of a dry powder inhaler composed of PLGA large porous particles encapsulating gentamicin sulfate. Adv Pharm Bull. 2019;9:255–61.
30. Yang Y, Bajaj N, Xu P, Ohn K, Tsifansky MD, Yeo Y. Development of highly porous large PLGA microparticles for pulmonary drug delivery. Biomaterials. 2009;30:1947–53.
31. Chvatal A, Ambrus R, Party P, Katona G, Jójárt-Laczkovich O, Szabó-Révész P, Fattal E, Tsapis N. Formulation and comparison of spray dried non-porous and large porous particles containing meloxicam for pulmonary drug delivery. Int J Pharm. 2019;558:68–75.
32. N'Guessan A, Fattal E, Chapron D, Gueutin C, Koffi A, Tsapis N. Dexamethasone palmitate large porous particles: a controlled release formulation for lung delivery of corticosteroids. Eur J Pharm Sci. 2018;113:185–92.
33. Liao Q, Yip L, Chow MYT, Chow SF, Chan HK, Kwok PCL, Lam JKW. Porous and highly dispersible voriconazole dry powders produced by spray freeze drying for pulmonary delivery with efficient lung deposition. Int J Pharm. 2019;560:144–54.
34. Peng T, Zhang X, Huang Y, Zhao Z, Liao Q, Xu J, Huang Z, Zhang J, Wu C, Pan X, Wu C. Nanoporous mannitol carrier prepared by non-organic solvent spray drying technique to enhance the aerosolization performance for dry powder inhalation. Sci Rep. 2017;7:46517.
35. Zhang X, Yue X, Cui Y, Zhao Z, Huang Y, Cai S, Wang G, Wang W, Hugh S, Pan X, Wu C, Tan W. A systematic safety evaluation of nanoporous mannitol material as a dry-powder inhalation carrier system. J Pharm Sci. 2020;109:1692–702.
36. Tse JY, Koike A, Kadota K, Uchiyama H, Fujimori K, Tozuka Y. Porous particles and novel carrier particles with enhanced penetration for efficient pulmonary delivery of antitubercular drugs. Eur J Pharm Biopharm. 2021;167:116–26.
37. Varun N, Ghoroi C. Engineered inhalable micro-balloon shaped drug particles for carrier-free dry powder inhalation (DPI) application. Powder Technol. 2022;408:117705.
38. Pham D, Fattal E, Ghermani N, Guiblin N, Tsapis N. Formulation of pyrazinamide-loaded large porous particles for the pulmonary route: avoiding crystal growth using excipients. Int J Pharm. 2013;454:668–77.
39. Zhou M, Shen L, Lin X, Hong Y, Feng Y. Design and pharmaceutical applications of porous particles. RSC Adv. 2017;7:39490.

40. Healy AM, Amaro MI, Paluch KJ, Tajbar L. Dry powders for oral inhalation free of lactose carrier particles. Adv Drug Deliv Rev. 2014;75:32–52.
41. Chakravarty P, Famili A, Nagapudi K, Al-Sayah MA. Using supercritical fluid technology as a green alternative during the preparation of drug delivery systems. Pharmaceutics. 2019;11:629.
42. Usmani OS, Roche N, Jenkins M, Stjepanovic N, Mack P, Backer WD. Consistent pulmonary drug delivery with whole lung deposition using the aerosphere inhaler: a review of the evidence. Int J Chron Obstruct Pulmon Dis. 2021;16:113–24.
43. Rabe KF. GFF MDI for the improvement of lung function in COPD –a look at the PINNACLE-1 and PINNACLE-2 data and beyond. Expert Rev Clin Pharmacol. 2017;10:685–98.
44. Rennard S, Fogarty C, Reisner C, Fernandez C, Fischer T, Golden M, Rose E, Darken P, Tardie G, Orevillo C. Randomized study of the safety, pharmacokinetics, and bronchodilatory efficacy of a proprietary glycopyrronium metered-dose inhaler in study patients with chronic obstructive pulmonary disease. BMC Pulm Med. 2014;14:118.
45. Ferguson GT, Hickey AJ, Dwivedi S. Co-suspension delivery technology in pressurized metered-dose inhalers for multi-drug dosing in the treatment of respiratory diseases. Respir Med. 2018;134:16–23.
46. Kumar R, Mehta P, Shankar KR, Rajora MAK, Mishra YK, Mostafavi E, Kaushik A. Nanotechnology - assisted metered - dose inhalers (MDIs) for high - performance pulmonary drug delivery applications. Pharm Res. 2022;39:2831. https://doi.org/10.1007/s11095-022-03286-y.
47. Taylor G, Warren S, Dwivedi S, Sommerville M, Mello L, Orevillo C, Maes A, Martin UJ, Usmani OS. Gamma scintigraphic pulmonary deposition study of glycopyrronium/ formoterol metered dose inhaler formulated using co-suspension delivery technology. Eur J Pharm Sci. 2018;111:450–7.
48. Quinn D, Seale JP, Reisner C, Fischer T, Golden M, Thomas M, Tardie G, Orevillo C. A randomized study of formoterol fumarate in a porous particle metered-dose inhaler in patients with moderate-to-severe COPD. Respir Med. 2014;108:1327–35.
49. Zhu L, Li M, Liu X, Jin Y. Drug-loaded PLGA electrospraying porous microspheres for the local therapy of primary lung cancer via pulmonary delivery. ACS Omega. 2017;2:2273–9.
50. Kanojia N, Singh S, Singh J, Sharma N, Grewal AS, Rani L, Thapa K, Arora S. Recent advancements and applications of inhalable microparticles based drug delivery systems in respiratory disorders. Biointerface Res Appl Chem. 2021;11:10099–118.
51. Mangal S, Gao W, Li T, Zhou Q. Pulmonary delivery of nanoparticle chemotherapy for the treatment of lung cancers: challenges and opportunities. Acta Pharmacol Sin. 2017;38:782–97.
52. Kemeny NE, Niedzwiecki D, Hollis DR, Lenz HJ, Warren RS, Naughton MJ, et al. Hepatic arterial infusion versus systemic therapy for hepatic metastases from colorectal cancer: a randomized trial of efficacy, quality of life, and molecular markers (CALGB 9481). J Clin Oncol. 2006;24:1395–403.
53. Guerin C, Olivi A, Weingart JD, Lawson HC, Brem H. Recent advances in brain tumor therapy: local intracerebral drug delivery by polymers. Investig New Drugs. 2004;22:27–37.
54. Markman M, Cleary S, Pfeifle C, Howell SB. Cisplatin administered by the intracavitary route as treatment for malignant mesothelioma. Cancer. 1986;58:18–21.
55. Duvillard C, Romanet P, Cosmidis A, Beaudouin N, Chauffert B. Phase 2 study of intratumoral cisplatin and epinephrine treatment for locally recurrent head and neck tumors. Ann Otol Rhinol Laryngol. 2004;113:229–33.
56. Markman M. An update on the use of intraperitoneal chemotherapy in the management of ovarian cancer. Cancer J. 2009;15:105–9.
57. Markman M, Reichman B, Hakes T, Rubin S, Lewis JL Jr, Jones W, et al. Evidence supporting the superiority of intraperitoneal cisplatin compared to intraperitoneal carboplatin for salvage therapy of small- volume residual ovarian cancer. Gynecol Oncol. 1993;50:100–4.
58. Wang J, Jansen JA, Yang F. Electrospraying: possibilities and challenges of engineering carriers for biomedical applications – a mini review. Front Chem. 2019;7:258.

59. Patel B, Gupta N, Ahsan F. Low-molecular-weight heparin (LMWH)–loaded large porous PEG-PLGA particles for the treatment of asthma. J Aerosol Med Pulm Drug Deliv. 2014;27:12–20.
60. He S, Gui J, Xiong K, Chen M, Gao H, Fu Y. A roadmap to pulmonary delivery strategies for the treatment of infectious lung diseases. J Nanobiotechnol. 2022;20:101.
61. Pham D, Grégoire N, Couet W, Gueutin C, Fattal E, Tsapis N. Pulmonary delivery of pyrazinamide-loaded large porous particles. Eur J Pharm Biopharm. 2015;94:241–50.
62. Weers JG, Bell J, Chan HK, et al. Pulmonary formulations: what remains to be done? J Aerosol Med Pulm Drug Deliv. 2010;2:S5–2.
63. Rowsell JLC, Yagli OM. Metal-organic frameworks: a new class of porous materials. Micropor Mesopor Mat. 2004;73:3–14.
64. Zhou Y, Niu B, Wu B, Luo S, Fu J, Zhao Y, Quan G, Pan X, Wu C. A homogenous nanoporous pulmonary drug delivery system based on metal-organic frameworks with fine aerosolization performance and good compatibility. Acta Pharm Sin B. 2020;10(12):2404–16.
65. Emmi G, Silvestri E, Squatrito D, et al. Thrombosis in vasculitis: from pathogenesis to treatment. Thromb J. 2015;13:1–10.
66. Mackman N, Bergmeier W, Stouffer GA, Weitz JI. Therapeutic strategies for thrombosis: new targets and approaches. Nat Rev Drug Discov. 2020;19:333–52.
67. Qi Y, Zhao G, Liu D, Shriver Z, Sundaram M, Sengupta S, Venkataraman G, Langer R, Sasisekharan R. Delivery of therapeutic levels of heparin and low-molecular-weight heparin through a pulmonary route. Proc Natl Acad Sci. 2004;101:9867–72.
68. Singh R, Dwivedi SP, Gaharwar US, Meena R, Rajamani P, Prasad T. Recent updates on drug resistance in mycobacterium tuberculosis. J Appl Microbiol. 2019;128:1547–67.
69. Contreras LG, Sung J, Ibrahim M, Elbert K, Edwards D, Hickey A. Pharmacokinetics of inhaled rifampicin porous particles for tuberculosis treatment: insight into rifampicin absorption from the lungs of Guinea pigs. Mol Pharm. 2015;12:2642–50.
70. Cai Y, Xu M, Yuan M, Liu Z, Yuan W. Developments in human growth hormone preparations: sustained-release, prolonged half-life, novel injection devices, and alternative delivery routes. Int J Nanomedicine. 2014;9:3527–38.
71. Kim HK, Chung HJ, Park TG. Biodegradable polymeric microspheres with "open/closed" pores for sustained release of human growth hormone. J Control Release. 2006;112:167–74.
72. Bosquillon C, Préat V, Vanbever R. Pulmonary delivery of growth hormone using dry powders and visualization of its local fate in rats. J Control Release. 2004;96:233–44.
73. Koushik K, Dhanda DS, Cheruvu NPS, Kompella UB. Pulmonary delivery of deslorelin: large-porous PLGA particles and HPßCD complexes. Pharm Res. 2004;21:1119–26.
74. Kuriakose A, Hu W, Nguyen KT, Menon JU. Scaffold-based lung tumor culture on porous PLGA microparticle substrates. PLoS One. 2019;14:e0217640.
75. Singh A, Malviya R, Sharma PK. Pulmonary drug delivery system: a novel approach for drug delivery. Curr Drug Ther. 2011;6:137–51.
76. Chaurasiya B, Zhao Y. Dry powder for pulmonary delivery: a comprehensive review. Pharmaceutics. 2021;13:31.
77. Pulivendala G, Bale S, Godugu C. Inhalation of sustained release microparticles for the targeted treatment of respiratory diseases. Drug Deliv Transl Res. 2020;10:339–53.
78. Labiris NR, Dolovich MB. Pulmonary drug delivery part I: physiological factors affecting therapeutic effectiveness of aerosolized medications. Br J Clin Pharmacol. 2003;56:588–99.
79. Kim CS. Physiological factors affecting lung deposition. J Aerosol Med Pulm. 2021;34:147–54.
80. Zara DL, Sun F, Zhang F, Franek F, Sivars KB, Horndahl J, Bates S, Brännström M, Ewing P, Quayle MJ, Petersson G, Folestad S, JRV O. Controlled pulmonary delivery of carrier-free budesonide dry powder by atomic layer deposition. ACS Nano. 2021;15:6684–98.
81. Liang Z, Ni R, Zhou J, Mao S. Recent advances in controlled pulmonary drug delivery. Drug Discov. 2015;20:380–9.
82. Gupta V, Davis M, Hope-Weeks LJ, Ahsan F. PLGA microparticles encapsulating prostaglandin E_1-hydroxypropyl-β-cyclodextrin (PGE_1-HPβCD) complex for the treatment of pulmonary arterial hypertension (PAH). Pharm Res. 2011;28:1733–49.

83. Sung JC, Padilla DJ, Garcia-Contreras L, VerBerkmoes JL, Durbin D, Peloquin CA, Elbert KJ, Hickey AJ, Edwards DA. Formulation and pharmacokinetics of self-assembled rifampicin nanoparticle systems for pulmonary delivery. Pharm Res. 2009;26:1847–55.
84. Kim I, Byeon HJ, Kim TH, Lee ES, Oh KT, Shin BS, Lee KC, Youn YS. Doxorubicin-loaded porous PLGA microparticles with surface attached TRAIL for the inhalation treatment of metastatic lung cancer. Biomaterials. 2013;34:6444–53.
85. Thananukul K, Kaewsaneha C, Opaprakasit P, Lebaz N, Errachid A, Elaissari A. Smart gating porous particles as new carriers for drug delivery. Adv Drug Deliv Rev. 2021;174:425–46.

Application of Numerical Simulation (CFD) to Probe Powder, Particles, and Inhalers

6

Tan Suwandecha and Teerapol Srichana

Abstract

Dry powder inhaler (DPI) is one of many pulmonary drug delivery systems that are commonly prescribed for the treatment of a variety of diseases. However, the efficiency of the current DPI is limited as a variable, and low portions of the dose reach the deep lungs. The complexity of a DPI platform that involves multiple physics domains, including aerodynamics and solid mechanics, is a significant obstacle to improving its performance and efficiency. Numerical simulation such as computational fluid dynamics (CFD), discrete element method, and finite element analysis, is beneficial for investigating the underlying aerosolization mechanisms and evaluating DPI design. Visualization of fluid flow and particle trajectory, particle interaction, and parametric optimization is enabled by numerical simulation. It opens the opportunities for computer-aided designs of DPI systems in formulation development, device development, and observation of patient factors on particle deposition and targeting aerosol to a particular area of the lungs. This chapter reviews the basic principles of CFD and related physics and the application of numerical simulation to study fluid dynamics in DPIs. The limitation and future trends of numerical simulations in aerosol delivery system research are also discussed.

T. Suwandecha
Drug and Cosmetics Excellence Centre and School of Pharmacy, Walailak University, Nakhon Si Thammarat, Thailand
e-mail: tan.su@wu.ac.th

T. Srichana (✉)
Drug Delivery System Excellence Centre and Department of Pharmaceutical Technology, Faculty of Pharmaceutical Sciences, Prince of Songkla University, Songkhla, Thailand
e-mail: teerapol.s@psu.ac.th

P. P. Mehta, V. Dhapte-Pawar (eds.), *Pulmonary Drug Delivery Systems: Material and Technological Advances*,
https://doi.org/10.1007/978-981-99-1923-9_6

Keywords

Dry powder inhaler · CFD · Aerosol · Numerical simulation · Discrete element method

6.1 Introduction

Pharmaceutical aerosol is a technique to deliver medications in the form of droplets, fine mist, and powders into the pulmonary airways. It can be used for the treatment of local or systemic diseases. There are four types of pharmaceutical aerosols, i.e., nebulizers, pressurized metered-dose inhalers (pMDI), soft mist inhalers (SMIs), and DPIs. Each aerosol system comprises three key components, i.e., micronized drug particle, carrier (lactose or propellant), and inhaler device. Nebulizers, pMDIs, and SMIs are liquid dosage forms, while DPIs are solid dosage forms. During the act of inhalation, the inhaler device delivers formulation in the form of droplets, fine mist, and fine particles due to energy provided by compressed gas, piezoelectric vibration, pressurized propellant systems, and inspiratory flow produced by the patient. Compared to all available aerosols, DPIs are most commonly preferred due to ease of storage, handling, and use. Also, DPIs has an excellent ability to deliver high-dose drugs (e.g., TOBI® Podhaler®), protein/peptides (e.g., Afrezza®), and other biological substance to pulmonary airways for various applications [1].

Unit dose, multiple unit doses, and reservoir-based inhalers are the three major types of DPIs available in the market. The unit dose system is usually hard gelatin or HPMC capsule filled with powder medicament, which is actuated by separation of cap/body (e.g., Rotahaler® Cipla) or piercing action of needle (Spinhaler® and Cyclohaler®). The unit dose system is also available in the form of a blister, i.e., Elpenhaler® by ELPEN Pharma. The multiple unit dose systems are usually composed of powder cartridge (Multihaler® Cipla), blister discs (Diskhaler® GSK), and 60's single blister strip (Accuhaler®, GSK) or 30's dual blister strip (Ellipta®, GSK) inhaler in which powder medicament dispersed by puncturing mechanism or separation of sealing foil by unique rotating gear during an act of inhalation. In reservoir DPIs (Symbicort® Turbuhaler® and Genuair®), the bulk powder was stored in the device compartment, and a fixed amount of powder releases upon actuation [2–4]. During inhalation maneuver, powder medicament follows four key processes, i.e., (i) fluidization/dispersion of powder, (ii) movement of drug/carrier complex (micro-aggregates), (iii) drug/carrier detachment, and (iv) drug deposition within pulmonary airways. Powder dispersion or deagglomeration efficacy is equally governed by formulation nature and inhaler device design. In most cases, inhaler devices significantly impacted powder dispersion/deagglomeration due to shear force turbulence generation, mechanical impaction between particles and the wall of the inhaler device, and inter-particle collision (Fig. 6.1) [5].

Most DPIs can deliver only about 30% of the loaded dose into the deep lungs. The particle with diameters <5 μm usually reaches the lower part airways and can be expressed as the fine particle fraction (FPF). Thus, to improve FPF several attempts have been made by researchers such as the use of particle engineering

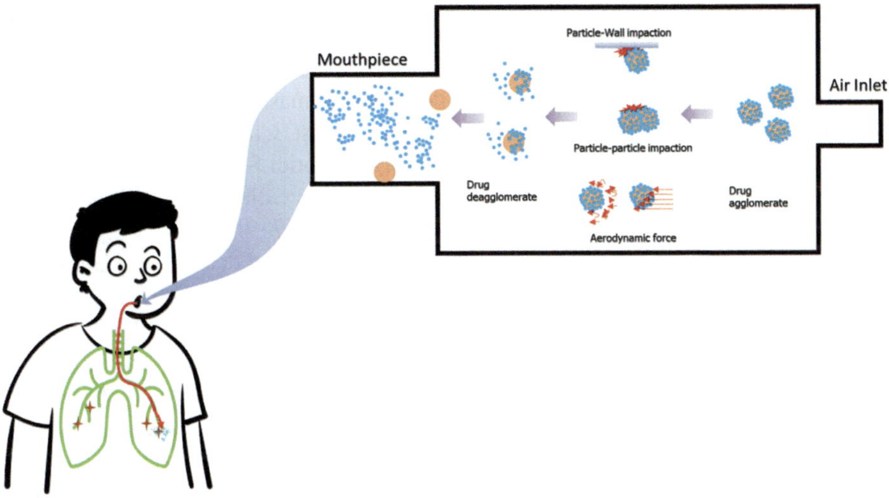

Fig. 6.1 The mechanism of powder delivery through inhaler device

methods, modification of existing inhaler devices, and use of a new powder dispersion method to aerosolize the drug formulation. Lactose fines (5–20%) or ternary agents (e.g., magnesium stearate) are also used to effectively disperse the cohesive micronized drug particle. Among these approaches, modification or development of new inhaler devices is appealing to maximize the FPF. However, reaching the goal is not simple. The aerosolization process involves two critical physics domains: aerodynamics and solid mechanics.

Additionally, the interaction between the two domains needs to be addressed as a powder–fluid interaction. The powder–fluid interaction is a complex, nonlinear problem and needs a fundamental understanding of complex geometries and mechanisms. Thus, a prediction of DPIs aerodynamic and biological efficiency only using empirical information is quite tricky. Moreover, in most cases, experimental analysis is always carried out at fixed parameters such as flow rate, pressure, and powder formulation. By considering all these challenges in the DPIs domain, computer-aided designs such as numerical analysis and multi-physics computational simulation were explored by various active researchers to probe the formulation/device interactions, particle/particle interactions, device components, and patient factors impacts on particle deposition and aerodynamic performance [6].

6.2 Computational Fluid Dynamics

A standard method for exploring the aerodynamic profiles in DPIs is CFD which may couple with particle tracking to observe the particle movement in the inhalers. The CFD modeling is a numerical model for analyzing and solving the problem of fluid flow that is governed by three fundamental laws, including (1) the law of conservation of mass, (2) Newton's second law of motion (transport of momentum), and (3) the first law of thermodynamics (Conservation of energy).

These three laws apply to a finite small control volume in a moving fluid. Partial differential equations (PDEs) are used to describe the mass, momentum, and energy transfers extensively developed by French physicist Claude-Louis Navier and Anglo-Irish physicist and mathematician George Gabriel Stokes. The collective of PDEs expresses the motion of viscous fluid substances called Navier–Stokes equations. The transport equations are described as follows [7–9]:

The law of conservation of mass—The mass is conserved in a closed system such as a controlled volume. It cannot be created or destroyed. The mass entering a controlled volume had to equal the mass escaping from the controlled volume. In other words, the difference between input and output is zero. For fluid flow, the mass is described in terms of the density change rate between the inlet and outlet flow (Eq. 6.1) [10].

$$\frac{\partial \rho}{\partial t} + \rho(\nabla \bullet \mathrm{u}) = 0 \tag{6.1}$$

where, ρ = fluid density, t = time, u = fluid velocity, and ∇=gradient operator.

The law of conservation of momentum—The conservation of momentum follows Newton's second law. The rate of increase of material momentum within the control volume is equal to the net rate at which momentum enters the control volume with the flowing fluid with the summation of the forces acting on the fluid in the control volume (Eq. 6.2) [10].

$$\rho \frac{\mathrm{Du}}{\mathrm{Dt}} = \rho g - \nabla \mathrm{p} + \mu \nabla^2 \mathrm{u} \tag{6.2}$$

where, Δp = surface force and μ = viscosity coefficient.

The law of conservation of energy—The conservation of energy follows the rate of change of energy inside the fluid element is equal to the net flux of heat into the element and the rate of work done on the element due to body and surface forces (Eq. 6.3) [10].

$$\rho \frac{\mathrm{DE}}{\mathrm{Dt}} = \frac{\partial(\rho E)}{\partial t} + \mathrm{V} \bullet (\rho E u) \tag{6.3}$$

where, E = energy per unit mass.

The velocity and pressure fields, as well as other vital characteristics, can be obtained by solving the Navier–Stokes equations in a coupled or segregated method. The most accurate method to solve the Navier–Stokes is the direct numerical simulation (DNS) which can capture all turbulent time and length scales. Unfortunately, this method is very computationally intensive. Although it is receiving more attention in fluid dynamics research, it is still only used in a restricted number of DPIs simulations. As a result, the more straightforward methods of relying on heuristic turbulence models are much more practical for DPI research and development.

These methods simplified the Navier–Stokes equations to perform at steady-state conditions based on RANS (Reynolds averaged Navier-Stokes) approach. The RANS approach breaks down the turbulence flow into average velocity flow and velocity fluctuations from turbulence. To close the RANS equation for solving, additional modeling is required. The models are known as turbulence models, such as standard k-ε, re-normalization group (RNG) k-ε, realizable k-ε, Wilcox's k-ω, and Reynolds stress model (RSM). These models provide a reasonable level of simulation fidelity while balancing the computing demands. The k-ε model consists of two variables: k, the turbulence kinetic energy, and ε (epsilon), the rate of dissipation of turbulence kinetic energy. The RNG k-ε improves over the standard k-ε model by renormalizing the Navier–Stokes equations to address the effects of more minor scales of motion, which may be improved accuracy in rotating flows. The realizable k-ε made improvement on the turbulent viscosity and transport equation for the dissipation rate resulting in significant improvement to analyze complex flow such as rotation, boundary layers under strong adverse pressure gradients, separation, and re-circulation.

The k-ω model has a similar concept to the k-ε model. However, the rate of dissipation of turbulence is replaced with ω (omega), the specific rate of dissipation of kinetic energy. The k-ω model is a low Reynolds number model that is useful for the flow with strong curvature, separated flows, and jets. The nonlinearity of this model results in more difficulty in converging than the k-ε model. The RSM, also known as the Reynolds Stress Transport Model, is a useful turbulent model. This model computes the Reynolds stress tensor directly by avoiding the eddy-viscosity hypothesis and resolves the complicated interactions within turbulent flows and the directional effects of the Reynolds stresses. The RSM provides higher simulation accuracy than k-ε and k-ω [11, 12].

Another alternative to the k-ε, k-ω, and RSM model is large-eddy simulation (LES). LES is designed to ignore the minor length scales by employing the low-pass filtering of the Navier–Stokes equations. This method nearly achieves the simulation accuracy of DNS while considerably reducing the computational cost. Many scholars have used LES in DPIs computer simulation in recent years. Some numerical investigations used Stress Blended Eddy Simulation (SBES) instead of a full LES simulation, resulting in a hybrid of the RANS and LES. These approaches agreed well with the particle image velocimetry data [13, 14].

6.3 Particles Tracking

The flow field profile alone cannot fulfill the DPI's dynamic aerosolization process. The numerical modeling of the particle motion in the air is characterized as a granular multiphase flow process. The multiphase flow can be implemented using either an Eulerian or Lagrangian approach by coupling the dispersed phase with the fluid flow model. The Eulerian approach assumes that the gas and solid phases can coexist and interpenetrate with each other in a confined element and the interface between them is unresolved. The solid phases are modeled as a second continuum.

The Navier–Stokes conservation equations for each phase are solved. The interaction inside the solid phase in granular flow may have sub-regimes such as elastic regime and plastic (inelastic) regime. The stresses in the solid phase are calculated by taking into account the random motion of the solid particles brought on by the collision of inelastic particles. The solid-phase stresses from the inelastic particle–particle collision are calculated using the random motion of solid particles. The mass and momentum exchange between the two phases is modeled in the conservation equations for each phase as source/sink terms [15, 16].

The Lagrangian particle tracking recognizes the discrete nature of particle phase and probable particle size distribution [6]. The assumption of the Lagrangian particle methods considers particles as point masses which is the Lagrangian parcel concept (LPC). The motion of particles in the fluid, including the change of particle location and the linear and angular components of particle velocity, is calculated using the relevant forces acting on the particle according to the Newtonian second law (Eq. 6.4) [17].

$$\frac{du_p}{dt} = F_D\left(u - u_p\right) + \frac{g_x\left(\rho_p - \rho\right)}{\rho_p} + F_x, F_D = \frac{18\mu}{\rho_p d_p^2}\frac{C_D \operatorname{Re}}{24} \tag{6.4}$$

where u is the fluid phase velocity, and u_p is the particle velocity, ρ is the density of the fluid and ρ_p is the density of the particle, g_x is the acceleration due to gravitation, F_x is an additional acceleration (force/unit particle mass) term, $F_D\ (u–u_p)$ is the drag force per unit particle mass, μ is the molecular viscosity of the fluid, d_p is the diameter of the particle, C_d is drag coefficient, and Re is Reynolds number.

This Lagrangian concept can be implemented in the Discrete Particle Method (DPM) and the DEM (Discrete Element Method). The DPM assumes that the collisions between particles occur instantaneous, one-to-one impaction, and have a concise time scale compared to the other time scale and the new particles' velocity. The possibility of particle deformation from the external force is ignored. Determination of a collision between the particles for the particle of interest is necessary for the DPM approach. The numerical efficiency of this model is considerably high. However, if the collision frequency is high, the time step size can be very short. Reducing the time step and using deterministic calculation will result in more numerical time-demanding depending on the mesh size. It is much more practical to implement a stochastic collision model where a virtual collision partner is created from the local population of the particle phase during each time step. The translational and angular velocity changes during the collision events were calculated from the virtual particle by solving the impulse equations with appropriate values for the restitution and friction coefficients [18]. The DEM differs from DPM as it includes multiple contacts between several particles. As a result, this model is well suited to particle-laden flows in which numerous particles have a chance to collide at the same time. Particle deformation is allowed in this model; thus, elastic/plastic deformation and frictional forces can be analyzed. Most particle tracking methods in inhaler research are one-way coupling that only calculates the effect of fluid dynamics force on the particle. On the other hand, two-way coupling account for the interaction between fluid and particles or vice versa [19].

The inter-particle interaction can be divided into four classes: elastic, elastic-adhesive, elastic-plastic, and elastic-plastic with adhesion, depending on the materials. For elastic contact, deformation is recoverable, and the Hertz contact model effectively characterizes the normal contact force. Based on a theoretical explanation of how smooth, elastic spheres deform when they come into frictional contact, the Hertz contact model generates both regular and shear forces. The Hertz model assumes that each particle in contact can be considered an elastic half-space, the strains are minor and within the elastic limit, and the surfaces are continuous, non-conforming are frictionless [20].

Several contact models were developed to overcome the limitation of the Hertz model. These models have been used in the explanation of dry powder agglomeration and deagglomeration, such as Derjaguin, Landau, Verwey, and Overbeek (DLVO), Johnson-Kendall-Roberts (JKR), and Derjaguin, Muller and Toporov (DMT). The JKR model extended the Hertz model and was adapted in many for modeling the adhesion between particles [21]. This model implies that particle diameter is significantly greater than the practical length scale across which the van der Waals adhesion force is affected, allowing the adhesion force to be assumed to act exclusively inside the flattened contact zone between two colliding particles. When the particles are not in contact, no adhesion force acts between them before or after the collision. Because both elastic and adhesive forces determine the size of the contact region, these forces are combined nonlinearly to provide the total contact/adhesion force on any pair of particles in touch with each other [22]. Different CFD and particle tracking methods can be applied to investigate and characterize the phenomena related to dry powder drug delivery, as shown in Table 6.1.

Table 6.1 CFD modeling and particle tracking in dry powder drug delivery research

Target of interest	CFD method	Turbulent model	Particles tracking	References
Aerolizer® DPI	RANS	k–ω SST	DPM	[23–26]
Aerolizer® DPI	RANS	k–ω SST	DPM	[27]
Aerolizer® DPI	RANS	RSM	DEM	[28]
Modified Cyclohalers DPI	RANS	RNG k-ε	DPM	[29]
Turbuhaler DPI	RANS + LES	k–ω SST	DPM	[30, 31]
Inhalator and Cyclohaler DPI	RANS	RNG k-ε	DPM	[32]
Standardized entrainment tube	RANS	k–ω SST	DPM	[33–35]
Twincer® DPI	RANS	Realizable k-ε	DPM	[36]
Entrainment tube	RANS	Standard k-ε	Eulerian-Eulerian	[37]
L-shaped channel	LES	–	DEM	[38]
Human airway bifurcation model	DNS	–	DPM	[39]

6.4 Application of Numerical Simulations

The numerical simulation of the DPI device and human airways was conducted in the early 2000s'. The flow field in an idealized human oropharynx was studied using RANS approaches. The simulation analyzed both laminar and turbulent models [40]. The simulation on commercial DPI devices, the Aerolizer® and Rotahaler® DPI, was first performed by Coates et al. (2004). They found that slight modification of the inhaler geometry can largely impact the DPI performance. From their findings, other research groups pay more attention to CFD approaches in DPIs developments [24]. The computational analysis of DPIs includes various domains such as inhaler device geometry, formulation nature, agglomerate breakup mechanism, and device/patient interface [35]. Some applications of numerical simulation (CFD) to probe powder, particles, and inhalers are illustrated as follows.

6.4.1 Simulation of the Airflow Field and Particle Motion Inside the Inhaler and Human Airways

The most basic form of numerical simulation for DPIs and human airways is to simulate the airflow field with particle trajectory. The air velocity, turbulence kinetic energy, and the fluid dynamic force acting on particles, particle trajectory, and particle impaction can be collected and visualized. Coates et al. (2007) investigated the effect of multiple iterations of Aerolizer® device modification on airflow patterns and particle trajectory. They used standard RANS with k–ω SST turbulence model and DPM particle tracking. The particle tracking was performed as one-way coupling post-processing using the CFD flow field to calculate the force acting on particles. The particle deformation and inter-particle interaction are not included in the numerical model. The device design modifications were grid structure, mouthpiece length, capsule, air inlet size, and mouthpiece geometry. Excellent agreement was found between the simulation and experimental axial and tangential flow profiles through the devices at the flow rate of 60 and 120 liters per minute (LPM). Velocity profiles from CFD simulation showed how the flow field created in the inhaler mouthpiece was influenced by the inhaler grid (Fig. 6.2). The decreased voidage of the grid resulted in intensified integral scale strain rate in the mouthpiece. Highly turbulent flows directly affected the frequency of particle–mouthpiece collision and increased powder retention inside the mouthpiece. This phenomenon reduced overall inhaler performance. The mouthpiece length and variations in capsule size did not affect the aerosolization performance. However, completely removing the capsule from the device reduced the turbulence level and diminished the performance. Interestingly, the size and number of the capsule hole dramatically increased the performance of the inhaler, especially with the four 0.6 mm holes. In addition, reducing air inlet size increased FPF at the flow rate of 30 and 45 LPM by increasing the turbulence intensity, capsule spinning velocity, and particle impaction velocities. Contradictory, narrowing the air inlet ports to one-third of the original size at the flow rate of 60 LPM decreased FPF due to lower particles released from the

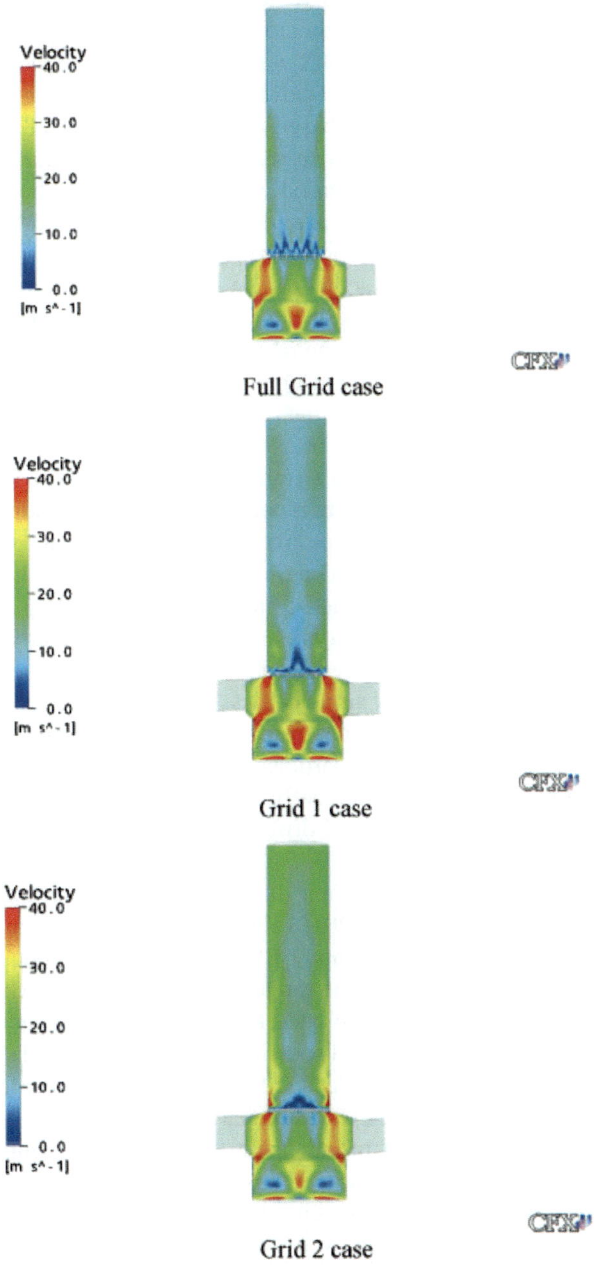

Fig. 6.2 Velocity profiles from CFD simulation of Aerolizer® device (Reprint from [24] with permission by Ellis Horwood)

device before the effective dispersion [23–26]. A nearly similar approach was employed to determine the effect of the circular chamber height of Aerolizer® on the performance of the DPI without the capsule in the chamber. The simulation was performed using RANS with low Reynolds number standard k-ω and Lagrangian DPM particle tracking based on the flow rate of 28.3 LPM. As the height of the circulation chamber increased, the velocity also increased. The particle deposition study showed that increased circulation chamber height leads to higher levels of deposition within the inhaler device due to higher turbulence generation [41].

Early CFD simulations were performed at constant inhalation flow rates, which did not resemble the actual patient's inhalation patterns. Thus, Kim and Chen (2009) investigated the effects of inhalation patterns on particle deposition in the human airways. Different inhaling rates with breath-holding patterns were investigated in a three-dimensional human upper airway model extending from the mouth to the third generation of the trachea. They used standard RANS with k-ω turbulence model and DPM particle tracking. The wall surface was assumed to trap all particles that collided with it and terminate the tracking. The sinusoidal inhalation patterns and velocity profiles were modeled at the inhaler outlet. They validated the simulation data with various experimental data for bifurcations under laminar and transient flows and showed good agreement. They found that multi-curvature geometry greatly affected the particle trajectory. The secondary flow was found to intensify at the first bend at the pharynx and relaxed after the larynx. The airflow velocity and particle size affected the particle deposition fraction. Short inhalation time resulted in much higher peak air velocities, such as 1 s and 5 s inhalation led to peak velocities of 15 m/s and 3 m/s, respectively. The high peak velocity resulted in a high deposition of 3–5 μm particles at the larynx regions. On the other hand, 1–2 μm particle deposition was rarely affected by the inhalation patterns. Slow inhalation for 3–5 seconds reduced drug loss in the upper airways and improved deep lung delivery, especially with 3–5 μm particles [42].

A numerical simulation of Turbuhaler®, a multiple-dose DPI, was conducted by Milenkovic et al. (2013) [30]. The simulation was based on RANS with k–ω SST turbulence model and LES. The study was modeled with steady-state airflow with pressure drop in the range of 800 to 8800 Pa, which resulted in flow rates of 20–70 LPM. The particle tracking was a Lagrangian approach without collision-induced breakage or redispersion effects. The particle injected into the simulation was single-sized particles (1–10 μm) or with a mean diameter of 2.2 and 4.5 μm. The velocity profiles were comparable for LES and k–ω SST. Thus, they decided to use the k–ω SST turbulence model to reduce computational time. The particle deposition inside the device was related to the particle size. The 5 μm particle was mainly deposited on the circular chamber surface. Whereas the small particles in the range of 1–2 μm were primarily deposited in the helical and mouthpiece regions. The CFD simulation with the experimental results was in good agreement. The relationship between the pressure drop and flow rate was predicted with the k–ω SST turbulence model. CFD successfully predicted the aerosolization outcome at various flow rates (30–70 LPM) with the inspiratory volume of 2 and 4 L and particle size of 2 μm [30]. Further refinement of this study was extended with the dynamic flow profile. The FPF results were close to the experimental data in other works [31].

Zhou et al. (2013) carried out a CFD simulation on Aerolizer® Foradile®. They use RANS with k–ω SST turbulence and DEM particle tracking. The inter-particle forces, such as the contact force, damping force, and van der Waals forces, were considered in DEM. They extended the simulation beyond the inhaler device to the USP induction port that was used in the in vitro experiment for aerosol particle size characterization. They inspected the effect of the mouthpiece length, air inlet size, and grid. The mouthpiece length alteration showed no difference in the airflow patterns, and it was confirmed with the in vitro aerosolization data. The CFD revealed that the air inlet size reduction significantly increased turbulence, and it was reflected in the experimental FPF results. For example, the FPF increased from 10% to 15% when reducing the air inlet size by one-third at a flow rate of 30 LPM. More voidage grid resulted in the swirling airflow pattern in the mouthpiece and increased drug deposition in the USP induction port [43].

Wong et al. (2010 and 2011) thoroughly explored the standardized entrainment tubes and CFD with Lagrangian particle tracking to probe the aerosolization consequence. They investigated the influence of grid structure, wall impaction, and turbulence on aerosol performance and the break-up of a model formulation. The model formulation is composed of primary mannitol particles (3.91 μm) and particles agglomerates (642.80 μm). The aerosolization experiments were performed using custom-designed entrainment venturi tubes and impactor apparatus (15–140 LPM; flow rate). The experimental and simulation analysis revealed that the impaction process is mainly responsible for initiating the large particle agglomerate break-up [33–35].

Suwandecha et al. (2014) examined the effect of turbulent kinetic energy (TKE) on inhaler performance. The numerical analysis on Cyclohaler® and Inhalator® was performed using RANS with standard k-ε turbulence model and DPM particle tracking. Three carrier-particle size ranges with a mean diameter of 33, 68, and 73 μm was used. The dispersion flow rates were 30, 60, and 90 LPM. Each particle-wall impaction event was recorded and calculated for the probability of carrier-drug particle agglomerate deagglomeration. The numerical and experimental analysis showed that TKE and FPF had a significant direct correlation [32]. The study was extended to develop a new DPI device from Cyclohaler® and Rotahaler®. Several design modifications of Cyclohaler® and Rotahaler® were critically evaluated with the help of simulation and experimental analysis. The simulation approach used here was similar to previously published data [32]. The grid in the inhaler highly impacted the performance of the inhaler by reducing the turbulence within the mouthpiece [29].

Various commercial inhalers' performances were carefully evaluated by numerical and experimental analysis. Primarily, Donovan et al. (2012) modeled the carrier particle (32, 108, and 275 μm) trajectories using two commercial DPIs (Aerolizer® and HandiHaler®). The aerosolization flow rate was set at 60 LPM. They discovered that both carrier-particle size and inhalation device design had a strong influence on the particle–inhaler collisions, which was an important mechanism for enhancing the inhaler dispersion performance. The Aerolizer® followed impaction-based deagglomeration and its aerosolization outcome was carrier-particle size dependent. Large carrier particles (275 μm) with Aerolizer® showed smaller mass median aerodynamic diameter (MMAD) values. Whereas, HandiHaler® aerosolization outcome was not dependent on carrier-particle size and impaction mechanism [44].

The numerical analysis of the effect of turbulence, air flow characteristic, gravity, drag force, lift force, and pressure gradient inside the inhaler on the carrier-particle motion was performed by Sommerfeld and Schmalfuß (2015). An open-source CFD simulation code (OpenFOAM) was used to run the simulation. RANS with k–ω SST turbulence was used to simulate the airflow. A standard swirl-flow inhaler device, such as Aerolizer®, was explored. The model particles were the monodisperse particle with a mean diameter of 50, 110, and 500 μm and Rosin-Rammler size distribution with a mean diameter of 500 μm. The flow rate was 100 LPM. They found that when the particle size increased, the average particle velocities decreased dramatically and the particle residence time in the DPI increased. However, it had a positive effect on the wall collision frequency due to higher inertia force. It was determined that the critical parameter in Lagrangian simulations of particle motion within inhaler devices was the transverse lift forces generated by shear flow and relative particle rotation [27].

Lee et al. (2018) compared the Activair® (Korea United Pharm. Inc., Korea) to a modified spiral mouthpiece design using numerical calculation, particle image velocimetry, and Andersen cascade impactor. Activair® is a low-resistance multi-dose DPI. The RANS with k-ω turbulence model and DPM particle tracking were employed for numerical simulation. The experiment was performed at the flow rate of 28.3, 60, and 90 LPM. The model particle was Lactohale® 200 (72 μm) mixed with micronized arformoterol tartrate (2.5 μm) and micronized budesonide (2.5 μm) at a ratio of 97.96:0.04:2.00. The CFD results revealed that the non-spiral device had a linear streamline along the longitudinal axis of the inhaler, resulting in the vortex between the longitudinal axis of the inhaler and the wall of inhalers. The swirling motion of the airflow exiting the spiral channel led to the formation of radial velocity in the spiral mouthpiece. The particle residence time in the spiral design was significantly higher than in the Activair® design. In addition, regardless of the airflow rate, the spiral mouthpiece provided more uniform particle velocity and swirl flow. The spiral mouthpiece will significantly improve DPI performance in terms of drug particle dispersion, allowing for a more constant dose distribution [45].

Kopsch et al. (2018) demonstrated a quite different approach to CFD simulation. Three different acrylic entrainment modules were assembled. Three different lactoses were used as model dry powder particles. Primarily, Respitose® SV003 (59.18 μm), Lactohale® 100 (119.35 μm), and Lactohale® 200 (76.82 μm) were used in the present study. The flow modeling was done using RANS with standard k-ε turbulence model. The particle tracking was performed using the Eulerian-Eulerian approach instead of the Eulerian-Lagrangian approach. The inhalation profile was a transient flow rate profile. The CFD results were validated with high-speed camera imaging at 10000 frames per second. The entrainment of both powders Respitose® SV003 and Lactohale® 100 were comparable between experiment and CFD predictions. However, the entrainment process of Lactohale® 200 was found to be unreliable in both the experiment and CFD. The physical properties, mainly the morphology of lactose particles, are critical as they showed a significant impact on the results of the simulation. The strong cohesion force between Lactohale® 200 particles significantly affected the entrainment process. As compared to Lactohale® 200, the Respitose® SV003 and Lactohale® 100 had narrow particle size distribution and good flowability as they are produced by the sieving method. The cohesive interaction of particles and fine fractions was not studied in the present study, and therefore extra experimental and numerical efforts are needed to probe the nature of Lactohale® 200 [37].

The complex fluid and particle dynamics in capsule-based DPI were observed by Almeida et al. (2022) for the development of high-dose inhaled therapeutics. The low and high-resistance Cyclohaler® DPI variation, as well as formulation adjustments that affect the adhesive bonding between the active pharmaceutical ingredient (API)/API and API/excipient particles, were examined in high-dose formulation systems (10 mg or more API). The CFD-DEM method was SST κ-ω turbulence model with Lagrangian particles tracking approach. The interaction between the fluid and solid phase was modeled by two-step CFD-DEM coupling. The first step was transferring fluid flow characteristics to the DEM solver and computing the fluid forces acting on particles. The porosity values and momentum source terms were mapped and transferred back to the CFD solver and applied to the solution at the next time step. The results between the simulation and experiment showed good agreement that capsule discharge was driven by the centrifugal acceleration and air flow entering the capsule, which showed similar trends between the low and high-resistance devices. The particle dynamics after leaving the capsule clearly visualized the segregation effect (Fig. 6.3). The flow field from the high-resistance device resembled the cyclone separator, while the low-resistance device did not. Additionally, the formulation with high API/lactose adhesion and API/API cohesion tended to aerosolize poorly in high-resistance devices, whereas low-resistance devices resulted in better aerosolization [46].

The consistency of results between numerical analysis and in vitro experiments was an interesting issue due to numerical analysis is highly dependent on the model and input data to provide an equivalence prediction of aerosol delivery system performance to the in vitro experiment. Ignjatovic et al. (2021) conducted an experiment to compare the aerosolization performance of a solid lipid microparticle DPI (SLM-DPI), the CFP-DPM simulation to three in vitro assessment methods including the twin stage impinger (TSI), fast screening impactor (FSI), and next generation impactor (NGI). Five formulations with different process parameters, such as mixing time and ratio of lipid, poloxamer, and salbutamol sulfate, were used in this experiment. The aerodynamic and micrometric properties of SLM were measured and used as input parameters for the simulation. The inhaler used in this study was a low-resistance unit dose RS01® device. The particle-wall stickiness model was also included in the CFD-DEM model. The results from in silico and in vitro were

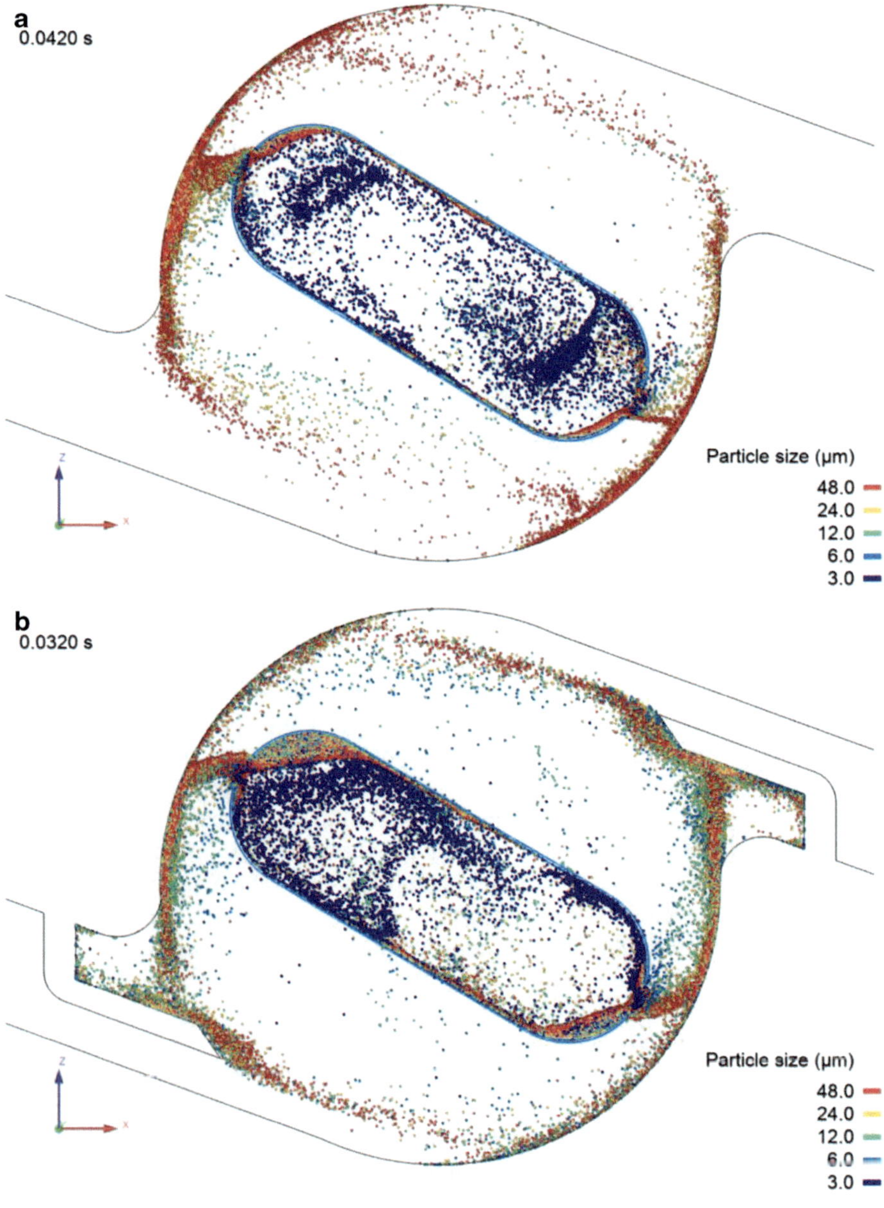

Fig. 6.3 Particle flow inside the capsule of modified Cyclohaler® with low resistance (**a**) and high resistance (**b**) (Reprint from [24] with permission by Elsevier B.V.)

Table 6.2 Comparison of CFD-DPM prediction on dispersion performance in five formulations [47]

Formulation	Composition	Method	d_{v90} (μm)	CFD-DPM prediction parameter	
				% EF	% FPF
F1	1% SS, 0.4% P188	8 min mix TIn 89°C	8.84 ± 0.37	= FSI, =NGI and = TSI	<FSI, <NGI and < TSI
F2	1% SS, 1.5% P188	2 min mix TIn 89°C	8.51 ± 0.06	= FSI, <NGI and < TSI	<FSI, <NGI and < TSI
F3	2% SS, 1.5% P188, Tre	2 min mix TIn 89°C	7.10 ± 0.03	>FSI, >NGI and > TSI	<FSI, <NGI and < TSI
F4	1% SS, 1.5% P188, Tre	2 min mix TIn 89°C	12.21 ± 0.06	= FSI, =TSI and < NGI	<FSI, <NGI and < TSI
F5	1% SS, 1.5% P188	2 min mix TIn 80°C	6.63 ± 0.04	>FSI, >NGI and > TSI	>FSI, >NGI and > TSI

SS Salbutamol sulfate, *P188* Poloxamer 188, *Tre* Trehalose, *TIn* inlet temperature of spray drying, *%EF* emitted fraction, *FSI* Fast Screening Impactor

compared using correlation coefficient, root mean square error, and normalized root mean square error. It was found that the CFD-DPM emitted fraction result was mostly comparable to the three in vitro methods. However, the CFD-DPM failed to correctly predict emitted fraction in a formulation containing trehalose due to the particle-wall stickiness model may not mimic the real situation. For fine particle fraction, CFD-DPM prediction was lower in comparison to experimental values (Table 6.2) [47].

6.4.2 Simulation of the Inter-Particle Forces and deagglomeration

Due to the limitation of simple particle tracking, the inter-particle forces need to be addressed to accurately reflect the dynamic process of powder deagglomeration inside the DPI device and the particle deposition inside the human respiratory tracts. The modeling of inter-particle forces is essential for completely mimicking the real physical phenomenon of dry powder formulation and aerosolization process.

Thornton et al. (1996) used the JKR model with DEM to create an agglomerate consisting of monodisperse 1000 particles with a radius of 100 μm. The agglomeration of particles was imposed with a centripetal gravity field after they were initially scattered around a designated area. Later, the surface energy was added in tiny increments to reach the necessary level up to 3.0 J/m^2. A stable agglomeration was formed by the gradual removal of the centripetal gravity field. Although the JKR model with DEM produced stable agglomerates that mimicked those in actual drug powder, however, it is very complex to understand how agglomeration took place DPIs. Re-attachment of dispersed tiny particles most likely causes new agglomerates to develop during an actual inhaling process [48].

On the other hand, the actual particle agglomeration process was studied by Afkhami et al. (2018 and 2019). The model particle was a spherical particle with a diameter of 50 μm. The particle agglomerate formation experiment was performed in a rectangular flow channel with width, height, and length of 0.25 m, 0.004 m, and 0.126 m, respectively. The particle was randomly placed inside the flow channel. The flow direction followed the length of the flow channel with a mean streamwise velocity of 4.5 m/s. They reported how fluid turbulence and particle surface energy impacted agglomeration. Their findings showed that the turbulent flow dominated particle mobility, which resulted in particle-particle interactions, and there was a positive link between particle agglomeration and particle surface energy. There were two distinct zones with elevated particle agglomeration: (i) near-wall regions due to high particle concentration and (ii) high turbulence regions close to the walls due to the shearing impact of the flow at the no-slip boundary [49]. They extended the investigation to find the effect of turbulence intensity and particle cohesiveness on the agglomeration behavior in particle-fluid flow. The results indicated that the intensity of the flow turbulence had a strong influence on the rate of agglomeration. When dealing with particles that have high surface energy, an increase in the turbulence intensity of the flow directly improved the agglomeration rate. In contrast, the agglomeration of particles with low surface energy decreased as the intensity of the flow turbulence increased [50].

The carrier-based DPI inter-particle agglomerates have also been simulated as it is the first stage in fine particle dispersion modeling. A DEM simulation of carrier-based formulation agglomeration was performed. Primarily, a carrier particle (26 to 52 μm) was created, and drug particles ($n = 242$ and 2.5 μm) were arbitrarily placed across the carrier. Afterward, drug particles were programed to move toward the center of the carrier at a specific velocity until they stuck to the carrier surface and attained a stable position. The influence of adhesion (0.1–0.6 mJ/m^2) and air velocity (1.0–4.0 m/s) was systematically studied in the present case. Simulation analysis showed that dispersion number (number of drug particles detached from carrier surface) improves with enhancing air velocity and decreases with increasing the adhesion force. In summary, the DPI outcome was mainly controlled by the balance of the adhesive and removal forces [51, 52].

Yang et al. (2015) also looked at how electrostatic force affected the mixing performance of a model formulation. The mixture of carrier (35 μm) and API particle (2.5 μm) was used to performed the simulation analysis. A cubic container (105–210 μm; length) was used for mixing. The physical properties of lactose were assumed for both the carrier and API particles. Moreover, pre-charging the carrier and API particles with positive and negative charges, respectively, allowed for the consideration of electrostatic interactions between the particles. It was demonstrated that as vibrational velocity, amplitude, and frequency increased, the contact number decreased. The contact number grew as the charge increased and decreased as the mixing container size increased. Yang et al. (2015) initially concluded that electrostatic force was the dominant force in the mixing procedure. It contradicted the previously proposed mechanism that described the van der Waals as the main contributor to the mixing process. Thus, the long- and short-range adhesive forces had to be studied while developing the mixing protocol [52].

Simulation of the inter-particle force and powder agglomerates can be used to analyze the deagglomeration process during the inhalation. Six different loose aggregates were assembled from 100 μm particles with a number of particles from 250 to 8000 which resulted in an aggregates diameter of 0.764 mm–2.34 mm. The interface energy between particles in aggregate assembly was 0.1 J/m^2. However, one loose aggregate assembly was modeled with interface energies from 0.1 to 1.0 J/m^2. The flow condition for the numerical examination of the aerodynamic dispersion of the aggregates was a uniform flow field. The aggregate may be viewed as a carrier-free powder. The JKR model was used to determine the effect of particle cohesion and contact force, which was closely associated with aggregate surface energy and diameter. There was a relative particle-fluid velocity threshold in their simulation results, and once this barrier was exceeded, dispersion occurred swiftly and approached a completely scattered condition. The behavior of the aggregations was determined by the ratio of the aggregation diameter to the primary particle diameter. Above the threshold velocity, the aggregation broke up if the size ratio was small. However, a larger size ratio caused the dispersion behavior to switch to a much slower stripping of the surface because the force on the aggregate's surface took longer to propagate to the inside [53–55].

The agglomerates-wall impaction was a major mechanism of deagglomeration because the impaction force exerted by the particle can be significantly higher than that created by fluid flow. Thornton et al. (1999) modeled a dense spherical agglomeration with a mean diameter of 1.11 μm composed of auto-adhesive particles hitting orthogonally with a target wall. They found that rebound, fracture, or shattering of agglomerates depended on the magnitude of the force that was directly related to the impact velocity. The plastic deformation of the microstructure inside the impact zone absorbed the majority of the input energy. Interestingly, the agglomeration strength was not largely affected by pre-existing defects [56]. Moreover, several parameters related to agglomeration and deagglomeration phenomenon were also cautiously investigated by the same research group [57–59]. The studies were conducted based on process engineering problems that focused on the primary particle size range from 16 to 60 μm. This size range was larger than a typical aerosol particle for inhalation. The agglomeration of 5000 primary particles with a median diameter of 20 μm was simulated to understand the impact with a wall at a velocity range of 0.5 m/s–2.0 m/s. It was found that the impact velocity had an inverse relationship to the deagglomerate fragment size. The portion of fine fragments also had an inverse relationship to the solid fraction. In addition, the contact density determined the mode of deagglomeration. The low contact density agglomerate resulted in deagglomeration by disintegration, while the fractured mechanism dominated deagglomeration in a higher contact density agglomerate [57–59]. The mathematical model related to agglomeration and deagglomeration phenomenon was demonstrated by Tong et al. (2009). They used 3D DEM to simulate the deagglomeration of granular materials. The agglomerate was treated as an assembly of discrete particles with a size of 5 μm. Initially, 5000 mono-dispersed, non-overlapping particles in a spherical region were randomly generated. The particles relocated toward the space center as a result of the centripetal force. The process was repeated until all of the particles reached their stable positions, at which point practically all of the kinetic energy of individual particles were stabilized and dissipated. The agglomeration then collided with a target wall at various velocities and angles. To generate statistically reliable results, each impact was simulated four times with the agglomerate rotated randomly. According to the DEM results, the formed agglomerates were not strongly bound together and experienced plastic deformation on impact before being dispersed into small pieces. The damage ratio of agglomerates rose as the impact velocity increased. The impact angle dependency was investigated from 10° to 90° angle. An impact angle of 45° showed the best agglomerates break-up performance, while an impact angle of 10° and 90° showed poor agglomerates break-up performance. The shallow angle of 10° provided less force distribution to the agglomerate structure, while the impact angle of 90° generated large magnitudes of force but less covering area compared to the 45° angle [60].

Tong et al. (2010) studied the DPI powder aerosolization mechanism using CFD with the DEM technique. The CFD was used to observe the aerodynamic flow profile inside the inhaler, while DEM was used to model the deagglomeration from the fluid force. The simulation was used to predict the agglomeration behavior of different particle sizes (d_{50}_2.60–4.14 µm) in a cyclonic flow. The cyclonic flow field mimicked the flow characteristic found in the swirling chamber of the commercial DPI, Aerolizer®. The results showed that particle-wall impact dominated powder dispersion, with internal shearing caused by air movement playing just a minor role [61]. The powder deagglomeration in various customized impaction throats was investigated by Tong et al. (2011). The throats were joined from two tubes with three different angles of 15°, 45°, or 90°. A total of five throats were modeled, as shown in Fig. 6.4. They found improvement in dispersion performance for customized impaction throats with two angles. The first angle provided impaction that caused the initial break-up of the agglomerate, whereas the second produced more fine particles smaller than 5 µm [62].

Furthermore, the CFD and DEM technique was used to examine the mechanisms of particle dispersion in the Cyclohaler® device. The swirling chamber wall of the Cyclohaler® was a major collision surface for the agglomerates, as multiple significant collision events were observed inside the swirling chamber. However, the de-agglomerated fragments after the impaction were found as large fragments without

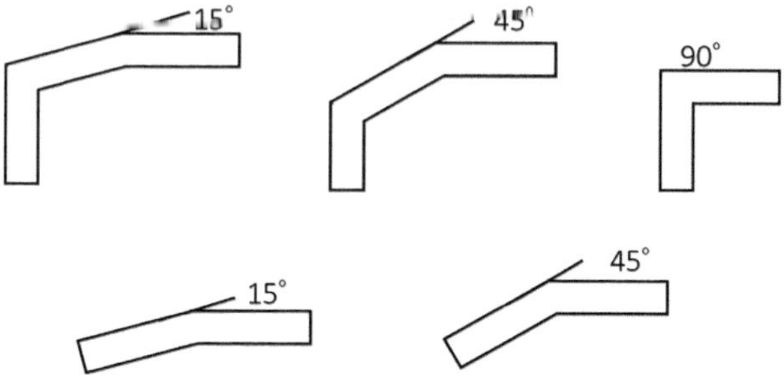

Fig. 6.4 Schematics of impaction throats (modified from Tong et al. [62] with permission by Elsevier B.V)

creating much fine particle fraction. The collision with the grid was important and discovered as the primary cause of the FPF [63].

Breuer and Khalifa (2019) tried to improve the breakup models for compact powder agglomerates exposed to turbulent flows using a four-way coupled Euler–Lagrange hard-sphere approach applied in combination with the large-eddy simulation technique. The model agglomerates were formed from the primary particles varied between 0.97, 2.47, and 5.08 μm. The simulation was conducted under the main hypothesis that when the associated stresses exceed the agglomerate's strength, it will break up. Three stresses expose to agglomerates; (i) drag stress, (ii) rotary stresses, and (ii) turbulent stresses (variations in fluid velocities inside a turbulent flow). Pressure and velocity gradients which were caused by velocity differences on opposite sides of the agglomerates might lead to deformation or break-up of the agglomerates. Drag-induced stress was the stress from fluid drag force and an inertia force on a cord of agglomerate when it suddenly faced steady flow with a uniform fluid velocity. The rotary stress was caused by the rotation of agglomerates, which may translate to the centrifugal force acting on the largest cross-sectional area of the agglomerate. The post-breakup treatment was implemented to model the kinetics of fragments after the occurrence of breakup. The simulation of fluid flow was performed at 25 LPM. It was found that the drag stress was the main mechanism followed by the rotary stress for deagglomeration, while breakup due to turbulence does not occur for the low Reynolds number case studied. The numerical simulation was extended to the high-Re case. The main mechanism of deagglomeration was also the drag stress; however, the breakup of agglomerates by turbulent stresses showed a marginal role. The inter-particle collisions or particle-wall collisions had some important roles in the breakup by drag stresses and rotary stresses [64]. Although this research was not aimed at DPI, it still provided valuable information for the further development of DPI.

Until recently, simulation was used to probe the interaction between the agglomerates and the wall of the inhaler device. However, the interaction between the agglomerates was not thoroughly investigated. Thus, Tong et al. (2016) cautiously used the CFD-DEM approach to simulate the collision between two mannitol agglomerates. Two identical sphere mannitol agglomerates (51 μm) were introduced separately with an initial velocity the same as the flow velocity (10–30 m/s) at the feeding points through the left and right inlets of T-shape pipes. The results showed that the collision process of two agglomerates could be divided into two stages. The first stage was the restructuring stage which occurred in a very short period. The second stage was the fragmentation stage which lasted a more extended period. The FPF values were, directly and indirectly proportional to airflow rate and agglomerate strength. Furthermore, increasing collision angle was also responsible for increased FPF values and fragment size [65].

Dizaji et al. (2020) investigated the collision of two agglomerates in a shear flow. They used a more sophisticated four-way coupled CFD-DEM approach. The fluid phase was resolved using a high-order finite difference method, while the particles were simulated using an adhesive soft-sphere DEM. The fluid flow was not resolved to the individual particle scale. The fluid forces acting on the particles were treated as a smooth body force field, the overall forces acting on numerous particles in a particular location. The computations considered both formations of agglomerates in a turbulent flow and deagglomeration. The deagglomeration was examined in detail on the collision process that occurs when one or two agglomerates are taken and placed in a plane shear flow. It was meant to mimic a relatively small segment of the entire turbulent flow. It was found that depending on the adhesion parameter and the offset distance to the agglomeration radius of the gyration ratio, the collision could result in agglomerate merging, bouncing, and fragmentation [66].

van Wachem et al. (2020) investigated the particle–particle and particle–wall interactions using DEM and the discrete fragmentation model (DFM). The agglomerates were prepared using 2 μm primary particles resulting in five different sizes of agglomerates ranging from 10 to 50 μm. The numerical simulation was performed with micro- and macro-scale. The micro-scale simulation included collision between pair of agglomerates and agglomerate-wall collision. The simulations of agglomerate–agglomerate collisions revealed that the fraction of disaggregated primary particles were directly related to a function of the dimensionless collision intensity and particle properties. They also found that at low impact intensity, the agglomerate stuck and remained on the wall. In contrast, the agglomerate broke up at a higher intensity, and most primary particles bounced off the wall. The modeling fitted with micro-scale collision. The macro-scale simulation modeled the agglomerates as single-parent particles. When an agglomerate collision occurred, the remaining volume of agglomerates was calculated using a fraction of broken-off small fragments. The force distribution after the break-up was determined and assigned to primary fragments. They introduced a novel dimensionless parameter to capture the attributes of the agglomeration and the collision that incorporated Webber number, number of contacts, and elastic adhesion index. The macro-scale simulation with DFM, probability density framework, and dimensionless parameter fitted well with micro-scale simulation without resolving the dynamic characteristic of each individual primary particle [67].

For the carrier-based deagglomeration process, Yang et al. (2014) used a CFD-DEM approach to investigate the dispersion of a carrier-based agglomeration in a uniform airfield. The carrier particle in the size range of 26–52 μm was covered with 2.5 μm drug particles. The carrier-drug particle agglomerate was placed in the airflow of 1.0–4.0 m/s. The side of the agglomerate that faced in-flow was labeled as the upstream side, while the opposite side was labeled as the downstream side. They discovered that the fine API particles on the downstream side of the carrier-particle agglomerate were directly taken off when airflow was introduced. The API particles in the middle sections of the agglomerate were transferred to the downstream side before they began to separate from the carrier surface. On the other hand, the majority of the API particles on the upstream side were still intact [51]. The CFD-DEM technique was later used to investigate the interaction and effect of a carrier-based agglomeration and collision surface. Two dispersion stages were observed that were rapid dispersion and stabilization. Particle dispersion caused by particle-wall impaction could be well represented using the cumulative Weibull distribution function, which is controlled by the balance between the impact energy and adhesion energy.

The effect of the fine particle on the carrier and the transfer of fine particles during drug–carrier agglomerates collision was examined using DEM analysis. The carrier-fine particle model was to deposit 3 μm fine lactose particles on the 200 μm mannitol carrier particle to form three layers with an even distribution on a spherical section of a carrier particle. The experiment revealed that the kinetic energy derived from impact velocity up to 1.2 m/s was directly related to the fraction of fine particle transfer before being reduced at higher impact velocities. The interface energies within the fines and between the fines and carrier, had a negative effect on the fine particle transfer. In addition, fine particle fragments were reorganized after a collision, and those reattached to another carrier exhibited higher tensile strength [68]. The influence of the fine drug particle loading on the aerosolization performance after impaction with the target wall was observed by Tong et al. (2017). The simulation was performed using CFD-DEM with various carrier-drug mass ratios. The model formulation was polystyrene carriers (121.8 μm) with salbutamol sulfate particles (2.1 μm) at three different ratios (1:5, 1:10, and 1:100). The numerical analysis was done at the flow velocity of 20 m/s and impact velocity of 15 m/s. The result showed no change in aerodynamic outcome (e.g., FPF values) with an increasing drug loading until a critical threshold was surpassed. Increasing the carrier-drug mass ratio could improve aerosolization performance in the formulation with small carrier particles. Moreover, the wall characteristic was also important to investigate [28].

Ariane et al. (2018) studied the dispersion process of drug particles from an inhaler. Mainly, the effect of the carrier rotation (±100,000 rad/s), angle of impact (between 5° and 90°), and impact velocity (1 and 20 m/s) were examined. The DEM approach was used to simulate the fluid dynamic effects. The 3D model was utilized with a single carrier particle of 100 μm diameter and 882 drug particles of 5 μm diameter to study the dispersion process. Results showed that agglomeration rotation played a significant role in dispersion if the rotational agglomerate kinetic energy was less than the translational velocity. The dispersion ratio was slightly affected by an elastic or sticky wall [69].

The detachment rate of the fine drug particles from the carrier particles during the agglomerate-wall impaction was calculated using a fully resolved Lattice–Boltzmann method (LBM). Initially, a micro-scale model based on LBM was developed, which featured a 100 μm spherical carrier particle randomly covered with a monolayer of small 5 μm spherical powder. The wall impaction velocity was set at 10 m/s with a variable angle of 1°–90°. They found that increasing the impact angle from 1° to 20° improved the lift-off probability of drug particles from 27.92% to 50%. Beyond 20° impact angle, the lift-off probability of drug particles reached a plateau. The remaining drug particle that was not lift-off was probably detached by a sliding or rolling mechanism [70]. The effect of surface roughness on the detachment of fine particles from carrier surfaces was investigated using CFD or DEM. The drug particle size was kept constant (5 μm). The roughness of carrier particles was modeled as a half-sphere patch on the surface with a radius of 1.25 μm or 1.56 μm, and the length between half sphere was varied from 3 to 5 μm. They found that the lift-off, sliding, and rolling mechanisms played an essential role in the detachment of particles from rough surfaces. The rolling detachment correlated very well with

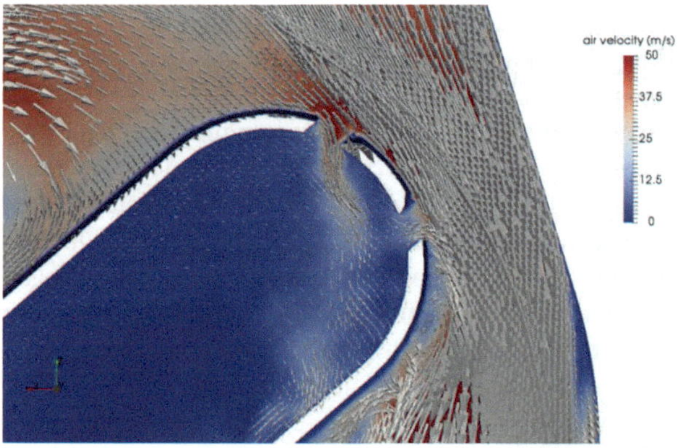

Fig. 6.5 CFD simulation of the airflow through two of the capsule holes at a capsule angle of 45° in a plane normal to the primary axis (Reprint from [24] with permission by Elsevier B.V.)

asperity contact distance, while the drag and lift force had an inverse relationship with the contact distance. At a small contact distance, the drag force was quite large compared to what occurred in the case of a spherical particle on a plane wall [71].

To date, there have been no computational or experimental studies performed to investigate collisions between capsules and the inhaler. Thus, Benquea and Khinast (2019) studied the motion of a capsule in the swirl chamber by high-speed image analysis. The high-speed images were used to determine the rate of capsule-inhaler collisions, the collision velocity, and the collision locations. The resulting capsule motion information was then used as an input in particle simulations to study the effect on the particle discharge from the capsule. Specifically, CFD simulations were used to study the effect of drag force and inertial forces on particles within the rotating capsule. A transparent Aerolizer® device (Plastiape S.p.A) and size 3 capsule containing 25 mg powder (salbutamol 200 mcg/capsule) were used in this study. The numerical analysis was performed using LES and DEM. Powder cohesiveness was simulated using a macroscopic elastoplastic adhesive (MEPA) contact model. The capsule was set to rotate at a constant speed of 4500 rpm. The particle image velocimetry was used to validate and confirm the simulation results. The CFD revealed the airflow velocity profile inside the capsule of up to about 10 m/s, and the local peak velocity was 35 m/s close to the capsule holes (Fig. 6.5). The minor air flow path escaping the capsule was through the tangential hole of the capsule to the adjacent air flow filed outside the capsule. There was some airflow redirection inside the capsule that facilitated particle mixing and fluidization. In addition, the carrier particles inside the capsule, which were relatively large compared to the drug particles, were dominantly affected by the inertial force. In contrast, the smaller drug particles inside the capsule received significant force from the drag from the airflow [72].

6.4.3 Modeling of Advanced DPI Designs and Formulations

Recently, the pulmonary delivery of high-dose medication using DPI has gained more attention. It is very challenging due to aerosolizing such high amounts of powder using the patient's limited inhalation force. The high-dose DPIs were

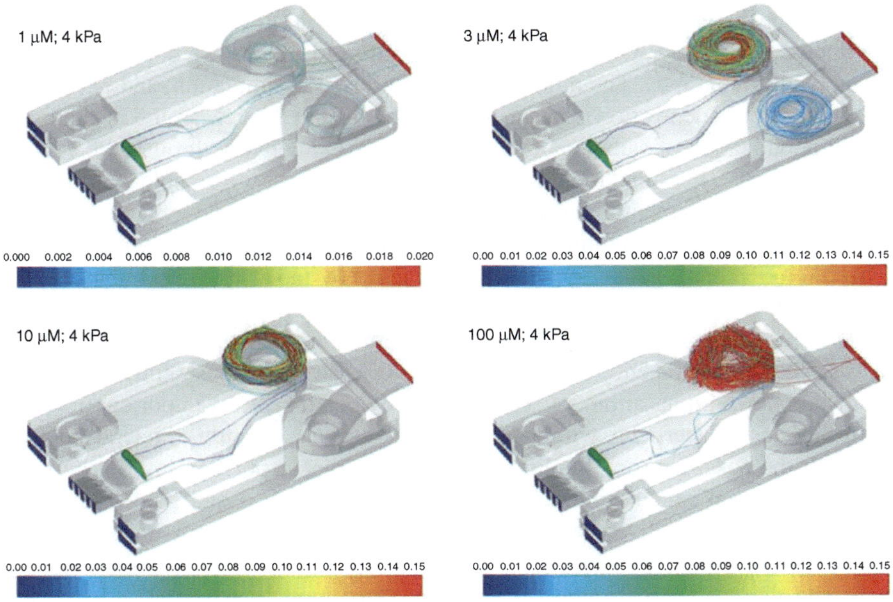

Fig. 6.6 Particle residence time in Twincer™ from CFD simulation for particles size of 1 μm, 2 μm, 10 μm, and 100 μm (Reprint from [24] with permission by Oxford University Press)

usually formulated without carriers or in a low quantity of carriers. Fine drug particles are very adhesive and challenging to disperse efficiently. Thus, innovative designs of the inhaler are necessary.

De Boer and team proposed a new design based on the cyclone separator principle to deliver high-dose antibiotics, colistimethate sodium. They named it air classifier technology that was implemented in Twincer™ DPI [73–76]. The device design modifications were tested numerically and experimentally. The air classifier in the Twincer™ device is constituted by two parallel circular chambers with three airflow channels each. The CFD analysis was performed using realizable k-ε with disperse particle model. It was found that the air inlet bypass contributed little effect to the classifier airflow characteristics. They also found that small amounts of lactose particles (150–200 μm) in the formulation played a significant role in the device's performance. The simulation visualized the retention time of the particles in the classifiers. The small particles escaped the classifiers as earlier as 0.006 s. In contrast, the large lactose particles retained much longer (Fig. 6.6). The large lactose carrier acted as a sweeper to reduce the fine particles-inhaler wall adhesion and improved the drug delivery performance. Additionally, the pressure drop across the device had a practically low impact on the aerosolization performance [36].

An interesting new method of delivering aerosol to the lungs is nose-to-lung (N2L) aerosol delivery which is useful for administration to infants and children due to the fact that infants and small children are generally nasal breathers. Bass et al. (2019) established a CFD method for simulating N2L aerosol administration to an infant based on comparisons with concurrent in vitro experiments. The 3D geometries of nasal airways from 3 to 6-month-old infants were obtained using computed tomography. The models were used for CFD simulation, and 3D printed for in vitro drug deposition experiments. The aerosol in this research was generated using nebulization of salbutamol sulfate solution and delivery through the ventilation interface. The simulation was performed using a low Reynold number k-ω model. It was found that the polyhedral meshing of the nasal cavity model allowed faster convergence and required fewer control volumes. The turbulent flow was discovered in the nasal geometry, which was caused by input jets from the nasal cannula interface. However, because of the small particle size, turbulent dispersion had little influence on deposition. Based on experimental and CFD simulation results, aerosol delivery efficiency was higher than 90% via the delivery tube, nasal cannula, and newborn nasal model. This result provided a piece of important information that aerosol particles with MMAD of about 1.5 μm could efficiently deliver through the nasal to reach the lungs with low inertial deposition loss in the nasal cavity and nasal cannula interface [77]. The initial method and guideline for the simulation of N2L delivery were adapted to develop the DPI–patient interfaces for children. The interfaces were designed to have two models: the mouth–throat interface and the nose–throat interface. Each interface model has several design variances by the gradual and rapid expansion of geometry, adding rod array, and teardrop-shaped baffle. The CFD simulation and particle tracking were achieved using DPM and low Reynold number k-ω with user-defined near-wall function. Among various design variations, the best mouthpiece interface design was the design with wide expansion tapped to the outlet with rod array 343 configuration close to the capillary inlet. It reduced the drug loss in the extrathoracic to approximately 5%. For nose-throat delivery, the nasal cannular with wide expansion tapped to the outlet with rod array 343 configuration close to the capillary inlet equipped with an extended capillary [78].

Kopsch and team conducted a series of research to determine the feasibility of creating a DPI device that efficiently operated independently of patient inhalation force. The simulation was based on a two-phase Eulerian–Eulerian approach where the drug particles were modeled as a second continuous phase, and a volume fraction was added to track the local fraction of the drug phase. The airflow inside the DPI device was split into two paths that were air bypass channel from the inlet to the mouthpiece and airflow through the entrainment compartments. The adaptive bypass element in the shape of an elastically deformable flap vale was introduced to the DPI device. When applied pressure decreases are considerable, the flap should bend to allow for a larger flow rate than would otherwise be possible. Initially, the flap's tip is "pre-loaded" against the channel's upper barrier. Pressure on the flap's surface pushes it down when a pressure decrease is delivered across the channel. Low pressures will not be enough to overcome the pre-loading. The CFD simulation with fluid-structure interaction was conducted to simulate the behavior of the variable bypass element. The parametric optimization methodology was used to determine the optimal layout of the device. The quantify performance of the bypass design was determined by two cost functions, which were the similarity of drug

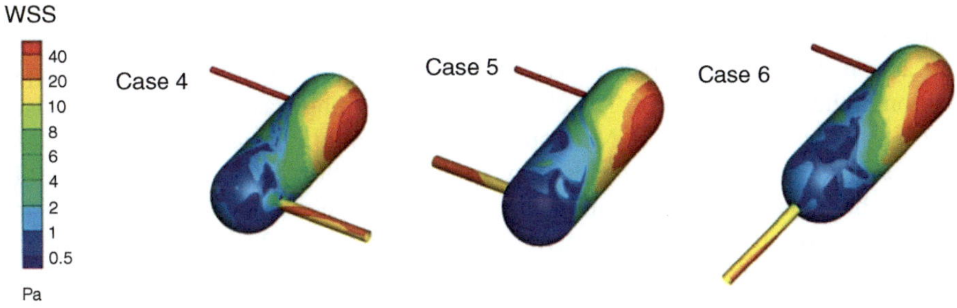

Fig. 6.7 Contours of WSS distribution on DAC unit surface in Case 1–6 (Reprint from [24] with permission by Elsevier B.V)

release with different inhalation profiles and the comparison of release to a target function. The fluid-structure simulation and the experimental results showed good agreement and were able to analyze the behavior of a flap valve variable bypass. The variable flow resistance DPI device using a variable bypass mechanism was able to maintain a higher flow rate compared to what would be achievable with fixed flow resistances DPI [37, 79–82].

Dai et al. (2021) studied the effects of device design on drug delivery efficiency for an active DPI named dose aerosolization and containment (DAC) units. The DAC devices mainly consist of the inlet capillary, containment volume (capsule size 0), and outlet capillary. The effects of piercing aperture location, reservoir length, and inhaler inlet flow rate on emitted dose and particle deposition were evaluated. The CFD simulation in each device was resolved by the low Reynolds number k-ω turbulence model with near-wall corrections. The inlet flow rates were 2–5 LPM, resulting in outlet flow rates of 15, 30, and 45 LPM. The albuterol sulfate (1 to 20 μm) particles were used in study. The particle tracking was performed using the Lagrangian approach. The flow field parameters were extracted from CFD, such as turbulent kinetic energy, the volume-averaged value of turbulent kinetic energy, and wall shear stress. They suggested that the offset between the inlet and outlet orifice and the high flow rate increased emitted dose. These effects were clearly visualized from the contour render of the wall shear stress of Case 4–6 design (Fig. 6.7). The reservoir inside the DPIs device provides an area for the strong jet at the DAC unit's outlet to relax and disseminate, resulting in improved delivery efficiency in the mouth and throat. The higher deposition in the mouth and throat correlated to the larger particle size and higher inhalation flow rate. At a flow rate of 30 LPM, 10 and 100% particles of 1 and 10 μm size were deposited in the mouth and throat region, respectively [83].

6.4.4 Modeling of Powder Deposition with Patient Factors

Patient factors are also important to DPI performance in real-world usage. The inter-patient variation and disease states can affect the inhalation maneuver and the dispersion and deposition of aerosol particles. Therefore, it is challenging to ensure that the delivered dose of dry powder inhalation is consistent and absorbs or exerts its action. The inhalation pattern was an important factor affecting the particle deposition in the airway passage. The effect of breath holding was studied in comparison to the steady flow condition. Medical imaging was used to create the airways model (CT scan). Inhalation was numerically simulated at flow rates of 28.3 and 60.0 L/min, which matched the inhalation conditions in patients with bronchial asthma and healthy participants, respectively. A total of 4,538,820 fluticasone particle (D_{50} = 5 μm) was introduced into the mouth of the airways model. Many of the particles ingested were deposited in the airways as a result of breath holding during inhalation, while the bulk was deposited in the mouth/throat area as a result of exhaling without breath holding. The turbulence caused by retaining one's breath aided particle accumulation in the airways. As a result, one of the most important factors influencing the therapeutic impact of formulations is breath holding [84].

Benque and Khinast (2022) conducted research on the effect of the inter-patient variability on the dispersion in Aerolizer® inhaler using CFD-DEM simulation. The inhalation profiles of patients with asthma aged 18–55 years, children with asthma aged 5–17 years, and COPD patients over the age of 55 years were taken from the literature [85]. The cohesive polydisperse lactose particles (30–95 μm) were used. The simulation results revealed that small carrier particles (30.6 μm) were found to strongly follow the airflow and reach higher velocities of 18–20 m/s. The inhalation profile simulation created varied air velocities and caused the particle velocities to vary vastly. However, the constant flow rate (60 or 100 LPM) simulation resulted in uniform velocities of particles due to the flow simulation being considered as the pseudo-steady condition. The frequency of particle acceleration incidents at high magnitudes (>10^7 m/s^2) seemed to compare with the peak inhalation flow (PIF). The particle acceleration incidents in the swirl chamber at high magnitudes, i.e., 1.6×10^7–2.5×10^7 m/s^2 are comparable with 10 and 2 times for normal adults (93.7 LPM) and children (71.3 LPM) PIF, respectively. The effect of the inhalation profile on the distribution of the cumulative impulse proved to be minor. In contrast, the parameters of the inhalation profile had a considerable influence on the peak force distribution [86]. The study further explored the effects of gravity on the motion of the capsule, powder dispersion, and emission with various patient inhalation profiles. The initial powder bed configuration was obtained from high-speed camera recordings of the powder pattern inside a capsule that was installed inside a transparent Aerolizer® at a flow rate of 50, 70, and 100 LPM. They found that the initial powder bed configuration and the capsule position had a strong effect on the powder aerosolization. The gravity and initial flow profile had remarkable interaction in the first hundred milliseconds of inhalation. The low rotation rate of a capsule (850 rpm) in children with asthma profiles could not produce enough centrifugal force (1.034×10^{-13} N) to overcome the gravitational force (2.047×10^{-12} N). That caused an uneven distribution of powder and the initial release of the carrier-particle clusters from the capsule. When the capsule holes are aligned with the inhaler's air inlets, airflow into the capsule prevents powder release [87].

CFD was used in a study of cystic fibrosis (CF) patients' inspiratory maneuvers on lung deposition of PulmoSphere™ tobramycin inhalation powder (MMAD 3.0–3.8 μm) that was administered with Podhaler™. The 12 patients (12–45 years old) were selected from the observational study (https://trialsearch.who.int/Trial2.aspx?TrialID=NTR5212). An inhalation profile recorder was used to record the airflow profile at the patient's residence. The forced expiratory volume in 1 second of the subjects ranged from 35 to 110%. Inhalation profiles were divided into three categories: instructed slow, uninstructed, and instructed fast inhalation. The 3D lung models were created using chest CT scans of CF patients. When compared to uninstructed or instructed fast inhalation, CFD analysis revealed that instructed slow inhalation of tobramycin inhalation powder resulted in significantly higher tobramycin concentrations in big and small airway models. The tobramycin deposition in small airways was around 700 ng/ml for uninstructed or instructed fast inhalation compared to 1200 ng/ml for instructed slow inhalation [88].

In addition to the inhalation profile, the head position of the patient also affected the particle deposition pattern in the airways. The physiologically realistic mouth-throat geometries were scanned using MRI and employed as a CFD model in three head positions (head-straight, head-up, and head-left). An arbitrary cylindrical inhaler mouthpiece with 4 mm radius and 1200 mm length was used. The mouthpiece up to 600 mm was inserted into the mouth-throat region. A total number of 250,000 rigid spherical particles with diameters diameter of 0.5–15 μm was uniformly released at the inlet of the mouth-throat with an inhalation flow rate of 15, 30, and 60 LPM. The mouthpiece of an inhaler and the lips were imitated at three different angles (i.e., 0°, 15°, and 30°). Head posture was discovered to have no effect on particle deposition in the oropharyngeal area. Whereas the inhaler mouthpiece angle, particularly 15° -orientation at a modest flow rate, significantly enhanced drug loss in the oropharyngeal area. The flow field structures at the mouth area are obviously affected by the inlet direction, which in turn affects particle trajectory. Because of the powerful jet flow, direct inertial impact was discovered along the symmetry plane of the mouthpiece and oral cavity. The deposition far from the symmetry plane, on the other hand, was caused by centrifugal inertial impaction. The presented experimental and simulation analysis suggested that future inhalers should have components that force users to hold the inhaler at the proper orientation angle [89].

The highly variable aerosol deposition of the preterm infant was rarely characterized. Preterm infants often do not have fully formed nasal meatus structures or curved turbinate flow passageways, which grow and become more noticeable with age. Bass et al. (2022) utilized CT scans of a 28-week-old preterm infant and some part of the 6-month-old infant's airways to form a full model from nostrils through the nasal cavity, nasopharynx, larynx, and a section of the trachea. The nasal prong interface was used to deliver aerosol to the preterm infant nose-throat model. Eight different prong configurations included (i) single- and dual-prong configurations, (ii) external and internal prong types, and (iii) internal prongs at three insertion depths. For the internal prong, the insertion depth varied. The drug formulation was a spray-dried spherical salbutamol sulfate excipient enhanced growth formulation with a mean diameter of 100–900 nm [90]. The CFD model validation showed that experimental data and numerical analysis results were in good agreement. The drug loss in the nose-throat area with the single and dual internal prong with 5 mm insertion cases and a flow rate of 1.85 LPM from in vitro experiment was 25%, and 24%, respectively, and from CFD analysis was 26% and 34%, respectively. The MMAD was 1.9 and 2.63 μm for the single and dual internal prong cases, respectively. The turbulent jet effect originating from the prong was clearly visualized and observed by the CFD simulation. For example, strong chaotic flow behavior was observed through the nasopharyngeal and laryngeal regions by using the single external prong, while the dual external prong reduced the flow rates entering the nose-throat region. The prong insertion depths of 2–5 mm seldom affected the aerosol deposition in the nose-throat model, whereas the single prong with 5 mm insertion remarkably caused the deposition [91].

Prediction of particle deposition in the realistic human airways was a major step in the numerical analysis of the aerosol delivery system. However, the deposition alone does not ensure that the drug was appropriately absorbed and distributed as expected. Vulović et al. (2018) developed a computational method to couple the CFD and physiologically based pharmacokinetic (CFD-PBPK) modeling. For the CFD part, Aerolizer® was chosen as a model inhaler, and the low Reynold number k-ω model and DPM were used for the flow simulation and particle tracking. The physicochemical and aerodynamic properties of amiloride HCl DPI (20 mg) and nebulized solution (1.91 mg) were used for CFD-PBPK coupling simulation. The drug particles in the formulations were prepared using jet milling or spray drying. The median particle size of four jet-milled formulations (JM1-JM4) was 1.71 μm, 2.07 μm, 2.62 μm, and 2.96 μm, respectively. The median particle size of the spray-dried formulation (SD) was 2.89 μm. The inhalation flow rate was fixed at 28.3 LPM. The % FPF of JM1-JM4 and SD was 39.78, 41.25, 45.67, 50.62, and 50.19, respectively. Concentration-time profiles of inhaled amiloride were predicted from the pulmonary absorption parameters and drug deposition pattern. Data from prior studies and the oral dosage form were used to validate the physiologically based pharmacokinetics (PK) model. Peak concentration (C_{max}), estimated time to peak concentration (T_{max}), and area under the curve (AUC) were all comparable to in vivo data for PBPK validation. The C_{max}, T_{max}, and AUC from the 20 mg immediate release capsule simulation were 40.70 ng/ml, 2.44 h, and 761.58 ng·h/ml, respectively, and from in vivo study were 30.20 ng/ml, 2.63 h, and 757.14 ng h/ml, respectively. The extrathoracic deposition was considered for inhalation dosing in accordance with the oral dosage form. The lung volume, blood flow, drug tissue-to-plasma partition coefficient, and blood-to-plasma concentration ratio were used to estimate the pulmonary absorption coefficient. According to the PK simulation, approximately 50% of the amiloride that was breathed was absorbed through the lungs, with the remaining amount being absorbed from the gastrointestinal tract. The predicted C_{max} of JM1-JM4 and SD was directly proportional to the pulmonary absorption percentage. Specifically, the pulmonary absorption of amiloride dry powders i.e. JM2 and JM4 was 39% and 44% while the simulation predicted C_{max} was 24 ng/ml and 33 ng/ml. Similarly, the C_{max}, T_{max}, and AUC of simulated nebulized solution were 1.35 ng/ml, 0.48 h, and 2.79 ng h/ml, respectively compared to in vivo study, i.e., 1.33 ng/ml, 0.52 h, and 17.61 ng h/ml, respectively [92].

6.5 Future Trend of Numerical Simulations in Aerosol Delivery System Research

The use of computational analysis has considerably increased in recent years. The CFD simulations such as LES and DNS are more feasible in the present situation to investigate the complex drug/device combination. The LES and DNS can resolve a much more accurate simulation of the turbulent flow field than RANS. The particle behavior simulation is far more complicated than the flow field, including near-realistic dynamic particle interactions and the coupling effect between the solid phase and the fluid phase. The useful simulation strategy that has been proposed is multi-scale modeling, where the micro-scale and macro-scale are analyzed separately. However, there is still room for great improvement in setting up the simulation environment to mimic the real physical and chemical phenomena during the aerosolization process and drug particle outcome in the airways. These processes include the growing effect of the particle in the high humidity of human airways and particle deposition patterns in special populations such as neonatal, pediatric, and children. A higher level of simulation fidelity expects to extend the application in the aerosol delivery system design, including:

- Targeting the deposition in the specific area of the airways seems possible as the numerical analysis can predict the deposition pattern from the inhaler to the lower part of the airways. The parametric assessment and optimization for inhaler devices and dry powder formulation can be used for pulmonary drug targeting. This application was well studied by Das et al. (2018). The CFD-optimized delivery system could target the upper airways of children for nearly 80% [93].
- Numerical analysis is a valuable tool during formulation development. Spray drying is one of the most common manufacturing methods to develop DPIs formulation. The CFD technique can optimize the processing parameters of the spray dryer. The significant difference between the simulation in the aerosol delivery system and the spray-drying process is that the heat transfer problem had to be resolved with CFD and particle simulation. The CFD simulation of Buchi® spray dyer system showed promising potential as it was able to predict drying parameters to produce particles with the MMAD and ED comparable to the experimental results. These measurements yielded excellent linear correlations. An efficient tobramycin excipient enhanced growth formulation was achieved by the CFD-based spray drying optimization, which produced 100–900 nm spherical particles [94, 95].
- The coupling of CFD simulations with pharmacokinetics simulation is very useful for aerosol drugs aiming for systemic absorption. The drug deposition pattern in the airways can be transferred to a pharmacokinetics model such as physiologically based pharmacokinetic and calculated for the pharmacokinetics profile

such as plasma drug concentration. This application has been demonstrated in many environmental inhalation risk assessments and inhaler studies [92, 96–98].

- The fluid structural interactions (FSI) simulation also needs to be improved. Many design ideas for the inhaler device comprised mechanical elements that can dynamically interact with the airflow, such as a flap and capsule. The flutter-based dispersion mechanism was proposed and demonstrated by Selvam et al. (2010) [99].
- The advent of data science, such as machine learning (ML), can be applied to numerical simulation research. The parametric optimization using CFD and ML was demonstrated by Kotteda et al. (2020). The TensorFlow machine learning framework was trained and validated with the dataset from CFD. The optimized parameter was re-evaluated with CFD [100].
- The US Food and Drug Administration (FDA) recently published draft guidance on inhaler products, including DPI, pMDI, SMI, and nasal spray. (https://fda.report/media/130621/5+Locally-Acting+Orally-Inhaled+and+Nasal+Drug+Products.pdf). FDA has explored the potential of CFD modeling to understand and predict the performance of orally inhaled and nasal drug products. The CFD-PBPK models had the possibility to predict deposition, tissue concentration, blood concentration, and pharmacodynamic outcome after administration. New drug applicants might use this in silico data to waive BE study [101].

6.6 Summary

CFD is a valuable tool for developing the aerosol delivery system that can numerically explore the airflow field, particle trajectory, agglomeration, deagglomeration, and particle deposition inside the inhaler and extend to the human airway model. The application in developing an aerosol delivery system, such as those with high-dose, nose-to-lung route, pediatric patient compatibility, and independence from patients' breathing efforts, has been discussed carefully in the chapter. The most popular numerical approach in aerosol delivery system research is RANS and Lagrangian particle tracking, while many research groups are progressing to reach the feasibility of LES. Although computational modeling has several advantages for developing DPI devices, it still has some limitations when compared to conventional devices. A lot of dynamic efforts, such as developing an advanced mathematical model and extension of the numerical simulation beyond drug deposition to probe biological response, will be needed to construct the most practical numerical tool for effective pulmonary drug delivery research. In the near future, the in silico method and computer-aided pharmaceutical aerosol development will play a significant role in both the research and regulation of new drugs.

References

1. Hickey AJ. Emerging trends in inhaled drug delivery. Adv Drug Deliv Rev. 2020;157:63–70. https://doi.org/10.1016/j.addr.2020.07.006.
2. Xiroudaki S, Schoubben A, Giovagnoli S, Rekkas DM. Dry powder inhalers in the digitalization era: current status and future perspectives. Pharmaceutics. 2021;13(9) https://doi.org/10.3390/pharmaceutics13091455.
3. Rospond B, Krakowska A, MuszyŃSka B, Opoka W. The history, current state and perspectives of aerosol therapy. Acta Pharma. 2022;72(2):225–43. https://doi.org/10.2478/acph-2022-0017.
4. Ye Y, Ma Y, Zhu J. The future of dry powder inhaled therapy: promising or discouraging for systemic disorders? Int J Pharm. 2022;614:121457. https://doi.org/10.1016/j.ijpharm.2022.121457.
5. Anthony JH, Sandro RR. Pharmaceutical inhalation aerosol technology, third edition. Drugs and the pharmaceutical sciences. Boca Raton, Florida: CRC Press; 2019.
6. Sommerfeld M, Cui Y, Schmalfuß S. Potential and constraints for the application of CFD combined with Lagrangian particle tracking to dry powder inhalers. Eur J Pharm Sci. 2019;128:299–324. https://doi.org/10.1016/j.ejps.2018.12.008.
7. Ansys Inc. Ansys Fluent theory guide. Cannonsburg, PA, USA: ANSYS, Inc; 2013.
8. Ansys Inc. Chapter 1: Basic fluid flow. In: ANSYS Inc., ANSYS Fluent theory guide. Cannonsburg, PA, USA: ANSYS Inc.; 2013.
9. Tu J, Yeoh GH, Liu C. Chapter 3–Governing equations for CFD—fundamentals. In: Tu J, Yeoh GH, Liu C, editors. Computational fluid dynamics. Burlington: Butterworth-Heinemann; 2008. p. 65–125.
10. Tu J, Yeoh GH, Liu C. Appendix a–full derivation of conservation equations. In: Tu J, Yeoh GH, Liu C, editors. Computational fluid dynamics. Burlington: Butterworth-Heinemann; 2008. p. 410–3.
11. Ansys Inc. Chapter 4: Turbulence. In: Ansys Inc, editor. ANSYS Fluent theory guide. Cannonsburg, PA, USA: ANSYS Inc.; 2013.
12. Tabatabaian M. Turbulence modeling–approaches and models. In: Tabatabaian M, editor. CFD module. Multiphysics modeling series. Dulles, Virginia: Mercury Learning & Information; 2015.
13. Fletcher DF, Chaugule V, Gomes dos Reis L, Young PM, Traini D, Soria J. On the use of computational fluid dynamics (CFD) modelling to design improved dry powder inhalers. Pharm Res. 2021;38(2):277–88. https://doi.org/10.1007/s11095-020-02981-y.
14. Alexopoulos AH, Milenkovic J, Kiparissides C. An integrated computational model of powder release, dispersion, breakage, and deposition in a dry powder inhaler. In: Kraslawski A, Turunen I, editors. Computer aided chemical engineering. Elsevier; 2013. p. 139–44.
15. ANSYS Inc. 17.5. Eulerian model theory. Ansys Fluent theory guide. USA: ANSYS Inc; 2011. p. 537–88.
16. Schellander D. Eulerian granular phase modelling. In: Schellander D, editor. CFD simulations of particle laden flows: particle transport and separation. Hamburg, Germany: Anchor Academic Publishing; 2014. p. 11–31.
17. Ansys Inc. Chapter 16: Discrete phase. In: Ansys Inc, editor. ANSYS Fluent theory guide. Cannonsburg, PA, USA: ANSYS Inc.; 2013.
18. Schellander D. Lagrangian discrete phase modelling. In: Schellander D, editor. CFD simulations of particle laden flows: particle transport and separation. Hamburg, Germany: Anchor Academic Publishing; 2014. p. 32–44.
19. Antonio AM. The combined finite-discrete element method. Hoboken, NJ: Wiley; 2004.
20. Nan W, Goh WP, Rahman MT. Elasto-plastic and adhesive contact: an improved linear model and its application. Powder Technol. 2022;407:407. https://doi.org/10.1016/j.powtec.2022.117634.

21. Yang J, Wu CY, Adams M. Numerical modelling of agglomeration and Deagglomeration in dry powder inhalers: a review. Curr Pharm Des. 2015;21(40):5915–22. https://doi.org/10.2174/1381612821666151008150845.
22. Johnson KL, Kendall K, Roberts AD, Tabor D. Surface energy and the contact of elastic solids. Proceedings of the Royal Society of London A Mathematical and Physical Sciences. 1971;324(1558):301–13. https://doi.org/10.1098/rspa.1971.0141.
23. Coates MS, Chan H-K, Fletcher DF, Raper JA. Effect of design on the performance of a dry powder inhaler using computational fluid dynamics. Part 2: air inlet size. J Pharm Sci. 2006;95(6):1382–92. https://doi.org/10.1002/jps.20603.
24. Coates M, Fletcher D, Chan H, Raper JA. A comparative study of two marketed pulmonary drug delivery devices using computational fluid dynamics. In: Dalby R, Byron P, Peart J, Suman J, Farr S, editors. Respiratory drug delivery IX. Horwood International Publishing; 2004. p. 821–4.
25. Coates MS, Fletcher DF, Chan HK, Raper JA. The role of capsule on the performance of a dry powder inhaler using computational and experimental analyses. Pharm Res. 2005;22(6):923–32. https://doi.org/10.1007/s11095-005-4587-y.
26. Coates MS, Chan HK, Fletcher DF, Chiou H. Influence of mouthpiece geometry on the aerosol delivery performance of a dry powder inhaler. Pharm Res. 2007;24(8):1450–6. https://doi.org/10.1007/s11095-007-9262-z.
27. Sommerfeld M, Schmalfuß S. Numerical analysis of carrier particle motion in a dry powder inhaler. J Fluids Eng. 2015;138(4) https://doi.org/10.1115/1.4031693.
28. Tong Z, Zhong W, Yu A. CFD-DEM study the effect of carrier-drug mass ratio on the aerosolisation process in original and modified dry powder inhalers. Springer Proceedings in Physics 2017. p. 957–965.
29. Suwandecha T, Wongpoowarak W, Srichana T. Computer-aided design of dry powder inhalers using computational fluid dynamics to assess performance. Pharm Dev Technol. 2016;21(1):54–60. https://doi.org/10.3109/10837450.2014.965325.
30. Milenkovic J, Alexopoulos AH, Kiparissides C. Flow and particle deposition in the Turbuhaler: a CFD simulation. Int J Pharm. 2013;448(1):205–13. https://doi.org/10.1016/j.ijpharm.2013.03.004.
31. Milenkovic J, Alexopoulos AH, Kiparissides C. Deposition and fine particle production during dynamic flow in a dry powder inhaler: a CFD approach. Int J Pharm. 2014;461(1–2):129–36. https://doi.org/10.1016/j.ijpharm.2013.11.047.
32. Suwandecha T, Wongpoowarak W, Maliwan K, Srichana T. Effect of turbulent kinetic energy on dry powder inhaler performance. Powder Technol. 2014;267:381–91. https://doi.org/10.1016/j.powtec.2014.07.044.
33. Wong W, Fletcher DF, Traini D, Chan HK, Crapper J, Young PM. Particle aerosolisation and break-up in dry powder inhalers 1: evaluation and modelling of venturi effects for agglomerated systems. Pharm Res. 2010;27(7):1367–76. https://doi.org/10.1007/s11095-010-0128-4.
34. Wong W, Fletcher DF, Traini D, Chan HK, Crapper J, Young PM. Particle aerosolisation and break-up in dry powder inhalers: evaluation and modelling of impaction effects for agglomerated systems. J Pharm Sci. 2011;100(7):2744–54. https://doi.org/10.1002/jps.22503.
35. Wong W, Fletcher DF, Traini D, Chan H-K, Young PM. The use of computational approaches in inhaler development. Adv Drug Deliv Rev. 2012;64(4):312–22.
36. De Boer AH, Hagedoorn P, Woolhouse R, Wynn E. Computational fluid dynamics (CFD) assisted performance evaluation of the Twincer™ disposable high-dose dry powder inhaler. J Pharm Pharmacol. 2012;64(9):1316–25. https://doi.org/10.1111/j.2042-7158.2012.01511.x.
37. Kopsch T, Murnane D, Symons D. Computational modelling and experimental validation of drug entrainment in a dry powder inhaler. Int J Pharm. 2018;553(1–2):37–46. https://doi.org/10.1016/j.ijpharm.2018.10.021.
38. van Wachem B, Thalberg K, Remmelgas J, Niklasson-Björn I. Simulation of dry powder inhalers: combining micro-scale, meso-scale and macro-scale modeling. AICHE J. 2017;63(2):501–16. https://doi.org/10.1002/aic.15424.

39. Stylianou FS, Sznitman J, Kassinos SC. Direct numerical simulation of particle laden flow in a human airway bifurcation model. Int J Heat Fluid Flow. 2016;61:677–710. https://doi.org/10.1016/j.ijheatfluidflow.2016.07.013.
40. Heenan AF, Matida E, Pollard A, Finlay WH. Experimental measurements and computational modeling of the flow field in an idealized human oropharynx. Exp Fluids. 2003;35(1):70–84. https://doi.org/10.1007/s00348-003-0636-7.
41. Tuteric T, Vulovic A, Cvijic S, Ibric S, Filipovic N. Effect of circulation chamber dimensions on aerosol delivery efficiency of a commercial dry powder Inhaler Aerolizer®. Proceedings–2017 IEEE 17th International Conference on Bioinformatics and Bioengineering, BIBE 2017. p. 555–8.
42. Kim J, Chen JS. Effective of inhaling patterns on aerosol drug delivery: CFD simulation. ASME International Mechanical Engineering Congress and Exposition, Proceedings. 2009:83–91.
43. Zhou QT, Tong Z, Tang P, Citterio M, Yang R, Chan HK. Effect of device design on the aerosolization of a carrier-based dry powder inhaler–a case study on aerolizer® foradile®. AAPS Journal. 2013;15(2):511–22. https://doi.org/10.1208/s12248-013-9458-6.
44. Donovan MJ, Kim SH, Raman V, Smyth HD. Dry powder inhaler device influence on carrier particle performance. J Pharm Sci. 2012;101(3):1097–107. https://doi.org/10.1002/jps.22824.
45. Lee HJ, Kwon IH, Lee HG, Kwon YB, Woo HM, Cho SM, et al. Spiral mouthpiece design in a dry powder inhaler to improve aerosolization. Int J Pharm. 2018;553(1–2):149–56. https://doi.org/10.1016/j.ijpharm.2018.10.036.
46. Almeida LC, Bharadwaj R, Eliahu A, Wassgren CR, Nagapudi K, Muliadi AR. Capsule-based dry powder inhaler evaluation using CFD-DEM simulations and next generation impactor data. Eur J Pharm Sci. 2022;175 https://doi.org/10.1016/j.ejps.2022.106226.
47. Ignjatovic J, Austersic T, Cvijic S, Bodic A, Duris J, Ibric S, et al. Comparative assessment of computational vs. in vitro methods for the estimation of dry powders for inhalation emitted fraction. BIBE 2021–21st IEEE International Conference on BioInformatics and BioEngineering, Proceedings 2021.
48. Thornton C, Yin KK, Adams MJ. Numerical simulation of the impact fracture and fragmentation of agglomerates. J Phys D Appl Phys. 1996;29(2):424–35. https://doi.org/10.1088/0022-3727/29/2/021.
49. Afkhami M, Hassanpour A, Fairweather M, Njobuenwu DO. Fully coupled LES-DEM of particle interaction and agglomeration in a turbulent channel flow. Comput Chem Eng. 2015;78:24–38. https://doi.org/10.1016/j.compchemeng.2015.04.003.
50. Afkhami M, Hassanpour A, Fairweather M. Effect of Reynolds number on particle interaction and agglomeration in turbulent channel flow. Powder Technol. 2019;343:908–20. https://doi.org/10.1016/j.powtec.2018.11.041.
51. Yang J, Wu C-Y, Adams M. Three-dimensional DEM–CFD analysis of air-flow-induced detachment of API particles from carrier particles in dry powder inhalers. Acta Pharm Sin B. 2014;4(1):52–9. https://doi.org/10.1016/j.apsb.2013.11.003.
52. Yang J, Wu CY, Adams M. DEM analysis of the effect of electrostatic interaction on particle mixing for carrier-based dry powder inhaler formulations. Particuology. 2015;23:25–30. https://doi.org/10.1016/j.partic.2014.12.007.
53. Calvert G, Hassanpour A, Ghadiri M. Mechanistic analysis and computer simulation of the aerodynamic dispersion of loose aggregates. Chem Eng Res Des. 2011;89(5):519–25. https://doi.org/10.1016/j.cherd.2010.08.013.
54. Calvert G, Hassanpour A, Ghadiri M. Analysis of aerodynamic dispersion of cohesive clusters. Chem Eng Sci. 2013;86:146–50. https://doi.org/10.1016/j.ces.2012.06.035.
55. Calvert G, Ghadiri M, Dyson M, Kippax P, McNeil-Watson F. The flowability and aerodynamic dispersion of cohesive powders. Powder Technol. 2013;240:88–94. https://doi.org/10.1016/j.powtec.2012.07.003.

56. Thornton C, Ciomocos MT, Adams MJ. Numerical simulations of agglomerate impact breakage. Powder Technol. 1999;105(1–3):74–82. https://doi.org/10.1016/S0032-5910(99)00120-5.
57. Mishra BK, Thornton C. Impact breakage of particle agglomerates. Int J Miner Process. 2001;61(4):225–39. https://doi.org/10.1016/S0301-7516(00)00065-X.
58. Kafui KD, Thornton C. Numerical simulations of impact breakage of a spherical crystalline agglomerate. Powder Technol. 2000;109(1-3):113–32. https://doi.org/10.1016/S0032-5910(99)00231-4.
59. Thornton C, Liu L. How do agglomerates break? Powder Technol. 2004;143-144:110–6. https://doi.org/10.1016/j.powtec.2004.04.035.
60. Tong ZB, Yang RY, Yu AB, Adi S, Chan HK. Numerical modelling of the breakage of loose agglomerates of fine particles. Powder Technol. 2009;196(2):213–21. https://doi.org/10.1016/j.powtec.2009.08.001.
61. Tong ZB, Yang RY, Chu KW, Yu AB, Adi S, Chan HK. Numerical study of the effects of particle size and polydispersity on the agglomerate dispersion in a cyclonic flow. Chem Eng J. 2010;164(2):432–41. https://doi.org/10.1016/j.cej.2009.11.027.
62. Tong ZB, Adi S, Yang RY, Chan HK, Yu AB. Numerical investigation of the de-agglomeration mechanisms of fine powders on mechanical impaction. J Aerosol Sci. 2011;42(11):811–9. https://doi.org/10.1016/j.jaerosci.2011.07.004.
63. Tong ZB, Zheng B, Yang RY, Yu AB, Chan HK. CFD-DEM investigation of the dispersion mechanisms in commercial dry powder inhalers. Powder Technol. 2013;240:19–24. https://doi.org/10.1016/j.powtec.2012.07.012.
64. Breuer M, Khalifa A. Refinement of breakup models for compact powder agglomerates exposed to turbulent flows considering relevant time scales. Comput Fluids. 2019;194 https://doi.org/10.1016/j.compfluid.2019.104315.
65. Tong Z, Zhong W, Yu A, Chan H-K, Yang R. CFD–DEM investigation of the effect of agglomerate–agglomerate collision on dry powder aerosolisation. J Aerosol Sci. 2016;92:109–21. https://doi.org/10.1016/j.jaerosci.2015.11.005.
66. Dizaji FF, Marshall JS, Grant JR. Collision and breakup of fractal particle agglomerates in a shear flow. J Fluid Mech. 2019;862:592–623. https://doi.org/10.1017/jfm.2018.959.
67. van Wachem B, Thalberg K, Nguyen D, Martin de Juan L, Remmelgas J, Niklasson-Bjorn I. Analysis, modelling and simulation of the fragmentation of agglomerates. Chem Eng Sci. 2020;227:115944. https://doi.org/10.1016/j.ces.2020.115944.
68. Nguyen D, Rasmuson A, Thalberg K, Björn IN. The exchange of fines between carriers in adhesive particle mixing: a study using DEM simulation. Powder Technol. 2016;288:266–78. https://doi.org/10.1016/j.powtec.2015.10.048.
69. Ariane M, Sommerfeld M, Alexiadis A. Wall collision and drug-carrier detachment in dry powder inhalers: using DEM to devise a sub-scale model for CFD calculations. Powder Technol. 2018;334:65–75. https://doi.org/10.1016/j.powtec.2018.04.051.
70. Cui Y, Sommerfeld M. The modelling of carrier-wall collision with drug particle detachment for dry powder inhaler applications. Powder Technol. 2019;344:741–55. https://doi.org/10.1016/j.powtec.2018.12.067.
71. Cui Y, Sommerfeld M. Lattice–Boltzmann simulations for analysing the detachment of micron-sized spherical particles from surfaces with large-scale roughness structures. Particuology. 2022;61:47–59. https://doi.org/10.1016/j.partic.2021.05.004.
72. Benque B, Khinast JG. Understanding the motion of hard-shell capsules in dry powder inhalers. Int J Pharm. 2019;567 https://doi.org/10.1016/j.ijpharm.2019.118481.
73. De Boer AH, Hagedoorn P, Gjaltema D, Goede J, Frijlink HW. Air classifier technology (ACT) in dry powder inhalation: part 1. Introduction of a novel force distribution concept (FDC) explaining the performance of a basic air classifier on adhesive mixtures. Int J Pharm. 2003;260(2):187–200. https://doi.org/10.1016/S0378-5173(03)00250-3.
74. De Boer AH, Hagedoorn P, Gjaltema D, Goede J, Frijlink HW. Air classifier technology (ACT) in dry powder inhalation: part 3. Design and development of an air classifier family for

the Novolizer® multi-dose dry powder inhaler. Int J Pharm. 2006;310(1–2):72–80. https://doi.org/10.1016/j.ijpharm.2005.11.030.

75. De Boer AH, Hagedoorn P, Gjaltema D, Goede J, Frijlink HW. Air classifier technology (ACT) in dry powder inhalation: part 4. Performance of air classifier technology in the Novolizer® multi-dose dry powder inhaler. Int J Pharm. 2006;310(1–2):81–9. https://doi.org/10.1016/j.ijpharm.2005.11.029.
76. De Boer AH, Hagedoorn P, Gjaltema D, Goede J, Kussendrager KD, Frijlink HW. Air classifier technology (ACT) in dry powder inhalation part 2. The effect of lactose carrier surface properties on the drug-to-carrier interaction in adhesive mixtures for inhalation. Int J Pharm. 2003;260(2):201–16. https://doi.org/10.1016/S0378-5173(03)00264-3.
77. Bass K, Boc S, Hindle M, Dodson K, Longest W. High-efficiency nose-to-lung aerosol delivery in an infant: development of a validated computational fluid dynamics method. J Aerosol Med Pulm Drug Deliv. 2019;32(3):132–48. https://doi.org/10.1089/jamp.2018.1490.
78. Bass K, Longest W. Development of dry powder inhaler patient interfaces for improved aerosol delivery to children. AAPS PharmSciTech. 2020;21(5):157. https://doi.org/10.1208/s12249-020-01667-3.
79. Kopsch T, Murnane D, Symons D. A personalized medicine approach to the design of dry powder inhalers: selecting the optimal amount of bypass. Int J Pharm. 2017;529(1–2):589–96. https://doi.org/10.1016/j.ijpharm.2017.07.002.
80. Kopsch T, Murnane D, Symons D. Air flow entrainment of lactose powder: Simulation and experiment: IUTAM Bookseries 2019. p. 107–117.
81. Kopsch T, Murnane D, Symons D. Optimizing the entrainment geometry of a dry powder inhaler: methodology and preliminary results. Pharm Res. 2016;33(11):2668–79. https://doi.org/10.1007/s11095-016-1992-3.
82. Kopsch T, Murnane D, Symons D. Development of an adaptive bypass element for passive entrainment flow control in dry powder inhalers. Proc Inst Mech Eng C J Mech Eng Sci. 2019;233(15):5201–13. https://doi.org/10.1177/0954406219845416.
83. Dai W, Huang F, Yu J, Li R, Tong Z. Numerical study of effects of device design on drug delivery efficiency for an active dry powder inhaler. J Aerosol Sci. 2021;157 https://doi.org/10.1016/j.jaerosci.2021.105801.
84. Kadota K, Imanaka A, Shimazaki M, Takemiya T, Kubo K, Uchiyama H, et al. Effects of inhalation procedure on particle behavior and deposition in the airways analyzed by numerical simulation. J Taiwan Inst Chem Eng. 2018;90:44–50. https://doi.org/10.1016/j.jtice.2017.11.008.
85. Azouz W, Chetcuti P, Hosker H, Saralaya D, Chrystyn H. Inhalation characteristics of asthma patients, COPD patients and healthy volunteers with the Spiromax® and Turbuhaler® devices: a randomised, cross-over study. BMC Pulm Med. 2015;15(1):47. https://doi.org/10.1186/s12890-015-0043-x.
86. Benque B, Khinast JG. Estimating inter-patient variability of dispersion in dry powder inhalers using CFD-DEM simulations. Eur J Pharm Sci. 2021;156 https://doi.org/10.1016/j.ejps.2020.105574.
87. Benque B, Khinast JG. Carrier particle emission and dispersion in transient CFD-DEM simulations of a capsule-based DPI. Eur J Pharm Sci. 2022;168 https://doi.org/10.1016/j.ejps.2021.106073.
88. Meerburg JJ, Andrinopoulou ER, Bos AC, Shin H, Van Straten M, Hamed K, et al. Effect of inspiratory maneuvers on lung deposition of tobramycin inhalation powder: a modeling study. J Aerosol Med Pulm Drug Deliv. 2020;33(2):61–72. https://doi.org/10.1089/jamp.2019.1529.
89. Stylianou FS, Angeli SI, Kassinos SC, Svensson M. The effect of flow rate, head position, and inhaler orientation on the airflow and particle deposition in an mri-based mouth-throat geometry. 10th International Symposium on Turbulence and Shear Flow Phenomena, TSFP 2017.
90. Son Y-J, Longest PW, Tian G, Hindle M. Evaluation and modification of commercial dry powder inhalers for the aerosolization of a submicrometer excipient enhanced growth (EEG) formulation. Eur J Pharm Sci. 2013;49(3):390–9. https://doi.org/10.1016/j.ejps.2013.04.011.

91. Bass K, Momin MAM, Howe C, Aladwani G, Strickler S, Kolanjiyil AV, et al. Characterizing the effects of nasal prong interfaces on aerosol deposition in a preterm infant nasal model. AAPS PharmSciTech. 2022;23(5):114. https://doi.org/10.1208/s12249-022-02259-z.
92. Vulović A, Šušteršič T, Cvijić S, Ibrić S, Filipović N. Coupled in silico platform: computational fluid dynamics (CFD) and physiologically-based pharmacokinetic (PBPK) modelling. Eur J Pharm Sci. 2018;113:171–84. https://doi.org/10.1016/j.ejps.2017.10.022.
93. Das P, Nof E, Amirav I, Kassinos SC, Sznitman J. Targeting inhaled aerosol delivery to upper airways in children: insight from computational fluid dynamics (CFD). PLoS One. 2018;13(11):e0207711. https://doi.org/10.1371/journal.pone.0207711.
94. Longest PW, Farkas D, Hassan A, Hindle M. Computational fluid dynamics (CFD) simulations of spray drying: linking drying parameters with experimental Aerosolization performance. Pharm Res. 2020;37(6):101. https://doi.org/10.1007/s11095-020-02806-y.
95. Longest W, Hassan A, Farkas D, Hindle M. Computational fluid dynamics (CFD) guided spray drying recommendations for improved aerosol performance of a small-particle antibiotic formulation. Pharm Res. 2022;39(2):295–316. https://doi.org/10.1007/s11095-022-03180-7.
96. Yoo SJ, Ito K. Assessment of transient inhalation exposure using in silico human model integrated with PBPK-CFD hybrid analysis. Sustain Cities Soc. 2018;40:317–25. https://doi.org/10.1016/j.scs.2018.04.023.
97. Andersen M, Sarangapani R, Gentry R, Clewell H, Covington T, Frederick CB. Application of a hybrid CFD-PBPK nasal dosimetry model in an inhalation risk assessment: an example with acrylic acid. Toxicological Sci. 2000;57(2):312–25. https://doi.org/10.1093/toxsci/57.2.312.
98. Longest PW, Bass K, Dutta R, Rani V, Thomas ML, El-Achwah A, et al. Use of computational fluid dynamics deposition modeling in respiratory drug delivery. Expert Opin Drug Deliv. 2019;16(1):7–26. https://doi.org/10.1080/17425247.2019.1551875.
99. Selvam P, McNair D, Truman R, Smyth HDC. A novel dry powder inhaler: effect of device design on dispersion performance. Int J Pharm. 2010;401(1–2):1–6. https://doi.org/10.1016/j.ijpharm.2010.07.056.
100. Kotteda VMK, Badhan A, Kumar V. Parametric optimization of a dry powder inhaler. American Society of Mechanical Engineers, Fluids Engineering Division (Publication) FEDSM 2020.
101. US Food and Drug Administrationl. FY2018 GDUFA science and research report: locally-acting orally-inhaled and nasal drug products 2018.

Chitosan-Based Particulates Carriers for Pulmonary Drug Delivery

7

Piyush Pradeep Mehta and Vividha Dhapte-Pawar

Abstract

Chitosan (CH), a naturally occurring linear polysaccharide, is a well-known carrier for drug delivery due to its biodegradable and biocompatible nature. Additionally, CH and its synthetic derivatives have been widely explored as cargos in nano- and microparticulate drug delivery systems for different routes of administration owing to its distinct, pH sensitivity, low toxicity, and controlled drug release properties. It has also been established as a pulmonary drug carrier in the therapy of various infectious diseases. The success of CH and its synthetic derivatives as a pulmonary drug carrier is a vital interplay of their mucoadhesive nature, permeation properties, and site-specific delivery potential. In the present chapter, we have systematically reviewed the pulmonary applications of CH-based novel drug delivery systems considering their aerosolization potential and in vivo respiratory dispersion performance. Moreover, different studies discussing CH cytotoxicity potential with the help of in vitro human cell lines are also elaborated. Lastly, a few critical concerns linked to the applicability of CH carriers in the real world are proposed to provide insights for future investigations.

P. P. Mehta
Global Respiratory R&D, Cipla R&D Center, Mumbai, Maharashtra, India
e-mail: piyu053@gmail.com

V. Dhapte-Pawar (✉)
Department of Pharmaceutics, Poona College of Pharmacy, Bharati Vidyapeeth
(Deemed to be University), Pune, Maharashtra, India
e-mail: dhapte.vv@bharatividyapeeth.edu

P. P. Mehta, V. Dhapte-Pawar (eds.), *Pulmonary Drug Delivery Systems: Material and Technological Advances*,
https://doi.org/10.1007/978-981-99-1923-9_7

Keywords

Chitosan · Linear polysaccharide · Biodegradable polymer · Amphiphilic particles · Mucoadhesive particles and chitosomes

7.1 Introduction

Pulmonary drug delivery is a useful approach to attain both local and systemic action for a drug substance [1–4]. The large surface area (80–120 m^2), thin epithelial membrane, high vascularization, easy transcytosis in the alveolar region, and low enzymatic activity are prominent features that make the lung an ideal drug delivery platform for the management of both local (e.g., asthma, cystic fibrosis, tuberculosis (TB), and emphysema) as well as systemic (e.g., diabetes, cancer, influenza virus, and Parkinson's) disorders [5–8]. Targeted drug delivery, rapid onset of action, superior therapeutic action, lower systemic toxicity, and sustained drug release profile over other conventional dosage forms (e.g., tablets and capsules) are the key benefits of pulmonary drug delivery [9–12]. Moreover, noninvasive and organ-specific pulmonary drug delivery ensures lower systemic adverse effects than oral or intravenous delivery systems [13, 14]. Pulmonary drug delivery has the potential to achieve an equivalent therapeutic action to oral or parenteral delivery systems at lower doses (5–500 μg) [15]. Dry powder inhalers (DPIs) [16, 17], pressurized metered-dose inhalers (pMDIs) [18], soft mist inhalers (SMIs) [7], and nebulizers are used to deliver drug to pulmonary airways. Sugars (e.g., lactose (LAC), mannitol (MAN), and trehalose) and propellants (e.g., hydrofluorocarbon) were used as a carrier or diluent in aerosols. Carrier properties have a significant impact on aerodynamic and biological behavior of a drug substance. The well-balanced interplay between micronized drug substance (0.5–5 μm) and carrier molecule plays an essential role in successful pulmonary drug delivery [19, 20]. Similar to conventional carriers, polymer, co-polymer, and matrix have been used to formulate pulmonary drug delivery systems [21, 22].

Various natural and synthetic biodegradable polymers have been studied and analyzed for the pulmonary delivery of different drugs [23]. Polymers like hyaluronic acid (HA) [24, 25] sodium caseinate [26, 27], poly(lactic-co-glycolic) acid (PLGA) [28], poly (ethylene glycol) (PEG) [29], dextran [30], chitosan (CH) [31], and polyethyleneimine (PEI) [32] have been thoroughly investigated for pulmonary drug delivery. Nano and micron-sized carriers developed using biodegradable polymeric matrix offered several distinctive advantages such as high drug loading, sustained release profile, distinctive surface morphology, modified surface roughness, appropriate aerosolization pattern, and superior biological performance in pulmonary drug delivery [33, 34]. The HA-like polymer can design and develop multifunctional formulations containing therapeutic protein/peptides, genetic material (DNA, siRNA, RNA), and antibiotics [35–38]. Likewise, dextran and its derivatives, i.e., dextran propionate succinate and acetylated dextran, can be used to tailor pH-sensitive, biocompatible, and effective targeted drug delivery systems [39, 40]. PLGA is one of the most versatile polymers commonly used to design novel drug delivery systems, i.e., nanoparticles (NPs), microparticles (MPs), and microspheres

[41–43]. However, the hydrophobic nature of PLGA restricts the proper encapsulation and successive deep pulmonary delivery of hydrophilic molecules [44]. This setback can be controlled by linking the hydrophilic PEG molecules to the PLGA matrix. This linkage has opened a window for designing various PLGA-PEG co-polymers for encapsulating both hydrophobic and hydrophilic therapeutic agents [45].

Similarly, CH is also a multipurpose, naturally occurring, cationic, linear polysaccharide. CH is thoroughly explored for designing various drug delivery systems [46]. CH can efficiently enclose both hydrophobic and hydrophilic molecules because of its unique amphiphilic nature [47]. CH and its derivatives are excellent polymers for pulmonary delivery of anti-microbial, anti-cancer, phytoconstituents, and anti-infective agents [48]. Formulation scientists also accept CH due to its non-toxic nature, biodegradability, and better biocompatibility with respiratory cells [49]. Furthermore, absorption, permeation enhancement, and mucoadhesive properties are important advantages of CH over other biodegradable polymers [50, 51]. CH possesses a promising capability to transform the morphology of carrier particles [52, 53] thereby, revising the dispersion and deagglomeration pattern of particles leading to an excellent aerodynamic deposition profile within the pulmonary airways. Besides, CH has natural immunogenicity which is lacking in other polymers which permit its use as an adjuvant for vaccine delivery into the pulmonary airways [49, 54]. Consequently, here, we provide an overview of CH and its essential derivatives in pulmonary drug delivery, covering its applications, aerodynamic outcomes, toxicological issues, and commercialization challenges.

7.2 Chitosan in Pulmonary Drug Delivery

CH (Fig. 7.1) is a well-known polymer and the second most accessible natural polymer after cellulose [55]. It is a linear polysaccharide derived through deacetylation of naturally occurring chitin and therefore includes an N-acetyl glucosamine moiety [49]. It is a semicrystalline polymer, and the extent of deacetylation mainly governs its crystallinity [56]. CH (1526.5 g/mol) is insoluble in aqueous solutions at neutral pH; but, it is soluble in aqueous acidic media (pH <5) as the primary amino groups ($-NH_2$) are protonated ($-NH_3^+$) and thus, acquires a positive charge [49, 57]. The positively charged CH offers good mucoadhesive properties [58, 59] that assist the adhesion between the delivery system and mucosa of pulmonary epithelial cells to offer the sustained release profile for loaded substances [60, 61]. The cationic CH enhances the drug absorption pattern by opening the continuous tight intercellular junctions of the airway epithelium [62, 63]. Additionally, CH eases transepithelial drug transport within the alveolar vicinity. CH itself holds good bactericidal properties [64, 65]. CH is most active on the cell surface of bacteria or fungi. The interaction between bacterial cells and CH is governed mainly by electrostatic interaction. The positive charge of a protonated amino group of CH interacts with the negative charge molecules on the surface of bacterial cells. Eventually, the permeabilization of the bacterial cell surface causes the leakage of intracellular matter, which ultimately leads to the death of bacterial cells [66]. Few recent scientific reports have shown that CH can penetrate the plasma membrane of bacterial cells and affect

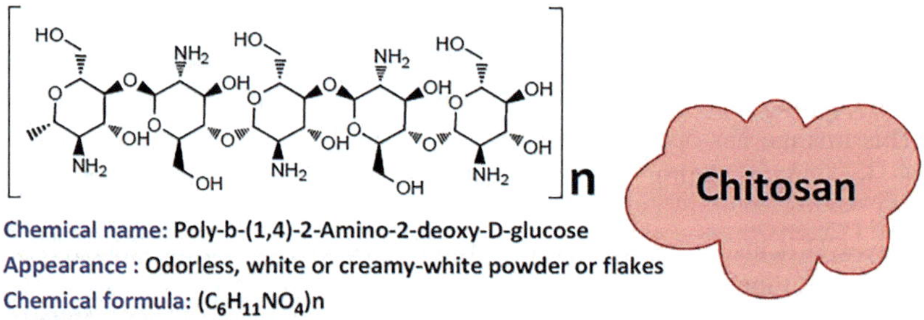

Fig. 7.1 Important physical, chemical, and biological features of chitosan

DNA expression by binding with its nucleic acid [66]. Moreover, CH is prone to depolymerization in the lysozyme, one of the most vital anti-microbial enzymes present in the airways. Hence, several researchers and material chemists explored CH to deliver different anti-bacterial substances into the respiratory tract [67]. Furthermore, according to hydrolysis time and degree of deacetylation, CH offers different physicochemical and biological properties (Table 7.1). These fundamental traits further assist the airway delivery of anti-cancer agents, phytochemicals, anti-microbial substances, proteins/peptides, and genetic materials against life-threatening pulmonary complications [48, 68, 69]. As per the above-discussed literature, it is apparent that CH has plentiful drug delivery applications as a carrier, and a few such exciting drug delivery examples are thoroughly described in the below segment.

7.2.1 Role of Chitosan in Carrier Engineering

Carrier active sites and surface roughness, both are important attributes in designing DPI [70–72]. The well-balanced interplay between the active sites and surface roughness precisely depicts the aerosolization performance (Fig. 7.2). As showed in Fig. 7.2 micronized drug particles will attach and remain within the asperities of the carrier with a micro metered roughness while micronized drug particles (Fig. 7.2a) will simply adsorb on top of the surface of a smooth-surface carrier (Fig. 7.2b). Both types of roughness own a moderately high number of active sites. In case of nanometer roughness (Fig. 7.2c), only a few asperities will be accessible by

Table 7.1 Summary of chitosan derivatives with fundamental physicochemical and biological properties [48]

Chitosan derivative	Physicochemical properties	Biological properties
Cationic CH	• Highly positive charge • Good film-forming capacity • Compatible with different cations • Develop complexes with anions. • Enhanced water solubility at pH < 9	• Biodegradable • Enhanced intestinal permeation • Improved mucoadhesive property • Possess anti-bacterial and anti-fungal properties • Potent hydroxyl radical scavenger
N-acyl-CH	• Hygroscopic nature • Form amphiphilic hydrogel • Possess excellent water-absorption and retention capacity • Commonly used as an anti-cancer drug carrier	• Biodegradable • Possess anti-bacterial properties • Good transfection efficacy • Easy endocytic uptake • Cytocompatible with chondrocytes
O-acyl-CH	• Possess excellent organic soluble properties • Good absorptive properties • Mucoadhesive in nature	• Fungicidal in nature • Possess good cellular permeability • Endorse enzymatic hydrolysis
N-Carboxyalkyl/ aryl-CH	• Possess excellent gel/ film-forming ability • Water soluble in nature • Own chelating capacity	• Antioxidant in nature • Possess anti-mutagenic properties
O-Carboxyalkyl-CH	• Assist in improving adhesive properties	• Possess anti-bacterial properties • Improved blood compatibility • Better chondrocytes adhesion
Thiolated CH	• Enhanced mucoadhesiveness • Possess cohesive network • Excellent in situ gelling properties	• Biodegradable • Possess enzyme inhibitory activity • Enhanced permeation properties
Sugar linked CH	• Water soluble in nature • Specific to cell/virus/bacteria as a function of sugars moiety	• Possess good intercellular recognition and adhesion
Sulfated CH	• Water soluble in nature	• Possess anti-viral and anti-microbial properties • Possess anti-obesity via anti-adipogenesis inhibition • Capable of stopping human malignant melanoma cell adhesion • Possess anti-coagulant and hemagglutination inhibitory properties

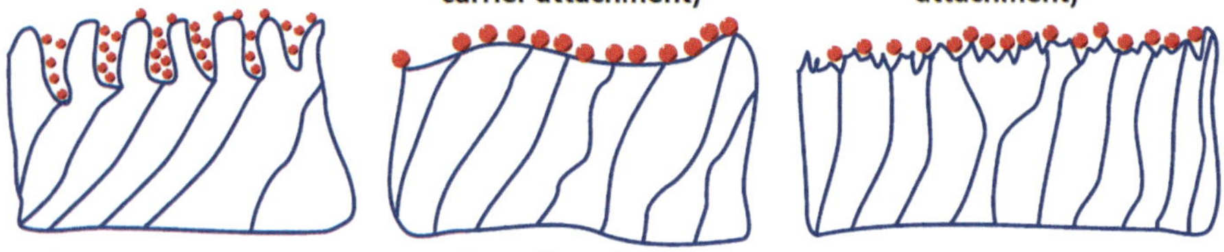

Fig. 7.2 Graphic demonstration of coarse lactose carrier (50–160 μm; blue color surface) surface roughness and micronized drug (0.5–6 μm; red color dots) attachment

micronized drug particles on carrier surface. The number of active sites positioned in the sequence of micrometer roughness > smooth surface > nanometer roughness. Hence, carrier particles with nanometer roughness possess the lowest numbers of active sites and which provide the DPI formulations with an option of a satisfying drug loading and facile drug detachment [73, 74]. Briefly, regulating surface roughness is the preferred option to develop DPIs with improved aerosolization performance. Consequently, Huang and co-workers (2018) formulated CH-based binary carriers (CBBC) of nanometer roughness using MAN by the spray-drying method. Micronized budesonide (BUD; 4.14 μm) was mixed with CBBC (1:10 w/w) manually for 5 min to develop DPI. Spherical-shaped CBBC displayed average roughness (<20 nm) and particle size (100.51 μm) apt for pulmonary delivery. During aerodynamic analysis, BUD-CBBC showed 1.42-fold higher fine particle fraction (FPF) than commercial BUD formulation (Pulmicort®) using Turbospin™ at 60 L/min. During in vitro cell lines studies using adenocarcinoma human alveolar basal epithelial cells (A549) and human lung cancer cell line (Calu-3) by MTT assay, CBBC showed no major cytotoxicity and thus evolved as a safe carrier for BUD delivery. In an in vivo pharmacokinetic study using DP-4R dry powder insufflator, BUD-CBBC showed a marked improvement in area under the curve $(AUC)_{0\text{-}8h}$ (2.29-fold) and maximum serum concentration (C_{max}) (4.68-fold) compared to commercial BUD formulation (Pulmicort®) at the same dose (800 μg/kg). Moreover, BUD-CBBC displayed a significant improvement in the half-life ($t_{1/2}$) and mean residence time ($MRT_{0\text{-}8h}$), representing better pulmonary retention. In the present case study, the difference in the molecular migration rates of CH and MAN produce surface roughness. Particularly, low molecular weight MAN owns better solvent affinity and lower viscosity compared to the high molecular weight CH. Therefore, MAN experience low resistance when scattered within the solution, migrates speedily to the outer layer compared to CH and as a result generates a heterogeneous surface with nanometer roughness. To sum up, CH-based binary carriers possess the good potential to regulate the surface roughness and aerosolization performance of DPIs [75].

The physical mixture of solid NPs with micron-sized carrier particles is an attractive alternative to formulate DPIs with improved aerodynamic performance. Dispersion of solid NPs on surface of micron-sized carrier particles will prevent NPs from agglomeration and NPs will also prevent coarse accumulation during mixing and storage. The concept of nano-on-microparticles was well explored by Alhajj et al. (2020) by developing CH NPs and LAC-PEG 3000 MPs (LM-PEG$_{3000}$ MPs) using spray drying. Subsequently, CH NPs and LM-PEG$_{3000}$ MPs (1:9 w/w) were physically mixed using a vortex blender (40 Hz for 30 min) to improve the inhalation performance. CH NPs formulated using high molecular weight (310000–375,000 Da) CH [HMW-CH] showed lower particle size (347.27 nm), polydispersity index (0.42), and zeta potential (20.17 mV) as compared to low molecular weight (20000–50,000 Da) CH [LMW-CH] 610.03 nm, 0.63 and 27.47 mV, respectively. The reduction in mean particle size of HMW-CH NPs mainly was ascribed to lower concentration (0.05%) and the ability of HMW-CH long chains to hold particles more efficiently. Besides, the reduction in the zeta potential of HMW-CH NPs was primarily attributed to the lower degree of deacetylation compared to LMW-CH. Well-defined, spherical-shaped CH NPs showed lower surface roughness (0.5 nm) values. Plate-shaped rough LM-PEG$_{3000}$ MPs showed particle size (5.43 μm) and surface roughness (127.21 nm) apt for pulmonary delivery. CH NPs loaded LM-PEG$_{3000}$ MPs aerodynamic analysis was performed by cascade impactor using a breath-activated DPI device at a flow rate of 48 L/min. HMW-CH-based NPs loaded LM-PEG$_{3000}$ MPs showed lower mass median aerodynamic diameter (MMAD) of 3.85 μm with FPF of 34.42%, while LMW-CH-based NPs loaded LM-PEG$_{3000}$ MPs displayed MMAD of 4.13 μm with FPF of 30.83%. As observed in the present study, CH NPs with a higher level of circularity, large extent of zeta potential, and spherical shape experience less inter-nanoparticulate aggregation and have largely interacted with MPs that assist to deliver the NPs to lower regions of airways. Principally, physical mixing of NPs with LAC MPs mutually reduces their agglomeration through surface adsorption phenomenon. Consequently, nano-on-microparticles offered one of the most effective pulmonary drug delivery options [76].

Carvedilol (CAV) is a unique antihypertensive molecule with activity against both α- and β-adrenergic receptors. CAV has limited bioavailability (~20%) because of its first-pass metabolism with log P and pKa of 2.79 and 7.8 (weak base), respectively. It shows that CAV has adequate lipophilicity to pass through the airway membranes. It satisfies the criterion for the development of pulmonary drug delivery. Varshosaz and co-workers fabricated polymeric NPs of poly (ethylene-co-vinyl acetate) [PEVA] coated with CH for pulmonary delivery of CAV. Initially, PEVA NPs were formulated by O/W solvent evaporation technique and coated with mucoadhesive CH (0.2%) polymer. Subsequently, they were spray-dried using MAN and LAC carriers to obtain dry powder formulations. Uniform, spherical-shaped CH-coated CAV PEVA NPs showed mean particle size, zeta potential, and entrapment efficacy of 530.5 nm, 39.2 mV, and 87.5%, respectively. High lipophilicity of CAV resulted in superior entrapment efficiency values. When assessed using the mucin adsorption technique, CH-coated CAV PEVA NPs exhibited good mucoadhesive properties (>40%). In an in vitro release study using PBS (pH 7.4), NPs showed a biphasic release pattern with initial fast release followed by slower drug release up to 20 h. However, in an acidic medium, 0.1 N HCl containing 2% Tween 20 (pH 1.2) showed faster drug release due to protonation of amino moieties of CH in acidic pH, which caused hydration of the film coating leading to rapid CAV release. Spray-dried MAN-based powders showed particle size and bulk density of 4.2 μm and 0.39 g/cm^3, respectively. In comparison, spray-dried LAC-based powders displayed particle size and bulk density of 3.1 μm and 0.46 g/cm^3, respectively. In aerodynamic assessment using Easyhaler® at 60 L/min, MAN-based powders exhibited emitted dose (ED) and FPF of 90.49% and 58.82%, respectively, with an aerodynamic diameter of 1.4 μm. In contrast, LAC-based powders showed ED and FPF of 90.69% and 21.26%, respectively, with an aerodynamic diameter of 2.82 μm. Spray-dried MAN showed superior FPF compared to spray-dried LAC particles due to the cohesion of individual particles to form larger aggregates. Particularly, spray-dried MAN showed twofold lower aggregated particles compared to spray-dried LAC particles. Overall, the lower density and aerodynamic diameter of spray-dried MAN-based powder significantly improved the FPF of CAV. Even if the preliminary aerodynamic analysis showed suitable outcomes, NPs loaded spray-dried carriers need to experience vigorous clinical evaluation [77].

7.2.2 Engineered Nicotine Particles for Pulmonary Delivery

Tobacco smoking and nicotine dependence is a widespread public health crisis across the globe and cause death to >7 million people per year [78]. The injurious effects of tobacco smoking are mainly because of the by-products of tobacco combustion; for instance, tar ingredients and carbon monoxide rather than nicotine [79]. In view of that, researchers tried to develop sustained release nicotine formulation for pulmonary delivery. Sustained release nicotine formulation helps to mimic the effects of tobacco smoking and notably reduces cravings and withdrawal symptoms [79]. Nicorette® (15 mg) and Nicotrol® (10 mg) are the currently available inhalers for pulmonary delivery of nicotine; but these inhalers deliver nicotine into the buccal region, not in the alveoli, resulting in reducing plasma maximum concentration and delayed time to attain maximum concentration for beneficial action [79, 80].

Suitably, Wang et al. (2017) formulated nicotine hydrogen tartrate (NHT) loaded CH NPs using the W/O emulsion method and thoroughly studied them for aerodynamic properties. Particularly, NHT: CH NPs were formulated at three different concentrations, i.e., 1:1, 2:1, and 3:1. Afterward, NHT-CH nanosuspension was freeze-dried to obtain respirable nano-powders. NHT: CH NPs showed mean particle sizes of 198.30, 204.40, and 411.20 nm for 1:1, 2:1, and 3:1 concentration, respectively. Low concentration (1:1) NHT: CH NPs showed fine spherical NPs with smooth surfaces while high concentration (2:1 and 3:1) NHT:CH NPs displayed surface roughness, indicating that a significant amount of NHT was collected on the CH NPs surface. NHT: CH NPs demonstrated % entrapment efficiency of 77.50, 79.80, and 87.70% for 1:1, 2:1, and 3:1 concentration, respectively.

Thermal and X-ray diffraction studies displayed that NHT was converted to an amorphous form and present as the nano-dispersion in CH NPs. Freeze-dried NHT: CH (3:1) nano-powders exhibited good powder flowability compared to other two formulations. In an in vitro release study using PBS, NHT: CH NPs (1:1 and 2:1) formulations followed Fickian diffusion drug release while NHT:CH NPs (3:1) formulations showed non-Fickian diffusion owing to the higher concentration of NHT in the CH NPs. During aerodynamic analysis, freeze-dried NHT: CH (3:1) nano-powders showed a higher fine particle dose (FPD) 3.18 mg compared to NHT:CH (2:1 and 1:1) nano-powders i.e., 2.50 and 1.71 mg, respectively using Rotahaler® at 60 L/min for 5 sec [79]. Based on the outcome of the present study Wang and research group also investigated the impact of CH NPs surface charge on DPI aerosolization pattern. The NHT: CH NPs were positively charged, and the amount of positive charge was reduced with increasing NHT concentration in the NPs. NHT: CH NPs demonstrated surface charge of 19, 14.20, and 12.2 mV for 1:1, 2:1, and 3:1 concentration, respectively. The NHT molecule has a negative surface charge because of the presence of the tartaric acid salt thus the positive charge of the CH amino groups was declined with increased concentration of NHT in the formulation. The surface morphology analysis of freeze-dried NHT:CH nano-powders showed presence of micro-agglomerates of NPs with mean particle size (<5 μm) appropriate for pulmonary delivery. Aerodynamic performance of freeze-dried NHT:CH nano-powders was compared with NHT:CH (10%) loaded on Inhalac® 120 coarse LAC formulation. During aerodynamic assessment using Rotahaler® at 60 L/min, freeze-dried NHT:CH nano-powders demonstrated % FPF 39.7, 33.6, and 32.4% for 1:1, 2:1, and 3:1 concentration, respectively. Data suggest that the intrinsic surface positive charge of NHT:CH NPs enhance the repulsive forces among the NPs and showing in improved FPF with increased surface charge. Moreover, NHT:CH (10%) loaded on Inhalac® 120 coarse LAC formulation showed % FPF 51.2, 53.1, and 55.5% for 1:1, 2:1, and 3:1 concentration, respectively. LAC-based formulation showed higher FPF values with increasing NHT concentration, as large LAC carriers (100 μm) help to reduce the cohesive forces among the CH NPs and enhance the de-agglomeration. The data suggest the mechanism of NHT:CH NPs aerodynamic dispersion without large LAC carriers is dependent on the surface charge of the NPs [81]. Further to understand the clinical relevance of the NHT:CH (3:1) freeze-dried nano-powders formulation Wang and research group (2020) studied the locomotor activity of C57BL/6 mice. In-vitro aerodynamic profile of the NHT:CH (3:1) freeze-dried nano-powders formulation (50 mg of NHT) was initially studied using the nose-only inhalation device at the flow rate and exposure time of 0.9 L/min and 5 min. Set flow rate and exposure time achieved optimal delivery of NHT with a total recovered dose and deposition percentage of 27.15 mg and 17.75%, respectively. During the in-vivo locomotor activity using C57BL/6 mice treated with NHT:CH (0.88 mg) and NHT (0.59 mg) showed similar locomotor activity as compared to nicotine (0.5 mg/kg) subcutaneous injection. Moreover, no significant differences were noted during a histological examination for untreated and freeze-dried nano-powders formulation-treated lung tissue samples. Briefly, freeze-dried nano-powders showed that reformulation did not change the therapeutic potential of NHT [82]. Still further detailed investigation of the impact of NHT:CH NPs on pulmonary cells over a longer period is needed.

7.2.3 Chitosan Particles for Pulmonary Delivery of Antibiotic Agents

Conventional delivery systems are unable to deliver an anti-microbial agent to the site of action since they are unable to remain adherent to the site of action or occasionally, they must deliver the drug indirectly via blood to the alveolar region. To address the challenges of rapid drug clearance and reduced drug deposition within the airway Mukhtar et al. (2020) designed nanotechnology-based biocompatible DPI for effective pulmonary delivery of anti-microbial agent isoniazid (INH). Initially, INH-loaded CH/HA and thiolated-CH/HA (TCH/HA) nanosuspensions were fabricated using a cross-linker sodium-tripolyphosphate (TPP) by ionic gelation method. Later nanosuspensions were freeze-dried to obtain respirable nano-powders. CH/HA nano-powders showed small aggregates with a broad particle size distribution owing to the electrostatic attraction between TPP and nano-powders. In contrast, TCH/HA nano-powders showed smooth-surfaced uniformly scattered particles. CH/HA and TCH/HA nano-powders showed a bulk density of 0.118 and 0.162 g/cm^3, respectively. Both CH/HA and TCH/HA nano-powders displayed a controlled release profile of INH (48 h) and followed the Higuchi release model. During the permeability study, TCH/HA nano-powders showed 1.5-fold higher INH permeation than CH/HA nano-powders. In aerodynamic analysis, TCH/HA nano-powders showed 1.13-fold superior INH FPF than CH/HA nano-powders at 60 L/min. Besides, both CH/HA and TCH/HA nano-powders exhibited more than 70% INH deposition within the acinar region which is the most affected by tuberculosis. CH/HA and TCH/HA nano-powders were also compatible with A549 cells and showed no cytotoxicity. Briefly, potential of combination of biocompatible and bioadhesive polymers showed enhanced drug dispersion and contact time with the lung mucosal surface [83].

Bedaquiline (BDQ) received accelerated approval in 2012 by US FDA for treatment of resistant TB, more widespread in India, China, and Eastern Europe [84]. BDQ is a highly hydrophobic (Log P 6.37) diarylquinoline molecule that inhibits the proton pump of ATP synthase, an enzyme needed for energy synthesis in bacteria [85]. BDQ tablets (Sirturo® by Johnson and Johnson's; 400 mg/day for 2 weeks followed by 200 mg thrice a week for about 22 weeks) are currently prescribed for the treatment of TB [86]. Therefore, target-specific, pulmonary delivery of BDQ was explored by Rawal et al. (2018) to shorten duration of treatment and reduce the side effects of the long-term anti-tubercular therapy. Initially, BDQ-loaded CH (BDQ-CH) nanosuspension was formulated using the ionic gelation method and thoroughly optimized by a two-level three-factor factorial design. Afterward, BDQ-CH nanosuspension was freeze-dried (BDQ-CH NPs) and mixed with LAC pre-blend to obtain a DPI. The nanosuspension sonication time, CH, and TPP concentration significantly impacted particle size and zeta potential of BDQ-CH nanosuspension. The optimized batch displayed particle size and zeta potential of 109.70 nm and 36 mV, respectively. In in vitro release studies, BDQ-CH NPs showed slow and sustained BDQ release in phosphate buffer saline (PBS; pH 7.4). BDQ-CH NPs showed only 65% BDQ release within 10 h, while BDQ alone showed 100% release within 14 h. In addition, BDQ-CH NPs displayed superior FPF (28.27%) values compared to conventional LAC-based BDQ DPI (15.38%) using Rotahaler® device at 28.3 L/min. During in vitro (macrophage J774 cell lines) and in vivo toxicity (acute and chronic) studies, BDQ-CH NPs displayed a better safety profile than conventional LAC-based DPI and BDQ oral solutions. During in vivo pharmacokinetic study, BDQ-CH NPs displayed a marked 1.24 and 1.80-fold higher AUC compared to conventional LAC-based BDQ DPI and the BDQ oral solution, respectively. Moreover, BDQ-CH NPs showed a marked 1.27 and 1.65-fold superior C_{max} with reduced dosing frequency compared to conventional LAC-based BDQ DPI and the BDQ oral solution, respectively. So, fabricated CH-based NPs open a new avenue for designing therapeutic options for the effective treatment of pulmonary TB [86]. Prothionamide (PTI) is a thioamide molecule with anti-tubercular activity. PTI is a second-line drug prescribed to treat multidrug-resistant TB. It is available in conventional tablet (250 mg) formulation and its erratic absorption and frequent administration limit its long-term clinical application. Accordingly, Debnath et al. (2018) developed CH-coated PTI NPs using ionic gelation method followed by freeze drying to achieve targeted delivery with reduction in dosing frequency. Spherical-shaped CH-PTI NPs exhibited particle size and aerodynamic particle size of 314.37 nm and 1.76 μm, respectively, suitable for effective pulmonary delivery. In an in vitro release study using simulated lung fluid, CH-PTI NPs displayed initial burst release followed by sustained release up to 24 h (96.91%). During the biological assessment, CH-PTI NPs showed that CH effectively maintained the PTI concentration above minimum inhibitory concentration for 12 h after single-dose administration (2 μg/mL) and enhanced the PTI residency in the pulmonary airways for more than 24 h. The biological and aerodynamic assessment successfully demonstrated the dose and dosing frequency reduction in the management of TB [87].

Ciprofloxacin hydrochloride (CIP) is a second-generation fluoroquinolone antibiotic that has superb activity against pulmonary pathogens and shows concentration-dependent killing making it well suited for inhaled drug delivery [88]. CIP-loaded poly-ε-caprolactone NPs (CIP-PLC-NPs) were fabricated by solid-in-oil-in-water (s/o/w) emulsion solvent evaporation method for pulmonary administration. Additionally, the CIP-PLC-NPs surface was modified using CH to improve the entrapment efficacy. CIP-PLC-NPs were further converted into nanocomposite MPs by a freeze drying method using MAN as carrier material to achieve better aerodynamic performance. CIP-PLC-NPs showed particle size, positive zeta potential, and encapsulation efficiency of 489 nm, 62 mV, and 8.26%, respectively. Spherical and smooth surface nanocomposite MPs displayed suitable aerodynamic performance with a MMAD of 5.41 μm and FPF of 20.23% using Aerolizer® at 28 L/min. During an anti-microbial study using *Escherichia coli* (ATCC 25922), *Klebsiella pneumoniae* (RSKK 574), and *Pseudomonas aeruginosa* (ATCC 27853) by cup plate technique, nanocomposite MPs showed 93.33, 85.36, and 65.23% anti-microbial efficacy, respectively. During 6-month storage, nanocomposite MPs effectively retained their structural properties and anti-microbial activity. This investigation shows that CIP preserved its anti-microbial potency when integrated into the particulate systems. Briefly, CIP particulate DPI shows as an effective, simple to use, and well-tolerated alternative [89].

7.2.4 Chitosan Functionalized Particles for Pulmonary Delivery of Biological Materials

CH oligomers are used in many fields of drug delivery. CH and its derivatives are well known to enhance the absorption of synthetic molecules across the mucosal barriers such as intestinal and nasal mucosae. Similarly, Yamada et al. (2005) studied the role of CH oligomers in pulmonary absorption of peptide. Particularly, different CH oligomers, i.e., CH dimer (340 Da), tetramer (663 Da), hexamer (985 Da), and water-soluble CH (1800 Da) were used to study the pulmonary absorption of interferon-α (INF-α) in rats. Following intratracheal administration (1×10^7 IU/kg), CH oligomers significantly increased serum INF-α concentration. Among the different CH oligomers, CH hexamer (0.5% w/v) was more efficient in improving the pulmonary absorption of INF-α with a marked 2.6-fold higher AUC value of INF-α as compared with intramuscularly administered INF-α. High molecular weight (22–96 kDa) CH polymers fail to improve the pulmonary absorption of INF-α owing to their poor water solubility. Besides, leakage of protein, lactate dehydrogenase, and bronchoalveolar lavage fluid assessment showed that CH oligomers did not cause any injury to the rat pulmonary tissues. Thus, the present study represented that the judicious selection of CH oligomers is essential for enhancing pulmonary absorption of therapeutically active proteins and peptides [90]. Zhang et al. (2016) developed CH hexamer (0.5% w/v) based mucoadhesive formulation for pulmonary delivery of peptide hormone, i.e., calcitonin. Developed mucoadhesive formulation exhibited higher calcitonin absorption due to mucoadhesion between negative mucus membrane and positive CH hexamer. The maximum calcitonin absorption is caused by CH hexamer compared to other CH oligomers, i.e., dimer and tetramer. It was credited to the optimum charge ratio of hexamer-calcitonin and self-aggregation capability of CH hexamer. Moreover, compared to CH tetramer and dimer, CH hexamer showed higher mucoadhesion potential due to the longer chains and more positive charges due to amino groups in its structure. Thus, CH hexamer has a superior tendency to interpenetrate and entangle with the mucus protein, resulting in stronger mucoadhesive interactions between the positive CH oligomers and the negative mucin glycoproteins. Additionally, hypoglycemic effects of formulation from pulmonary delivery were found to be higher than oral delivery. Briefly, results suggest the advantage of CH oligomers in pulmonary delivery of peptides over the traditional drug delivery route [91].

Parathyroid hormone (PTH) is a peptide hormone involved in bone formation, accelerate bone growth, and enhance bone density when it is administrated in as low dose regularly. PTH subcutaneous injection (Forteo®; Eli Lilly and Company) is currently prescribed for the postmenopausal women who have osteoporosis. But 24 months of the treatment cycle (20 μg/day) brings a great pain and inconvenience to the patient. Thus, Hong et al. (2018) carefully design noninvasive pulmonary route for PTH delivery to improve patient compliance and adherence to long-term therapy. PTH-loaded CH oligosaccharide (1:15 w/w; PTH-CSO) dry powders were formulated using an aqueous solution of a novel green supercritical fluid system equipped with an aqueous solution hydrodynamic cavitation mixer-based atomizer for improved pulmonary delivery of PTH. Well-defined spherical PTH-CSO MPs displayed well-controlled particle size (1 μm) and high loading efficacy (82.40%) suitable for deep pulmonary delivery. Circular dichroism confirmed the structural integrity of co-precipitated PTH-CSO MPs. At the same time, physicochemical analysis, i.e., thermo-gravimetric analysis and X-ray diffraction of PTH-CSO, verified the amorphous state of MPs. During aerodynamic analysis, PTH-CSO MPs showed satisfactory FPF (63.51%) and MMAD (1.7 μm) values using HandiHaler® at 60 L/min. Briefly, the proposed green processing scheme was found to be substantially apt for developing respirable drug/polymer composite MPs without adding any organic solvents [92].

Gene therapy is one of the most capable therapeutic alternatives for curative compared to symptomatic treatment [93]. The success in gene therapy depends on the efficient delivery of nucleic acid into the target cells. CH has good potential for DNA complexation and is helpful as non-viral vectors for gene delivery. Okamoto and co-workers formulated CH-plasmid DNA (CH-pDNA) complex powders using MAN as a carrier by a supercritical carbon dioxide/ ethanol admixture process. Rectangular shape CH-pDNA complex with positively charged polymer amine (N) groups to negatively charged nucleic acid phosphate (P) groups (N/P) ratio of 5 displayed yield of 82.2% and particle size of 12.2 μm. During in vivo luciferase activity using the mouse, after pulmonary administration of CH-pDNA complex, luciferase activity in mouse lung was increased compared to CH-pDNA solutions. CH-pDNA complex powder showed improved luciferase activity (2700%) at the end of 9 h than CH-pDNA solutions, while sodium acetate-based pDNA powders showed only 305% luciferase activity at the end of 9 h. This study suggested that CH complexes are helpful for pulmonary gene delivery because CH controlled the degradation of plasmid. During the supercritical process, the use of CH improved the yield of powder and CH complex powder improved the luciferase activity compared to plasmid DNA powder without CH and plasmid DNA solutions with or without CH [94].

7.2.5 Chitosan Decorated Microparticulate Systems

Losartan potassium (LP) is a non-peptide angiotensin II receptor antagonist, commonly prescribed in the management of hypertension. LP has established substantial attention in treatment of pulmonary hypertension allied with pulmonary vein stenosis. LP is a BCS class 3 drug with low permeability, short half-life (1.5–2.5 h), variable oral bioavailability (33%), and first-pass metabolism. Consequently, sustained release, non-oral delivery system such as pulmonary drug delivery is a right approach for LP delivery. But LP presents few physicochemical challenges with water-soluble carriers (e.g., polysaccharides) owing to its high aqueous solubility as water-soluble carriers hinder its encapsulation and sustained release profile. Therefore, Elmowafy and Soliman (2019) fabricated LP-loaded CH and dextran sulfate microplex (LP-CH/DS-MC) by a spray-drying method to obtain sustained release non-oral delivery systems. CH (positively charged) and DS (negatively charged) polymers allow to form a stable polyelectrolyte complex due to favorable strong ionic interactions. Additionally, an amphiphilic amino acid leucine was used as a dispersibility enhancer, and it also reduces the adhesion of spray-dried powder to the cyclone. The spray drying process offered a good yield (90.53%) with lower moisture content (2.16%). Micrometric robust LP-CH/DS-MC (1:0.5:0.5) showed higher encapsulation (75%) with a particle size of 1.2 μm. Moreover, LP-CH/DS-MC exhibited a superior swelling index (2.60) and zeta potential (−2.28 mV) apt for intrapulmonary bio-adhesion. In an in vitro release study using simulated lung fluid, LP-CH/DS-MC particles showed sustained release characteristics for 24 h and follow the Higuchi model ($R^2 = 0.946$). Irregularly dimpled and wrinkled-shaped LP-CH/DS-MC particles displayed a higher ED (95.95%) and a FPF (74.00%) using Aerolizer® at 60 L/min. In the in vivo acute inhalation toxicity studies wherein animals were treated with LP-CH/DS-MC particles, neither the integrity of the alveolar cell was affected, nor was any inflammation response trigged compared to animals treated with saline. The physicochemical characterization, outcomes of in vitro release, good powder flow properties, apt mucoadhesive potentials, and adequate aerodynamic performance showed LP microplex potential as an inhalation system. However, sustained release, non-oral delivery microplex systems need further checks for bench-to-bedside translation [95].

Microspheres are a promising class of the microparticulate systems. Being small in size, smooth surface, and large surface to volume ratios, microspheres are normally utilized for controlled release of hydrophobic molecules. Use of different biodegradable polymers makes microspheres suitable for high drug loading, sustained drug release, and targeted drug delivery. Simply, microspheres are model carriers for many drug delivery systems and pulmonary drug delivery is no exception to same [96, 97]. Consequently, Zhang et al. (2013) formulated biocompatible theophylline/carboxymethyl CH microspheres by spray drying for pulmonary drug delivery. Smooth surface spherical microspheres showed narrow particle sizes with improved drug loading efficiency. The in vitro hemolysis assessment of microspheres showed low hemolytic indices (<5%) and confirmed the biocompatibility of this novel material for lung delivery. During cell viability study using fibroblasts, microspheres showed concentration-dependent toxicity. However, no toxicity was reported at a lower concentration. The in vivo implant experiment using Sprague-Dawley rats indicated that the inflammatory reaction gradually decreased and disappeared without any significant difference from an operative suture. Additionally, during the in-vivo tissue biocompatibility study, after implanting microspheres (60 mg) in the skeletal muscle of rats, no significant hepatotoxicity and renal toxicity were observed. Moreover, microspheres did not show any hemoglobin synthesis-related toxicity in histopathological analysis with a lower degree of inflammatory response. Thus, spray-dried CH microspheres were potential biocompatible carriers for the pulmonary drug system [98].

7.2.6 Chitosan-Coated Nanoparticulate Systems

Dendrimers are nanosized, hyperbranched 3D polymeric structure composed of three key fragments, i.e., a central core, branched layer, and terminal functional group. They present a novel class of polymeric molecules with unique physicochemical properties and multiple functionalities [99]. Dendrimers provide distinctive platforms for imaging, diagnostics, theranostics, targeting, and drug delivery applications. Recently, dendrimers displayed huge clinical potential and advantages in pulmonary drug delivery [100]. Considering physicochemical and biological potential of dendrimers, Restani and co-workers fabricated CH-coated dendrimer-based pulmonary delivery of ibuprofen (IBP). Biodegradable IBP-loaded POxylated poly-urea dendrimers coated with CH were formulated by supercritical-fluid-assisted spray drying for lung delivery. Engineered nano-in-micro dry powders showed a spherical shape with a narrow size distribution between 1.4 and 5.5 μm and 0.9 g/cm^3 density. IBP-loaded engineered nano-in-micro dry powders showed improved solubility, good swelling kinetics, and sustained release profile at physiological pH. In an in vitro study using MTT assay, no significant cytotoxicity effect on the metabolic activity of A549 cells was observed at the lowest concentration (1×10^{-3} m) up to 120 h. In aerodynamic assessment using Aerolizer® device at 60 L/min, DPI demonstrated lower MMAD (<3 μm) and higher FPF values (<45%). The good swelling index guarantee the NPs sustained release and superior cellular uptake with promising biophysical properties of the dendrimers for IBP pulmonary delivery in both systemic and local actions. In summary, hybrid nano-in-micro formulations displayed promising applications for deep lung delivery in the upcoming time [101].

Yang et al. (2008) formulated CH-modified PLGA NPs for pulmonary delivery of paclitaxel (PTX). PTX-loaded PLGA NPs were prepared using a solvent evaporation method and coated with 0.7% w/v CH under sonication followed by magnetic stirring overnight. In double-distilled water, CH-modified PTX PLGA NPs showed particle size, zeta potential, and entrapment efficacy of 280.62 nm, 31.83 mV, and 69.62%, respectively. In contrast, in fresh mouse (CDF1), plasma CH-modified PTX PLGA NPs showed particle size and zeta potential of 2670.51 nm and 20.92 mV, respectively. An increase in particle size and zeta potential in plasma medium was accredited to electrical interaction between positively charged NPs and negatively charged plasma proteins. Aggregates were readily dissociated when agitated gently for 5 min indicating that the electronic interaction was transient and weak. Amorphous CH-modified PTX PLGA NPs displayed spherical shape morphology with a smooth surface. During the in vitro release study, CH-modified PTX PLGA NPs showed biphasic release behavior, i.e., initial faster release followed by a gradual release in PBS and plasma medium. PLGA NPs followed erosion-controlled and diffusion-based models in plasma and PBS medium, respectively; hence PLGA NPs showed higher cumulative release in plasma than PBS. During an in vitro uptake study using A549 cells, PLGA NPs showed good internalization. Most prominently, in an in-vivo study using lung-metastasized mice model, CH-modified PTX PLGA NPs showed an 18.5-fold higher lung distribution index of PTX than Taxol™ (10 mg/kg). The quick formation of NPs aggregates in contact with plasma proteins during blood circulation and electrical interaction-mediated enhanced uptake across the pulmonary endothelial cells was mainly responsible for higher lung distribution of PTX [102]. Taking together the hopeful advantages shown by CH-coated nanoparticulate platforms and the possibility to solve clinical challenges, CH possess the huge potential to build the next generation of pulmonary drug delivery systems.

7.2.7 Chitosan Containing Nebulized Particles

Like DPIs and MDIs, nebulizers also possess a very good potential to delivery drug molecule to the pulmonary airways. To deliver a drug using nebulizer, the drug was first dispersed into an aqueous medium and using a dispersing force (jet of gas or ultrasonic waves) a mist of droplets was generated in the range of an inhalable size. A variety of nebulizers such as spinning disc, jet, and ultrasonic nebulizers are available to delivery drug to the lung. Spinning disc, jet, and ultrasonic nebulizers use centrifugal force, high-pressure gas, and high-frequency vibrations (piezoelectric crystal), respectively, to generate a mist of droplets apt for inhalation [103, 104]. In most of the clinical cases, nebulizers showed efficient drug localization, and drug deposition pattern with suitable aerodynamic diameter [105]. Thus, nebulizer is an encouraging option for delivery drug to the pulmonary airways to mitigate airway inflammation, asthmatic conditions and allow protection from further exacerbations.

Ferulic acid (FA) is a US FDA-approved phytophenolic acid, with a strong, free radicle scavenging activity and widespread pharmacological actions. Its immunomodulatory action is known to ease various allergic complications such as Th1/Th2-mediated inflammation in the pulmonary airways. Accordingly, it can be used as a drug for effective management of pulmonary diseases such as asthma. However, low bioavailability, poor permeability, and short half-life of 30 min limit its therapeutic applicability to a great level. Its poor biopharmaceutical properties can be surpassed using CH-based mucoadhesive nanocarrier and surface-functionalized with hyaluronic acid to confirm transportation and protection of the FA across the pulmonary region. Thus, FA-containing HA-engineered CH NPs (HA-FA CH-NPs) were formulated and aerosolized using a vibrating mesh nebulizer to manage asthma effectively. Spherical-shaped HA-FA CH-NPs showed particle size, zeta potential, entrapment efficiency, and loading capacity of 164.2 nm, 24.00 mV, 65.00%, and 18.50%, respectively. During the aerosolization study, HA-FA CH-NPs displayed satisfactory MMAD of 1.81 μm using collision nebulizer (8 mL NPs) for 10 min at 0.5 L/min. HA-FA CH-NPs displayed safety during the biological assessment and efficiently controlled prophylaxis mitigated inflammation, pulmonary airway hypersensitivity, and remodeling in ovalbumin-induced female BALB/c mice models. HA surface engineering enhanced the thermal stability of CH-based NPs and allowed specific binding to epithelial cells compared to non-engineered CH-based NPs. The transportation of CH nanocarriers is mediated by absorptive transcytosis and is appropriate for transmucosal drug delivery. As well, HA functionalization modifies the surface chemistry of nanocarriers to assist immune stealth properties, transmucosal transportation, and better interaction with the pulmonary milieu. Thus, CH (as a permeation enhancer) in HA's presence effectively loosens the tight junctions of the pulmonary airway epithelium to ease both transcellular and paracellular drug delivery [106].

Colloidal carriers such as liposomes help to modify pharmacokinetics, protect the encapsulated molecule from enzymatic attack and increase the drug bioavailability [11]. But the clinical application of liposomes is restricted by their stability in biological milieu. To surmount this issue, the coating of liposomes vesicles with suitable polymer like CH is usually appreciated. CH coating not only increase liposomes vesicles stability in the biological milieu but also provide mucoadhesive properties to increase bioavailability. Consequently, CH-xanthan gum coated liposomes (chitosomes) for pulmonary delivery of rifampicin (RIF) were developed by Manca and co-workers and thoroughly investigated for pulmonary delivery. RIF-loaded multilamellar vesicles were fabricated by hydrogenated soy phosphatidylcholine (P90H) and soy phosphatidylcholine (P50) using the thin-film hydration method. RIF liposomes coating was performed using simple antisolvent addition method. RIF: P90H: P50 (10:30:30) coated with CH-xanthan gum (1:0.5) [RIF-30] showed mean particle size, zeta potential and entrapment efficacy of 2429 nm, 18 mV, and 83%, respectively, while RIF: P90H: P50 (10:60:60) coated with CH-xanthan gum (1:0.5) [RIF-60] showed mean particle size, zeta potential and entrapment efficacy of 2354 nm, 15 mV, and 79%, respectively. Both chitosomes vesicles showed smooth and compact surface, spherical shape morphology suitable for pulmonary delivery. For aerodynamic assessment of chitosomes, vesicles were nebulized using Markos Mefar MB2 air-jet nebulizer equipped with Nebula compressor (Markos Mefar), which delivers 3 mL of a sample containing 300 mg of RIF at a flow rate of 7 L/min. Nebulized RIF-30 chitosomes showed FPD and FPF of 1260 μg and 22.93%, respectively, whereas nebulized RIF-60 chitosomes displayed FPD and FPF of 2320 μg and 37.77%, respectively. Liposomes coating with the appropriate ratio of CH-xanthan gum showed better resistance to aerosolization, capable of delivering a higher amount of drug to airways [107].

7.2.8 Chitosan-Based Carriers for Intranasal Delivery

Stimulation of pulmonary immunity plays a crucial function in preventing the distribution of *Mycobacterium tuberculosis* during the initial infection stage. Moreover, recent research associated with mucosal immunity suggests that local mucosal immune responses are much more effective than systemic immune responses to reduce pathogens that infect via mucous membranes [108]. Thus, Ai et al. (2013) fabricated a natural cationic polysaccharide mucosal gene carrier, i.e., CH complexed pHSP65pep gene to stimulate specific pulmonary immunity. CH complexed pHSP65pep gene showed narrowed particle size distribution with a hydrodynamic radius of 400 nm. During in vivo studies using *Mycobacterium bovis* BCG challenge, BALB/c mice were infected intranasally for 4 weeks after the last immunization, i.e., 4 doses of 12.5 μg of the complex showed significant production of secretory immunoglobulin A in lung lavage fluid and noticeably improved pulmonary CD^{4+}, CD^{8+}, and IFN-γ^{+} T cell responses as compared with naked gene vaccine. CH also has immune-stimulating potential, which is involved in increasing accumulation and activation of macrophages and polymorphonuclear cells [109]. Thus, four doses of intranasal vaccine efficiently stimulated both humoral as well as cellular immunity in the pulmonary airways [110].

7.2.9 Cancer Detection Assay

Diagnostic maneuvers play a vital role in the characterization of disease progression in living objects. Nowadays, most of the diagnostic techniques involve use of nanotechnology or novel polymer-based devices to visualize and probe disease progression and biological processes noninvasively [111]. CH-based particles such as glycol CH NPs [112] and CH-sensors [113] were currently used as a diagnostic and imaging agent. Similarly, National Taiwan University Hospital developed the pH-responsive CH-based cell culture model to evaluate the cell detachment ratio to correlate the overall survival in lung cancer patients. The pH-responsive CH-based cell culture model assists in developing an alternative technique for early diagnosis and accurate prognosis in cancer therapeutics. The purified patient-derived and isolated tumor cells will be analyzed to express surface markers using flow cytometry. In the present situation, the sponsor recruits the subjects for the study [114]. CH like biomaterial-based cell culture models will be helpful for early diagnosis as well as accurate prognosis in cancer therapeutics. Like drug delivery, CH showed huge potential in the field of molecular imaging and diagnosis.

7.3 Future Viewpoint

CH is a classic polymer in pharmaceutical, biological applications because of its unique physicochemical properties and biocompatible attributes. CH and its derivatives-based NPs (e.g., nanogel, nanospheres, and SLN) and MPs (e.g.,

Table 7.2 Summary of chitosan-based delivery systems for management of pulmonary diseases

Formulation	Method	Particle properties	Outcome	References
Nanoparticulates drug delivery systems				
Prothionamide	Ionic gelation followed by freeze drying	Spherical and 314.37 nm	NPs showed in-vitro sustained release and high FPF values (82%)	[87]
Bedaquiline	Ionic gelation followed by freeze drying	109.70 nm	NPs showed a marked 1.80 fold improvement in bioavailability compared to oral solution	[86]
Hyaluronic acid decorated and ferulic acid loaded CH NPs	Ionic gelation followed by freeze drying	Spherical and 164.2 nm	Novel NPs ease the pulmonary inflammation and hypersensitivity in ovalbumin-induced female BALB/c mice	[106]
Isoniazid hybrid nanoplex	Ionic gelation followed by freeze drying	Smooth-surface uniform NPs	Controlled in-vitro release with 1.13-fold superior FPF compared to nano-powder	[83]
Isoniazid	Ionic gelation followed by spray drying	Spherical and 449.10 nm	Sustained release up to 6 days and high ED (> 90%)	[115]
Isoniazid and rifampicin	Ionic gelation followed by spray drying	Spherical and 230.00 nm	Sustained release with superior pulmonary deposition using nebulizer system	[116]
Rifampicin	Ionic gelation followed by probe sonication	Spherical and 124.10 nm	Sustained release NPs with FPF of 33% and superior pulmonary pharmacokinetic compared to oral solution	[117]
Rifampicin octanoyl CH NPs	Double emulsion solvent evaporation	Smooth spherical and 253 nm	Excellent aerosolization profile with FPF of 43.27% using a jet nebulizer	[118]
Rifampicin	Emulsion solvent evaporation	Spherical compact and < 215 nm	Co-polymer offer controlled release profiles with higher drug loading efficacy	[119]
Isoniazid and rifampicin nanogel	Emulsion cross-linking process	Spherical and < 130 nm	Sustained release with higher drug deposition within the pulmonary airways	[120]

(continued)

Table 7.2 (continued)

Formulation	Method	Particle properties	Outcome	References
Nicotine	W/O emulsion followed by freeze drying	Spherical and <500 nm	After aerosol delivery NPs attained optimal nicotine delivery and satisfactory in vivo locomotor activity	[82]
Ciprofloxacin	O/W emulsification followed by freeze drying	Smooth, spherical and 489 nm	Novel NPs showed FPF of 20.23% and retained structural and anti-microbial properties during long-term 6-month storage	[89]
Folic acid conjugated gemcitabine CH/PEG NPs	Ionic gelation	184.30 nm	Sustained and pH-sensitive gemcitabine release with improved cellular uptake and half-life	[121]
Raloxifene	Single emulsion solvent evaporation	Multilayer NPs and 210.6 nm	Multilayer NPs showed a higher level of cytotoxicity against A549 cells with enhanced half-life and potential of raloxifene	[122]
Paclitaxel	Solvent evaporation method and magnetic stirring	280.62 nm	Electrical interaction-mediated enhanced uptake across lung endothelial cells and 18.5-fold higher lung distribution index compared to Taxol™ at the same dose in lung-metastasized mice model showed NPs fitness in pulmonary delivery	[102]
Paclitaxel solid lipid NPs	Nanoprecipitation	Spherical and 250 nm	Sustained release up to 3 days and inhaled NPs showed a marked a 7- and 32-folds improvement in lung drug concentrations compared to inhaled Taxol® following 1 h and 6 h of delivery	[123]

Table 7.2 (continued)

Formulation	Method	Particle properties	Outcome	References
Doxorubicin and paclitaxel hollow nanospheres	Electrostatic adsorption	Spherical	Nanospheres showed better cytotoxicity against A549 cells with good inhibition of cell proliferation and effective cell apoptosis due to synergistic action	[124]
pHSP65pep gene	Supercritical fluid	The hydrodynamic radius of 400 nm	Intranasal administered (4 doses of 12.5 μg) complex gene carriers stimulate humoral and cellular immunity in the pulmonary airways	[110]
Interferon-α	–	–	Hexamer (0.5% w/v) showed 2.6-fold improved pulmonary absorption compared with intramuscularly administered INF-α without any damage to pulmonary tissues	[90]
Microparticulates drug delivery systems				
Isoniazid	Spray drying	<4 μm	MPs showed a controlled release profile with good ex vivo mucoadhesive potential and no cytotoxic effect against alveolar macrophages	[125]
Losartan microplex	Spray drying	Irregularly dimpled shaped and 1.2 μm	Microplex offered controlled release characteristics for 24 h and a respirable fraction of 74% at 60 L/min	[95]
Budesonide/ MAN binary carrier	Spray drying	Spherical shaped and 100.51 μm	Binary carrier showed a marked 2.29-fold improvement in bioavailability compared to commercial product at the same dose (800 μg/kg)	[75]

(continued)

Table 7.2 (continued)

Formulation	Method	Particle properties	Outcome	References
Levofloxacin Microspheres	Spray drying	Swellable microspheres	Microspheres present immediate drug release properties with a higher (90%) emitted dose	[126]
Doripenem	Ionic gelation followed by spray drying	Porous and <6 μm	CH MPs offered a controlled release profile with a high emitted dose (98%) and were effective for local pulmonary diseases such as pneumonia	[127]
Paclitaxel/ quercetin NPs co-loaded in CH microspheres	Ionic gelation followed by spray drying	Spherical and 3.37 μm	Microspheres offer a controlled release profile and apt aerodynamic diameter (1.80 μm) with superior plasma drug concentrations and half-life	[128]
Cisplatin microspheres	Emulsification with ionic gelation	Spherical and 5.20 μm	Sustained release up to 12 h and showed suitable aerodynamic diameter (2.71 μm) with superior IC_{50} value as compared to drug alone	[129]
Docetaxel microspheres	W/O emulsification	Smooth, spherical and 9.60 μm	During the biological assessment, microspheres achieved maximum drug concentration within lung tissues	[130]
Parathyroid hormone	Supercritical fluid system equipped with hydrodynamic cavitation mixer-based atomizer	Spherical and 1 μm	Amorphous MPs showed a % FPF of 63.51% at 60 L/min	[92]
Dapsone Microcapsules	Pre-emulsification followed by the rotor-stator system, high pressure homogenizer and spray drying	Spherical globules and 430 nm	Microcapsules offer controlled release profiles up to 24 h and higher FPF (50%) values	[131]

Table 7.2 (continued)

Formulation	Method	Particle properties	Outcome	References
Theophylline/ carboxymethyl CH microspheres	Spray drying	Smooth surface spherical microspheres	In-vivo studies confirmed material biocompatibility with lower tissue toxicity and lower inflammatory response in histopathology analysis	[98]
CH-plasmid DNA complex powders	Supercritical carbon dioxide/ ethanol admixture	Rectangular shape and 12.2 μm	During in vivo luciferase activity after pulmonary administration, CH-pDNA complex powder showed 8.85-fold improved luciferase activity as that of sodium acetate-based pDNA powders	[94]
Ibuprofen	Supercritical-fluid-assisted spray drying	Spherical and < 5.5 μm	Biodegradable dry powders demonstrated lower MMAD (< 3 μm) and higher FPF values (< 45%) using Aerolizer® (60 L/min)	[101]
Rifampicin chitosomes	Thin-film hydration RIF: P90H: P50 (10:30:30)	Smooth, compact and spherical with 2.4 μm	Chitosomes nebulized using Markos Mefar MB2 air-jet nebulizer equipped with Nebula compressor (Markos Mefar) showed FPD and FPF of 1260 μg and 22.93%	[107]
Rifampicin chitosomes	Thin-film hydration RIF: P90H: P50 (10:60:60)	Smooth, compact and spherical with 2.3 μm	Chitosomes nebulized using Markos Mefar MB2 air-jet nebulizer equipped with Nebula compressor (Markos Mefar) showed FPD and FPF of 2320 μg and 37.77%	[107]

(continued)

Table 7.2 (continued)

Formulation	Method	Particle properties	Outcome	References
CH-coated carvedilol loaded PEVA NPs	O/W solvent evaporation technique and MAN spray drying	4.2 μm	Microcomposites exhibited ED and FPF of 90.49% and 58.82%, respectively, using Easyhaler® at 60 L/min with an aerodynamic diameter of 1.4 μm	[77]
CH-coated carvedilol loaded PEVA NPs	O/W solvent evaporation technique and LAC spray drying	3.1 μm	Microcomposites exhibited ED and FPF of 90.6% and 21.26%, respectively, using Easyhaler® at 60 L/min with an aerodynamic diameter of 2.82 μm	[77]

microplex, microspheres, and microcapsules) systems were designed systematically and studied judiciously for the management of pulmonary diseases (Table 7.2). Researchers showed that by varying degrees of deacetylation and the molecular weight of CH, suitable nano or micro-carrier could be tailored according to the expected mean particle sizes and zeta potential. Using glutaraldehyde or tripolyphosphate, CH NPs fabricated via cross-linking method improved mucoadhesive properties with higher cellular uptake upon pulmonary delivery. Insulin and terbutaline sulfate encapsulated CH NPs showed sustained drug release profile, whereas CH-based dry powders of salbutamol sulfate, clindamycin, and beclomethasone dipropionate showed promising controlled release profiles (Fig. 7.3).

Additionally, a dry powder system containing beclomethasone dipropionate (hydrophobic), terbutaline sulfate (hydrophilic), leucine (aerosolization enhancer), and CH as a drug release modifier displayed promising sustained release of both actives from a single formulation. Nanometer surface roughness (<20 nm), binary carriers, and CH NPs loaded LAC-PEG-3000 MPs showed satisfactory aerosolization performance and therapeutic outcome. In few cases, drug encapsulation within CH NPs and MPs led to minimum drug systemic toxicity and extended biological half-life of the drug. Moreover, CH has multiple functional moieties (-NH and -OH), and researchers have explored them thoroughly to develop pulmonary carriers. Moreover, the cationic amino group at the C_2 position of CH was also utilized to develop inhalable micro-complexes via electrostatic interaction with the anionic groups of different polyions (e.g., dextran, carrageenan, and HA). Besides, cationic CH has also been used to develop pulmonary vaccines due to its ability to activate the immune system directly. These novel CH-based carriers were meticulously explored and investigated as pulmonary carriers against the conventional LAC-based carrier systems.

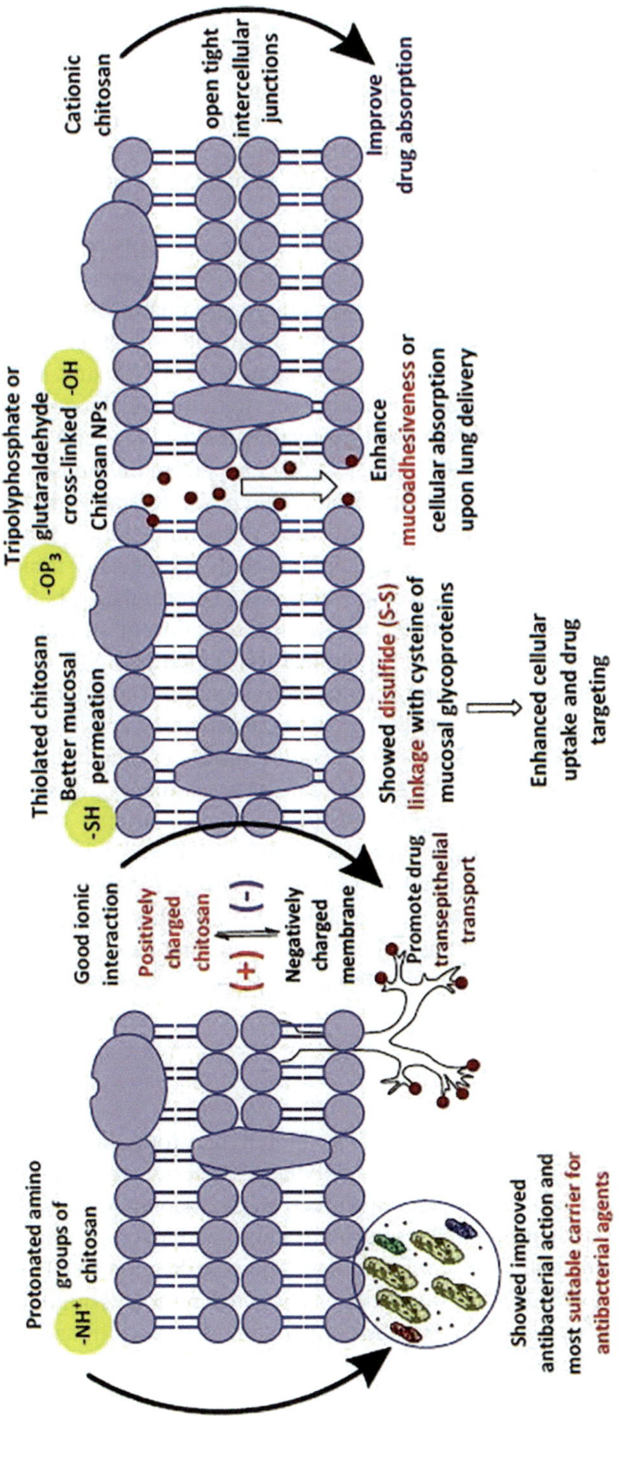

Fig. 7.3 Interaction of chitosan with pulmonary membrane

Several types of CH molecules from numerous origins eventually acquired different physicochemical properties and thus created complications for regulatory bodies to set harmonized assessment standards. In addition, to deliver drug substances at the target site or to modulate biological properties, biopolymer CH was generally utilized with other synthetic polymers and excipients. However, using synthetic polymers may lead to adverse effects in the human body, mainly during chronic therapy. So, extra attention is required while exploring synthetic polymers as modifier agents in CH-based delivery systems.

Moreover, CH-based novel inhalable carriers are relatively new, and thus, in-depth analysis is essential together with powder dispersion and de-agglomeration characterization, dose uniformity and in vivo pulmonary pharmacokinetics. Suitable dissolution apparatus equipped with an aerosol particle collection system can be adopted to understand the in vitro release kinetics of CH NPs and MPs. In most investigations, the CH formulation's safety, i.e., toxicity and biocompatibility, is described based on a short-term exposure protocol while the long-term effects remain unattended. Further, translation of biological outcomes from rat/mouse lung models to humans is challenging owing to the considerable physiological variation between the airways of animals and humans. In most cases, the administration routes explored for animal models are not fit for human application. It is especially tricky to assess and validate the pulmonary efficacy of the designed CH formulation before thinking about them for clinical investigations. Hence, the selection of suitable pulmonary models will assist in accomplishing the clinical translation is highly essential. Cell culture inserts organized cascade impactor in such situations is a better alternative for evaluating inhaled CH formulation during the initial screening and long-term studies with reproducible conditions. Besides, cell culture inserts or membrane-loaded cascade impactors can also be used for a more realistic assessment of dissolution kinetics, drug transport and cellular uptake of drug particles. Briefly, such robust yet straightforward in vitro and in silico models potentially offer a platform to study the underlying mechanisms by which inhaled CH-based medicines interact with the pulmonary airways and offer good reproducibility. In the future, various judiciously designed CH formulation systems and sophisticated analytical techniques would result in bench-to-bedside translational research for inhaled CH-based medicines.

7.4 Conclusion

This chapter resourcefully addresses CH as a biopolymer for pulmonary drug delivery. CH and its derivatives are proficient polymeric materials for pulmonary delivery of anti-cancer drugs, antibiotics, anti-infective, phytocompounds and biological substances owing to their mucoadhesive, anti-microbial, biocompatible, biodegradable, and permeation enhancement properties. The modification of CH structure in a broader perspective aid in improving its physical behavior and biological properties as a lung delivery vehicle. Also, the conjugation of various ligands with CH is vital in designing efficient lung anti-cancer systems. However, a few issues like limited pulmokinetics data, biosafety information and toxicological challenges can restrict CH and its derivatives application in the clinic. Thus, in the upcoming time, a collaborative effort between scientists from multiple fields, including formulation, materials science, biotechnology, toxicology and clinical settings, would potentiate the translation of CH-based novel particulate systems into commercially viable inhalable products.

References

1. Weber S, Zimmer A, Pardeike J. Solid lipid nanoparticles (SLN) and nanostructured lipid carriers (NLC) for pulmonary application: a review of the state of the art. Eur J Pharm Biopharm. 2014;86(1):7–22.
2. Mehta PP, Dhapte-Pawar VS. Novel and evolving therapies for COVID-19 related pulmonary complications. Am J Med Sci. 2021;361(5):557–66.
3. Mehta P. Imagine the superiority of dry powder inhalers from carrier engineering. J Drug Deliv. 2018;2018:5635010.
4. Valente SA, Silva LM, Lopes GR, Sarmento B, Coimbra MA, Passos CP. Polysaccharide-based formulations as potential carriers for pulmonary delivery–a review of their properties and fates. Carbohydr Polym. 2022;277:118784.
5. Mehta PP, Bothiraja C, Kadam SS, Pawar AP. Potential of dry powder inhalers for tuberculosis therapy: facts, fidelity and future. Artif Cells Nanomed Biotechnol. 2018;46(sup3):S791–806.
6. Mehta PP. Dry powder inhalers: a concise summary of the electronic monitoring devices. Ther Deliv. 2021;12(1):1–6.
7. Mehta PP, Pawar AP, Mahadik KR, Kadam SS, Dhapte-Pawar V. Dry powder coating techniques and role of force controlling agents in aerosol. Polymer Coatings: Technology and Applications. 2020:41–74.
8. Mehta PP. Dry powder inhalers: a brief overview of the drug detachment techniques. Ther Deliv. 2020;11(3):139–43.
9. Dhanjal DS, Mehta M, Chopra C, Singh R, Sharma P, Chellappan DK, Tambuwala MM, Bakshi HA, Aljabali AA, Gupta G, Nammi S. Novel controlled release pulmonary drug delivery systems: current updates and challenges. In modeling and control of drug delivery systems 2021 (pp. 253-272). Academic Press.
10. Mehta PP, Bothiraja C, Mahadik KR, Kadam SS, Pawar AP. Phytoconstituent based dry powder inhalers as biomedicine for the management of pulmonary diseases. Biomed Pharmacother. 2018;108:828–37.
11. Mehta PP, Ghoshal D, Pawar AP, Kadam SS, Dhapte-Pawar VS. Recent advances in inhalable liposomes for treatment of pulmonary diseases: concept to clinical stance. J Drug Deliv Sci Technol. 2020;56:101509.
12. Mehta PP. Dry powder inhalers: upcoming platform technologies for formulation development. Ther Deliv. 2019;10(9):551–4.
13. Depreter F, Pilcer G, Amighi K. Inhaled proteins: challenges and perspectives. Int J Pharm. 2013;447(1–2):251–80.
14. Ho DK, Nichols BL, Edgar KJ, Murgia X, Loretz B, Lehr CM. Challenges and strategies in drug delivery systems for treatment of pulmonary infections. Eur J Pharm Biopharm. 2019;144:110–24.
15. Al-Obaidi H, Granger A, Hibbard T, Opesanwo S. Pulmonary drug delivery of antimicrobials and anticancer drugs using solid dispersions. Pharmaceutics. 2021;13(7):1056.
16. Mehta PP, Kadam SS, Pawar AP. Effect of USP induction ports and modified glass sampling apparatus on aerosolization performance of lactose carrier-based fluticasone propionate dry powder inhaler. J Drug Deliv Sci Technol. 2020;58:101794.

17. Mehta PP, Kadam SS, Pawar AP. Influence of modified induction port, modified DUSA assembly and device air-inlet geometry on the aerosolization pattern of a dry powder inhaler. J Drug Deliv Sci Technol. 2020;55:101416.
18. Kumar R, Mehta P, Shankar KR, Rajora MA, Mishra YK, Mostafavi E, Kaushik A. Nanotechnology-assisted metered-dose inhalers (MDIs) for high-performance pulmonary drug delivery applications. Pharm Res. 2022;1-25:2831.
19. Zhou QT, Morton DA. Drug–lactose binding aspects in adhesive mixtures: controlling performance in dry powder inhaler formulations by altering lactose carrier surfaces. Adv Drug Deliv Rev. 2012;64(3):275–84.
20. Ramachandran V, Murnane D, Hammond RB, Pickering J, Roberts KJ, Soufian M, Forbes B, Jaffari S, Martin GP, Collins E, Pencheva K. Formulation pre-screening of inhalation powders using computational atom-atom systematic search method. Mol Pharm. 2015;12(1):18–33.
21. Mehta PP, Dhapte-Pawar V. Role of surfactants in pulmonary drug delivery. In: In green sustainable process for chemical and environmental engineering and science. Academic Press; 2022. p. 559–77.
22. Mehta PP, Dhapte-Pawar VS. Repurposing drug molecules for new pulmonary therapeutic interventions. Drug Deliv Transl Res. 2021;11(5):1829–48.
23. Vinjamuri BP, Kotha AK, Kolte A, Haware RV, Chougule MB. Polymer applications in pulmonary drug delivery. In: Applications of polymers in drug delivery. Elsevier; 2021. p. 333–54.
24. Vasvani S, Kulkarni P, Rawtani D. Hyaluronic acid: a review on its biology, aspects of drug delivery, route of administrations and a special emphasis on its approved marketed products and recent clinical studies. Int J Biol Macromol. 2020;151:1012–29.
25. Zillen D, Beugeling M, Hinrichs WL, Frijlink HW, Grasmeijer F. Natural and bioinspired excipients for dry powder inhalation formulations. Curr Opin Colloid Interface Sci. 2021;101497
26. Zhou QT, Leung SS, Tang P, Parumasivam T, Loh ZH, Chan HK. Inhaled formulations and pulmonary drug delivery systems for respiratory infections. Adv Drug Deliv Rev. 2015;85:83–99.
27. Elzoghby AO, El-Fotoh WS, Elgindy NA. Casein-based formulations as promising controlled release drug delivery systems. J Control Release. 2011;153(3):206–16.
28. Zhu L, Li M, Liu X, Du L, Jin Y. Inhalable oridonin-loaded poly(lactic-co-glycolic)acid large porous microparticles for in situ treatment of primary non-small cell lung cancer. Acta Pharm Sin B. 2017;7(1):80–90.
29. Deshmukh M, Kutscher HL, Gao D, Sunil VR, Malaviya R, Vayas K, Stein S, Laskin JD, Laskin DL, Sinko PJ. Biodistribution and renal clearance of biocompatible lung targeted poly(ethylene glycol) (PEG) nanogel aggregates. J Control Release. 2012;164(1):65–73.
30. Waters RC, Hochhaus G. Characterization of a dextran-budesonide prodrug for inhalation therapy. Eur J Pharm Sci. 2019;129:58–67.
31. Mohammadi Z, Eini M, Rastegari A, Tehrani MR. Chitosan as a machine for biomolecule delivery: a review. Carbohydr Polym. 2021;256:117414.
32. Xie Y, Kim NH, Nadithe V, Schalk D, Thakur A, Kılıç A, Lum LG, Bassett DJP, Merkel OM. Targeted delivery of siRNA to activated T cells via transferrin-polyethylenimine (Tf-PEI) as a potential therapy of asthma. J Control Release. 2016;229:120–9.
33. Savla R, Minko T. Nanotechnology approaches for inhalation treatment of fibrosis. J Drug Target. 2013;21(10):914–25.
34. Gulati N, Chellappan DK, MacLoughlin R, Dua K, Dureja H. Inhaled nano-based therapeutics for inflammatory lung diseases: recent advances and future prospects. Life Sci. 2021;285:119969.
35. Mallakpour S, Azadi E, Hussain CM. Chitosan, alginate, hyaluronic acid, gums, and β-glucan as potent adjuvants and vaccine delivery systems for viral threats including SARS-CoV-2: a review. Int J Biol Macromol. 2021;182:1931–40.

36. Tripodo G, Trapani A, Torre ML, Giammona G, Trapani G, Mandracchia D. Hyaluronic acid and its derivatives in drug delivery and imaging: recent advances and challenges. Eur J Pharm Biopharm. 2015;97(Pt B):400–16.
37. Huang G, Huang H. Hyaluronic acid-based biopharmaceutical delivery and tumor-targeted drug delivery system. J Control Release. 2018;278:122–6.
38. Luo Z, Dai Y, Gao H. Development and application of hyaluronic acid in tumor targeting drug delivery. Acta Pharm Sin B. 2019;9(6):1099–112.
39. Hu Q, Lu Y, Luo Y. Recent advances in dextran-based drug delivery systems: from fabrication strategies to applications. Carbohydr Polym. 2021;264:117999.
40. Maity S, Saha S, Sarkar S, Kumar P, Gautam AK, Sonkar AB. Dextran-based nanomaterials in drug delivery applications. In: Biopolymer-based nanomaterials in drug delivery and biomedical applications. Academic Press; 2021. p. 293–312.
41. Li X, Jiang X. Microfluidics for producing poly (lactic-co-glycolic acid)-based pharmaceutical nanoparticles. Adv Drug Deliv Rev. 2018;128:101–14.
42. Swider E, Koshkina O, Tel J, Cruz LJ, de Vries IJM, Srinivas M. Customizing poly(lactic-co-glycolic acid) particles for biomedical applications. Acta Biomater. 2018;73:38–51.
43. Fredenberg S, Wahlgren M, Reslow M, Axelsson A. The mechanisms of drug release in poly(lactic-co-glycolic acid)-based drug delivery systems–a review. Int J Pharm. 2011;415(1–2):34–52.
44. Feng SS, Mu L, Win KY, Huang G. Nanoparticles of biodegradable polymers for clinical administration of paclitaxel. Curr Med Chem. 2004;11(4):413–24.
45. Shen TW, Fromen CA, Kai MP, Luft JC, Rahhal TB, Robbins GR, DeSimone JM. Distribution and cellular uptake of PEGylated polymeric particles in the lung towards cell-specific targeted delivery. Pharm Res. 2015;32(10):3248–60.
46. Elieh-Ali-Komi D, Hamblin MR. Chitin and chitosan: production and application of versatile biomedical nanomaterials. Int J Adv Res (Indore). 2016;4(3):411–27.
47. Peers S, Montembault A, Ladavière C. Chitosan hydrogels incorporating colloids for sustained drug delivery. Carbohydr Polym. 2021;118689
48. Rasul RM, Tamilarasi Muniandy M, Zakaria Z, Shah K, Chee CF, Dabbagh A, Rahman NA, Wong TW. A review on chitosan and its development as pulmonary particulate anti-infective and anti-cancer drug carriers. Carbohydr Polym. 2020;5(250):116800.
49. Islam N, Ferro V. Recent advances in chitosan-based nanoparticulate pulmonary drug delivery. Nanoscale. 2016;8(30):14341–58.
50. Mohammed MA, Syeda JTM, Wasan KM, Wasan EK. An overview of chitosan nanoparticles and its application in non-parenteral drug delivery. Pharmaceutics. 2017;9(4):53.
51. Elgadir MA, Uddin MS, Ferdosh S, Adam A, Chowdhury AJK, Sarker MZI. Impact of chitosan composites and chitosan nanoparticle composites on various drug delivery systems: a review. J Food Drug Anal. 2015;23(4):619–29.
52. Shukla SK, Mishra AK, Arotiba OA, Mamba BB. Chitosan-based nanomaterials: a state-of-the-art review. Int J Biol Macromol. 2013;59:46–58.
53. Cheung RC, Ng TB, Wong JH, Chan WY. Chitosan: an update on potential biomedical and pharmaceutical applications. Mar Drugs. 2015;13(8):5156–86.
54. Bivas-Benita M, van Meijgaarden KE, Franken KL, Junginger HE, Borchard G, Ottenhoff TH, Geluk A. Pulmonary delivery of chitosan-DNA nanoparticles enhances the immunogenicity of a DNA vaccine encoding HLA-A*0201-restricted T-cell epitopes of mycobacterium tuberculosis. Vaccine. 2004;22(13–14):1609–15.
55. Jiménez-Gómez CP, Cecilia JA. Chitosan: a natural biopolymer with a wide and varied range of applications. Molecules. 2020;25(17):3981.
56. Francis Suh JK, Matthew HWT. Application of chitosan-based polysaccharide biomaterials in cartilage tissue engineering: a review. Biomaterials. 2000;21:2589–98.
57. Kalia S. Biodegradable green composites. John Wiley & Sons; 2016. p. 1–380.
58. Yamamoto H, Kuno Y, Sugimoto S, Takeuchi H, Kawashima Y. Surface-modified PLGA nanosphere with chitosan improved pulmonary delivery of calcitonin by mucoadhe-

sion and opening of the intercellular tight junctions. J Control Release. 2005;102(2): 373–81.
59. Dhawan S, Singla AK, Sinha VR. Evaluation of mucoadhesive properties of chitosan microspheres prepared by different methods. AAPS PharmSciTech. 2004;5(4):e67.
60. Goyal AK, Garg T, Rath G, Gupta UD, Gupta P. Development and characterization of nanoembedded microparticles for pulmonary delivery of anti-tubercular drugs against experimental tuberculosis. Mol Pharm. 2015;12(11):3839–50.
61. Learoyd TP, Burrows JL, French E, Seville PC. Sustained delivery by leucine-modified chitosan spray-dried respirable powders. Int J Pharm. 2009;372(1–2):97–104.
62. Murata M, Nakano K, Tahara K, Tozuka Y, Takeuchi H. Pulmonary delivery of elcatonin using surface-modified liposomes to improve systemic absorption: polyvinyl alcohol with a hydrophobic anchor and chitosan oligosaccharide as effective surface modifiers. Eur J Pharm Biopharm. 2012;80(2):340–6.
63. Thanou M, Verhoef JC, Junginger HE. Chitosan and its derivatives as intestinal absorption enhancers. Adv Drug Deliv Rev. 2001;50(Suppl 1):S91–101.
64. Raafat D, Sahl HG. Chitosan and its antimicrobial potential–a critical literature survey. Microb Biotechnol. 2009;2(2):186–201.
65. Yilmaz AH. Antibacterial activity of chitosan-based systems. Functional Chitosan. 2020:457–89.
66. Li J, Zhuang S. Antibacterial activity of chitosan and its derivatives and their interaction mechanism with bacteria: current state and perspectives. Eur Polym J. 2020;138:109984.
67. Kaczmarek MB, Struszczyk-Swita K, Li X, Szczęsna-Antczak M, Daroch M. Enzymatic modifications of chitin, chitosan, and Chitooligosaccharides. Front Bioeng Biotechnol. 2019;7:243.
68. Singh B, Maharjan S, Cho KH, Cui L, Park IK, Choi YJ, Cho CS. Chitosan-based particulate systems for the delivery of mucosal vaccines against infectious diseases. Int J Biol Macromol. 2018;110:54–64.
69. Gulati N, Dua K, Dureja H. Role of chitosan based nanomedicines in the treatment of chronic respiratory diseases. Int J Biol Macromol. 2021;185:20–30.
70. Mehta PP, Kadam SS, Pawar AP. Exploring the impact of extrinsic lactose fines, a USP modified sampling device and modified centrifuge tube on the delivered dose uniformity and drug detachment performance of a fluticasone propionate dry powder inhaler. J Drug Deliv Sci Technol. 2020;57:101681.
71. Mehta PP, Bothiraja C, Kadam SS, Pawar AP. Probing the influence of lactose fines, USP modified induction port and modified DDU apparatus on aerodynamic behavior of fluticasone propionate dry powder inhaler. New J Chem. 2019;43:17327.
72. Mehta PP. Probing the effect of USP induction port geometry on aerodynamic performance of dry powder inhalers. Drug Delivery to the Lungs. 2020;31
73. Littringer EM, Mescher A, Schroettner H, Achelis L, Walzel P, Urbanetz N. Spray dried mannitol carrier particles with tailored surface properties–the influence of carrier surface roughness and shape. Eur J Pharm Biopharm. 2012;82(1):194–204.
74. Kawashima Y, Serigano T, Hino T, Yamamoto H, Takeuchi H. Effect of surface morphology of carrier lactose on dry powder inhalation property of pranlukast hydrate. Int J Pharm. 1998;172(1):179–88.
75. Huang Y, Huang Z, Zhang X, Zhao Z, Zhang X, Wang K, Ma C, Zhu C, Pan X, Wu C. Chitosan-based binary dry powder inhaler carrier with nanometer roughness for improving in vitro and in vivo aerosolization performance. Drug Deliv Transl Res. 2018;8(5):1274–88.
76. Alhajj N, Zakaria Z, Naharudin I, Ahsan F, Li W, Wong TW. Critical physicochemical attributes of chitosan nanoparticles admixed lactose-PEG 3000 microparticles in pulmonary inhalation. Asian J Pharmaceutical Sci. 2020;15(3):374–84.
77. Varshosaz J, Taymouri S, Hamishehkar H. Fabrication of polymeric nanoparticles of poly (ethylene-co-vinyl acetate) coated with chitosan for pulmonary delivery of carvedilol. J Appl Polym Sci. 2014;131(1)

78. https://www.who.int/news-room/fact-sheets/detail/tobacco#:~:text=Key%20facts,exposed%20to%20second%2Dhand%20smoke.
79. Wang H, George G, Bartlett S, Gao C, Islam N. Nicotine hydrogen tartrate loaded chitosan nanoparticles: formulation, characterization and in vitro delivery from dry powder inhaler formulation. Eur J Pharm Biopharm. 2017;113:118–31.
80. Cipolla D, Gonda I. Inhaled nicotine replacement therapy. Asian J Pharmaceutical Sci. 2015;10(6):472–80.
81. Wang H, George G, Islam N. Nicotine-loaded chitosan nanoparticles for dry powder inhaler (DPI) formulations–impact of nanoparticle surface charge on powder aerosolization. Adv Powder Technol. 2018;29(12):3079–86.
82. Wang H, Holgate J, Bartlett S, Islam N. Assessment of nicotine release from nicotine-loaded chitosan nanoparticles dry powder inhaler formulations via locomotor activity of C57BL/6 mice. Eur J Pharm Biopharm. 2020;154:175–85.
83. Mukhtar M, Pallagi E, Csóka I, Benke E, Farkas Á, Zeeshan M, Burián K, Kókai D, Ambrus R. Aerodynamic properties and in silico deposition of isoniazid loaded chitosan/thiolated chitosan and hyaluronic acid hybrid nanoplex DPIs as a potential TB treatment. Int J Biol Macromol. 2020;165(Pt B):3007–19.
84. Mahajan R. Bedaquiline: first FDA-approved tuberculosis drug in 40 years. Int J Appl Basic Med Res. 2013;3(1):1–2.
85. Sarathy JP, Gruber G, Dick T. Re-understanding the mechanisms of action of the anti-mycobacterial drug Bedaquiline. Antibiotics (Basel). 2019;8(4):261.
86. Rawal T, Patel S, Butani S. Chitosan nanoparticles as a promising approach for pulmonary delivery of bedaquiline. Eur J Pharm Sci. 2018;124:273–87.
87. Debnath SK, Saisivam S, Debanth M, Omri A. Development and evaluation of chitosan nanoparticles based dry powder inhalation formulations of Prothionamide. PLoS One. 2018;13(1):e0190976.
88. McShane PJ, Weers JG, Tarara TE, Haynes A, Durbha P, Miller DP, Mundry T, Operschall E, Elborn JS. Ciprofloxacin dry powder for inhalation (ciprofloxacin DPI): technical design and features of an efficient drug–device combination. Pulm Pharmacol Ther. 2018;50:72–9.
89. Topal GR, Devrim B, Eryilmaz M, Bozkir A. Design of ciprofloxacin-loaded nano-and microcomposite particles for dry powder inhaler formulations: preparation, in vitro characterisation, and antimicrobial efficacy. J Microencapsul. 2018;35(6):533–47.
90. Yamada K, Odomi M, Okada N, Fujita T, Yamamoto A. Chitosan oligomers as potential and safe absorption enhancers for improving the pulmonary absorption of interferon-alpha in rats. J Pharm Sci. 2005;94(11):2432–40. https://doi.org/10.1002/jps.20454.
91. Zhang H, Huang X, Sun Y, Xing J, Yamamoto A, Gao Y. Absorption-improving effects of chitosan oligomers based on their mucoadhesive properties: a comparative study on the oral and pulmonary delivery of calcitonin. Drug Deliv. 2016 Sep 1;23(7):2419–27.
92. Hong DX, Yun YL, Guan YX, Yao SJ. Preparation of micrometric powders of parathyroid hormone (PTH1–34)-loaded chitosan oligosaccharide by supercritical fluid assisted atomization. Int J Pharm. 2018;545(1–2):389–94.
93. Stern M, Geddes DM, Alton E. Taking stock of gene therapy for cystic fibrosis. Respir Res. 2000;1:78–81.
94. Okamoto H, Nishida S, Todo H, Sakakura Y, Iida K, Danjo K. Pulmonary gene delivery by chitosan–pDNA complex powder prepared by a supercritical carbon dioxide process. J Pharm Sci. 2003;92(2):371–80.
95. Elmowafy E, Soliman ME. Losartan-chitosan/dextran sulfate microplex as a carrier to lung therapeutics: dry powder inhalation, aerodynamic profile and pulmonary tolerability. Int J Biol Macromol. 2019;136:220–9.
96. Mehta P. Dry powder inhalers: a focus on advancements in novel drug delivery systems. J Drug Deliv. 2016;2016:1.
97. Mehta P. Imagine the superiority of dry powder inhalers from carrier engineering. J Drug Deliv. 2018;2018:1.

98. Zhang WF, Zhao XT, Zhao QS, Zha SH, Liu DM, Zheng ZJ, Li WT, Zhou HY, Yan F. Biocompatibility and characteristics of theophylline/carboxymethyl chitosan microspheres for pulmonary drug delivery. Polym Int. 2014;63(6):1035–40.
99. Korake S, Shaikh A, Salve R, Gajbhiye KR, Gajbhiye V, Pawar A. Biodegradable dendritic Boltorn™ nanoconstructs: a promising avenue for cancer theranostics. Int J Pharm. 2021 Feb 1;594:120177.
100. Mehta P, Kadam S, Pawar A, Bothiraja C. Dendrimers for pulmonary delivery: current perspectives and future challenges. New J Chem. 2019;43(22):8396–409.
101. Restani RB, Silva AS, Pires RF, Cabral R, Correia IJ, Casimiro T, Bonifácio VD, Aguiar-Ricardo A. Nano-in-micro POxylated Polyurea dendrimers and chitosan dry powder formulations for pulmonary delivery. Part Part Syst Charact. 2016;33(11):851–8.
102. Yang R, Yang SG, Shim WS, Cui F, Cheng G, Kim IW, Kim DD, Chung SJ, Shim CK. Lung-specific delivery of paclitaxel by chitosan-modified PLGA nanoparticles via transient formation of microaggregates. J Pharm Sci. 2009;98(3):970–84.
103. Ali M. Pulmonary drug delivery. In handbook of non-invasive drug delivery systems 2010 Jan 1 (pp. 209-246). William Andrew Publishing.
104. Hyers BM, Spencer C, Eisenkraft JB. Humidification and filtration. In anesthesia equipment 2021 Jan 1 (pp. 183-192). WB Saunders.
105. Hertel SP, Winter G, Friess W. Protein stability in pulmonary drug delivery via nebulization. Adv Drug Deliv Rev. 2015 Oct 1;93:79–94.
106. Dhayanandamoorthy Y, Antoniraj MG, Kandregula CAB, Kandasamy R. Aerosolized hyaluronic acid decorated, ferulic acid loaded chitosan nanoparticle: a promising asthma control strategy. Int J Pharm. 2020;591:119958.
107. Manca ML, Manconi M, Valenti D, Lai F, Loy G, Matricardi P, Fadda AM. Liposomes coated with chitosan-xanthan gum (chitosomes) as potential carriers for pulmonary delivery of rifampicin. J Pharm Sci. 2012;101(2):566–75.
108. Yuki Y, Kiyono H. Mucosal vaccines: novel advances in technology and delivery. Expert Rev Vaccines. 2009;8:1083–97.
109. Kumar M, Behera AK, Lockey RF, Zhang J, Bhullar G, De La Cruz CP, Chen LC, Eong LKW, Huang SK, Mohapatra SS. Intranasal gene transfer by chitosan-DNA nanospheres protects BALB/c mice against acute respiratory syncytial virus infection. Hum Gene Ther. 2002;13:1415–25.
110. Ai W, Yue Y, Xiong S, Xu W. Enhanced protection against pulmonary mycobacterial challenge by chitosan-formulated polyepitope gene vaccine is associated with increased pulmonary secretory IgA and gamma-interferon (+) T cell responses. Microbiol Immunol. 2013 Mar;57(3):224–35. https://doi.org/10.1111/1348-0421.12027.
111. Agrawal P, Strijkers GJ, Nicolay K. Chitosan-based systems for molecular imaging. Adv Drug Deliv Rev. 2010 Jan 31;62(1):42–58.
112. Lee S, Kang SW, Ryu JH, Na JH, Lee DE, Han SJ, Kang CM, Choe YS, Lee KC, Leary JF, Choi K. Tumor-homing glycol chitosan-based optical/PET dual imaging nanoprobe for cancer diagnosis. Bioconjug Chem. 2014 Mar 19;25(3):601–10.
113. Nasution TI, Nainggolan I, Hutagalung SD, Ahmad KR, Ahmad ZA. The sensing mechanism and detection of low concentration acetone using chitosan-based sensors. Sensors Actuators B Chem. 2013;177:522–8.
114. Evaluation of the Cell Detachment Ratio on pH-responsive Chitosan as a Prognostic Factor in Lung Cancer. NCT04218188. https://clinicaltrials.gov/ct2/show/record/NCT04218188.
115. Pourshahab PS, Gilani K, Moazeni E, Eslahi H, Fazeli MR, Jamalifar H. Preparation and characterization of spray dried inhalable powders containing chitosan nanoparticles for pulmonary delivery of isoniazid. J Microencapsul. 2011;28(7):605–13.
116. Garg T, Rath G, Goyal AK. Inhalable chitosan nanoparticles as antitubercular drug carriers for an effective treatment of tuberculosis. Artificial Cells Nanomed Biotechnol. 2015;44(3):1–5.
117. Rawal T, Parmar R, Tyagi RK, Butani S. Rifampicin loaded chitosan nanoparticle dry powder presents an improved therapeutic approach for alveolar tuberculosis. Colloids Surf B Biointerfaces. 2017;154:321–30.

118. Petkar KC, Chavhan S, Kunda N, Saleem I, Somavarapu S, Taylor KMG, Sawant KK. Development of novel octanoyl chitosan nanoparticles for improved rifampicin pulmonary delivery: optimization by factorial design. AAPS PharmSciTech. 2018;19(4):1758–72.
119. Rajan M, Raj V. Formation and characterization of chitosan-polylacticacidpolyethylene glycol-gelatin nanoparticles: a novel biosystem for controlled drug delivery. Carbohydr Polym. 2013;98(1):951–8.
120. Wu T, Liao W, Wang W, Zhou J, Tan W, Xiang W, Cai X. Genipin crosslinked carboxymethyl chitosan nanogel for lung-targeted delivery of isoniazid and rifampin. Carbohydr Polym. 2018;197:403–13.
121. Wang F, Wang Y, Ma Q, Cao Y, Yu B. Development and characterization of folic acid-conjugated chitosan nanoparticles for targeted and controlled delivery of gemcitabinein lung cancer therapeutics. Artificial Cells Nanomed Biotechnol. 2017;45(8):1530–8.
122. Almutairi FM, Abd-Rabou AA, Mohamed MS. Raloxifene-encapsulated hyaluronic acid-decorated chitosan nanoparticles selectively induce apoptosis in lung cancer cells. Bioorg Med Chem. 2019;27(8):1629–38.
123. Rosière R, Van Woensel M, Gelbcke M, Mathieu V, Hecq J, Mathivet T, Wauthoz N. New folate-grafted chitosan derivative to improve delivery of paclitaxel-loaded solid lipid nanoparticles for lung tumor therapy by inhalation. Mol Pharm. 2018;15(3):899–910.
124. Tao L, Jiang J, Gao Y, Wu C, Liu Y. Biodegradable alginate-chitosan hollow nanospheres for codelivery of doxorubicin and paclitaxel for the effect of human lung cancer A549 cells. Biomed Res Int. 2018;2018:1–11.
125. Oliveira PM, Matos BN, Pereira PAT, Gratieri T, Faccioli LH, Cunha-Filho MSS, Gelfuso GM. Microparticles prepared with 50-190 kDa chitosan as promising non-toxic carriers for pulmonary delivery of isoniazid. Carbohydr Polym. 2017;174:421–31.
126. Gaspar MC, Sousa JJS, Pais AACC, Cardoso O, Murtinho D, Serra MES, Olivier J-C. Optimization of levofloxacin-loaded crosslinked chitosan microspheres for inhaled aerosol therapy. Eur J Pharm Biopharm. 2015;96:65–75.
127. Yildiz-Pekoz A, Akbal O, Tekarslan SH, Sagirli AO, Mulazimoglu L, Morina D, Cevher E. Preparation and characterization of doripenem-loaded microparticles for pulmonary delivery. J Aerosol Med Pulm Drug Deliv. 2018;31(6):347–57.
128. Liu K, Chen W, Yang T, Wen B, Ding D, Keidar M, Zhang W. Paclitaxel and quercetin nanoparticles co-loaded in microspheres to prolong retention time for pulmonary drug delivery. Int J Nanomedicine. 2017;12:8239–55.
129. Menon M, Singh D, Lohade A, Parmar J, Hegde D, Soni P, Samad A. Development of chitosan-based dry powder inhalation system of cisplatin for lung cancer. Indian J Pharm Sci. 2012;74(6):521–6.
130. Wang H, Xu Y, Zhou X. Docetaxel-loaded chitosan microspheres as a lung targeted drug delivery system: in vitro and in vivo evaluation. Int J Mol Sci. 2014;15(3):3519–32.
131. Ortiz M, Jornada DS, Pohlmann AR, Guterres SS. Development of novel chitosan microcapsules for pulmonary delivery of dapsone: characterization, aerosol performance, and in vivo toxicity evaluation. AAPS PharmSciTech. 2015;16(5):1033–40.

Multifunctional Cyclodextrins Carriers for Pulmonary Drug Delivery: Prospects and Potential

8

Piyush Pradeep Mehta and Vividha Dhapte-Pawar

Abstract

Cyclodextrin (CD) is one of the most explored complexing agents in the pharmaceutical sector to improve different therapeutic agents' solubility, stability, and bioavailability. Additionally, CDs and CDs-derivatives are easily tailored to fabricate diverse multifunctional carrier systems such as micelles, liposomes, niosomes, polymeric nanoparticles, lipid nanoparticles, and porous microparticles to circumvent the problems associated with both drug substances as well as carrier engineering. Moreover, the recent development in materials sciences, nanotechnology, molecular biology, and supramolecular chemistry greatly influenced the pharmaceutical application of CDs. The basis of this chapter is to provide insights into the potential of these multifaceted complexing agents for pulmonary delivery. This chapter summarizes and interprets scientific data on inhaled CDs-based carriers, which have been published in recent years. Initially, this chapter provides basic details on CDs and CD derivatives, followed by miscellaneous practical applications in pharmaceutical industries. In the subsequent section, the pulmonary applications of CDs are cautiously discussed with particular emphasis on aerosolization fate, powder dispersion behavior, pharmacokinetic outcomes, and long-term stability performance. The last section discusses the prospects and potential for translating inhaled CDs-based carrier systems. The bottom-line marks CD-based carriers for the growth of pulmonary medicines in the upcoming years.

P. P. Mehta
Global Respiratory R&D, Cipla R&D Center, Mumbai, Maharashtra, India
e-mail: piyu053@gmail.com

V. Dhapte-Pawar (✉)
Department of Pharmaceutics, Poona College of Pharmacy, Bharati Vidyapeeth (Deemed to be University), Pune, Maharashtra, India
e-mail: dhapte.vv@bharatividyapeeth.edu

P. P. Mehta, V. Dhapte-Pawar (eds.), *Pulmonary Drug Delivery Systems: Material and Technological Advances*,
https://doi.org/10.1007/978-981-99-1923-9_8

Keywords

Cyclodextrins · Complexing agents · "Host-guest" association · Hydroxypropyl-β-CD · Sulfobutylether-β-CD and cyclic oligosaccharides

8.1 Introduction

Cyclodextrin (CD) is a group of synthetic cyclic oligosaccharides derived from the enzymatic degradation of starch [1]. They are composed of a macrocyclic ring of 6–8 glucose subunits linked by α-1,4 glycosidic bonds, and the characteristic truncated cone-shaped interior comprises a dimensionally stable hydrophobic cavity [2]. Based on glucose subunits involved in CD structure, they are further classified as α-CD (6 units), β-CD (7 units), and γ-CD (8 units) [3]. CDs fit into the family of cage molecules, and the unique cyclic configuration of CD molecules presents a hydrophilic outer surface and hydrophobic internal cavity, thus forming the "inclusion complex" [4]. Their remarkable structural features assist in forming a "host-guest" type relationship that helps to revise the intrinsic physicochemical properties of a guest molecule [4, 5]. The successful application of CD is mainly based on a balanced "host-guest" association [4]. Accordingly, several researchers thoroughly explored various methods to form stable drug/CD complexes such as kneading, grinding, co-precipitation, neutralization, solution, and slurry methods [6, 7]. Moreover, a few techniques such as drug ionization, complex-in-complex, salt formation, the acid/base ternary complexes, polymer complexes, solubilization of CD aggregates, and a combination of two or more methods are also used judiciously by researchers to improve the complexation efficiency of the drug/CD mixture [6, 8]. Additionally, several scientists used CD's highly adaptable chemical nature to synthesize mono- or poly-substituted CD derivatives to tune their complex formation abilities and improve guest molecule's physicochemical and biological properties [9]. Chemically modified CDs mainly include the anionic sulfobutylether-β-CD (SBE-β-CD) sodium salt, randomly substituted neutral CDs such as 2-Hydroxypropyl-β-CD (HP-β-CD), 2-Hydroxypropyl-γ-CD, and methyl-β-CD [10]. Table 8.1 showed commonly used CD molecules in pharmaceutical formulations.

CDs (Fig. 8.1) andv their derivatives have miscellaneous practical applications in various industries such as chemistry (e.g., analytical chemistry, agro-chemistry, click chemistry, supramolecular, and macromolecular chemistry) [11–13], chromatography [14], catalysis [15], biotechnology [16], cosmetics [17], foods [18, 19], enzyme technology [20], biomedical (e.g., pharmacy and medicines) field [21], adhesives [22], plastics [23, 24], decontamination, toiletries [25, 26], and textile industries [27]. Particularly in terms of pharmaceutical applications, highly purified CDs are available. Globally almost 40 different CDs-based pharmaceutical products are employed in about 14 formulations, such as different types of tablets (conventional, sublingual, and chewing tablets), capsules, injectable solutions, nasal sprays, dermal products, eye drop solutions, and suppositories [28–30]. In most pharmaceutical products, CDs are utilized as complexing agents to enhance the aqueous solubility of hydrophobic

Table 8.1 Summary of commonly used cyclodextrin molecules in pharmaceutical formulations

Cyclodextrin	Empirical formula and molecular weight (g/mol)	Melting point (°C)	Lethal dose (LD_{50}; g/kg)	Commercial products
α-CD	$C_{36}H_{60}O_{30}$ and 972	250–260	Rat (IV): 0.79	• Solution for injection: Caverject 10 mcg by Pfizer ltd. and Viridal duo 10 mcg/mL by Advanz pharma.
β-CD	$C_{42}H_{70}O_{35}$ and 1135	255–265	Rat (IV): 1.00 Rat (Oral): 18.8	• Sublingual tablet: Nicorette Microtab 2 mg by McNeil products ltd.
γ-CD	$C_{48}H_{80}O_{40}$ and 1297	240–245	Rat (IV): 4.00 Rat (Oral): 8.00	• Na.
HP-β-CD	$C_{63}H_{112}O_{42}$ and 1541.5	262–269	–	• Solution for injection: AKIS 75 mg/ml by Flynn Pharma Ltd. • Tablet: MAVENCLAD 10 mg by Merck.
SBE-β-CD	$C_{70}H_{119}Na_7O_{56}S_7$ and 2242.05	–	Rat (IV): 2.00 Rat (Oral): 2.00	• Powder for infusion: VFEND 200 mg powder by Pfizer Ltd. • Solution for injection: ABILIFY 7.5 mg/mL by Otsuka Pharmaceuticals (UK.) Ltd.

molecules, augment the bioavailability of a drug substance, and extent their physicochemical stability [31]. In each case, CD application varies according to the nature of the drug (e.g., solubility, crystalline nature, and stability) and type of final dosage form (e.g., tablet, powder, suspension, or parenteral solution) [32, 33].

CDs are utilized to substitute organic solvents in parenteral and topical formulations, to act as absorption enhancers for nasal delivery, to improve oral bioavailability of BCS Class II and some BCS Class IV drugs, to modify the dose-to-solubility ratio in ophthalmic formulation, to overcome gastrointestinal irritation, improve drug accessibility at the mucosal surface for large lipophilic molecules and increase dermal availability of drugs [6, 34]. Additionally, recent pre-clinical and clinical investigations have shown that CDs can be explored for enhancing the drug delivery potential of almost any type of formulation [35, 36]. Besides, recent advancements in material sciences, nanotechnology, biomaterials, particle engineering, and molecular and cell biology have triggered rapid growth in the segment of CDs-based novel drug delivery systems (NDDS). Developing various CDs-based NDDS such as nanoparticles (NPs) [37], nanocomposites [38], nanoaggregates [39], solid lipid NPs [40], liposomes [41, 42], microencapsulation [43], microparticles (MPs), and microspheres [44] have further presented many vital benefits. Recognizing the unique potential of CDs in drug delivery and recent developments in various

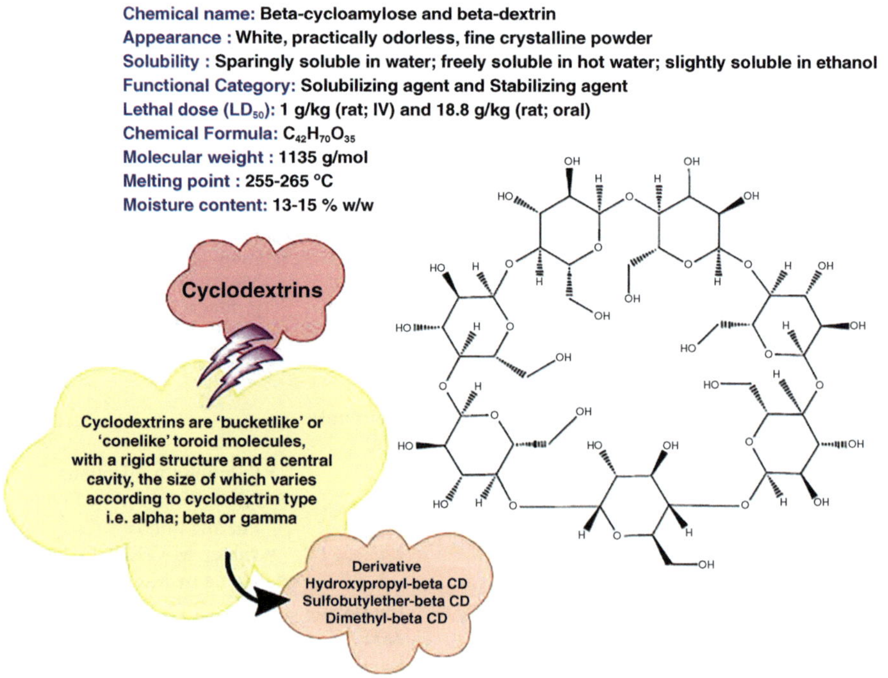

Fig. 8.1 Cyclodextrins' critical physical, chemical, and biological features (The structure of Betadex (β-CD) is shown)

scientific fields, few active formulation scientists have developed CDs-based inhaled dosage forms [45]. Consequently, we represent the various vital applications of CDs for pulmonary drug delivery and the evolution of CDs-based pulmonary technologies via in-depth analysis of the scientific sources.

8.2 Cyclodextrins in Pulmonary Drug Delivery

Pulmonary delivery systems are generally explored for local (e.g., asthma and chronic respiratory disease) and systemic (e.g., diabetes and hypertension) diseases [46–50]. Pulmonary drug delivery effectively avoids the first-pass metabolism and offers a suitable therapeutic response with better patient compliance [51, 52]. However, the efficacy of pulmonary drug delivery can be limited by a slow drug dissolution rate and poor aqueous solubility [53]. Further, the mucociliary clearance in the upper airways and macrophages in the alveoli region eliminate insoluble particles from the pulmonary airways [53, 54]. CDs can offer better opportunities in pulmonary drug delivery by improving the solubility, dissolution kinetics, and physicochemical stability of poorly water-soluble and chemically unstable substances. Further, it can reduce mucociliary clearance, improve drug absorption, and offer beneficial local therapeutic action. Besides, inclusion complexes can be easily

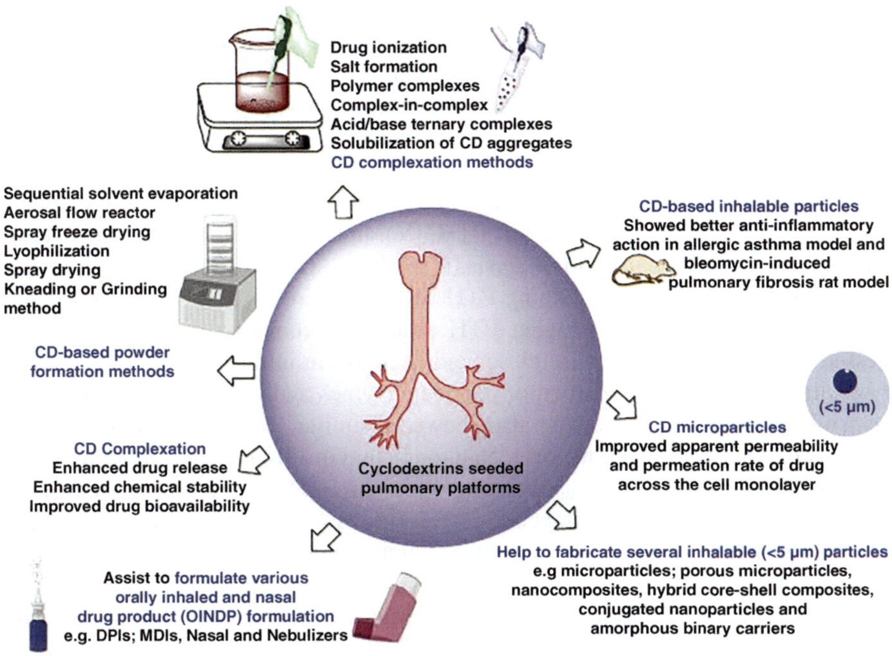

Fig. 8.2 Application spectrum of the cyclodextrins in pulmonary drug delivery

transformed into the inhalable solid form using several scalable drying techniques such as spray drying [55–57], freeze-drying [58], and spray freeze drying [59] (Fig. 8.2). Therefore, in the following segments, the most recent achievements in the domain of CD-based inhalable dosage forms with particular importance on their aerodynamic and biological functions are described exhaustively.

8.2.1 Cyclodextrins Functionalized Particles for Pulmonary Delivery of Biological Molecules

Parathyroid hormone (PTH) is an endogenous polypeptide hormone (4118 Da) that plays a key role in calcium and phosphate homeostasis regulation, new bone formation, regulation of bone growth, and improves mineral density. Troubles coupled with long-term treatment cycles, day-to-day injection, low oral bioavailability, are the main motives for investigating alternative drug delivery routes for PTH. In this manner, co-administration of polypeptide hormone with an absorption enhancer through the pulmonary route is an appealing approach. Accordingly, Poursina et al. (2015) developed porous inhalable PTH MPs using HP-β-CD as an absorption enhancer by the spray freeze drying method to improve systemic bioavailability of PTH. During formulation optimization HP-β-CD (0.04% w/w) and citrate buffer preserved the PTH physicochemical and conformational stability. Porous spherical-shaped particles showed a mean particle size of 15.95 μm. In aerosolization study, PTH MPs showed fine particle fraction (FPF) and emitted dose (ED) of 14.2% and 37.2%, respectively, using Cyclohaler® at 60 L/min. PTH MPs containing 10% w/w leucine showed % FPF of 91.1%. During the in vivo study after pulmonary delivery using an insufflator in Sprague Dawley rats, PTH MPs (20 μg PTH) showed satisfactory bioavailability of 47.25% with a marked 1.33 and 1.32-fold improvement in $t_{1/2}$ (half-life) and MRT (mean residence time), respectively, compared to the Cinnopar® (0.1 mL PTH, IV). In comparison to the in vitro aerodynamic data, during in vivo analysis, PTH MPs showed higher bioavailability compared to the leucine-containing PTH MPs. Additionally, PTH MPs processed using citrate buffer showed higher PTH absorption compared to PTH MPs processed in water mainly because citric acid decrease in intracellular pH which prevents the proteolytic degradation of peptide enzymes and improve the paracellular transport of peptides. Citric acid also showed a similar absorption pattern for biologicals like human growth factor-1, calcitonin, and insulin by pH-regulating mechanisms. The pulsatile pharmacokinetic pattern of PTH MPs demonstrated better therapeutic outcomes than intravenous PTH. Briefly, inhalable porous MPs offered an effective alternative to developed PTH formulation for the management of osteoporosis [60].

Tewes et al. (2011) fabricated synthetic peptide calcitonin, i.e., salmon calcitonin (SCT) MPs using stabilizing excipients such as polyethylene glycol (PEG) and HP-β-CD. The covalent binding of PEG with peptide help to increase its half-life and reduce immunogenicity while HP-β-CD assists to maintain the high glass transition temperature (Tg) and ultimately regulate the thermal properties of the formulation. Particularly, SCT-MPs were formulated using linear PEG (10 kDa) and HP-β-CD (1:1) and spray dried to obtain inhalable dry powder. Spray dried, spherical, and rough surface SCT-MPs showed mean particle size and loading efficacy of 2.84 μm and 93.3%, respectively. In aerodynamic analysis, SCT-MPs showed FPF and mass median aerodynamic diameter (MMAD) of 24.4% and 7.3 μm, respectively, using Handihaler®. While SCT-MPs fabricated using branched PEG showed FPF and MMAD of 19.4% and 10.4 μm, respectively, using Handihaler®. SCT-MP showed good, controlled permeability performance using a human lung cancer cell line (Calu-3) owing to the ability of HP-β-CD to act as permeability enhancers either by enhancing the drug's apparent solubility, protecting SCT against enzymes, and temporarily regulating the pulmonary membrane. Furthermore, during pharmacokinetics analysis, Wistar rats in a two-compartment model insufflated SCT-MPs (100 μg/kg) exhibited a marked 4.76 and 2.3-fold improvement in time to reach maximum plasma concentration (t_{max}) and absolute bioavailability, respectively, as compared to nebulized SCT solution. Like in-vitro aerodynamic data, during pharmacokinetics study SCT-MPs fabricated using liner PEG showed a marked 1.46-fold higher bioavailability compared to the SCT-MPs fabricated using branched PEG at the same dose. Superior pharmacokinetics results are mainly attributed to the ability of PEG and HP-β-CD to raise local viscosity which eventually slows down clearance and/or degradation mechanisms involved in SCT elimination and contributes to the increase in SCT bioavailability [61].

Current therapies in diabetes management involve the use of intensive insulin regimens to attain better control of blood glucose levels. Inhaled insulin offers an alternate therapy for management of diabetes as part of a basal/bolus approach in subjects who are unable to use pre-prandial insulin injections. Ungaro and co-workers (2009) developed large porous particle-based inhaled insulin using poly (lactide-co-glycolide) (PLGA) by double emulsion method. Additionally, HP-β-CD was used as an alternative osmotic agent to retain protein integrity and prevent protein precipitation/aggregation/inactivation compared to salts (e.g., calcium and sodium chloride). Moreover, HP-β-CD imparts the desired bulk properties with a highly porous structure and low density (0.1 g/mL) to insulin particles. Formulated insulin particles showed actual loading and encapsulation efficiency of 35.4 IU/100 mg and 59%, respectively. Spherical and well-porous insulin particles showed a volume mean diameter of 26.2 μm. During aerodynamic analysis, insulin particles showed FPF and MMAD of 61.3% and 5.02 μm, respectively, using Turbospin® at 60 L/min. During the in vitro release study, insulin particles displayed a biphasic release pattern with an initial burst followed by a sustained release for 10 days. Confocal microscopic investigations using fluoresceine isothiocyanate labeled insulin particles in male Wistar rats showed good uptake of insulin in the alveolar region. During in vivo biological testing using streptozotocin-induced diabetic rats, intratracheally administered insulin particles showed a significant reduction in blood glucose level over a longer time than insulin solutions administered via pulmonary route while dosing schedule (4 IU/kg). Overall, biodegradable large porous particles were effective carriers to deliver protein like insulin at the desired site of action [62]. However, issues related to local toxicity, long-term accumulation, and immunogenicity need to be addressed in the future.

Cyclosporin A (CSA) is a natural cyclic non-ribosomal polypeptide immunosuppressant derived from the fungus *Beauveria nivea*. CSA is well known for its immunosuppressant, anti-fungal, and anti-asthmatic activities. CSA showed a clinical advantage in patients with chronic severe steroid-dependent asthma. But oral administration of CSA to asthmatic patients showed limited therapeutic benefits owing to its systemic adverse effects [63]. To avoid adverse effects, the direct administration of CSA to the pulmonary airways was carefully investigated by Matilainen et al. (2006). Particularly, CSA: HP-β-CD (1:269 w/w) complex was formulated using the phase-solubility system and subjected to freeze-drying to obtain an inhalable solid powder. Additionally, lactose (LAC): CSA (346:1 w/w) and LAC: HP-β-CD/CSA (14:1 w/w) formulations were fabricated to probe the aerodynamic properties of the CSA. During an aerodynamic study using Taifun® inhaler device, CSA: LAC, HP-β-CD/CSA, and HP-β-CD/CSA: LAC exhibited respirable fraction and MMAD of 19.8, 33.0, 34.6% and 3.1, 2.7 and 2.7, respectively. Powder stored (32 days, 40 °C, 75% RH) in Taifun® inhaler device showed respirable fraction of 23.6% (CSA: LAC), 28.5% (HP-β-CD/CSA) and 34.3% (HP-β-CD/CSA: LAC) with MMAD of 2.8 (CSA: LAC), 3.1 (HP-β-CD/CSA), and 2.6 μm (HP-β-CD/CSA: LAC), respectively. HP-β-CD/CSA loaded on LAC carrier effectively retained the respirable fraction and MMAD on storage. Briefly, HP-β-CD/

CSA loaded on LAC carrier retained the physical stability of complex and showed dosing repeatability of CSA on storage [64].

CSA low oral bioavailability was mainly attributed to its poor aqueous solubility (0.04 mg/gm) and large molecular weight (1202.61 g/mol) [65]. Thus, Fukaya and co-workers logically developed CSA inclusion complexes using α-CD, dimethyl-β-cyclodextrin (DM-β-CD), glucosyl-α-CD (Gα-CD) and maltosyl-α-CD (Mα-CD) to improve aqueous solubility and comprehend pulmonary fate of same. CSA and CD's inclusion complexes were fabricated using kneading method and subsequently pulverized to obtain particle size suitable for inhalation. CSA inclusion complexes showed a marked 40, 80, 130, and 570-fold improvement in CSA solubility using α-CD, Mα-CD, Gα-CD, and DM-β-CD, respectively. CSA-Mα-CD showed highest ciliary activity of the trachea (ciliostatic potential) in the frog's palate model compared to other inclusion complexes. The growing number of eosinophils in the lung is a distinguishing trait of asthma and it plays a key role in the pathogenesis of asthma. During the in vivo investigation using rats after administration using a nose-only inhalation chamber, CSA-Mα-CD complex (contain 49.7% CSA) showed effective inhibition of allergen-induced eosinophil accumulation for a longer duration at a 9-times lower dose than CSA alone. Compared to CSA alone, CSA-Mα-CD complex showed improved pharmacological effect as complex, comparatively more stable in the aqueous mucosal layer. The dissolved complex stays at the absorption site and then releases CSA more gradually at the site [66].

Recombinant human growth hormone (RHGH) is a single-domain, non-glycosylated globular protein (22 kDa) indicated for treatment of patients suffering from inadequate secretion of endogenous growth hormone or adults with growth hormone deficiency. Presently, RHGH is delivered via parenteral route, i.e., 2–3 injections per week. Accordingly, to establish needle-free RHGH therapy, Jalalipour et al. (2008) fabricated spray dried RHGH powder using DM-β-CD and protein stabilizing agent (LAC and zinc) for pulmonary delivery. Spray dried, dimple-shaped RHGH-DM-β-CD (1:100) complex showed particle size and surface roughness of 2.3 μm and 55.6 nm, respectively. During aerodynamic analysis, HGH-DM-β-CD displayed FPF and dispersibility of 34.89% and 86.96% using Spinhaler® at 60 L/min. In an in vivo study, RHGH-DM-β-CD complex showed a marked 3.38, 3.0, and 2.98-fold improvement in maximum serum concentration (C_{max}), absolute bioavailability, and area under curve (AUC) compared to RHGH alone at the same dose (0.1 mg), respectively. Particularly, DM-β-CD concentration played a significant role in the absorption and aerosolization performance of RHGH [67]. Overall, the pulmonary airway is a vital route for delivery of hormones, proteins, and peptides since it presents an excellent therapeutic efficacy and better patient compliance.

8.2.2 Engineered Cyclodextrin Particles for Pulmonary Delivery of Phytoconstituents

Resveratrol (REV) is a polyphenol, stilbene, and a member of resorcinols mainly sourced from grapes. REV is a phytoalexin with antioxidant and chemo-preventive activities. Recent investigations showed that REV effectively scavenges oxygen-derived free radicals and prevents the release of inflammatory cytokines from alveolar macrophages in COPD [68, 69]. Despite these therapeutic activities, the clinical translation of REV is mainly hindered due to its first-pass metabolism and limited bioavailability. Further, as a polyphenolic molecule, RES is highly prone to phase II metabolism and only a small amount of REV can reach the pulmonary airways using traditional oral dosage forms [70, 71]. To tackle these limitations, Kong et al. (2020) developed a respirable REV: HP-β-CD inclusion complex and systematically studied it for pulmonary pharmacokinetic properties. REV: HP-β-CD complex was administered to rats in four different ways, i.e., bolus tail vein injection (10 mg/kg), oral gavage (50 mg/kg), intratracheal administration using orotracheal intubation (20 mg/kg), and inhalable administration using nose-only inhalation exposure system at a flow rate of 12 L/min (2.3 mg/kg). After pulmonary administration, the systemic exposure of REV was significantly improved compared to oral delivery. The absolute bioavailability was in the following order: intratracheal (92.95%) > inhalable (76.31%) > oral gavage (16.68%). Intratracheal and oral gavage showed t_{max} of 0.13 and 0.37 h, respectively, which indicates that REV was absorbed more quickly through the transpulmonary route than gastrointestinal route. Additionally, REV: HP-β-CD complex showed quick and superior (100 times) accumulation of REV in pulmonary airways compared to oral delivery at early time points. Moreover, the metabolic investigation using rat liver and lung microsomes showed that the metabolic activity of pulmonary airways was much lower than the liver. Thus, the improved pulmonary pharmacokinetic properties of REV were mainly attributed to HP-β-CD's ability to improve the solubility and stability of REV within the biological milieu [72]. All these results showed that administration of REV via pulmonary route is an efficient way to prevent extensive pre-systemic metabolism and deliver more drugs into the desired site of action.

Pneumonia is a common disease known by lung parenchyma inflammations, mainly triggered by infections of diverse pathogens such as bacteria, fungi, viruses, mycoplasma, and chlamydia. Andrographolide (AG) is the major bioactive ingredient of *Andrographis paniculate* and is well known for its multiple therapeutic actions such as antibacterial, anti-inflammatory, and antioxidant. However, AG is very poorly water-soluble with low bioavailability, leading to poor therapeutic action. In addition, the mechanism of anti-pneumonia effect of AG was not well studied [73, 74]. With this prospect, Zhang et al. (2017) developed inhalable AG-β-CD MPs to fight against resistant *Staphylococcus aureus* pneumonia. Initially, AG-β-CD suspension was formulated and subjected to lyophilization to obtain powder. AG inclusion and drug loading efficiency were found to be 63.92% and 9.61%, respectively. AG-β-CD MPs showed grouped bricks-like morphology with particle size suitable for pulmonary delivery (<5 μm). AG-β-CD MPs displayed a faster and superior drug release profile than AG powder in simulated lung fluid. AG-β-CD MPs showed FPF and ED 81.83% and 96.08%, respectively. Additionally, during in vivo anti-pneumonia activity using bacterial-pneumonic rat models after intratracheal administration (Penn Century aerosol micro sprayer, USA), AG-β-CD MPs (1 mg of AG) and AG powder alone (10 mg) showed satisfactory bacterial inhibition with inflammation easing by regulating the immune responses. In contrast, penicillin sodium (240 mg/mL) killed bacteria without significant immune regulation. Pneumonia leads to lung injury thus treating lung injury is an important step in pneumonia therapy. AG-β-CD MPs strongly attenuated the lung injury with a little local hemorrhage compared to AG powder. Briefly, immunoregulation of AG, solubility enhancement of β-CD, and locally accelerated AG dissolution from β-CD inclusion allows the inhalable AG-β-CD MPs to effectively treat *S. aureus* pneumonia [73].

Tea tree oil (TTO) is an essential oil acquired using ethanol extraction or steam distillation of the leaves and terminal branchlets of *Melaleuca alternifolia*. TTO is a natural, non-contaminating, non-toxic, biodegradable, and seldom resistant antimicrobial phytoconstituent. Yet, its volatilization, strong stimulation, instability under light or oxygen, allergic reactions, and low water solubility restrict its beneficial pharmacological action. Accordingly, Li et al. (2017) developed TTO: β-CD inclusion complex (TTO-β-CD) to improve its physical properties and tested the same for anti-pneumonia potential. TTO-β-CD inclusion complex was fabricated using grinding method. Initially, β-CD was ground with water and carefully added to the solution of ethanol: TTO (1:1, v/v). The mixture was ground, dried under vacuum (30 °C for 2 h), and passed through a 180-mesh sieve to obtain powder. TTO-β-CD MPs showed tapped density and D_{50} of 0.40 g/mL and 8.85 μm, respectively. Rough surface irregularly shaped TTO-β-CD MPs showed aerodynamic diameter and FPF of 5.59 μm and 51.22%, respectively. During an in vivo study using rat models, i.e., fungal (*Candida albicans*) and bacterial (*Acinetobacter baumannii*) pneumonia, TTO-β-CD MPs exhibited superior and similar anti-pneumonic effects compared to TTO alone and positive control fluconazole and penicillin, respectively. TTO-β-CD MPs efficiently protect the host by regulating the level of neutrophils, leukocytes, and modulating the immune response. TTO-β-CD MPs downregulated cyclooxygenase 2 (COX 2) and inflammatory cytokines such as TNF-a, IL-6, and IL-1b in infected rat lung model. Briefly, β-CD inclusion not only improved the physical properties and aerodynamic performance of TTO but also retained the anti-pneumonia potential of TTO. Thus, TTO-β-CD MPs are a promising approach for the effective management of bacterial and fungal pneumonia [75].

Fisetin (FIS) is a bioactive flavonoid commonly found in the smoke tree, fruits, vegetables, and nuts. FIS possesses a variety of therapeutic values such as anti-inflammatory, antioxidant, neuroprotective, antiproliferative, and apoptotic activities. The low aqueous solubility (10.45 μg/mL), high lipophilicity (log P 3.2), and poor oral bioavailability (44.1%) hinder its effective drug delivery [76]. Accordingly, Mohtar et al. (2017) developed FIS: SBE-β-CD complex to improve its physicochemical properties and explore the full potential of FIS in pulmonary disease management. Initially, FIS: SBE-β-CD (2:1) complex was fabricated by co-solvent technique and subjected to spray drying to obtain MPs of suitable aerodynamic properties. FIS-SBE-β-CD MPs showed pitted surface morphology and a mean particle size of 3.14 μm with acceptable residual solvent limits. During aerodynamic analysis using Aerolizer® device, FIS-SBE-β-CD MPs showed FPF and ED of 88.08% and 32.49%, respectively, at 60 L/min. The addition of leucine (20% w/w) showed a marked 2.3-fold enhancement in FPF. FIS-SBE-β-CD MPs displayed good antioxidant activity in the 2,2-diphenyl-1-picrylhydrazyl (DPPH) assay signifying that the processing technique preserved the antioxidant activity of FIS. Furthermore, FIS-SBE-β-CD MPs exhibited a higher effective concentration (EC_{50}) of 75.86% in MTT assay using A549 cells [77]. In a nutshell, FIS: SBE-β-CD complex enhanced the solubility of FIS and spray dried MPs could deliver satisfactory FIS concentration to the desired region of airways for better therapeutic outcomes. Additionally, a synergistic combination of FIS with suitable anticancer molecules can be explored for pulmonary delivery using the SBE-β-CD [77].

8.2.3 Respirable Cyclodextrin Antibiotics-Loaded Complex

Clofazimine (CFZ) is a phenazine molecule with anti-inflammatory and anti-mycobacterial actions. Recently published data showed that CFZ demonstrated good inhibitory activity against multi-drug-resistant tuberculosis (TB) strains and with first- or second-line anti-TB drugs showed a significant reduction in the bacterial count, indicating its capacity to reduce the duration of TB treatment [78–80]. Additionally, the World Health Organization (WHO) suggests the use of oral CFZ in combination with other drugs for drug-resistant TB treatment. These points indicate toward CFZ as a budding candidate for pulmonary TB therapy. However, its low aqueous solubility (0.3 μg/mL), high lipophilicity (Log P; 7.66), and high gastric degradation, resulted in suboptimal delivery to the pulmonary airways. For these motives, de Castro et al. (2020) prepared CFZ: β-CD complex with enhanced CFZ water solubility for regulating dissolution kinetics in airway mucus and aerodynamic performance. CFZ: β-CD (1:7) complex was formulated using a solution method containing purified water and ethanol: water (7:3) mixture. Subsequently, prepared mixture was spray dried in the presence of leucine (15%) to obtain rough MPs with less hygroscopic nature and fewer agglomerates. CFZ: β-CD MPs (2.18 μm) showed tapped density and surface area of 0.26 g/cm^3 and 6.1 m^2/g, respectively. Using the high resistance HandiHaler® inhaler, CFZ: β-CD MPs showed a superior FPF (51.56%) with lower MMAD (2.9 μm). During in vitro cytotoxicity study using Calu-3 cells, CFZ: β-CD MPs showed >90% cell viability at the highest CFZ concentration (25 μM) which indicated high tolerability of Calu-3 cells for the CFZ: β-CD MPs, promoting its pulmonary delivery. In the in vitro permeability study using Calu-3 cells, CFZ: β-CD MPs exhibited maximum CFZ retention in the apical compartment. High CFZ concentration directly available at the site of infection would provide a more efficient therapy, while less systemic CFZ biodistribution and toxicity. Briefly, inhaled MPs containing β-CD and leucine denote an appropriate system to deliver CFZ to pulmonary airways [81].

Combination therapy is the regular medication for TB, which includes the oral administration of two to four antibiotic drugs, in large doses (600–900 mg), over a period of 6 months [80]. Poor patient compliance/adherence and multi-drug resistance development are the key challenges allied to combination therapy. Therefore, to tackle these unmet needs, Kadota et al. (2016) designed a pulmonary dosage form that combined isoniazid (INH) and rifampicin (RIF) drugs into a single formulation to ensure both safety and rapid relief for the patient. Initially, INH (100 mg) and RIF (200 mg) and HP-β-CD (1000 mg) were completely dissolved in water and ethanol (99.5%) mixture with subsequent spray drying to obtain inhalable combined MPs. Porous spray dried combined MPs showed satisfactory drug content (INH; 97.1% and RIF; 92.1%) with a mean particle size of 6.50 μm. Non-aggregating spray dried HP-β-CD: INH (2.04 μm) and HP-β-CD: RIF (7.4 μm) MPs exhibited satisfactory drug content, i.e., 91.4% and 130.9%, respectively. Spray dried, combined MPs displayed higher surface area (25.1 m^2/g) compared to other spray dried sugar-based combined MPs due to porous nature. In the aerodynamic analysis using Jethaler® dual chamber device, HP-β-CD: INH showed satisfactory FPF (3.13%) and ED (92.6%) values while HP-β-CD: RIF MPs presented acceptable FPF (38.3%) and ED (99.3%) values. Spray dried combined HP-β-CD MPs showed a marked 1.66 and 1.70-fold improvement in FPF compared to spray dried combined LAC MPs. Additionally, combined HP-β-CD MPs showed ED above 98% for both drugs. The porous morphology and improved surface area of HP-β-CD MPs were mainly accountable for better aerodynamic performance. Specifically, INH particles were encased by porous RIF particles with improved aerosolization properties. Briefly, a single-dosage powder formulation, consisting of two anti-TB, i.e., hydrophobic, and hydrophilic molecules was formulated for successful pulmonary delivery [82].

Similarly, Basha and co-researchers (2019) developed levofloxacin (LEV) and RIF-loaded β-CD tethered curdlan (CUD) NPs for pulmonary delivery. CUD (linear β-1, 3 glucan), a water-insoluble polymer of glucose, is identified by dectin-1 receptor expressed on macrophages it assists in effective internalization of the drug-loaded particles [83]. Basically, to improve the drug loading efficacy and synchronized sustained release of LEV and RIF β-CD complex were fabricated using solvent evaporation method and crosslinked with CUD NPs using an epichlorohydrin crosslinking agent. CUD NPs (226 nm) showed spherical morphology while β-CD conjugated CUD NPs (523 nm) showed a diamond shape. LEV-β-CD-CUD NPs, RIF-β-CD-CUD NPs, and LEV/RIF-β-CD-CUD NPs showed positive zeta potential of 12.11, 14.31, and 13.45 mV, respectively. Conjugated particles displayed a sustained drug release pattern up to 80 h in phosphate buffer saline (PBS; pH 7.4). Particularly, during in vitro release study, LEV-β-CD-CUD NPs displayed an initial burst release, with 80% of the LEV released within the first 8 h while in LEV/RIF-β-CD-CUD NPs, LEV release was sustained. LEV sustained release was mainly ascribed to the presence of RIF which hindered LEV release from the conjugated NPs. The slow and sustained release of RIF and LVX is important to reduce dosing frequency. In the in vitro antibacterial study, RIF-β-CD-CUD NPs, LEV-β-CD-CUD NPs, and LEV/RIF-β-CD-CUD NPs showed a minimum inhibitory concentration of 20.83, 46.88, and 23.44 μg/mL against *M. smegmatis*. During in vitro cytotoxicity study using Abelson leukemia virus-transformed cell line (RAW 264.7 macrophages) and L929 fibroblasts cells, conjugated NPs showed 85% cell viability at all concentrations in both the cell lines. Conjugated NPs effectively kill more than 95% of *M. smegmatis* dwelling inside the macrophages in 4 h. Furthermore, the uptake of conjugated NPs was 1.8 times superior by the macrophage cells through dectin-1 receptor compared to the fibroblast cells. Concisely, a novel curdlan-β-CD NPs emerged as one of the capable alternatives for the controlled intracellular release of hydrophobic drugs with fruitful pulmonary outcomes [84].

8.2.4 Pulmonary Delivery of Anti-inflammatory Agent-Loaded Cyclodextrin Complex

Roflumilast (ROF) is a member of benzamides, prescribed for the treatment of asthma and COPD. ROF is a water insoluble (0.52 mg/L), potent phosphodiesterases-4 (PDE-4) inhibitor, and an anti-inflammatory agent. PDE-4 is the key isoenzyme present in the cells concerned in COPD [85]. Therefore, ROF is utilized to lessen the number of outbreaks of COPD symptoms. ROF tablet (Daliresp®; AstraZeneca) is prescribed for COPD treatment in a dose of 500 μg and maximum single orally tolerated dose is 1000 μg. Even though being effective in treating COPD, oral ROF create mood or behavior problems including anxiety, weight loss, insomnia, depression, and even suicide. For these concerns, inhaled ROF is considered as an important option for safe COPD therapy. Additionally, few pre-clinical studies in ovalbumin-sensitized and -challenged BALB/c mice (500 μg/kg, inhaled ROF) [86] and allergen-challenged Brown Norway rats (600 μg/kg, intratracheal ROF) [87] emphasized that the administration of ROF by pulmonary route enhanced lung function and therapeutic index with safety. Accordingly, Mahmoud et al. (2018) formulated emulsified spray dried alginate MPs for the pulmonary delivery of ROF. Sodium alginate, calcium β-glycerophosphate, and β-CD were used as particle forming, crosslinking agent, and carrier, respectively. Spherical ROF-β-CD MPs showed yield, entrapment efficiency, particle size, and zeta potential of 65.26%, 108.76%, 3.13 μm, and − 6.75 mV, respectively. Aerodynamic assessment of ROF-β-CD MPs showed a 1.89 and 3.75-fold improvement in FPD and FPF, respectively, compared to ROF alone. Additionally, the smaller MMAD value (2.47 μm) for ROF-β-CD MPs compared to ROF ensured more effective deposition of MPs in pulmonary airways. ROF-β-CD MPs displayed more prominent inhibitory action on A549 cells viability with half maximal inhibitory concentration (IC_{50}) value of 24.68 μg/mL and higher inhibitory effect ($p < 0.05$) on pro-inflammatory cytokines, i.e., interleukin-6, interleukin 10, and tumor necrosis factor-α as compared to the ROF. Therefore, ROF-β-CD MPs hold good potential to regulate tumorigenesis as well as tumor-associated inflammation. Moreover, ROF-β-CD MPs (500 μg/actuation) displayed superior sustained broncho-dilatation potential in healthy human subjects as compared to salbutamol (90 μg/actuation; Ventolin® HFA, GSK). Thus, spray dried alginate crosslinked β-CD MPs displayed a promising alternative for successful delivery of anti-inflammatory agent like ROF [88]. Likewise, Suzuki et al. (2018) formulated spray dried MPs using HP-β-CD to probe their suitability in pulmonary delivery of ROF. Wrinkled or raisin-like ROF-HP-β-CD (1:9% w/w) MPs showed particle size, surface area, and drug loading of 2.63 μm, 3.45 m^2/g, and 94.9%, respectively. During X-ray powder diffraction (XRD) and thermal analysis,

ROF-HP-β-CD MPs exhibited characteristic amorphous halo and higher Tg (194.5 °C), respectively, which confirmed its stable physical nature. ROF-HP-β-CD MPs showed ED, FPF, and MMAD of 79.67%, 56.42%, and 3.40 μm, respectively, using high resistance device HandiHaler® at 60 L/min. The improved aerodynamic performance was mainly attributed to wrinkled-shaped raisin-like MPs. Mainly, wrinkled-shaped morphology offers lower particle-particle contact due to numerous surface irregularities and breaking of inter-particulate contacts. During transepithelial permeability analysis using Calu-3 cell monolayers, ROF-HP-β-CD MPs showed a satisfactory apparent permeability coefficient of 8.73 x 10^{-6} cm/s. During 8-week storage (25 °C and 60% RH), ROF-HP-β-CD MPs effectively retained the aerodynamic properties. In summary, physicochemical, in vitro permeability and aerosolization studies specified the potential application of HP-β-CD MPs for pulmonary delivery of ROF [89].

Oral ROF tablet with inhaled formoterol (FF), showed a substantial synergistic action by inhibition of eosinophilia, with no significant pharmacokinetic interactions in the therapy of asthma. However, there is no ROF, and FF combined inhaled formulation is available for treatment of asthma or COPD. A combined inhaled formulation of ROF and FF helps to improve adherence with therapeutic outcomes compared to monotherapy [90, 91]. Accordingly, Suzuki and co-workers (2020) developed ROF/FF-HP-β-CD (2:1:1663 w/w) MPs using spray drying method. ROF/FF-HP-β-CD MPs (20 mg) contains 25 and 12 μg of ROF and FF, respectively for effective therapeutic action, per 20 mg of MPs. ROF/FF-HP-β-CD MPs exhibited 93.9% and 100% drug loading for ROF and FF, respectively. ROF/FF-HP-β-CD MPs (2.70 μm) showed lower bulk density (0.21 g/cm^3) and high surface area (4.96 m^2/g). XRD analysis indicated a characteristic amorphous halo for spray dried MPs while thermal testing revealed higher Tg values for ROF (202.57 °C) and FF (235.49 °C), which indicates its good physical stability. ROF/FF-HP-β-CD MPs displayed satisfactory FPF of 52.86% (ROF) and 53.64% (FF) with >80% ED for both APIs using high resistance device HandiHaler® at 60 L/min. Satisfactory aerodynamic results were mainly attributed to wrinkled or raisin-like MPs since it offers lower inter-particulate cohesion than spherical particles with smooth surfaces, thus enhancing powder flow properties and aerosolization performance of the formulation. During in vitro permeability investigations, MPs showed high permeability for ROF and FF using Calu-3 epithelial monolayer, indicating that the ROF/FF-HP-β-CD MPs had good potential for lung absorption when delivered as a DPI. During in vivo studies using the experimental allergic asthma model, ROF/FF-HP-β-CD MPs demonstrated reduced airway hyperresponsiveness due to decreased airway collagen fiber content and eosinophil infiltration. Additionally, ROF/FF-HP-β-CD MPs exhibited no recrystallization of APIs when exposed to high relative humidity. Briefly, ROF/FF-HP-β-CD MPs displayed better aerosolization and anti-inflammatory performance compared to ROF-HP-β-CD MPs. These results suggest better applicability of the ROF/FF combination DPI for pulmonary delivery [92].

Inhaled corticosteroids such as budesonide (BUD, e.g., Symbicort Turbuhaler® and Pulmicort Flexhaler®) are most prescribed in treatment of asthma and COPD to control airway inflammation and improve patient pulmonary functions. High permeability and low lung residence time in airways are the leading reasons associated with suboptimal therapeutic response of corticosteroid therapy [93–95]. Accordingly, to regulate the lung permeability and residence time of BUD, Dufour and research group developed BUD: HP-β-CD inclusion complex. Specifically, BUD-HP-β-CD MPs were formulated using bi-fluid nozzle-equipped spray dryer system. BUD-HP-β-CD MPs showed a deflated-ball-like shape with a mean particle size of 3.11 μm and bulk density of 0.11 $g.cm^3$. During aerodynamic assessment using Aerolizer® inhaler at 100 L/min BUD-HP-β-CD MPs showed 1.67-fold improvement in FPF compared to the traditional LAC-based BUD formulation. During in vitro permeability studies using Calu-3 cells, BUD-HP-β-CD (HP-β-CD 5 mM) showed lower permeability with sustained release of BUD within the lung parenchyma than free BUD. In Calu-3 cells BUD showed permeability in the following order; BUD alone > BUD: HP-β-CD 1 mM > BUD: HP-β-CD 2 mM > BUD: HP-β-CD 5 mM. Higher the BUD complexed fraction showed the lower the permeability across the Calu-3 cell layer, indicating that the flux is related to the free BUD. Moreover, during the in vivo study using ovalbumin-induced asthma mice model, BUD-HP-β-CD MPs showed a similar anti-inflammatory response at 2.5 times lower BUD concentrations than the BUD alone. In nutshell, the study showed that CDs-based DPIs can be used to regulate aerosolization properties and permeability of BUD-like corticosteroids [96]. Kinnarinen et al. (2003) developed inhalable BUD: γ-CD complex (BUD-γ-CD) using a precipitation technique. BUD-γ-CD-loaded LAC (1:15; 10 μg of BUD) MPs showed acceptable aerodynamic performance with a respirable fraction of 35% as compared to BUD/γ-CD formulation prepared by dry mixing. Additionally, following 30-day storage within the Taifun® multi-dose DPI under high humidity (75%), BUD-γ-CD retained its aerosolization performance (31%) [97].

As we know, the pulmonary delivery of drugs in the form of CD complexes presents various benefits by improving the local therapeutic action of drug via enhanced solubility, dissolution, stability, and aerodynamic qualities. Thus, beclomethasone dipropionate (BDP) and γ-CD complex were formulated to improve the local therapeutic action of BDP. Particularly, BDP: γ-CD (2:5) complex was fabricated using ethanol/water (75:25) solvent mixture by spray drying to acquire the inhalable MPs with acceptable aerodynamic characteristics. Collapsed raisin-like MPs showed 51% yield and unimodal particle size distribution with more than 95% MPs below 3.71 μm suitable for pulmonary deposition. Additionally, spray dried MPs were mixed with carriers, i.e., LAC and trehalose to modify aerosolization characteristics of BDP: γ-CD complex. BDP/γ-CD-LAC and BDP/γ-CD-trehalose exhibited 6.32 and 5.71% moisture content, respectively. BDP/γ-CD-LAC and BDP/γ-CD-trehalose showed respirable fractions of 35.2 and 34.2%, respectively, using Rotahler device at 60 L/min. BDP/γ-CD-LAC and BDP/γ-CD-trehalose revealed a marked 2.79 and 2.72-fold improvement in respirable fraction, respectively, compared to commercial products (Beclotaide Rotacaps 100 μg). Spray dryer inlet temperature had a statistically significant ($p < 0.05$) impact on the aerodynamic performances of the obtained complex. BDP/γ-CD complex obtained with the inlet

(70 °C), outlet (50 °C) temperature, and solution flow of 5 mL/min showed optimum aerodynamic performance. During aerodynamic analysis, the respirable fraction was directly proportional and inversely proportional to outlet temperature and solution flow rate. The optimization of the spray dryer parameter helps to provide powder with low agglomeration and adhesive properties, thus offering better dispersion of the formulation and attainment of higher ED (> 40%) values. However, the main drawback of the developed DPI was the high % of BDP retained in the device and capsule due to the moisture content of the formulations. Thus, further efforts are needed to control the device and capsule powder retention [98]. Similarly, Prednisolone (PRD) and fludrocortisone acetate (FLA) inhalable fine powders were formulated using a complexing agent, HP-β-CD and concurrently coated with flow enhancer leucine by the continuous-mode aerosol flow reactor to improve solubility and aerodynamic qualities of both corticosteroids. The PRD-HP-β-CD and FLA-HP-β-CD showed drug loading efficiency of 14.6 and 4.4% w/w, respectively. Formulated MPs (<5 μm) showed rough surface morphology owing to the leucine deposition on the outer surface. During an in vitro dissolution study using a modified diffusion cell, FLA-HP-β-CD showed superior dissolution kinetics as compared to PRD-HP-β-CD in buffer solution (pH 7.3). During aerodynamic assessment using Easyhaler® inhaler, PRD-HP-β-CD, and FLA-HP-β-CD showed FPF of 54 and 67%, respectively, while using Twister® inhaler, PRD-HP-β-CD, and FLA-HP-β-CD showed FPF of 70 and 62%, respectively, at a flow rate of 55 L/min. The difference in % FPF cannot be described by the resistance of the inhalers as they are similar (0.033–0.036 kPa $(L\ min^{-1})^{-1}$). The higher % FPF of Twister® is understood by its chamber geometry, where powder agglomerates collide several times to deagglomerate before leaving the device. On the other hand, the powder from Easyhaler® is delivered as a cake which disperses, resulting in fine particles so it relies on a fast-developed peak inspiratory flow rate more than in Twister®. Additionally, HP-β-CD improved both corticosteroids' permeation rate in differentiated Calu-3 cell monolayer. Furthermore, during toxicity assessments using Calu-3 cells, both powders displayed no significant toxicity and reactive oxygen species induction [99]. Briefly, increase drug dissolution and permeation through the Calu-3 cell monolayer was mainly credited to HP-β-CD engineered particles and offered a good particle engineering method to develop DPI with superior aerodynamic functions.

8.2.5 Cyclodextrins Functionalized Nano-in-Micro Particles

The particle size of the drug and carrier are the functional parameters in the development of stable and aerodynamically effective DPIs. To regulate drug deposition patterns by controlling particle size, NP-embedded MPs (NEMs) were developed by Lebhardt and co-workers (2011) wherein MPs are easily disintegrated to deliver NPs at their expected site of action. Specifically, a modified solvent displacement technique was used to formulate tetramethyl rhodamine-labeled NPs (TMR-NPs) and Nile Red-labeled NPs (NR-NPs). Subsequently, surfactant-free stabilization of NPs was achieved by co-spraying with α-CD (carrier). TMR-NPs (164 nm) and

NR-NPs (165 nm) showed a negative zeta potential of 40 and 39.0 mV, respectively. Stabilized co-spray dried NPs showed re-dispersible particles smaller than 200 nm. The small crowns of α-CD molecules act as spacers between the NPs. Additionally, the rigid structure of α-CD and fixed orientation of the hydroxyl groups (-OH) for hydrogen (-H) bonding further assist the interaction between α-CD and NPs surface and stability. Partially collapsed spherical particles with a rough surface showed MMAD of 4.3 μm using high resistance HandiHaler® device at 39 L/min. Furthermore, during cell culture experimentation using promonocytic cell line (U 937), redispersed NPs showed fewer interactions and better uptake with macrophages than polymeric PLGA MPs. In conclusion, developed NEMs possess good potential and allow practical pulmonary application of NPs in high concentrations using complexing agent [100]. In conclusion, NEMs are one of the suitable carriers for development of DPIs for pulmonary drug delivery.

Similarly, Kamal and co-workers developed NEMs for co-delivery of genistein (GNS) and all-trans retinoic acid (ATRA) for lung cancer therapy. Initially, hybrid lipid core-protein shell NPs were fabricated by sequential emulsification followed by the zein nanoprecipitation method to assist co-encapsulation of the hydrophobic substances, i.e., GNS and ATRA. Next, the zein shell gathered onto the lipid core which acts as a platform for decorating the surface of NPs via coupling of targeting ligands, ATRA and biotin, via carbodiimide chemistry. Here, biotin enhanced entry of NPs into cancer cells via membrane-receptor-mediated endocytosis and ATRA with improved internalization into the nucleus where GNS works by inhibition of topoisomerase II and PIP kinase. Additionally, hybrid lipid core-protein shell NPs were embedded into MPs by spray drying to ensure effective pulmonary drug delivery. Developed ATRA/GNS NPs displayed particle size and positive zeta potential of 206 nm and 37 mV, individually with 98.7 and 87.5% entrapment efficiency for ATRA and GNS, respectively. NEMs were fabricated using mannitol: HP-β-CD: leucine (1.5:1.5:1) with stearyl amine (0.73 mg/mL) showed a yield and re-dispersibility index of 71.8% and 0.94, respectively. Low-density and spherical-shaped NEMs verified helpful aerosolization performance with FPF and MMAD of 70.81% and 2.47 μm, respectively. During the in-vivo study in lung carcinoma-bearing mice after intratracheal administration (Penn Century aerosol microsprayer, USA), NEMs indicated superior therapeutic performance compared to i.v. NPs suspension. NEMs showed 19.5-fold increased caspase-3 expression compared to untreated mice bearing lung carcinoma. Additionally, better anticancer action of the inhaled NEMs was established by the substantial 19.3-fold decrease in numbers and diameters of metastatic foci and tumor biomarkers compared to untreated mice bearing lung carcinoma. Overall, the fabricated respirable dual-targeted NEMs co-loaded with ATRA and GNS offer a potential alternative for localized lung cancer therapy [101].

Non-reducing sugars, for example, trehalose and raffinose, are promising ingredients for protection of biomolecules. During drying, when water is eliminated hydrogen (-H) bonds are formed between non-reducing sugars and biomolecule, preserving the structural integrity of the biomolecule. Additionally, they form amorphous glass during drying, which offers a rigid matrix around the biomolecules to control and stabilize their motion. Briefly, they are effective stabilizers of biomolecules in the amorphous state. However, in most of the cases spray drying considerably inactivated the biomolecules, when processed in the presence of non-reducing sugars. However, after spray drying, biomolecules in the presence of CDs like HP-β-CD showed good control over the inactivation of biomolecules. Based on the above-mentioned investigations Amaro et al. (2015) formulated spray dried composite NEMs using raffinose: HP-β-CD (RAF: HP-β-CD) and trehalose: HP-β-CD (TRE: HP-β-CD), to enhance formulation stability and carrier system performance for oral inhalation. Initially, RAF: HP-β-CD and TRE: HP-β-CD were individually poured into a mixture of methanol: n-butyl acetate (4:1 v/v) to obtain spray dried NEMs. XRD analysis of spray dried NEMs showed amorphous nature and typical diffuse halo pattern, a characteristic of amorphous NEMs. The halo pattern in XRD and Tg in thermal analysis increased with increasing HP-β-CD concentration in the system. Spherical and porous RAF: HP-β-CD (40:60) showed mean particle size, specific surface area, and Tg of 1.9 μm, 91.52 m^2/g, and 120.8 °C, respectively. Likewise, TRE: HP-β-CD (40:60) presented mean particle size, specific surface area, and Tg of 1.5 μm, 102.20 m^2/g, and 153.7 °C, respectively. During in vitro deposition study RAF: HP-β-CD (40:60) displayed ED and FPF of 92.7 and 87.3%, respectively, using high resistance Handihaler® device. Likewise, TRE: HP-β-CD (40:60) exhibited ED and FPF of 88.9 and 83.8%, respectively. In a nutshell, the CD added into the spray drying system increase the water scavenger capability of the formulation along with an increase in the thermal stability by improving the Tg of the composite amorphous NEMs [102].

8.2.6 Cyclodextrin Complex as Stability Enhancers

LAC is the most frequently used carrier in DPI, while its further application was hindered by its hygroscopicity. LAC presents the risk of absorbing moisture from the environment owing to its amorphous state after spray drying, which would result in the crystallization of LAC and the structural alteration of carrier particles, leading to permanent aggregation with the formation of solid bridge, and thus adversely affecting the aerosolization performance of DPIs. Thus, anti-hygroscopicity should be imparted to confirm the aerosolization outcome of LAC carrier-based DPIs, which will ease its use in DPIs [103]. It is well-studied that HP-β-CD possessed low hygroscopicity and is commonly used as an anti-hygroscopicity agent [104]. Accordingly, Zhao et al. (2019) formulated HP-β-CD/LAC binary carriers using the spray drying method to resolve the hygroscopicity issue. Particularly, aqueous solution of HP-β-CD (40% w/w) and LAC (60% w/w) was spray dried to obtain respirable binary carriers. Spray dried LAC showed a spherical shape with smooth surfaces while binary carriers exhibited amorphous spheroid particles with corrugated surfaces. Higher Tg values for binary carriers (> 200 °C) confirmed the superior physical stability compared to the spray dried LAC (< 100 °C). During in vitro aerodynamic analysis, binary carriers showed significantly higher FPF (59.16%) and lowest MMAD (2.33 μm) compared to the spray dried LAC (FPF 15.95% and

MMAD 4.61 μm) using Turbospin™ device at 60 L/min which clarify the improvement of in vitro aerodynamic outcome. During storage, binary carriers exhibited 17.89-fold lower moisture uptake as compared to spray dried LAC. Furthermore, following 10-day storage under high humidity (75%), binary carriers effectively preserved their physicochemical and aerosolization performance [105]. A similar research group also formulated HP-β-CD-raffinose (60:40% w/w) binary carriers (HP-β-CD-RF) using the spray drying method to resolve the hygroscopicity-induced agglomeration concern. Jujube-like microstructure spheroid with a rough surface exhibited a particle size of 4.06 μm and lower bulk density (0.053 g/mL) appropriate for successful pulmonary drug delivery. Amorphous HP-β-CD-RF displayed a significant 1.5-fold improvement in Tg compared to spray dried raffinose. The in-vitro aerosolization study using Turbospin™ device at 60 L/min HP-β-CD-RF showed a 3.75-fold high FPF value compared to spray dried raffinose. Additionally, during 10-day storage, HP-β-CD-RF retained flow and aerodynamic properties as compared to spray dried raffinose. Molecular dynamics simulation showed that the large residual space of molecules in raffinose binary carriers is accountable for the low density and superior aerodynamic properties. Moreover, during the in vivo study after pulmonary delivery using DP-4R (Pen-century Inc., USA) in Sprague Dawley rats, HP-β-CD-RF showed significant 6.12 and 1.19-fold enhancement in lung bioavailability and mean resistance time of RF as compared to spray dried raffinose [106]. Based on the formation mechanism of binary carriers, LAC with high hygroscopicity was entrapped in the micelle-like aggregate formed by HP-β-CD and moisture availability of LAC was controlled by HP-β-CD aggregate with low hygroscopicity. The amorphous and physically stable binary carrier's formation mechanism calls for more experimental evidence with an understanding of supramolecular structures to validate the development of robust DPI carriers.

8.2.7 Cyclodextrin Engineered Nebulized Particles

SBE-β-CD is sulfobutyl ether derivative of β-CD. SBE-β-CD was developed with an objective to enhance the solubility, physicochemical stability, bioavailability or reduce volatility, irritation, smell, or taste of the drug [107]. Currently, SBE-β-CD is in 13 US FDA-approved products and several clinical molecules [108]. Accordingly, to tackle biopharmaceutical issues linked with polyphenol REV, Wang et al. (2020) formulated biodegradable REV: SBE-β-CD (1:10) complex NPs by single emulsion, solvent evaporation technique. SBE-β-CD complexation showed significant improvement (66-fold) in aqueous solubility of REV. Uniform, spherical shape NPs demonstrated particle size and zeta potential of 172.4 nm and – 38.3 mV, respectively. During aerosolization study, NPs (2 mL) were nebulized using PARI® LC plus breath-actuated nebulizer (15 L/min for 4 min) and nebulized NPs showed FPF and MMAD of 94.53% and 2.20 μm, respectively. In in vitro cellular uptake study using A549 cell lines, NPs showed better cellular internalization compared to REV alone. Thus, NPs can deliver extensively more REV intracellularly to exert its desired remedial outcome. Besides, cell-based investigations (apoptosis, cytotoxicity, cellular uptake, and antioxidant activity) showed enhanced therapeutic efficiency of NPs compared to REV alone. Moreover, prophylactic 3D spheroid investigation signified an intensified anticancer effect of NPs against A549 cells with 2.4-fold reduction in tumor volume compared to REV alone on the 15th day. In addition, Cell Titer-Glo luminescent cell viability assay showed a remarkable reduction (2.8-fold) in core tumoral mass compared to REV alone and proved the enhanced permeation of NPs. Briefly, formulated biodegradable REV: SBE-β-CD complex NPs represent a novel respirable delivery system to target non-small cell lung cancer. Yet pharmacokinetic and biodistribution-related studies are required [109].

Pulmonary fibrosis is a chronic disease with the characteristics of alveolar epithelial cell injury and fibroblast proliferation leading to the loss of normal functional tissue and destruction of lung elasticity. Tetrandrine (TET) is a bis-benzylisoquinoline alkaloid obtained from the root of *Stephania tetrandra*. TET is known to prevent fibroblasts proliferation, collagen synthesis, and effectively reverse fibrosis. Thus, Su and co-workers (2020) developed TET: HP-β-CD inclusion complex to treat pulmonary fibrosis. Initially, ethanol: TET solution was dropwise added to HP-β-CD aqueous solution and stirred carefully (60 °C for 6 h) to obtain homogeneous system. Obtained nanosuspension was subjected to freeze drying to obtain powder form. TET and inclusion complex powder (5 mg TET) were mixed with normal saline (100 mL) and subjected to high-pressure homogenizer (2 cycles of 1000 bar) to obtain nanosuspension. During in vitro drug release study using PBS (pH 7.4) TET powder and TET-HP-β-CD powder showed slow (5.69%) and high (70.84%) dissolution rate after 1 h, respectively. TET and TET-HP-β-CD nanosuspension displayed particle size and positive zeta potential of 553.4 nm, 21.2 mV and 630.3 nm, 24.1 mV, respectively. In aerodynamic analysis, nebulized TET-HP-β-CD nanosuspension 1.49-fold improvement in FPF compared to TET nanosuspension. During in vivo studies using the bleomycin-induced pulmonary fibrosis rat model, TET-HP-β-CD nanosuspension exhibited less inflammation due to synchronized protein expression in fibrosis development with a restricted accumulation of hydroxy-proline in the pulmonary airways. Additionally, after intratracheal administration, TET-HP-β-CD nanosuspension showed a marked 1.55-fold enhancement in bioavailability of TET compared to TET nanosuspension at the same dose. Differences in pharmacokinetic and pharmacodynamic outcomes are mainly attributed to differences in lung distribution of nebulized formulations. TET-HP-β-CD nanosuspension has greater FPF and reduced MMAD than TET nanosuspension, indicating that TET-HP-β-CD droplets have better dispersibility and uniformity giving rise to better deposition and complete absorption in pulmonary airways. Hence, pulmonary delivery of TET-HP-β-CD is an appealing alternative for the treatment of pulmonary fibrosis [110].

Asthma is a heterogeneous pulmonary disease mainly characterized by inflammation, airway remodeling, and bronchial hyperresponsiveness. Angiotensin-(1–7) [Ang-(1–7)], an endogen heptapeptide of renin-angiotensin system act by the Mas receptor and reduce inflammation and fibrosis in different pathophysiological conditions. Ang-(1–7) regulates eosinophilic inflammation and precludes chronic allergic airway inflammation. By knowing therapeutic importance of Ang-(1–7) Magalhães and co-workers (2020) developed HP-β-CD inclusion complex for effective pulmonary delivery of Ang-(1–7). Chronic allergic lung inflammation was induced in mice using ovalbumin. Inhaled Ang-(1–7)-HP-β-CD containing 4.5 μg of Ang-(1–7) was delivered using ultrasonic nebulizer for treatment. Inhaled Ang-(1–7)-HP-β-CD displayed a significant reduction in count of eosinophils, interleukin-5, chemokine CCL11 in pulmonary airways, and plasma immunoglobulin-E. Additionally, inhaled Ang-(1–7)-HP-β-CD successfully controlled the pulmonary remodeling by reducing the matrixins 9 and 12 expressions and collagen deposition. Briefly, clinical evidence indicates that HP-β-CD inclusion complex is helpful for the treatment of chronic respiratory disease [111].

As we reviewed, we found that CD inclusion complexes were widely used to improve the solubility, dissolution kinetics, bioavailability, and aerosolization of inhaled powder and particles. Additionally, CD complexes were also used for many drug applications focusing on either local or systemic delivery. Thus, Evrard et al. (2004) studied the short-term toxicity of inhaled CD carriers, i.e., HP-β-CD, γ-CD, and randomly methylated-β-CD (RAMEB) using C57BL/6 mice. Sterile, pyrogenic, and isotonic CD solutions were formulated using sterile water for injection and delivered using an ultrasonic nebulizer. During aerosolization assessment, aqueous solutions of CD carriers produce particle sizes suitable for pulmonary deposition. In short-term toxicity analysis, animals were exposed to HP-β-CD (75 mM), RAMEB (75 mM), and γ-CD (100 mM) aqueous solutions daily for 7 days using an ultrasonic nebulizer. Bronchoalveolar lavage, lung and kidney histopathology, and blood urea analysis showed that aqueous CD solutions are nontoxic. However, HP-β-CD and γ-CD inhalation showed a slight rise (*$p < 0.05$) in lymphocyte count compared to PBS within the bronchoalveolar lavage. The short-term safety of nebulized formulations clarifies the lung and renal function and integrity. A raise in lymphocyte counts in the bronchoalveolar lavage after CD inhalations need more clinical investigation [112].

8.3 Clinical Applications of Cyclodextrin Complex

8.3.1 CyPath Online: Biological Assay

The bioAffinity Technologies (USA) is one of the leading biological organizations that develop as well as apply diagnostics and therapeutics to fight against cancer. The bioAffinity Technologies has developed a novel biological assay (the CyPath® Early Lung Cancer Detection Assay) to distinguish between non-cancer and cancer cells in patient sputum samples. Recently, bioAffinity Technologies conducted an observational clinical trial (NCT03837600) to understand the efficacy of the use of methyl-β-CD in sputum sample processing and to conclude the cell count in sputum. This observational clinical trial is near to conclusion, and the bioAffinity Technologies is waiting for constructive outcomes [113].

8.3.2 Camptothecin Nanomaterials

Camptothecin (CPT) is an alkaloid obtained from the steam wood of *Camptotheca acuminata*. CPT selectively inhibits the nuclear enzyme DNA topoisomerase (type I) and is well studied for antineoplastic potential. Recently a few active researchers judiciously designed and developed a nano-pharmaceutical (CRLX101) product comprised of the anticancer agent CPT conjugated to a linear and hydrophilic CD-based polymer to treat small cell lung cancer. Recently, Lumos Pharma (NewLink Genetics Corporation) started an interventional (phases I and II) clinical trial (NCT00333502) to assess the safety, toxicity, and maximum tolerated dose (MTD) of CRLX101 when delivered intravenously to patients with advanced solid tumors. This clinical trial is near ending, and the NewLink Genetics Corporation is actively waiting for a helpful result. Additionally, the University of Chicago (USA) with the National Cancer Institute (USA) also performed a randomized phase II clinical trial (NCT01803269) to assess the effect of second-line treatment (CRLX101, IV) in subjects with recurrent small cell lung cancer. This study showed positive outcomes with the absence of disease progression or unacceptable toxicity [114]. Recently, the National Cancer Institute (USA) also planned a non-randomized interventional clinical trial (NCT02769962) to investigate the safety and maximum dose of CLRX101 and olaparib together in the treatment of small cell lung cancer.

8.4 Future Outlook

Even though more exciting formulation developments are still being investigated, the case studies described in the above section demonstrate several practical pulmonary applications that stand out from the variety of CD-based micro and nanosystems. Table 8.2 summarizes the CD-based inhaled formulations used to treat various pulmonary diseases. Various CD-based MPs and NPs were carefully developed and studied for pulmonary delivery of hormones, proteins/peptides, phytoconstituents, antibiotics, and anti-inflammatory agents. The CD-based MPs, engineered NPs, and core-shell systems were also fabricated to improve drug substances' physical stability and aerosolization characteristics. Additionally, a range of CDs such as HP-β-CD, maltosyl-α-CD, dimethyl-β CD, and SBE-β-CD were used to develop pulmonary drug delivery systems. At the same time, CDs were also coupled with various force-controlling agents (e.g., leucine), polysaccharides (e.g., curdlan gum), and sugars (e.g., LAC and mannitol) to improve the aerodynamic performance, drug deposition pattern, and physical stability during long-term storage. All the CDs-based formulations showed that the aerodynamic drug deposition pattern was acceptable regardless of inhaler device type and inspiratory flow rate. Moreover, the CD containing dry powders demonstrated several vital advantages, i.e., ease of handling, adequate stability, tailored surface properties, good flow properties, and ultimately high pulmonary drug deposition.

Among the different CDs, HP-β-CD was most used to develop inhalable MPs and showed enhanced biopharmaceutical properties with a substantial reduction in

Table 8.2 Cyclodextrin-based inhaled formulations for the treatment of various pulmonary diseases

Active agent	Method	Cyclodextrin	Particle properties	Outcome	References
Parathyroid	Spray freeze drying	HP-β-CD	Porous spherical-shaped and 15.95 μm	Porous MPs showed satisfactory FPF (14.20%) using Cyclohaler® at 60 L/min and significantly improved bioavailability compared to intravenous injection	[60]
Salmon calcitonin	Spray drying	HP-β-CD	Spherical rough surface and 2.84 μm	MPs displayed a marked 4.76 and 2.3-fold improvement in tmax and bioavailability, respectively, compared to nebulized calcitonin	[61]
Insulin	Double emulsion method	HP-β-CD	Low-density spherical and 26.2 μm	MPs displayed FPF of 61.3% using Turbospin® (60 L/min) and sustained release profile for 10 days with effective delivery within alveoli	[62]
Cyclosporin A	Freeze drying	HP-β-CD	–	Dry powder showed a 1.74-fold higher respirable fraction than LAC-based formulation using the Taifun® inhaler device	[64]
Cyclosporin A	Kneading method	Maltosyl-α-CD	2.76 μm	Complex showed efficient reserve of allergen-induced eosinophil accumulation for longer duration with 9-times lower dose as compared to drug alone	[66]
Recombinant human growth hormone	Spray drying	Dimethyl-β-CD	Rough (55.60 nm) dimply shaped and 2.3 μm	Dry powder showed good FPF (34.89%) using Spinhaler® (60 L/min) and marked 3.0-fold improvement in bioavailability compared to recombinant human growth hormone alone at the same dose (0.1 mg)	[67]
Resveratrol	Freeze drying	HP-β-CD	–	Complex showed higher absorption kinetics, bioavailability (92.95%), and superior (100 times) drug accumulation in pulmonary airways than oral delivery	[72]
Andrographolide	Lyophilization	β-CD	Grouped bricks-like morphology and <5 μm	Complex showed FPF and ED 81.83% and 96.08%, respectively, with superior in vivo anti-pneumonia activity	[73]
Tea tree oil	Grinding method	β-CD	Rough surface irregularly shaped and <5 μm	Complex showed satisfactory FPF (51.22%) with superior fungal (*Candida albicans*) and bacterial (*Acinetobacter baumannii*) pneumonia effects	[75]

(continued)

Table 8.2 (continued)

Active agent	Method	Cyclodextrin	Particle properties	Outcome	References
Fisetin	Co-solvent technique with spray drying	SBE-β-CD	Pitted surface morphology and 3.14 μm	Complex showed FPF and ED of 88.08% and 32.49% using Aerolizer® (60 L/min) with good EC_{50} (75.86) using A549 cells	[77]
Clofazimine	Spray drying	β-CD	Rough MPs	Complex displayed superior FPF (51.56%) with lower MMAD (2.9 μm) using HandiHaler® with maximum drug retention in the Calu-3 cell monolayer and apical compartment	[81]
Combine isoniazid and rifampicin	Spray drying	HP-β-CD	Porous and non-aggregated MPs	MPs showed satisfactory ED for isoniazid (92.60%) and rifampicin (99.30%) using Jethaler® dual chamber device	[82]
Levofloxacin and rifampicin	Solvent evaporation	β-CD and curdlan	Positively charged (13.45 mV) and diamond-shaped morphology	Sustained drug release (80 h) with a minimum inhibitory concentration of 23.44 μg/mL against *M. smegmatis*	[84]
Roflumilast	Spray drying	β-CD	Negatively charged (6.75 mV) and 3.13 μm	MPs showed 3.75-fold improvement in % FPF compared to drug alone with superior sustained broncho-dilatation activity in healthy human subjects as compared to salbutamol (90 μg/actuation)	[86]
Roflumilast	Spray drying	HP-β-CD	Wrinkled-shaped raisin-like MPs and 2.63 μm	MPs showed FPF of 56.42% using HandiHaler® (60 L/min) with an excellent apparent permeability coefficient using Calu-3 cell monolayers	[89]
Roflumilast and formoterol	Spray drying	HP-β-CD	Amorphous wrinkled-shaped MPs and 2.70 μm	MPs displayed satisfactory FPF for roflumilast (52.86%) and formoterol (53.64%) using HandiHaler® (60 L/min) with reduced airway hyperresponsiveness and better anti-inflammatory action in an allergic asthma model	[92]

Table 8.2 (continued)

Active agent	Method	Cyclodextrin	Particle properties	Outcome	References
Budesonide	Bi-fluid nozzle-equipped spray drying	HP-β-CD	Deflated ball-like shape and 3.11 μm	MPs showed 1.67-fold improvement in FPF compared to LAC formulation using Aerolizer® (100 L/min) with better anti-inflammatory potential at 2.5 times lower drug concentrations	[96]
Beclomethasone dipropionate	Spray drying	γ-CD	Collapsed raisin-like MPs	LAC and trehalose based spray dried MPs displayed respirable fractions of 35.2 and 34.2%, respectively using Rotahler® (60 L/min)	[98]
Prednisolone	Continuous-mode aerosol flow reactor	HP-β-CD and leucine	Rough surface morphology and <5 μm	MPs revealed FPF of 54 and 70% using Easyhaler® and Twister® at 55 L/min with improved permeation rate in differentiated Calu-3 cell monolayer	[99]
Fludrocortisone acetate	Continuous-mode aerosol flow reactor	HP-β-CD and leucine	Rough surface morphology and <5 μm	MPs showed FPF of 67 and 62% using Easyhaler® and Twister® at 55 L/min with an improved permeation rate in differentiated Calu-3 cell monolayer	[99]
Tetramethyl rhodamine-labeled NPs and Nile red-labeled NPs	Modified solvent displacement and spray drying	α-CD	Partially collapsed spherical NEMs	Rough surface NEMs exhibited MMAD of 4.3 μm using high resistance HandiHaler® (39 L/min)	[100]
Genistein and all-trans retinoic acid	Sequential solvent evaporation followed by nanoprecipitation and spray drying	HP-β-CD, leucine, and mannitol	Positively charged (37 mV) and 206 nm	Nanocomposites showed FPF and MMAD of 70.81% and 2.47 μm with superior therapeutic performance than intravenous NPs suspension in lung carcinoma-bearing mice	[101]
Raffinose pentahydrate and trehalose dihydrate	Spray drying	HP-β-CD	Amorphous and porous spherical NEMs with <5 μm	Co-spray dried NEMs displayed superior FPF (>80%) and ED (>90%) compared to spray dried sugars alone	[102]
LAC binary carriers	Spray drying	HP-β-CD	Amorphous corrugated surfaces spheroid and 4 μm	Binary carriers displayed 3.7-fold high FPF value and 17.89-fold lower moisture uptake as compared to spray dried LAC	[105]

(continued)

Table 8.2 (continued)

Active agent	Method	Cyclodextrin	Particle properties	Outcome	References
Raffinose binary carriers	Spray drying	HP-β-CD	Rough jujube-like microstructure spheroid and 4.06 μm	Amorphous binary carriers showed a 3.75-fold high FPF value using Turbospin™ (60 L/min) with significant 6.12 and 1.19-fold enhancement in lung bioavailability and mean resistance time compared to spay dried raffinose	[106]
Resveratrol	Single emulsion and solvent evaporation	SBE-β-CD	Negatively charged (38.3 mV) spherical shape and 172.4 nm	Nebulized NPs displayed FPF and MMAD of 94.53% and 2.20 μm, respectively, with an intensified anticancer effect against A549 cells	[109]
Tetrandrine	Freeze drying	HP-β-CD	Positively charged (24.1 mV) and 630.3 nm	Nebulized NPs showed 1.49-fold improvement in FPF compared to drug alone with lower inflammation in a bleomycin-induced pulmonary fibrosis rat model	[110]
Angiotensin (1–7)	–	HP-β-CD	–	The inhaled formulation showed a significant reduction in eosinophils, interleukin-5, chemokine CCL11 in pulmonary airways and plasma immunoglobulin-E with reduced matrixins 9 and 12 expressions and collagen deposition	[111]

the dose needed for therapy. HP-β-CD was coupled with linear PEG to control the permeability and aerodynamic performance of protein molecules. Thus, based on solubility and absorption kinetics, different CD inclusion complexes can be developed to expand the clinical application of various biological and genetic materials in treating pulmonary diseases. Additionally, in most corticosteroids containing formulations, significant improvements such as higher physical stability, superior respirable fraction, and controlled permeability through pulmonary epithelium in the therapy of inflammation linked with asthma can be easily achieved by using HP-β-CD-based spray dried powders. This approach looks hopeful for improving airway targeting with reduced adverse effects linked with high doses (>500 μg) of corticosteroids. Besides, this strategy can also be explored for different pulmonary pathologies where drug concentration and airways retention are limiting factors such as asthma, COPD, TB, and lung carcinoma. Shortly, several biofunctional substances, ligands, or ions such as biotin, folic acid, and hyaluronic acid can be covalently linked to the CDs moieties to design enzyme-responsive, supramolecular assemblies to achieve desirable cell/ tissue-targeting capacity. Additionally, these CD assemblies can be used to form simultaneous biological sensing and drug delivery platforms and regulate intra/intercellular communication for effective disease diagnosis and management. Furthermore, with the new advances in material science, molecular biology, and particle engineering, several novel types of CDs-based stimuli-responsive (e.g., heat, light, ionic strengths, and magnetic energy) supramolecular assemblies with superior drug targeting properties and appealing biological functions would be the upcoming future [115]. Undoubtedly CDs are promising carriers for pulmonary drug delivery, and the application of CDs-based materials will further extend to play an essential battle against pulmonary complications.

8.5 Conclusion

This chapter analyzed and summarized the research efforts in developing CD-based drug delivery systems and their outcome in treating pulmonary diseases by selecting a few representative examples. CDs applications as encapsulating agents for anti-inflammatory agents, as an additive in formulating NPs, as a template to design high drug loading porous particles, as adjuvants to stabilize proteins, peptides, and hormones, as carriers for phytoconstituents, and as dry powder stabilizers significantly highlight the versatile nature of CDs in pulmonary drug delivery. A wide range of in vitro drug release models advanced microscopic screening and aerodynamic assessment established physiochemical and aerosolization applications of CDs. In contrast, in vivo animal models and cell culture setups verified the biological advantages of CDs-based carrier materials. Furthermore, CDs-based carrier materials showed few promising outcomes in biological sensing, cellular communication, disease diagnosis, and treatment. Basically, in pulmonary drug delivery, CD-based carriers from a single to multi-component complex, from single- to multi-drug carrier platforms, and various dosage forms to tackle several devastating pulmonary diseases. With the help of progress in supramolecular chemistry, materials science, nanotechnology, a better understanding of CDs will undoubtedly bring a transition in their drug delivery trend. The bottom line, the importance of CDs-based pulmonary systems is still growing, and this movement is expected to persist in the coming time.

References

1. Crini G. Review: a history of cyclodextrins. Chem Rev. 2014;114(21):10940–75.
2. Martin J, Díaz-Montaña JE, Asuero GA. Cyclodextrins: past and present. In: Arora P, Dhingra N, editors. Cyclodextrin: a Versatile Ingredient; 2018. p. 1–43.
3. Wüpper S, Lüersen K, Rimbach G. Cyclodextrins, natural compounds, and plant bioactives-a nutritional perspective. Biomol Ther. 2021;11(3):401.
4. Del Valle EM. Cyclodextrins and their uses: a review. Process Biochem. 2004;39(9):1033–46.
5. Saokham P, Muankaew C, Jansook P, Loftsson T. Solubility of Cyclodextrins and drug/Cyclodextrin complexes. Molecules. 2018;23(5):1161.
6. Loftsson T, Jarho P, Másson M, Järvinen T. Cyclodextrins in drug delivery. Expert Opin Drug Deliv. 2005;2(2):335–51.
7. Aiassa V, Garnero C, Longhi MR, Zoppi A. Cyclodextrin multi-component complexes: pharmaceutical applications. Pharmaceutics. 2021;13(7):1099.
8. Patel M, Hirlekar R. Multicomponent cyclodextrin system for improvement of solubility and dissolution rate of poorly water-soluble drug. Asian J Pharmaceutical Sci. 2019;14(1):104–15.
9. Carneiro SB, Costa Duarte FÍ, Heimfarth L, Siqueira Quintans JS, Quintans-Júnior LJ, Veiga Júnior VFD, Neves de Lima ÁA. Cyclodextrin⁻Drug inclusion complexes: In Vivo and In Vitro approaches. Int J Mol Sci. 2019;20(3):642.
10. Braga SS. Cyclodextrins: emerging medicines of the new millennium. Biomol Ther. 2019;9(12):801.
11. Lee JU, Lee SS, Lee S, Oh HB. Noncovalent complexes of Cyclodextrin with small organic molecules: applications and insights into host-guest interactions in the gas phase and condensed phase. Molecules. 2020;25(18):4048.
12. Bai CC, Tian BR, Zhao T, Huang Q, Wang ZZ. Cyclodextrin-catalyzed organic synthesis: reactions, mechanisms, and applications. Molecules. 2017;22(9):1475.
13. Guo C, Xiao Y. Negatively charged cyclodextrins: synthesis and applications in chiral analysis-a review. Carbohydr Polym. 2021;256:117517.
14. Xiao Y, Ng SC, Tan TT, Wang Y. Recent development of cyclodextrin chiral stationary phases and their applications in chromatography. J Chromatogr A. 2012;1269:52–68.
15. Rousseau J, Menuel S, Rousseau C, Hapiot F, Monflier E. Cyclodextrins as porous material for catalysis. In: Organic Nanoreactors. Academic Press; 2016. p. 15–42.
16. Leemhuis H, Kelly RM, Dijkhuizen L. Engineering of cyclodextrin glucanotransferases and the impact for biotechnological applications. Appl Microbiol Biotechnol. 2010;85(4):823–35.
17. Numanoğlu U, Sen T, Tarimci N, Kartal M, Koo OM, Onyüksel H. Use of cyclodextrins as a cosmetic delivery system for fragrance materials: linalool and benzyl acetate. AAPS PharmSciTech. 2007;8(4):E85.
18. Petitjean M, García-Zubiri IX, Isasi JR. History of cyclodextrin-based polymers in food and pharmacy: a review. Environ Chem Lett. 2021:1–12.
19. Gonzalez Pereira A, Carpena M, García Oliveira P, Mejuto JC, Prieto MA, Simal GJ. Main applications of Cyclodextrins in the food industry as the compounds of choice to form host-guest complexes. Int J Mol Sci. 2021;22(3):1339.
20. Astray G, Mejuto JC, Simal-Gandara J. Latest developments in the application of cyclodextrin host-guest complexes in beverage technology processes. Food Hydrocoll. 2020;106:105882.
21. Bognanni N, Bellia F, Viale M, Bertola N, Vecchio G. Exploring charged polymeric Cyclodextrins for biomedical applications. Molecules. 2021;26(6):1724.

22. Sajadi S, Rojas P, Oppenheimer SB. Cyclodextrin, a probe for studying adhesive interactions. Acta Histochem. 2007;109(4):338–42.
23. Szente L, Fenyvesi É. Cyclodextrin-enabled polymer composites for packaging. Molecules. 2018;23(7):1556.
24. Liu Y, Sameen DE, Ahmed S, Wang Y, Lu R, Dai J, Li S, Qin W. Recent advances in cyclodextrin-based films for food packaging. Food Chem. 2021;370:131026.
25. Xiao Z, Zhang Y, Niu Y, Ke Q, Kou X. Cyclodextrins as carriers for volatile aroma compounds: A review. Carbohydrate Polymers. 2021;269:118292.
26. Mamusa M, Resta C, Sofroniou C, Baglioni P. Encapsulation of volatile compounds in liquid media: fragrances, flavors, and essential oils in commercial formulations. Adv Colloid Interf Sci. 2021;102544
27. Bezerra FM, Lis MJ, Firmino HB, Dias da Silva JG, Curto Valle RCS, Borges Valle JA, Scacchetti FAP, Tessaro AL. The role of β-Cyclodextrin in the textile industry-review. Molecules. 2020;25(16):3624.
28. Loftsson T, Brewster ME. Pharmaceutical applications of cyclodextrins: basic science and product development. J Pharm Pharmacol. 2010;62(11):1607–21.
29. Rincón-López J, Almanza-Arjona YC, Riascos AP, Rojas-Aguirre Y. When Cyclodextrins met data science: unveiling their pharmaceutical applications through network science and text-mining. Pharmaceutics. 2021;13(8):1297.
30. Rincón-López J, Almanza-Arjona YC, Riascos AP, Rojas-Aguirre Y. Technological evolution of cyclodextrins in the pharmaceutical field. J Drug Deliv Sci Technol. 2021;61:102156.
31. Zarzycki PK, Głód BK. Cyclodextrins-based nanocomplexes for encapsulation of bioactive compounds in food, cosmetics, and pharmaceutical products: principles of supramolecular complexes formation, their influence on the antioxidative properties of target chemicals, and recent advances in selected industrial applications. Encapsulations. 2016:717–67.
32. Loftsson T. Cyclodextrins in parenteral formulations. J Pharm Sci. 2021;110(2):654–64.
33. Loftsson T, Duchêne D. Cyclodextrins and their pharmaceutical applications. Int J Pharm. 2007;329(1–2):1–1.
34. Loftsson T. Cyclodextrins and the biopharmaceutics classification system of drugs. J Incl Phenom. 2002;44:63–7.
35. Tiwari G, Tiwari R, Rai AK. Cyclodextrins in delivery systems: applications. J Pharm Bioallied Sci. 2010;2(2):72–9.
36. Cyclodextrins used as excipients Report published in support of the 'Questions and answers on cyclodextrins used as excipients in medicinal products for human use' (EMA/CHMP/495747/2013). European Medicines Agency, 2017.
37. Gómez-Graña S, Pérez-Juste J, Hervés P. Cyclodextrins and inorganic nanoparticles: another tale of synergy. Adv Colloid Interf Sci. 2021;288:102338.
38. Bakshi PR, Londhe VY. Widespread applications of host-guest interactive cyclodextrin functionalized polymer nanocomposites: its meta-analysis and review. Carbohydr Polym. 2020;242:116430.
39. Garrido PF, Calvelo M, Blanco-González A, Veleiro U, Suárez F, Conde D, Cabezón A, Piñeiro Á, Garcia-Fandino R. The Lord of the NanoRings: Cyclodextrins and the battle against SARS-CoV-2. Int J Pharm. 2020;588:119689.
40. Shelley H, Babu RJ. Role of Cyclodextrins in nanoparticle-based drug delivery systems. J Pharm Sci. 2018;107(7):1741–53.
41. Gharib R, Greige-Gerges H, Fourmentin S, Charcosset C, Auezova L. Liposomes incorporating cyclodextrin–drug inclusion complexes: current state of knowledge. Carbohydr Polym. 2015;129:175–86.
42. Muñoz-Shuguli C, Vidal CP, Cantero-López P, Lopez-Polo J. Encapsulation of plant extract compounds using cyclodextrin inclusion complexes, liposomes, electrospinning and their combinations for food purposes. Trends Food Sci Technol. 2020;108:177–86.
43. Capelezzo AP, Mohr LC, Dalcanton F, de Mello JM, Fiori MA. β-Cyclodextrins as encapsulating agents of essential oils. Cyclodext-Versatile Ingred. 2018;

44. Mathapa BG, Paunov VN. Nanoporous cyclodextrin-based co-polymeric microspheres for encapsulation of active components. J Mater Chem B. 2013;1(29):3588–98.
45. Sou T, Bergström CAS. Contemporary formulation development for inhaled pharmaceuticals. J Pharm Sci. 2021;110(1):66–86.
46. Mehta PP, Dhapte-Pawar V. Role of surfactants in pulmonary drug delivery. In: Green sustainable process for chemical and environmental engineering and science. Academic Press; 2022. p. 559–77.
47. Mehta P. Imagine the superiority of dry powder inhalers from carrier engineering. J Drug Deliv. 2018;2018:1.
48. Jadhav P, Patil P, Bhagwat D, Gaikwad V, Mehta PP. Recent advances in orthogonal analytical techniques for microstructural understanding of inhalable particles: present status and future perspective. J Drug Deliv Sci Technol. 2022;103089
49. Mehta PP. Dry powder inhalers: a concise summary of the electronic monitoring devices. Ther Deliv. 2021 Jan;12(1):1–6.
50. Mehta PP. Dry powder inhalers: upcoming platform technologies for formulation development. Ther Deliv. 2019 Sep;10(9):551–4.
51. Tan ZM, Lai GP, Pandey M, Srichana T, Pichika MR, Gorain B, Bhattamishra SK, Choudhury H. Novel approaches for the treatment of pulmonary tuberculosis. Pharmaceutics. 2020;12(12):1196.
52. Mehta PP, Dhapte-Pawar VS. Repurposing drug molecules for new pulmonary therapeutic interventions. Drug Deliv Transl Res. 2021;11(5):1829–48.
53. Washington N, Washington C, Wilson CG. Pulmonary drug delivery. Physiological pharmaceutics. Barriers to drug absorption. 2nd ed. London, UK: Taylor & Francis; 2001. p. 221–47.
54. Henning A, Schneider M, Nafee N, Muijs L, Rytting E, Wang X, Kissel T, Grafahrend D, Klee D, Lehr CM. Influence of particle size and material properties on mucociliary clearance from the airways. J Aerosol Med Pulm Drug Deliv. 2010;23(4):233–41.
55. Mehta P, Bothiraja C, Kadam S, et al. Effect of USP induction ports, glass sampling apparatus, and inhaler device resistance on aerodynamic patterns of fluticasone propionate-loaded engineered mannitol microparticles. AAPS PharmSciTech. 2019;20:197.
56. Kamble R, Palve P, Mehta P. Preparation, and evaluation of amorphous olmesartan medoxomil with porous silica microparticles using spray-drying technique. J Adv Pharma Edu Res. 2014;4(4):444–50.
57. Alhajj N, O'Reilly NJ, Cathcart H. Designing enhanced spray dried particles for inhalation: a review of the impact of excipients and processing parameters on particle properties. Powder Technol. 2021 May 1;384:313–31.
58. Mehta P, Dhapte VV. Propulsive PAT paradigm: optimization of freeze drying process. Int J Pharm Sci Rev Res. 2014;28(2):240–6.
59. Vishali DA, Monisha J, Sivakamasundari SK, Moses JA, Anandharamakrishnan C. Spray freeze drying: emerging applications in drug delivery. J Control Release. 2019;300:93–101.
60. Poursina N, Vatanara A, Rouini MR, Gilani K, Rouholamini NA. Systemic delivery of parathyroid hormone (1-34) using spray freeze-dried inhalable particles. Pharm Dev Technol. 2017;22(6):733–9.
61. Tewes F, Gobbo OL, Amaro MI, Tajber L, Corrigan OI, Ehrhardt C, Healy AM. Evaluation of HPβCD-PEG microparticles for salmon calcitonin administration via pulmonary delivery. Mol Pharm. 2011;8(5):1887–98.
62. Ungaro F, di Villa d'E, Bianca R, Giovino C, Miro A, Sorrentino R, Quaglia F, La Rotonda MI. Insulin-loaded PLGA/cyclodextrin large porous particles with improved aerosolization properties: in vivo deposition and hypoglycaemic activity after delivery to rat lungs. J Control Release. 2009;135(1):25–34.
63. Nizankowska E, Soja J, Pinis G, et al. Treatment of steroid dependent bronchial asthma with cyclosporin. Eur Respir J. 1995;8:1091–9.
64. Matilainen L, Järvinen K, Toropainen T, Näsi E, Auriola S, Järvinen T, Jarho P. In vitro evaluation of the effect of cyclodextrin complexation on pulmonary deposition of a peptide, cyclosporin a. Int J Pharm. 2006;318(1–2):41–8.

65. Berton P, Mishra MK, Choudhary H, Myerson AS, Rogers RD. Solubility studies of cyclosporine using ionic liquids. ACS omega. 2019 May 1;4(5):7938–43.
66. Fukaya H, Iimura A, Hoshiko K, Fuyumuro T, Noji S, Nabeshima T. A cyclosporin a/maltosyl-alpha-cyclodextrin complex for inhalation therapy of asthma. Eur Respir J. 2003;22(2):213–9.
67. Jalalipour M, Najafabadi AR, Gilani K, Esmaily H, Tajerzadeh H. Effect of dimethyl-beta-cyclodextrin concentrations on the pulmonary delivery of recombinant human growth hormone dry powder in rats. J Pharm Sci. 2008;97(12):5176–85.
68. Ma BN, Li XJ. Resveratrol extracted from Chinese herbal medicines: a novel therapeutic strategy for lung diseases. Chinese Herbal Medicines. 2020 Oct 1;12(4):349–58.
69. Mehta P, Bothiraja C, Mahadik K, Kadam S, Pawar A. Phytoconstituent based dry powder inhalers as biomedicine for the management of pulmonary diseases. Biomed Pharmacother. 2018;108:828–37.
70. Trotta V, Lee WH, Loo CY, Haghi M, Young PM, Scalia S, Traini D. In vitro biological activity of resveratrol using a novel inhalable resveratrol spray-dried formulation. Int J Pharm. 2015 Aug 1;491(1–2):190–7.
71. Dimer FA, Ortiz M, Pohlmann AR, Guterres SS. Inhalable resveratrol microparticles produced by vibrational atomization spray drying for treating pulmonary arterial hypertension. J Drug Deliv Sci Technol. 2015 Oct 1;29:152–8.
72. Kong D, et al. Pulmonary administration of resveratrol/hydroxypropyl-β-cyclodextrin inclusion complex: in vivo disposition and in vitro metabolic study. J Drug Deliv Sci Technol. 2020;60(2020):101995.
73. Zhang T, Zhu L, Li M, Hu Y, Zhang E, Jiang Q, Han G, Jin Y. Inhalable Andrographolide-β-cyclodextrin inclusion complexes for treatment of Staphylococcus aureus pneumonia by regulating immune responses. Mol Pharm. 2017;14(5):1718–25.
74. Mehta P, Dhapte V. A comprehensive review on pharmacokinetic profile of some traditional Chinese medicines. New J Sci. 2016;2016:1–32.
75. Li M, Zhu L, Zhang T, Liu B, Lina D, Jin Y. Pulmonary delivery of tea tree oil-b-cyclodextrin inclusion complexes for the treatment of fungal and bacterial pneumonia. J Pharm Pharmacol. 2017;69:1458.
76. Mehta P, Pawar A, Mahadik K, Bothiraja C. Emerging novel drug delivery strategies for bioactive flavonol fisetin in biomedicine. Biomed Pharmacother. 2018 Oct 1;106:1282–91.
77. Mohtar N, Taylor K, Sheikh K, et al. Design and development of dry powder sulfobutylether-β-cyclodextrin complex for pulmonary delivery of fisetin. Eur J Pharm Biopharm. 2017;113:1–10.
78. Wang MG, Liu XM, Wu SQ, He JQ. Impacts of clofazimine on the treatment outcomes of drug-resistant tuberculosis. Microbes Infect. 2022;105020:105020.
79. Mashele SA, Steel HC, Matjokotja MT, Rasehlo SS, Anderson R, Cholo MC. Assessment of the efficacy of clofazimine alone and in combination with primary agents against mycobacterium tuberculosis in vitro. J Glob Antimicrob Resist. 2022;29:343–52.
80. Mehta P, Bothiraja C, Kadam S, Pawar A. Potential of dry powder inhalers for tuberculosis therapy: facts, fidelity and future. Artificial Cells, Nanomedicine, and Biotechnology. 2018;46(sup3):S791–806.
81. Ribeiro R, de Castro V, Todaro LC, da Silva RP, Simon A, do Carmo FA, de Sousa VP, Rodrigues CR, Sarmento B, Healy AM, Cabral LM. Development of inhaled formulation of modified clofazimine as an alternative to treatment of tuberculosis. J Drug Deliv Sci Technol. 2020;58:101805.
82. Kadota K, Senda A, Tagishi H, Ayorinde JO, Tozuka Y. Evaluation of highly branched cyclic dextrin in inhalable particles of combined antibiotics for the pulmonary delivery of anti-tuberculosis drugs. Int J Pharm. 2017;517(1–2):8–18.
83. Chen Y, Wang F. Review on the preparation, biological activities and applications of curdlan and its derivatives. Eur Polym J. 2020;141:110096.
84. Yunus Basha R, Sampath Kumar TS, Doble M. Dual delivery of tuberculosis drugs via cyclodextrin conjugated curdlan nanoparticles to infected macrophages. Carbohydr Polym. 2019;218:53–62.

85. Pinner NA, Hamilton LA, Hughes A. Roflumilast: a phosphodiesterase-4 inhibitor for the treatment of severe chronic obstructive pulmonary disease. Clin Ther. 2012 Jan;34(1):56–66.
86. Murad HA, Habib HS, Rafeeq MM, Sulaiman MI, Abdulrahman AS, Khabaz MN. Co-inhalation of roflumilast, rather than formoterol, with fluticasone more effectively improves asthma in asthmatic mice. Exp Biol Med. 2017;242(5):516–26.
87. Chapman RW, House A, Jones H, Richard J, Celly C, Prelusky D, Ting P, Hunter JC, Lamca J, Phillips JE. Effect of inhaled roflumilast on the prevention and resolution of allergen-induced late phase airflow obstruction in Brown Norway rats. Eur J Pharmacol. 2007;571(2–3):215–21.
88. Mahmoud AA, Elkasabgy NA, Abdelkhalek AA. Design and characterization of emulsified spray dried alginate microparticles as a carrier for the dually acting drug roflumilast. Eur J Pharm Sci. 2018;122:64–76.
89. Suzuki EY, Amaro MI, de Almeida GS, Cabral LM, Healy AM, de Sousa VP. Development of a new formulation of roflumilast for pulmonary drug delivery to treat inflammatory lung conditions. Int J Pharm. 2018;550(1–2):89–99.
90. Barnes PJ. Scientific rationale for using a single inhaler for asthma control. Eur Respir J. 2007 Mar;29(3):587–95.
91. Nelson HS, Chapman KR, Pyke SD, Johnson M, Pritchard JN. Enhanced synergy between fluticasone propionate and salmeterol inhaled from a single inhaler versus separate inhalers. J Allergy Clin Immunol. 2003;112(1):29–36.
92. Suzuki EY, et al. Effects of a novel roflumilast and formoterol fumarate dry powder inhaler formulation in experimental allergic asthma. Int J Pharm. 2020;588:119771.
93. Lambrecht BN, Hammad H. The immunology of asthma. Nature Immunol. 2014;16:45–56.
94. Holgate ST, Polosa R. The mechanisms, diagnosis, and management of severe asthma in adults. Lancet. 2006;368:780–93.
95. Derendorf H, Nave R, Drollmann A, Cerasoli F, Wurst W. Relevance of pharmacokinetics and pharmacodynamics of inhaled corticosteroids to asthma. Eur Respir J. 2006;28:1042–50.
96. Dufour G, Bigazzi W, Wong N, Boschini F, de Tullio P, Piel G, Cataldo D, Evrard B. Interest of cyclodextrins in spray-dried microparticles formulation for sustained pulmonary delivery of budesonide. Int J Pharm. 2015;495(2):869–78.
97. Kinnarinen T, Jarho P, Järvinen K, Järvinen T. Pulmonary deposition of a budesonide/gamma-cyclodextrin complex in vitro. J Control Release. 2003;90(2):197–205.
98. Cabral-Marques H, Almeida R. Optimisation of spray-drying process variables for dry powder inhalation (DPI) formulations of corticosteroid/cyclodextrin inclusion complexes. Eur J Pharm Biopharm. 2009;73:121–9.
99. Vartiainen V, et al. Aerosolization, Drug Permeation and Cellular Interaction of Dry Powder Pulmonary Formulations of Corticosteroids with Hydroxypropyl-β-Cyclodextrin as a Solubilizer. Pharm Res. 2017;34(1):25–35. https://doi.org/10.1007/s11095-016-2035-9.
100. Lebhardt T, Roesler S, Uusitalo HP, Kissel T. Surfactant-free redispersible nanoparticles in fast-dissolving composite microcarriers for dry-powder inhalation. Eur J Pharm Biopharm. 2011;78:90–6.
101. Kamel NM, et al. Inhalable dual-targeted hybrid lipid Nanocore-protein Shell composites for combined delivery of Genistein and all-trans retinoic acid to lung cancer cells. ACS Biomater Sci Eng. 2019;6(1):71–87.
102. Amaro MI, Tajber L, Corrigan OI, Healy AM. Co-spray dried carbohydrate microparticles: crystallisation delay/inhibition and improved aerosolization characteristics through the incorporation of hydroxypropyl-β-cyclodextrin with amorphous raffinose or trehalose. Pharm Res. 2015;32(1):180–95.
103. Della Bella A, Salomi E, Buttini F, Bettini R. The role of the solid state and physical properties of the carrier in adhesive mixtures for lung delivery. Expert Opin Drug Deliv. 2018 Jul;15(7):665–74.
104. Maeda H, Iga Y, Nakayama H. Characterization of inclusion complexes of betahistine with β-cyclodextrin and evaluation of their anti-humidity properties. J Incl Phenom Macrocycl Chem. 2016 Dec;86(3):337–42.

105. Zhao Z, Zhang X, Cui Y, Huang Y, Huang Z, Wang G, Liang R, Pan X, Tao L, Wu C. Hydroxypropyl-β-cyclodextrin as anti-hygroscopicity agent in amorphous lactose carriers for dry powder inhalers. Powder Technol. 2019;358:29–38.
106. Zhao Z, Huang Z, Zhang X, Huang Y, Cui Y, Ma C, Wang G, Freeman T, Lu XY, Pan X, Wu C. Low density, good flowability cyclodextrin-raffinose binary carrier for dry powder inhaler: anti-hygroscopicity and aerosolization performance enhancement. Expert Opin Drug Deliv. 2018;15(5):443–57.
107. Davis ME, Brewster ME. Cyclodextrin-based pharmaceutics: past, present and future. Nat Rev Drug Discov. 2004;3(12):1023–35.
108. Stella VJ, Rajewski RA. Sulfobutylether-β-cyclodextrin. Int J Pharm. 2020;583:119396.
109. Wang X, Parvathaneni V, Shukla SK, Kulkarni NS, Muth A, Kunda NK, Gupta V. Inhalable resveratrol-cyclodextrin complex loaded biodegradable nanoparticles for enhanced efficacy against non-small cell lung cancer. Int J Biol Macromol. 2020 Dec 1;164:638–50.
110. Su W, Liang Y, Meng Z, Chen X, Lu M, Han X, Deng X, Zhang Q, Zhu H, Fu T. Inhalation of Tetrandrine-hydroxypropyl-β-cyclodextrin inclusion complexes for pulmonary fibrosis treatment. Mol Pharm. 2020;17(5):1596–607.
111. Magalhães GS, Gregório JF, Ramos KE, Ribeiro AT, Baroni IF, Barcelos LS, Pinho V, Teixeira MM, Santos RA, Rodrigues-Machado MG, Santos MJ. Treatment with inhaled formulation of angiotensin-(1-7) reverses inflammation and pulmonary remodeling in a model of chronic asthma. Immunobiology. 2020;151957
112. Evrard B, Bertholet P, Gueders M, Flament MP, Piel G, Delattre L, Gayot A, Leterme P, Foidart JM, Cataldo D. Cyclodextrins as a potential carrier in drug nebulization. J Control Release. 2004;96(3):403–10.
113. Sputum labeling utilizing synthetic Meso-Tetra (4-Carboxyphenyl) Porphyrin (TCPP) for detection of lung cancer. NCT03837600. https://clinicaltrials.gov/ct2/show/NCT03837600?term=Cyclodextrin+pulmonary&draw=2&rank=2
114. Topotecan hydrochloride or cyclodextrin-based polymer-camptothecin CRLX101 in treating patients with recurrent small cell lung cancer. NCT01803269. https://clinicaltrials.gov/ct2/show/NCT01803269?term=Cyclodextrin+pulmonary&draw=2&rank=1
115. Zhang YM, Liu YH, Liu Y. Cyclodextrin-based multistimuli-responsive supramolecular assemblies and their biological functions. Adv Mater. 2020;32(3):18.

TPGS Functionalized Carriers: An Emerging Approach for Pulmonary Drug Delivery

9

Piyush Pradeep Mehta and Vividha Dhapte-Pawar

Abstract

Tocophersolan (D-α-tocopheryl polyethylene glycol succinate or TPGS) is a non-ionic surfactant fabricated by esterifying Vitamin E succinate with polyethylene glycol. Owing to its amphipathic nature and unique physicochemical attributes, TPGS offers multiple advantages in drug delivery, such as improving solubility, permeability, and bioavailability as well as assisting in designing a range of vesicular, micellar, and novel multi-particulate drug formulations. Moreover, it possesses nutritional values and has been grouped as a safe excipient by US FDA and EMA for topical, parenteral, and nasal formulations. These unique attributes make TPGS more apt for the development of novel carriers for successful pulmonary drug delivery. The present chapter aims to summarize various TPGS-based pulmonary drug delivery systems. The introductory segment focuses on the pharmacotherapy of chronic respiratory disorders and the function of nutritional agents, i.e., TPGS in respiratory disease management. Subsequent section deals with physicochemical attributes and applications of TPGS in the pharmaceutical sector. The third section thoroughly describes the impact of TPGS-based biomaterials on aerosolization performance. The section

P. P. Mehta
Global Respiratory R&D, Cipla R&D Center, Mumbai, Maharashtra, India
e-mail: piyu053@gmail.com

V. Dhapte-Pawar (✉)
Department of Pharmaceutics, Poona College of Pharmacy, Bharati Vidyapeeth (Deemed to be University), Pune, Maharashtra, India
e-mail: dhapte.vv@bharatividyapeeth.edu

P. P. Mehta, V. Dhapte-Pawar (eds.), *Pulmonary Drug Delivery Systems: Material and Technological Advances*,
https://doi.org/10.1007/978-981-99-1923-9_9

also highlights the biological consequences after intratracheal administration of TPGS-based delivery systems and in vivo anticancer activity using a lung cancer-bearing animal model. The last piece of the chapter highlights the opportunities and challenges allied to TPGS-based delivery cargos. Briefly, TPGS-based multimodal therapeutic delivery platforms offer promising outcomes in respiratory disease management.

Keywords

Tocophersolan · Vitamin E · Nonionic surfactant · P-glycoprotein inhibitor · Polymeric micelles · Lung cancer

9.1 Introduction

At the beginning of the twenty-first century, chronic respiratory diseases (CRD) remain among the most common life-threatening diseases, with more than 500 million cases globally, and figures are still escalating. CRD is an umbrella phrase mainly utilized to illustrate airway diseases, and two of the most common are asthma and chronic obstructive pulmonary disease (COPD) [1–4]. Asthma was mainly distinguished by recurrent attacks of breathlessness and wheezing, while COPD causes limitations in respiratory airflow. Asthma is a common disease among children [5]. Apart from these diseases, cystic fibrosis, lung cancer, tuberculosis, pneumonia, pulmonary hypertension, lower respiratory infections, and occupational pulmonary diseases are also causing significant impact on the healthcare system and call for urgent medical care with occasional hospitalization [6–10]. In addition, risk factors for CRDs contain physical inactivity, active, or passive tobacco smoking, indoor or outdoor air pollution, allergens and occupational chemicals, fumes, and dust [11]. Airways are also incessantly exposed to reactive oxygen (O^{2-} and H_2O_2) or nitrogen (NO and NO^{3-}) species and, therefore, highly susceptible to damage caused by oxidative stress [11, 12]. Further, several CRD patients are likely to elevate in coming years owing to air pollution, rising smoking habits, and aging populations in several countries [13].

According to the European Respiratory Society (ERS), the occurrence of CRDs and pathogenesis in most cases are also directly allied to diet and nutrition. In the recently published literature, ERS highlighted that a "balanced diet" with fruits, vegetables, and fish decreases the hazard of causing pulmonary diseases, mainly asthma, and COPD. Specifically, the Mediterranean diet, fruits and vegetables, and vitamins (A, D, and E) protect against wheezing or asthma, whereas whole-grain products, fibers, fruits, vegetables, and fish reduced the risk of newly detected COPD [14, 15]. Thus, CRDs prevention by the intake of dietary agents is measured to be a suitable strategy to diminish the rate of CRDs occurrence and the practice of routine pharmacotherapy and the avoidance of smoking [16, 17]. Interestingly, it is well apparent from the recent scientific literature that co-administration of pharmacotherapy with apt nutritional agents, such as d-α-tocopheryl polyethylene glycol-1000 succinate (TPGS) as a source of vitamin E could be a successful treatment approach for CRDs [18].

TPGS (Fig. 9.1) is a multifunctional, nonionic surfactant obtained from vitamin E [19]. Mainly, it is synthesized via esterification of vitamin E succinate with polyethylene glycol (PEG) 1000 [19]. It has a unique amphiphilic nature due to the lipophilic alkyl tail of vitamin E and hydrophilic PEG head [20]. The unique structural arrangement of TPGS provides many appealing physicochemical properties [21]. TPGS has been used in different capacities such as solubilizer [22], permeation enhancer [23], additive, absorption enhancer [24], emulsifier [25], stabilizer, plasticizer for film [26, 27], nutritional supplement [28] in several pharmaceutical dosage forms, cosmetics products [29], and dietary supplements [30]. TPGS as a biomaterial is explored for numerous delivery systems such as nanoparticles (NPs) [31], micelles [32], liposomes [33], nanocrystals [34], TPGS-drug conjugates [35], and TPGS-based prodrugs [30, 36]. Additionally, changes in the PEG molecular weight led to a significant impact on the physicochemical and biological properties such as hydrophile-lipophile balance, critical micelle concentration (CMC), P-glycoprotein (P-gp) inhibiting potential, and blood circulation time after intravenous administration of TPGS [37]. Briefly, such ease in modifications assists formulation scientists, and material chemists to develop robust drug delivery platforms for acceptable pharmacological effects [38]. Furthermore, TPGS is also approved by the US FDA as a pharmaceutically safe adjuvant for oral, topical, parenteral, nasal, and vaginal/rectal routes of administration [31]. Undoubtedly, TPGS has received growing consideration in novel drug delivery systems in the past few years. In this context, the

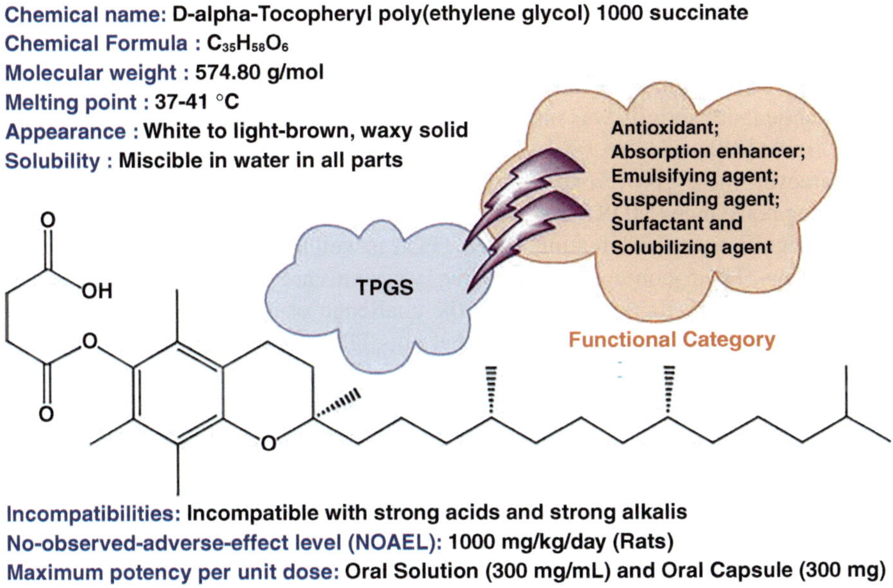

Fig. 9.1 Key physical, chemical, and biological features of TPGS

subsequent section will address the essential benefits of TPGS in the pulmonary drug delivery field to surmount the physicochemical, biological, and aerodynamic barriers.

9.2 TPGS in Pulmonary Drug Delivery

Several active researchers and research groups thoroughly explored TPGS in the pulmonary drug delivery system. It is an excellent solubilizer for hydrophobic anticancer, phytoconstituents, and antibiotic substances [39]. Its unique solubilization potential and suitable nano-carrier dilution stability further expanded the pulmonary applications of TPGS-based delivery systems [40]. In pulmonary drug delivery, TPGS can easily combine with other surfactants, polymers, block copolymers, and organic molecules to fabricate NPs, non-ionic vesicles, lipid vesicles, mixed micelles, microemulsion, microparticles (MPs), and mesophasic systems [31]. Lung is highly susceptible to oxidative damage and inflammatory actions caused by the external environment. Interestingly, in the pulmonary application, TPGS protects airways from various pollutants, peroxides, and free radicals generated in metabolic activities.

Additionally, it can work synergistically with other antioxidants and nutrients, such as vitamin C, selenium, and beta-carotene, to control peroxides or free radicals. Regarding its anti-inflammatory actions, TPGS effectively inhibits the enzyme lipoxygenase, which is mainly responsible for generating inflammation-causing leukotrienes [38]. Duhem and co-workers (2014) recently summarized that TPGS possess intrinsic anticancer potential. During anticancer activity assessment using human lung carcinoma cells rooted in nude mice, TPGS showed higher growth inhibition than tocopheryl succinate (TOS). TPGS showed more potent ROS generation, apoptosis, and growth inhibition than TOS even at a similar uptake rate within cells during in vitro analysis. This notable improvement in TPGS anticancer activity was mainly attributed to PEG interaction with cell membrane lipids [38].

Moreover, the P-gp overexpression is one of the active mechanisms involved in multidrug resistance (MDR) lung cancer treatment. Thus P-gp is a crucial target of clinical significance for defeating MDR. TPGS is well known for its P-gp inhibition action. Thus, TPGS-conjugated drug delivery systems are judiciously used as a platform in pulmonary therapy to address the challenge of MDR. Additionally, published recent scientific data showed that the TPGS-based novel drug delivery systems potentially reduced the dosage and frequency of the chemotherapeutic agents and remarkably controlled the undesirable adverse effects of long-term therapy with the enhancement of anticancer drugs within tumor cells [41]. Figure 9.2 showed vital applications of the TPGS in pulmonary drug delivery. This section will mainly focus on the potential use of the TPGS in pulmonary delivery systems as a drug carrier, formulation stabilizers, and adjuvants to improve the aerodynamic potential of drug moiety.

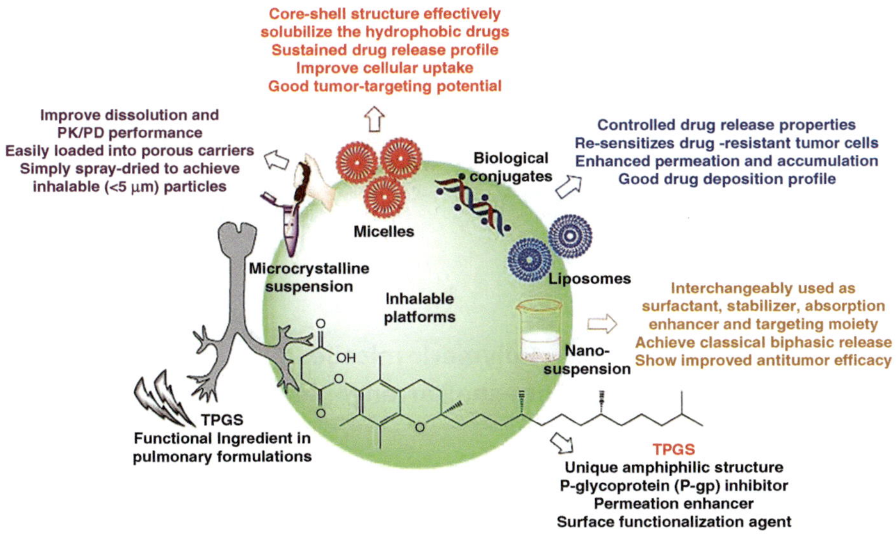

Fig. 9.2 TPGS-based promising inhalable platforms

9.2.1 TPGS Functionalized Polymeric Micelles

Polymeric micelles have achieved considerable attention in the last few years as a multifunctional nanotechnology-based drug delivery system. Polymeric micelles are characterized by a core-shell arrangement created from self-assembly of amphiphilic block copolymers in an aqueous solution. Polymeric micelles are shaped from self-aggregation of amphiphilic polymers with the hydrophobic part of the polymer on the inside (core) and hydrophilic on the outside (shell) part (Fig. 9.3) [42, 43]. As a result of this core-shell arrangement, polymeric micelles offer two main drug delivery benefits; (i) the hydrophobic core act as a solubilization depot for poorly water-soluble drugs and (ii) the hydrophilic shell offers protection from opsonin adsorption, which help to improve blood circulation time and bioavailability [42, 43]. The nano size, good solubilization potential and, the ease of formulation, make polymeric micelles fascinating carriers for different administration routes. They can produce a sustained and targeted drug release and regulate drug pharmacokinetic profile, with a reduction in adverse effects. Polymeric micelles have shown superior pharmacokinetic and pharmacodynamic outcomes in preclinical animal models and improved efficiency with a satisfactory safety profile for drugs [44]. Several polymeric micelles have reached the various clinical phases and are either in clinical trials or are approved for human use. The most studied administration route for polymeric micelles is the intravenous (injection/infusion) for cancer targeting followed by oral, and local (nasal, ocular, buccal) administration [45]. Similarly, various active material scientists and formulators were developed

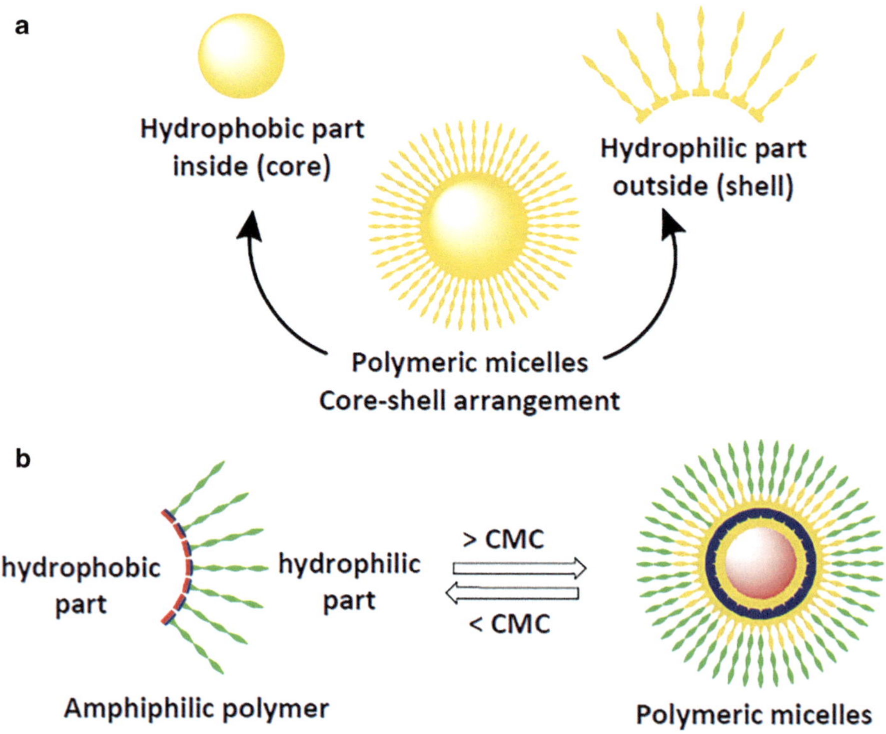

Fig. 9.3 Graphical illustration of polymeric micelles core-shell arrangement (**a**) and formation of polymeric micelles using amphiphilic polymers (**b**)

polymeric micelles using TPGS for pulmonary drug delivery. Consequently, some interesting TPGS-based polymeric micelles case studies are summarized below.

Paclitaxel (PTX) is a tetracyclic diterpenoid obtained from the bark of the Pacific yew tree, *Taxus brevifolia*. PTX is an important first-line chemotherapeutic agent which acts as inhibitor of cellular mitosis. It has proven efficiency against multiple cancers including prostate, ovarian, breast, and lung cancers. However, the clinical application of PTX is mainly restricted by its poor aqueous solubility, high lipophilicity (log P 3), and nonspecific pharmacokinetics in the systemic circulation. In this regard, nanostructured carriers were developed for successful delivery of PTX. Particularly, Zhang and Zhang (2015) formulated an amphiphilic copolymer monomethoxy PEG-poly(caprolactone) (MPEG-PCL) and chemically conjugated with TPGS carrier to shape polymeric micelles for pulmonary delivery of PTX [MPT-PTX micelles]. Hydrophobic PTX was loaded into PCL core while PEG shell efficiently avoids the nanocarriers from adsorption of protein and identification by the reticuloendothelial systems (RES) and results in prolonged blood circulation. TPGS was used to surpass the MDR by inhibiting P-gp overexpression. Uniform spherical-shaped MPT-PTX micelles showed particle size (110 nm) and zeta potential (26 mV) appropriate for dispersion stability and long in vivo circulation.

MPT-PTX micelles displayed a pH-sensitive sustained release profile, i.e., 50 and 36% of PTX release in pH 5 and 7.4, respectively, within 48 h. During cell uptake study using lung carcinoma epithelial (A549) cells, MPT-PTX micelles showed twofold higher uptake than PTX alone due to effective endocytosis progression. Additionally, MPT-PTX micelles effectively seized the human lung adenocarcinoma cell line (A549) at the G2/M phase and demonstrated a marked (20%) increase in the sub-G1 cell population. This indicated that PTX released from MPT-PTX micelles induced a mitotic arrest. In an in vivo pharmacokinetic study using Sprague-Dawley rats after intravenous administration (10 mg/kg), MPT-PTX micelles showed a marked fourfold improvement in area under curve (AUC) with prolonged blood circulation up to 24 h than PTX alone. Moreover, in the A549-based xenograft tumor model MPT-PTX micelles exhibited a marked two- and a fourfold reduction in tumor volume compared to PTX alone and saline group, respectively. Amphiphilic MPEG-PCL-TPGS block copolymer effectively shapes micelles by undergoing a spontaneous self-assembly process and three monomeric units of the copolymer (PEG, PCL, and TPGS) executed stipulated functions needed for lung cancer therapy. Briefly, experimental data suggest that MPT-PTX micelles efficiently inhibit tumor growth without causing any systemic adverse event [46].

Crizotinib (CZT) is an aminopyridine-based tyrosine kinase receptor inhibitor utilized for advanced non-small cell lung cancer therapy. However, CZT is not totally effective after frequent administrations since the patients become resistant to this drug. Furthermore, it can be anticipated that free CZT being a lipophilic molecule will have low aqueous solubility and poor blood circulation time. In this context, the delivery of CZT in nanosized carriers, i.e., amphiphilic polymeric micelles arise as a better solution to deliver CZT to the lung cancer tissues. Particularly, polylactic acid (PLA) and TPGS-based polymeric micelles were formulated by nanoprecipitation technique for pulmonary delivery of CZT. PLA-TPGS showed a CMC of 25 μg/mL. PLA forms the interior and carries the CZT, whereas TPGS exists on the periphery of the micelles. PLA-TPGS CZT micelles showed a particle size of 162 nm and displayed sustained release properties (96 h) in phosphate buffer saline (PBS). In addition, the absence of initial burst release verified the successful encapsulation of CZT within the PLA core of the micellar structure. In biocompatibility analysis using NCIH3122 lung cancer cells, PLA-TPGS polymeric backbone displayed satisfactory biocompatibility with cell viability of 90% at 100 μg/mL. During morphological imaging and Live/Dead assay, PLA-TPGS CZT micelles demonstrated better anticancer efficacy. Additionally, PLA-TPGS CZT micelles verified presence within the cytoplasmic region due to typical endocytosis-driven cellular uptake. Moreover, in annexin V/PI staining-based apoptosis assay, PLA-TPGS CZT micelles displayed a remarkable 1.6-fold improvement in apoptosis compared to CZT alone. Taken together, TPGS-modified PLA micelles present a potential pulmonary drug delivery system to enhance the chemotherapeutic efficiency of CZT in lung cancer treatment [47].

Docetaxel (DTX) is an antineoplastic agent that has a unique mechanism of action as an inhibitor of cellular mitosis. It plays a key role in the treatment of solid tumors including lung and breast cancer. Chemically DTX is a tetracyclic diterpenoid and semisynthetic side-chain analogue of PTX. To surmount the poor aqueous solubility of DTX (0.274 mg/L) and to decrease the systemic toxicity, nanostructured formulations such as polymeric micelles can be developed. But there is an issue that needs to be resolved to meet its successful clinical application, i.e., targeted delivery of DTX to lung cancer cells. The transferrin (TfR) receptor expression on cancer cells is high (approx. 100-fold) compared to the average expression in normal cells because of their quick proliferation rate and high iron requirement [48]. Additionally, a recent study showed that among the primary lung cancer cases, 19% of adenocarcinomas presented positive outcomes for TfR receptor-targeted delivery of micelles for lung cancer treatment [49]. Therefore, Singh et al. (2015) developed TfR-functionalized PLA-TPGS polymeric micelles for targeted pulmonary delivery of DTX. PLA-TPGS DTX micelles showed particle size, zeta potential, and entrapment efficacy of 135.3 nm, −8.64 mV, and 71.14%, respectively, while TfR functionalized PLA-TPGS DTX micelles showed particle size, zeta potential, and entrapment efficacy of 184.8 nm, 2.88 mV, and 86.63%, respectively. During the in vitro release study, TfR-PLA-TPGS DTX micelles displayed a sustained release profile for 72 h without an initial burst release compared to Docel™. TfR-PLA-TPGS DTX micelles sustained release property were mainly attributed to the micellar arrangement and strong interaction between micelles and DTX particles. During an in vitro cytotoxicity study using A549 cells, TfR-PLA-TPGS DTX micelles exhibited a significantly 72.98 and 8.12-fold lower half-maximal inhibitory concentration (IC_{50}) value compared to Docel™ and PLA-TPGS DTX micelles, respectively. The higher cytotoxicity of TfR-PLA-TPGS DTX micelles was owing to the transferrin receptor refereed endocytosis mechanism. Briefly, TPGS-based receptor-medicated targeted micelles exhibited better therapeutic potential in lung cancer treatment than commercial DTX injection [50].

Luteolin (LUT) is a tetrahydroxy flavone obtained from white mullein (*Verbascum lychnitis*). It showed several important therapeutic actions such as free radical scavenger, antioxidant, anti-inflammatory, and immune system modulator along with activity against various cancers. Still, LUT clinical application was mainly limited by its poor water solubility (0.14 g/L), high log P (2.4), low absorptive permeability, and high efflux rate via apical transporters such as multidrug resistance-associated proteins (MRPs) and P-gp. So, enhancement in the water solubility, membrane permeability, and reduction of the efflux phenomenon of LUT are vital for its effective clinical application. Accordingly, TPGS-based polymeric micelles are a right platform for effective LUT delivery. Thus, Yan et al. (2016) also formulated LUT-loaded long-circulating micelles using 1,2-distearoyl-snglycero-3-phosphoethanolamine--N-methoxy-PEG$_{-2000}$ (DSPE-PEG$_{2000}$) and TPGS for non-small cell lung cancer. Micelles containing LUT: DSPE-PEG$_{2000}$: TPGS (1: 4.5: 4.5) constituents showed satisfactory particle size (42.34 nm) and entrapment efficacy (81.73%). Spherical-shaped DSPE-PEG$_{2000}$-TPGS LUT micelles displayed sustained LUT release for more than 72 h without any initial burst release. During an in vitro cytotoxicity study using A549 cells, DSPE-PEG$_{2000}$-TPGS LUT micelles displayed a significantly threefold lower IC_{50} value compared to LUT alone. During in vivo antitumor testing using male BALB/c nude mice, DSPE-PEG$_{2000}$-TPGS LUT micelles showed superior antitumor action with tumor size less than 500 mm^3 and lowered toxic effects on normal tissues compared to LUT. Moreover, during in vivo imaging study, DSPE-PEG$_{2000}$-TPGS LUT micelles exhibited prolonged circulating (>24 h) properties with satisfactory targeted LUT delivery [51]. Briefly, TPGS-based polymeric micelles are a right platform for effective lung cancer therapy.

9.2.2 TPGS Decorated Mixed Micelles

Micelles are synthesized from single copolymers while mixed micelles are fabricated from a combination of two or more dissimilar block copolymers [52]. Mixed micelles are a practical and effective alternative to solve many of the issues allied with micelles without the need for complex synthetic schemes [53]. Mixed micelles synthesized from chemically diverse amphiphilic block copolymers can be used for various applications. In comparison to polymeric micelles, the mixed micellar showed various key advantages such as (i) enhance thermodynamic stabilities by lowering CMC; (ii) improve kinetic stabilities through stereo complexation, ionic interaction, hydrogen-bonding, and hydrophobic interactions; (iii) improved drug loading; (iv) allow better size control to avoid precipitation; and (v) ease of modifications such as cellular targeting and pH/temperature stimuli-responsive drug release [54]. Undoubtedly mixed micelle is a platform to widen the use of polymeric micelles as carriers in pulmonary drug delivery. Mixed micelles developed using TPGS for pulmonary drug delivery were summarized in the below section.

Baohuoside I (BI) is a well-known flavonoid obtained from *Herba epimedii*. It is well studied for several therapeutic activities such as antineoplastic, anti-inflammatory, and apoptosis stimulator. BI induces apoptosis in human non-small cell lung cancer cells through reactive oxygen species-mediated mitochondrial pathway. BI has low aqueous solubility, poor absorptive permeability, and high rate of efflux via apical efflux transporters such as MRPs and P-gp. These things limit the application of BI in the management of lung cancer therapy. Thus, BI-loaded TPGS, and non-ionic surfactant (solutol HS 15) mixed micelles were formulated using the thin-film hydration method. Well-defined, spherical-shaped BI: TPGS: solutol HS 15 (1:1.8:7.2) mixed micelles showed particle size and entrapment efficacy of 60.52 nm and 87.38%, respectively. BI mixed micelles showed typical sustained release properties with more than 87% BI release within 24 h. During an in vitro anticancer study using A549 cell lines, mixed micelles displayed more potent antiproliferative action at 2.60-fold lower IC_{50} as compared to BI alone. During an in vivo study using male BALB/c nude mice, BI mixed micelles displayed significantly superior antitumor action against cancer cells. During in vivo imaging study, mixed micelles exhibited efficient and targeted delivery using BI [55]. A similar research group also formulated BI mixed micelles using didecyldimethylammonium bromide (DDAB) and TPGS by thin-film hydration method to increase the aqueous solubility, membrane permeability and limit the efflux of BI. Additionally, hyaluronic acid (HA) was linked to the surface of the mixed micelle as a targeting ligand for the effective therapy of lung cancer. DDAB-TPGS BI mixed micelle showed particle size, zeta potential, and entrapment efficacy of 22.63 nm, 16.74 mV, and 88.36%, respectively. Uniformly distributed spherical HA-DDAB-TPGS BI mixed micelle displayed particle size, zeta potential, and entrapment efficacy of 31.42 nm, −1.25 mV, and 86.78%, respectively. During in vitro release study using PBS (pH 7.4), HA-DDAB-TPGS BI mixed micelle (82.43%) showed more sustained and superior BI release as compared to DDAB-TPGS BI mixed micelle (74.18%) for 24 h. During in vitro cytotoxicity study using A549 cells, HA-DDAB-TPGS BI showed a marked 4.51 and 1.89-fold lower IC_{50} values compared to BI alone and DDAB-TPGS BI, respectively, in 72 h incubation time. Moreover, HA-DDAB-TPGS BI showed a significant threefold reduction in tumor weight compared to BI alone at the same dose (10 mg/kg). Briefly, TPGS- and HA-based mixed micelles efficiently improved the BI lung tumor-targeting action [56]. Briefly, BI mixed micelles showed substantially improved antitumor activity, targeting efficacy, and can act as a potential delivery system to maximize the use of BI in lung cancer management.

Ginsenoside compound K (GCK) is a well-known phytoconstituent found in *Panax ginseng and Fusarium sacchari.* Chemically, GCK is a dammarane-type tetracyclic triterpenoid saponins and a key ingredient accountable for the pharmaceutical properties of ginseng. CGK is well known for its anticancer activities, and effectively inhibits pulmonary adenocarcinoma, gastric adenocarcinoma, proliferation of myeloid leukemia, and hepatoma cell lines [57]. However, its poor aqueous solubility, low permeability, and substantial P-gp efflux limit its pharmaceutical application. Therefore, Zhang et al. (2017) developed ascorbyl palmitate (AP) and TPGS-based mixed micelles to increase the solubility, targeting efficacy, and antitumor effect of GCK. Particularly, GCK-loaded AP-TPGS mixed micelles were formulated by thin-film hydration method in which amphiphilic AP has antioxidant activity with great self-aggregation potential while TPGS has good potential in improving the solubility and inhibiting P-gp efflux. Spherical shape, GCK mixed micelles showed particle size, zeta potential, and entrapment efficacy of 30.28 nm, −27.1 mV, and 91.34%, respectively. Mixed micelles showed a marked 41.56-fold improvement in GCK solubility. In the in vitro A549 cell model, GCK mixed micelles exhibited significant cell growth inhibition with G0/G1 phase cell cycle arrest, cell apoptosis, and adequate reserve of cell migration compared to GCK alone. In both in vivo near-infrared imaging and in vitro cellular fluorescence uptake, GCK mixed micelles showed enhanced cellular uptake and superior tumor-targeting properties. In in vivo study, GCK mixed micelles displayed a marked 7.25-fold increase in Bcl-2-associated X-protein/ B-cell lymphoma 2 (BAX/Bcl-2) ratio compared to GCK alone and confirmed that GCK mixed micelles promote cell apoptosis via improved accumulation in tumor. Furthermore, in the A549 xenograft mouse model, GCK mixed micelles exhibited a practical tumor-targeting efficacy primarily because of P-gp efflux inhibition activity of TPGS and apoptosis promotion [58]. Another research group also developed GCK containing TPGS, PEG, and PCL mixed micelles for successful management of lung cancer. Particularly, fixed molecular weight PEG-PCL (5000:5000) amphiphilic block copolymers and TPGS were used to shape the mixed micelles. The hydrophilic moiety of PEG prevented the absorption of micellar carriers by the reticuloendothelial system and enhanced drug stability and circulation while hydrophobic core of PCL regulates the water activity and relaxed the degradation rate of micelles. Spherical-shaped GCK mixed micelles formulated from thin-film hydration method showed particle size and entrapment efficiency of 53.07 nm and 94.60%, respectively, with 107.35-fold improvement in GCK water solubility compared to GCK. During the in vitro cytotoxicity study using A549 cells, GCK mixed micelles showed an IC_{50} value of 25.43 μg/mL. In an in vivo assessment, GCK mixed micelles (15 mg/kg) displayed superior tumor-targeting potential within 24 h with a 2.41-fold improved tumor inhibition rate than the positive control group [cisplatin (CIS); 2 mg/kg]. In vivo imaging study showed a longer maintenance time (>24 h) for mixed micelles in the tumor tissue and stronger fluorescence absorption at tumor tissue compared to other visceral organs confirmed the better-targeting ability of GCK mixed micelles [59]. However, further detailed clinical studies are needed to achieve bench-to-bedside research.

Oleanolic acid (OA) is a pentacyclic triterpenoid obtained from *Ophiopogon japonicus* and, Freziera. OA is one of the most prescribed triterpenoids in traditional Chinese medicine. OA has many pharmacological applications such as improvement of the body's defense systems, liver protection, and anti-inflammatory. OA is also well known for its anticancer activities. However, low aqueous solubility, poor permeability, and efflux have restricted its clinical application. Knowing this situation, Wu et al. (2016) thoughtfully developed OA-loaded TPGS and pluronic P105 mixed micelles (P105-TPGS OA) using the ethanol thin-film hydration method. In the present case, TPGS was used as an efflux inhibitor with a water-soluble amphiphilic molecular structure while P105 was used to capture hydrophobic OA and potent ability to sensitize multidrug-resistant tumors. Well-defined spherical-shaped P105-TPGS OA mixed micelles displayed particle size, zeta potential, and entrapment efficacy of 95.70 nm, −8.6 mV, and 93.60%, respectively. OA-mixed micelles showed typical sustained release properties for 120 h with only 40% OA release within the first 24 h in PBS (pH 7.4). During an in vitro cytotoxicity study using A549 cells, OA mixed micelles exhibited a 1.76-fold lower IC_{50} value compared to OA alone. Western blot and immuno-fluorescence investigation verified that the epithelial-mesenchymal transition was effectively reversed in OA mixed micelles treated A549 cells. Moreover, OA-mixed micelles mediate the anti-metastasis effect through extracellular signal-regulated kinase (ERK) signaling [60]. Briefly, P105-TPGS mixed micelles are a hopeful nano-drug delivery system for lung cancer treatment.

PTX is well-known anti-microtubule agent and is widely used for cancer therapy including non-small cell lung cancer (NSCLC). But the key disadvantage of PTX as a monotherapy is the quick development of drug resistance. PTX resistance is also credited to the intrinsically activated high levels of nuclear factor-κB (NF-κB). Additionally, superior levels of NF-κB protect the cancerous cells against programmed cell death in lung cancer. Thus, co-administration of PTX with a substance (e.g., parthenolide) that can suppress NF-κB activity can be used as an approach to surmount PTX-induced resistance. Parthenolide (PAR) acts by inhibiting the NF-κB DNA binding activity and avoiding the degradation of IκB-α and IκB-β (inhibitor of NF-κB) which is needed for the activation of NF-κB. However, co-administration of drugs is a very difficult task since each drug may have different solubility parameters. In this context, mixed micelles have received great attention due to their unique drug solubility and encapsulation properties. PTX and PAR (1:1) were entrapped into PEG_{2000}-DSPE micelles and mixed micelles were fabricated using TPGS by the film casting method. PTX-PAR mixed micelles showed particle size and entrapment efficacy of 12 nm and 97% (PTX) and 95% (PAR), respectively. In an in vitro release study using PBS (pH 7.4), PTX-PAR mixed micelles showed sustained release for both drugs almost for more than 72 h. During a cytotoxicity study using A549 cells and Taxol® resistant cell line (A549-T24), PTX-PAR mixed micelles showed lower IC_{50}, i.e., 64.15 and 128 nM, respectively. Moreover, mixed micelles displayed 79% and 70% cell death against A549 and A549-T24 cells, respectively. Furthermore, PEG_{2000}-DSPE and TPGS-based polymeric backbone effectively retained entrapment efficacy (>95%) and chemical stability during long-term 45 day storage. In summary, the increase in the sensitivity of the PTX-resistant cell line to the mixed micelles is attributed to a few key factors such as (i) co-administration of two drugs that act via two alternative pathways to avoid taxol resistance; (ii) presence of PEG shell on the surface favors direct interaction of the nano formulation with the cells ensuing in improved uptake via endocytosis; (iii) TPGS mixed micelles avoid P-gp mediated efflux; and (iv) inhibitory action of NF-κB by PAR. Briefly, the anticancer efficacy of PTX was considerably improved when co-administered with PAR using PEG_{2000}-DSPE and TPGS-based mixed micelles [61].

PTX mixed micelles were formulated using polylactide (PLA), PEG, and TPGS by the solvent evaporation method. Spherical-shaped PTX-mixed micelles showed entrapment efficacy, particle size, and zeta potential of 76.40%, 58.90 nm, and − 2.70 mV, respectively. PTX mixed micelles displayed a sustained release profile for 144 h, signifying the excellent stability needed for long-term circulation. PTX mixed micelles exhibited higher cytotoxicity against A549 cells with a 3.25-fold lower IC_{50} value than commercial PTX formulation. Additionally, in PTX-resistant A549 cells, PTX mixed micelles presented 57.31 and 3.84-fold lower IC_{50} values compared to PTX and PTX-TPGS formulation, respectively. During an in vivo study using the A549-based xenograft tumor model, PTX mixed micelles showed a marked threefold tumor growth inhibition compared to commercial PTX formulation. Moreover, in fluorescence resonance energy transfer (FRET) assessment, PTX mixed micelles showed presence within the tumor site to reveal anticancer activity [62]. All experimental results recommended that PTX mixed micelles act as a potential drug delivery platform for improved lung cancer therapy.

Piperine (PIP) is an alkaloid obtained from the roots and fruits of *Piperine nigrum*. PIP is well studied for various biological activities such as anti-inflammatory, anti-asthmatic, antimycobacterial, antioxidant, and immunomodulatory with intrinsic action against cancer cells. PIP is a potent inhibitor of P-gp and therefore reverses MDR in cancer cells. It has an important action on the drug metabolizing enzymes and acts as bioavailability enhancer for numerous chemotherapeutic molecules [63]. However, its low water solubility (0.04 mg/mL), poor stability, and quick metabolization have delimited its clinical use. Thus, PIP-loaded soluplus®-TPGS (4:1) mixed micelles were formulated using a thin-film hydration method. Uniform spherical shape mixed micelles showed particle size, zeta potential, and entrapment efficacy of 61.90 nm, −1.16 mV, and 90.90%, respectively. Lyophilized PIP mixed micelles exhibited good reconstitution in water with an accepted appearance. After 5 to 25-fold dilution, PIP mixed micelles retained particle size without any change in physicochemical properties. PIP mixed micelles showed sustained release behavior for 72 h compared to PIP alone. In cytotoxicity and cellular uptake study, PIP mixed micelles displayed better cytotoxicity potential with lower IC_{50} value and improved antitumor efficacy compared to physical mixture and PIP alone against A549 cells. During the pharmacokinetics study in male Sprague-Dawley rats after intravenous injection, PIP mixed micelles showed a marked 2.56 and 1.2-fold improvement in AUC and MRT compared to PIP alone at the same dose (3.2 mg/kg) [64]. In summary, the advantages of PIP mixed micelles contain prolonged circulation time, enhanced systemic stability, high cellular uptake, and superior anticancer effect.

Hinokiflavone (HF) is a biflavonoid obtained from *Platycladus orientalis* and *Selaginella tamariscina*. Chemically it is substituted apigenin, i.e., diflavonyl ether and acquires various therapeutic actions such as neuroprotective, antineoplastic, and antioxidant. However, its low aqueous solubility, poor stability, and limited bioavailability have restricted its clinical application. Thus, HF-loaded mixed micelles were formulated using soluplus (amphiphilic graft copolymer), TPGS, and dequalinium (DEQ; synthetic cationic surfactant) by thin-film hydration technique to enhance the solubilization efficacy and stability of HF. Soluplus and TPGS were used to improve the solution stability and inhibit the P-gp efflux, respectively, while double positively charged DEQ is used to improve the electrostatic interaction between the negatively charged membrane of cancer cells and micelles thus elevating the absorption of the mixed micelles by cancer cells. Additionally, due to double positive charge DEQ is selectively aggregated in the mitochondria of tumor cells to exert therapeutic action. HF: DEQ and TPGS: soluplus ratio was kept 1:1 and 1:4 to obtain small particle size with high encapsulation efficacy. HF mixed micelles showed particle size, zeta potential, and encapsulation efficacy of 67.83 nm, 2.34 mV, and 89.92%, respectively. During cytotoxicity study using CCK-8 assay, HF hybrid micelles showed a 2.47-fold lower IC_{50} value than HF alone. In mitochondrial targeting using ROS assay, HF hybrid micelles exhibited significantly improved ROS levels compared to HF alone. Additionally, in the mitochondrial membrane potential assay using the cationic lipophilic fluorochrome JC-1 method, HF hybrid micelles showed a noticeable 2.08-fold superior reduction in mitochondrial membrane potential compared to HF alone. In an in vivo antitumor study using A549-xenografted nude mice, HF hybrid micelles (80 mg/kg) displayed a better tumor inhibition rate (65%) as compared to HF alone (46%). HF hybrid micelles displayed superior antitumor activity by upregulating the Bax/Bcl-2 ratio and expression of Caspase-3 [65]. All these experimental outcomes showed that the novel DEQ containing HF mixed micelles had both strong tumor suppressor activity and little systemic toxicity.

Crizotinib (CRI) is an oral aminopyridine-based US FDA-approved drug for advanced NSCLC. CRI inhibits the anaplastic lymphoma kinase (ALK), hepatocyte growth factor receptor (c-MET) tyrosine kinases, and induces apoptosis via Caspase-3 activation. However, CRI is not fully effective after several administrations since patients develop quick resistance to CRI. Multidrug combination is a valuable option to surpass patient resistance to CRI. Thus, combination of palbociclib (PAL) and sildenafil (SIL) with CRI is explored to improved cytotoxic effect in cancer cells and regulate CRI resistance. Palbociclib (PAL) is a highly selective cyclin-dependent kinase 4/6 (CDK 4/6) inhibitor, that avoids pRb phosphorylation, therefore, resulting in G1 cell cycle arrest. Whereas sildenafil (SIL) is a phosphodiesterase 5 (PDE5) inhibitor and restrains several types of ATP-binding cassette drug-efflux transporters related to cancer MDR. Additionally, SIL combination with anticancer agents is beneficial as it can improve drug accumulation in cancer cells and upsurge the cytotoxicity of chemotherapeutics. Co-delivery of different chemotherapeutic agents using mixed micelles delivery system is a well-established approach. Thus, de Melo-Diogo et al. logically developed a TPGS-PLA mixed micelles system for simultaneous delivery of CRI, PAL, and SIL for effective treatment of advanced NSCLC. TPGS-PLA diblock copolymer was fabricated using the ring-opening polymerization method whereas triple drug loaded TPGS-PLA mixed micelles were prepared by solvent displacement technique. Blank TPGS-PLA mixed micelles showed a suitable mean particle size (172.2 nm) for therapeutic applications while small spherical-shaped triple drug-loaded mixed micelles showed a slight narrow mean particle size of 158.30 nm owing to stronger hydrophobic interactions within the interior of the micelle. TPGS-PLA mixed micelles showed satisfactory >60% (CRI) and > 80% (PAL and SIL) drug loading efficacy. During the in vitro release study, PBS (pH 7.4) mixed micelles exhibited a slightly faster release within the initial 12 h followed by a plateau stage and sustained release profile for 120 h. In the initial 12 h, PAL showed higher (60%) release as compared to CRI (30%) and SIL (40%) from micellar carriers. CRI-PAL-SIL mixed micelles (0.2 mg/mL) exhibited an improved cytotoxic effect against A549 cells as compared to a single (CRI) or dual (CRI-PAL) micelles. CRI-PAL dual combination displayed good cytotoxic action, however, cytotoxic effect of CRI-PAL-SIL combination exhibited a significant 1.78-fold decrease in A549 cells viability, indicating the active role of SIL in lung cancer therapy. Briefly, present investigation showed that the CRI-PAL-SIL combination encapsulated in TPGS-PLA mixed micelles accomplished a greater therapeutic outcome in lung cancer cells, emphasizing its potential in the treatment of NSCLC. Moreover, multidrug-loaded mixed micellar carriers can be further improved by targeting ligands specific for NSCLC to enhance selectivity toward lung cancer cells [66].

CIS is the platinum coordination complex and an alkylating agent used for the treatment of various solid tumors such as bladder, ovarian, and testicular cancers. Additionally, CIS is used as a first-line chemotherapeutic agent against head-and-neck, cervix, stomach, colon, and lung cancer. However, the therapeutic application of CIS is limited by its low bioavailability, unwanted side effects (such as chronic nephrotoxicity, peripheral neurotoxicity, and ototoxicity), and by the capacity of certain tumors to develop CIS resistance. Additionally, CIS is a small inorganic low molecular weight (300 g/mol) molecule and thus the retention time in the tumor is very short (i.e., CIS quickly enters into blood circulation) and hence the therapeutic outcome is inadequate [67]. The selective delivery of CIS to tumor cells using nanocarriers can substantially lessen the CIS chronic toxicity issue and thereby improve therapeutic efficiency. Accordingly, researchers focused on the development of TPGS hybrid micellar nanocarriers to increase therapeutic effects and decrease adverse effects linked with CIS delivery. Initially, polymeric CIS prodrug was synthesized using a one-step condensation polymerization method by diammine dichloro disuccinate platinum and ortho ester monomer. The CIS ester prodrug and TPGS (2:1) were co-assembled to form pH/redox dual-sensitive hybrid micelles. Spherical-shaped hybrid micelles showed a particle size of 38 nm with good physical stability in PBS for 5 days. During the in vitro release study, hybrid micelles showed 53.90 and 80.60% CIS release at pH 5.0 and 6.5 in 72 h, respectively, while the CIS release quantity was greatly inhibited at pH 7.4 (7.6%). The in vitro release study indicated that hybrid micelles were sensitive to mildly acidic conditions but remain stable in the neutral condition (pH 7.4). In an in vitro cytotoxicity study using A549 cells, hybrid micelles demonstrated significantly superior cytotoxicity of CIS due to TPGS-driven inhibition of efflux pumps. Hybrid micelles showed a 1.52-fold lower A549 cell viability compared to CIS alone. Moreover, during the in vitro growth inhibition rate using A549 multicellular spheroids, hybrid micelles displayed a noticeable 12-fold growth inhibition compared to CIS alone. Summing up, the outcomes attained in the present investigation evidently showed the potential of dual-sensitive hybrid micelles for pulmonary delivery of CIS [68].

The alpha V (αv) integrin (particularly αv-β3) is greatly expressed on tumor cells, tumor-associated blood vessels, and endothelial cells, however, weakly expressed in normal tissues thus making it a promising target for antitumor drug delivery based on selective toxicity [69]. The tripeptide arginine-glycine-aspartic sequence (RGD) showed good binding affinity toward integrin αv-β3 [70]. Additionally, in most of the cases, linear peptides undergo slow enzymatical degradation while small cyclic peptides exhibited resistance to enzymatic degradation and displayed long-term stability. Moreover, recent scientific data showed that nine-unit cyclic tumor-homing peptides with RGD sequence (iRGD) presented superior tumor-targeting activity because of their effective tumor penetration potential while traditional RGD peptides only carried the actives to the blood vessels. Particularly, iRGD acts via three-step progression, i.e., the RGD motif binds to αv integrins on tumor endothelium, induced a proteolytic cleavage and exposes a binding motif for neuropilin-1, which facilitates penetration into tissue and cells. Thus, conjugation to iRGD substantially enhanced the sensitivity of tumor-imaging vehicles and improved the activity of an antitumor molecule [71]. Thus, to improve the lung cancer tumor penetration and co-delivery of PTX/ survivin shRNA (shSur) Shen et al. (2013) fabricated iRGD peptide conjugated TPGS nanostructured carriers. The iRGD peptide was first linked with TPGS using acrylic chemistry and partially merged into pluronic P85-polyethyleneimine (P85-PEI)/TPGS complex NPs for co-delivering of PTX and shSur. Particularly, P85-PEI/TPGS-iRGD-TPGS loaded PTX/shSur mixed micelles were formulated by the thin-film hydration method. Uniform spherical mixed micelles showed particle size and positive surface charge of <160 nm and 30 mV, respectively. Mixed micelles displayed drug loading and entrapment efficacy 1.88% and 94.02%, respectively. During the in vitro release study, mixed micelles showed higher PTX (>85%) release within 72 h in the acidic medium. Mixed micelles displayed the lowest PTX release in the simulated extracellular environment as compared to the acidic medium. During an in vivo study using male Balb/c nude mice, mixed micelles considerably improved the accumulation of PTX/ shSur within the tumor environment with subsequent inhibition of survivin expression that produced potent cell apoptosis within the tumor milieu. Moreover, mixed micelles showed an apparent sixfold reduction in tumor volume compared to commercial PTX formulation. In a summary, iRGD mediated PTX/shSur co-delivery system was found to be a very useful approach for lung cancer therapy [72].

9.2.3 TPGS Decorated Liposomes

In the growing field of nanocarriers, liposomes are one of the most investigated nanostructured carriers for several key applications such as drug delivery, targeting, theragnostic, imaging, and diagnostics [73]. Structurally, liposomes are sub-micron (50–500 nm) spherical phospholipid vesicles fabricated using one or more lipid bilayers [74]. On basis of particle size and lipid bilayers, liposomes are grouped as small unilamellar vesicles (20–100 nm), large unilamellar vesicles (100 nm −1 mm), giant unilamellar vesicles (>1 mm), and multilamellar vesicles or multi-vesicular vesicles [75]. Particle size and lipid layer arrangement of liposomes showed a significant impact on the pharmacokinetics, circulation half-life, and overall biological outcomes [74]. Different types of drug delivery systems such as polymeric NPs, micelles, solid lipid NPs, nanostructured lipid carriers, microspheres, polymer–drug conjugates, and macromolecules (e.g., dendrimers) and lipid vesicles (e.g., liposomes and pro-liposomes) are also well studied for pulmonary drug delivery of several therapeutics [75–77]. Furthermore, few active researchers intensively explored TPGS in developing liposomes for pulmonary drug delivery. Accordingly, TPGS decorated liposomes for pulmonary drug deliveries were summarized in the below section.

Various drug delivery systems such as micelles and NPs are well reported for pulmonary delivery of GCK. However, Yang et al. (2016) developed GCK-containing liposomes modified with TPGS to enhance the membrane permeability, solubility, and anticancer action of GCK. GCK-TPGS liposomes were formulated using the thin-film hydration method and lyophilized to obtain dry powder. GCK-TPGS liposomes developed with a 3:7 (TPGS: phospholipid) ratio showed satisfactory drug loading efficiency (98.4%) and unimodal particle size (119.3 nm). GCK-TPGS liposomes showed good improvement in solubility of GCK (1.0 mg/

mL) due to the increase in hydrophobic interactions between the phospholipid and aromatic ring of TPGS. Smooth surface and spherical GCK-TPGS liposomes displayed controlled release properties for 24 h (32.1%) without any initial burst release. In an in vitro cytotoxicity study using A549 cells, GCK-TPGS liposomes exhibited significantly 1.52-fold lower cytotoxicity than GCK alone. GCK-TPGS liposomes showed a better hyper-sensitizing effect on A549 cells with enhanced cellular uptake. Moreover, in the in vivo antitumor assay using male athymic nude mice, GCK-TPGS-liposomes (15 mg/kg) achieved 1.65-fold superior antitumor efficacy compared to GCK alone. In in vivo study, the fluorescence signal at the tumor site was evidently detected and maintained for 24 h which confirmed that TPGS-modified liposomes delivered and retained GCK competently at tumor site. In summary, combined effect of the improvement of solubility, absorption, passive targeting by liposome, and the greater sensitization of MDR tumor cells by TPGS to GCK are the main reasons for the significant repression of tumor growth for TPGS-modified GCK liposome [78].

Luteolin (LUT) is a tetrahydroxy flavone found in various Chinese medicinal plants such as *Verbascum lychnitis, Carex fraseriana, Herba Schizonepetae*, and honeysuckle flower. Both isolated LUT and extracts from LUT-rich plants showed various key therapeutic properties such as anti-inflammatory, antimicrobial, and antioxidant activities [79]. LUT also presented antitumor activity in various cancers including breast cancer, cervical cancer, gastric cancer, hepatoma, and NSCLC. The mechanism of LUT antitumor activity depends on its capacity to impede proliferation, stimulate apoptosis of cancer cells, and avoid angiogenesis and metastasis. However, low water solubility (0.85 mg/mL), high lipophilicity (2.53), large volume of distribution, high clearance, and enterohepatic recirculation restrict LUT clinical applications [80]. Moreover, LUT is a substrate of MRPs refereed via P-gp. Therefore, to effectively regulate the biophysical properties and improve therapeutic outcomes of LUT, Li et al. (2016) formulated TPGS containing liposomes. Mainly, LUT-TPGS liposomes were formulated using yolk lecithin PC-98 T with a phosphatidylcholine, cholesterol, and TPGS (63:31:6) by film-dispersion method. Spherical-shaped LUT-TPGS liposomes exhibited unimodal particle size (176.2 nm) with zeta potential (−23.55 mV) and satisfactory entrapment efficacy (71.55%). LUT-TPGS liposomes showed a sustained release profile with a cumulative release of 46.88% after 120 h. LUT-TPGS liposomes (10 μg/mL) displayed superior cellular uptake by endocytosis and apoptosis through upregulating Bax/Bcl-2 ratio in A549 cells compared to LUT solution. The intravenous administration of LUT-TPGS liposomes (10 mg/kg) to tumor-bearing nude mouse model showed a significant 2.34-fold reduction in tumor volume compared to LUT solution. During in vivo imaging study using A549 inoculated nude mice, after tail vein injection of LUT-TPGS liposomes, LUT was mainly concentrated at the tumor site, representing that TPGS-functionalized liposomes delivered LUT directly into tumor and are retained for a longer period at tumor site (>24 h). Improved antitumor activity of LUT-TPGS liposomes was mainly ascribed to receptor-mediated endocytosis and the EPR effect. Briefly, experimental outcomes indicate that TPGS-functionalized liposomes are an excellent nanocarriers for chemotherapy and can be effectively explored carefully for treatment of lung cancer [33].

Apigenin (APG) is a 4′,5,7-trihydroxyflavone obtained from various fruits and vegetables for example nuts, oranges, tea, celery, and parsley. APG is a bioactive dietary flavone that plays a substantial role in the prevention and treatment of numerous health problems. APG displays antitumor effect in lung, liver, breast, and colorectal cancer via inducing autophagy, apoptosis, cell cycle arrest, and suppressing stem cell-like properties [81]. However, the clinical application of APG is mainly constrained because of its poor biopharmaceutical properties, such as low water solubility (183 mg/L), high lipophilicity (3.02), and poor absorption. Additionally, like most chemotherapeutic agents, APG develops resistance during chronic treatment, which result in therapy failure. These problems can be surmounted by using synergistic anticancer drug combinations which lessen drug resistance, decrease dose needed, and improve antitumor action. Tyroservatide (YSV) is a bioactive tripeptide obtained from the spleen of pigs. YSV (367.40 Da) is used in several clinical studies as a non-cytotoxic anticancer drug. YSV affects integrin-focal adhesion kinase signal transduction and inhibit histone deacetylase activity. YSV considerably prevents the proliferation, invasion of tumor cells, adhesion, and hamper the expression of angiogenesis-related factors at the protein and mRNA level [82]. YSV also displays antitumor effects in lung cancer [83]. Therefore, cocktail therapy, i.e., APG-loaded TPGS liposomes with YSV was logically developed to reduce chemoresistance and improve biological outcomes. APG-loaded TPGS liposomes were formulated using the modified thin-film hydration method. Small, uniform, and spherical-shaped APG-TPGS liposomes showed particle size and zeta potential of 118.60 nm and − 6.8 mV, respectively. During cellular uptake using A549 cells, APG-TPGS liposomes showed better intracellular localization with fair distribution throughout the cytoplasm. In vivo imaging studies using A549-bearing male nude mice showed that liposomes distributed the APG directly into the tumor site and were retained for a longer period (>12 h). During, in vitro cytotoxicity and in vivo antitumor studies, cocktail therapy exhibited acceptable outcomes with >50% tumor inhibition as compared to monotherapy. Notably, compared to monotherapy, cocktail therapy (2.7 μg/mL apigenin and 270 μg/mL YSV) showed higher cytotoxicity, caused better G2 arrest with A549 cancer cell invasion at a lower dose. Cocktail therapy (APG-TPGS liposomes with YSV) revealed remarkably superior pro-apoptotic actions in ex vivo and in vivo investigations compared to APG and YSV either alone or in combination. Thus, cocktail treatment needed further detailed clinical investigation to strengthen the lung cancer treatment [84].

9.2.4 TPGS Functionalized Nanostructured Carriers

Polymeric NPs have received a good attention owing to their ability to control physiochemical attributes, drug release kinetics, and precisely control biodistribution. PTX-loaded polymeric NPs were fabricated from poly (lactic-co-glycolic acid) (PLGA), TPGS, and soybean lecithin using a modified nanoprecipitation method.

Particularly, TPGS-functionalized PLGA-loaded PTX NPs were formulated using the innovative inverse-phase nanoprecipitation system to treat lung cancer. In a novel inverse-phase nanoprecipitation system, TPGS containing aqueous phase was gradually dropped into the organic phase holding biodegradable polymer (PLGA) and oil phase containing hydrophobic PTX molecule. Well-defined, spherical-shaped polymeric NPs displayed narrow particle size (<150 nm), zeta potential (−20 mV), and satisfactory entrapment efficacy (80%). During the in vitro release study, polymeric NPs displayed a typical sustained release profile for 200 h using PBS. As observed during, in vitro intracellular uptake investigation using lung cancer cells (A549), polymeric NPs showed significantly increased intracellular uptake with a strong inhibitory effect against A549 cells compared to Taxol® injection (500 μg of PTX). Furthermore, during in vivo activity, using A549 xenograft model polymeric NPs showed selective accumulation of PTX within the tumor for better antitumor efficacy after intravenous administration (10 mg/kg). Taken together, TPGS decorated PLGA NPs appeared as a potential drug delivery system of PTX for lung cancer treatment [85].

Similarly, Zhao et al. (2013) developed PTX-loaded polymeric NPs using novel biodegradable diblock copolymer by the modified solvent evaporation method. Mainly, novel biodegradable diblock aliphatic polyesters copolymer was developed using polylactide (PLA), polyglycolide (PGA), PCL, and D-α-tocopheryl polyethylene glycol (2000) succinate ($TPGS_{2k}$) as drug carrier for lung cancer therapy. In copolymer, PCL, PGA, and $TPGS_{2k}$ act as permeability enhancer, regulate NPs degradation, and offer good antitumor effects, respectively. PTX polymeric NPs showed particle size, zeta potential, entrapment efficacy of 202.1 nm, −31.2 mV, and 96.5%, respectively. Uniform spherical-shaped smooth surface PTX polymeric NPs exhibited a sustained release profile for 30 days due to the swelling and degradation mechanism of the novel biodegradable diblock aliphatic polyesters copolymer matrix. PTX polymeric NPs showed 4.77-fold higher cytotoxic activity against A549 cells than commercial PTX formulation. Furthermore, in A549 lung cancer xenograft bearing, nude mice model, PTX polymeric NPs (10 mg/kg) showed marked 2- and fivefold tumor volume reduction compared to commercial PTX formulation and saline group, respectively, followed by intratumoral injection. In summary, the biodegradable diblock aliphatic polyester amphiphilic copolymer acts as a promising polymeric backbone for a PTX delivery vehicle in lung cancer treatment [86].

Sun et al. (2014) used a nanoprecipitation technique to fabricate PTX-loaded PLGA NPs. TPGS was explored as an emulsifier to ease NPs formation in an ethanol–water system. In the NPs formation process, the organic solution containing PTX and PLGA was gradually dispersed into the aqueous solution containing TPGS. During dispersion activity, organic phase was evaporated slowly, PTX was encircled by PLGA and precipitated to form NPs. TPGS was concurrently layered around PLGA polymeric core. NPs showed a uniform particle size of 100 nm, zeta potential (−22 mV), with high encapsulation efficacy of >80%. Spherical-shaped NPs showed a sustained release profile for 96 h. Primarily, NPs effectively prevented the quick release of PTX from NPs within the lymphatic circulation and thus improved the PTX concentration within the tumor milieu. In a biodistribution study using the A549 xenograft mice model after intravenous administration, NPs showed strong passive tumor-targeting potential owing to the EPR effect. Also, in A549 xenograft, NPs (10 mg/kg) illustrated marked tumor volume reduction compared to commercial PTX formulation followed by intravenous injection. Briefly, PLGA and TPGS imparted better PTX cell uptake, release properties, and particle stability to the developed nanostructure [87].

Mesoporous silica-based NPs (MSN) have drawn much attention in biomedical field due to several excellent material attributes such as chemical and mechanical stability, easy of functionalization, and good in vitro and in vivo biocompatibility. Additionally, various intrinsic structural features of MSN such as high pore volume, tunable pore size, high specific surface area assist to improve the drug loading capacity, encapsulation efficiency and impart capacity to incorporate range of endogenous and exogenous stimuli, such as pH, temperature, light, remote magnetic actuation, and enabling controlled release kinetics for a wide range of molecules. Accordingly, Cheng et al. (2017) explored MSN to fabricate pH-responsive doxorubicin (DOX) delivery to treat lung cancer effectively. The pH-responsive DOX particles were fabricated using three steps particle engineering progression, i.e., initially, DOX was loaded on MSN via aqueous diffusion method then polydopamine (PDA) layer was formed onto the surface of DOX-MSN through an oxidative polymerization scheme and finally, TPGS was introduced on NPs using Michael addition reaction. The PDA and TPGS coating contribute to introduce pH sensitivity and disable drug resistance, respectively. Developed NPs showed particle size, zeta potential, and loading capacity of 221.2 nm, 2.6 mV, and 10.15%, respectively. Additionally, during quantitative examination using thermal gravimetric analysis PDA-TPGS-MNS showed a marked 2.12 and 1.63-fold higher weight loss compared to MSN and PDA-MSN, respectively, and confirmed the successful conjugation of PDA and TPGS over MSN. Moreover, PDA-TPGS-MNS exhibited surface area, pore volume, and pore size of 470.23 m^2/g, 0.51 cm^3/g, and 2.31 nm, respectively. A marked 1.99 and 2.5-fold reduction in surface area, and pore volume of PDA-TPGS-MNS compared to MSN demonstrated that DOX effectively engaged the mesoporous channels of silica while PDA and TPGS efficiently stuff the entrance of the mesoporous channels. Well-ordered, porous, ellipsoidal shape particles showed a classical biphasic release pattern with the initial burst release of DOX (24 h) followed by a sustained release for more than 140 h (47.7%). When the pH values aligned in the range of 5.0–6.5 to simulate the milieu of endosomes/lysosomes, DOX-loaded PDA-MSN and TPGS-PDA-MSN showed 51.60 and 47.70% DOX release due to pH-sensitive nature of PDA. In in vivo biodistribution and pharmacokinetic studies, decorated DOX NPs showed satisfactory DOX deposition within the lung tissue with improved antitumor efficacy due to long-term circulation properties and higher DOX accumulation at the tumor site via the EPR effect. During histopathology assessment, NPs displayed large necrotic areas in tumor and cell death triggered by DOX which suggested the improved therapy efficacy. Undoubtedly, MNS routed drug delivery platform offers a great potential for the treatment of lung cancer [88].

In line with above-discussed investigation, Hou et al. (2015) also fabricated pH-sensitive antitumor DOX prodrug NPs to manage lung cancer effectively. The pH-sensitive and chemo-photodynamic combination therapy was formed using three steps particle engineering process, i.e., initially, cis-aconitic anhydride-modified DOX NPs (CAD) were developed using a simple precipitation method followed by lyophilization then TPGS-conjugated CAD NPs (TCAD) were developed using same process and lastly, photosensitizer chlorin e6 (Ce6) was loaded into TCAD using self-assembly process. Developed NPs showed entrapment efficacy, drug loading, and particle size of 85.25%, 14.89%, and 160 nm, respectively. Lyophilized NPs showed uniform spherical shape and monodispersed behavior with critical aggregation concentrations of 23.40 μg/mL. NPs showed accelerated DOX and Ce6 release under mildly acidic conditions (pH 5.5) owing to the rapid hydrolysis of the acid-sensitive amide linkage. During an in vitro cell uptake study using A549 cells, NPs exhibited a considerably superior uptake of DOX (1.44-fold) than the DOX alone due to the efficient endocytosis process. During an in vivo experiment using female BALB/c nude mice, NPs showed enhanced tumor-targeting potential due to the EPR effect than free Ce6, thus improving fluorescence imaging. Moreover, during chemo-photodynamic combination therapy, NPs (5.86 mg/kg; DOX and 4 mg/kg; Ce6) along with near-infrared laser irradiation showed superior apoptosis and necrosis of tumor cells using A549 cells and considerably higher tumor growth inhibition in A549 lung cancer-bearing mice model. The DOX NPs prodrug-based chemo-photodynamic combination therapy showed potential outcomes in lung cancer treatment. However, more detailed experimental efforts are needed to translate this high-performance NPs prodrug to clinical settings in near future [89].

9.2.5 Spray Dried TPGS Particles for Pulmonary Delivery

Usually, drugs (e.g., corticosteroids, antifungals, and oligopeptides) utilized in the treatment of pulmonary diseases display low water solubility, poor bioavailability that affects the drugs' mean retention time in the lung tissue and subsequently their therapeutic efficiency. Development of NPs is considered as one of the most prevalent methods to overcome the obstacle linked to the low water solubility and poor bioavailability. However, in most of the cases, NPs show low physical or mechanical stability and high possibility of exhalation before airway deposition making their use difficult in pulmonary drug delivery. Therefore, to surpass these limits, the controlled agglomeration of NPs to micro-sized particles was suggested in recently published literature. Accordingly, Malamatari et al. (2015) formulated micro-sized NPs agglomerates for pulmonary delivery of model drug indomethacin (IND). IND is poorly water-soluble (0.937 mg/L) molecule with Log P of 4.27. Initially, two different amphiphilic non-ionic copolymers, i.e., Pluronic® F127 and F68 as well as TPGS were used to develop stable nanosuspension using wet bead milling by a planetary micro mill. Subsequently, developed nanosuspensions were solidified by spray-drying method using matrix-forming agents (i.e., mannitol and leucine) to shape respirable micron-sized NPs agglomerates. NPs exhibited mean particle size in the range of 200–400 nm while morphology analysis of NPs agglomerates showed spherical and composite structure that contained micron-sized clusters of NPs. IND showed a significant 2.27, 2.08, and 15.01-fold improvement in water solubility using 0.5% w/v Pluronic® F127, F68, and TPGS stabilizer, respectively, compared to IND aqueous solution. Pluronic® F127, F68, and TPGS NP-agglomerates exhibited drug loading efficacy of 94.3, 89.4, and 101%, respectively. Pluronic® F127, F68, and TPGS reconstituted nanosuspension presented mean particle sizes of 422, 417, and 468 nm, respectively. Additionally, during thermal analysis absence of endothermic peak in spray-dried NP-agglomerates for the incorporated stabilizers specified the existence of molecular dispersion. Moreover, after spray drying, NP agglomerates retained the original polymorphic form, i.e., γ-IND. During aerodynamic analysis using Cyclohaler® at 60 L/min, NP-agglomerates showed fine particle fraction (FPF) in the following range Pluronic® F68 (61.90%) > TPGS (57%) > Pluronic® F127 (48.90%). The satisfactory aerodynamic outcome was mainly attributed to matrix-forming agents. More specifically, matrix-forming agent enhances the aerosolization outcome by creating a coating on the particle surface and protects the individual particles as well as avoids particle-particle fusion. Additionally, leucine assists to improve the powder dispersibility and reduce the cohesiveness of NPs agglomerates. Thus, there is reduction in the induction port and pre-separator deposition of IND. Particularly, Pluronic® F127 NPs agglomerates with matrix-forming agent showed a marked 2.33-fold reduction in induction port and pre-separator IND deposition compared to equivalent formulation without matrix former [90]. Briefly, NP-agglomerates not only improve solubility but also enhance the aerodynamic performance of drug substances.

As discussed above lung delivery of NPs remains a challenge as they are prone to be exhaled quickly and are highly cohesive, which makes them tricky to aerosolize. Thus, innovative approach of porous mannitol carrier to delivered NPs was carefully explored by Leung et al. (2017). Particularly, porous mannitol carrier was developed using the ultrasonic spray-freeze-drying method for pulmonary delivery of model drug cyclosporine A (CSA) NPs. CSA is a hydrophobic cyclic peptide with low water solubility (27 μg/mL), Log P of 2.92 and limited bioavailability. Initially, CSA nanosuspension was developed using TPGS stabilizers by anti-solvent precipitation method via a four-stream multi-inlet vortex mixer. Subsequently, mannitol was dissolved into CSA nanosuspensions and spray-freeze-dried to yield inhalable porous particles. During precipitation progression, the stabilizers adsorb to the particle surface to prevent CSA NPs agglomeration by repulsive entropic force because of TPGS PEG tail. CSA nanosuspension showed an average diameter and polydispersity index of 160 nm and 0.08, respectively. Morphology analysis showed less void space in spray-freeze-dried MPs compared to spherical and porous blank MPs. During dynamic vapor sorption study, spray-freeze-dried MPs displayed good physical stability over a wide range (0–90%) of relative humidity. In in vitro dissolution study using a flow-through dissolution apparatus with phosphate buffer (pH 7.4 and 0.25% w/v Tween 80) spray-freeze-dried CSA MPs showed complete dissolution within 60 min compared to CSA alone. During aerodynamic analysis using Osmohaler® at 100 L/min, mannitol carrier concentration and density significantly impacted FPF of CSA. The FPF of CSA decreased from 63 to 44% as the mannitol concentration increased from 1 to 5% w/w. Porosity of MPs was improved as the mannitol concentration reduced from 5 to 1% w/w. This resembles the aerosol outcome of spray-freeze-dried CSA MPs, i.e., % FPF improved with reducing mannitol content. Additionally, spray-freeze-dried CSA MPs were stable for long-term storage, i.e., 24 months at 20 °C. Formulations stabilized using TPGS offered better dose uniformity and dispersion, indicating TPGS as a better stabilizing agent. Briefly, the porous mannitol-loaded CSA nanosuspension showed superior aerosol performance and suitable dissolution profiles. However, additional biological studies are required to evaluate the clinical fate of the formulation [91].

Cisplatin (CIS) is the prototype platinum coordination complex grouped under alkylating agents and utilized intravenously in the management of several types of cancer such as ovarian, cervical, bladder, lung, head, neck, and testicular cancer. Despite its effective antitumor actions, clinical application of CIS is mainly restricted by its adverse effects such as dose-limiting nephrotoxicity, anemia, electrolyte disturbances, gastrointestinal reactions, myelosuppression, and ototoxicity. Additionally, CIS is mainly excreted through urine (>90%) and is collected at distal and proximal nephrons causing severe tubular damage and acute to chronic renal deficiency in most of the patients. Therefore, aerosolized CIS chemotherapy is a more logical approach to improve therapeutic efficacy by increasing CIS concentration at the lung tumor site while limiting total body exposure. Initially, bulk CIS was dispersed in isopropanol (5.0% w/v) and subjected to high-speed homogenization (24,000 rpm for 10 min) with subsequent processing via high-pressure homogenizer (8000 rpm) to obtain CIS microcrystals. After that, stabilizer TPGS (0.1% w/v) was solubilized in isopropyl alcohol and added into CIS microcrystalline suspension and subjected to spray drying to obtain respirable solid powder. Electrothermal atomic absorption spectrometry and thermal gravimetric analysis showed high CIS content (95.6%) and lower moisture content (0.08%), respectively. CIS microcrystals (2.6 μm) presented a marked 7.2-fold lower particle size compared to bulk CIS (18.9 μm) using laser diffractometer method while spray-dried particles dispersed under simulated breathing conditions (100 L/min airflow) showed a particle size of 6.8 μm with 79.7% undersized (< 5 μm) particles. Spherical-shaped spray-dried particles displayed an average agglomeration index of 2.3. During aerodynamic analysis using Axahaler® at 100 L/min, spray-dried particles showed better fine particle dose (FPD; 9.84 mg) compared to spray-dried particles prepared with triglyceride stabilizer (e.g., tristearin; 4.47 mg) and lipid stabilizer (DSPE-PEG$_{2000}$; 5.02 mg). In an in vitro drug release study by dissolution vessel containing impactor disk assembly (50 rpm), spray-dried particles showed controlled CIS release for 24 h (>90%) using 400 mL modified simulated lung fluid. Briefly, present investigation verified the combined use of high-speed homogenization, high-pressure homogenizer, and spray-drying method to fabricate dried CIS microcrystals for pulmonary delivery [92]. Same research group also extended their efforts to understand the in vitro and in vivo aerodynamic performance of CIS microcrystal. Particularly, spray-dried solid lipid MPs (SLM) containing CIS microcrystals were formulated using three different biocompatible matrices, i.e., (i) tristearin SLM, (ii) tristearin with TPGS SLM, and (iii) tristearin with DSPE-PEG$_{2000}$ SLM to modify the particle surface attributes. TPGS SLM and DSPE SLM showed CIS content of 48.9 and 49.9%, respectively. During aerodynamic assessment, TPGS SLM and DSPE SLM showed FPF of 37 and 50% with mass median aerodynamic diameter (MMAD) of 2.4 and 2.3 μm, respectively. Besides, fabricated SLMs were mixed with spray-dried mannitol and leucine (10:1) matrix (1:25) using 3D blending process (4 h) to attain an adequate content uniformity. During in-vitro emission efficiency analysis using DP-4 M dry powder insufflator™ TPGS SLM and DSPE SLM exhibited emission mass of 87.6 and 77.6% w/w with emission doses of 104.8 and 198.3%, respectively. Based on in vitro data tristearin SLM and TPGS SLM were picked for comparative clinical investigation. During in vivo study using mice after a single pulmonary insufflation using DP-4 M dry powder insufflator™ tristearin SLM and TPGS SLM displayed good CIS lung recovery of 33 and 30%, respectively. TPGS SLM displayed a marked 26.96 and 10.88-fold improvement in peak serum concentration (C_{max}) and AUC $_{(10–48\ min)}$, respectively, compared to intravenously administered CIS microcrystalline suspension at the same dose (1.25 mg/kg). Moreover, tristearin SLM and TPGS SLM exhibited 10 and 17.77-fold higher targeting efficacy, respectively, compared to intravenously administered CIS microcrystalline suspension. Briefly, TPGS SLM showed good lung retention and controlled release properties. However, systemic and local toxicity, with therapeutic efficacy against lung tumor model needs to be assessed in near future to validate the safety of CIS DPI [93].

Atorvastatin (ATV) is a synthetic statin frequently prescribed in the treatment of hypercholesterolemia. Chemically, AVT is a dihydroxy monocarboxylic acid and potent inhibitor of 3-hydroxy-3-methylglutaryl CoA (HMG-CoA) reductase (HMGCR). It is well documented that HMGCR is overexpressed in human lung adenocarcinoma tissues compared with normal tissues. Hence, the use of AVT in treatment of lung cancer is a fruitful approach. However, low water solubility, high lipophilicity (6.63), low bioavailability (12%), and high pre-systemic clearance (> 80%) restrict AVT clinical applications. In this view, Ishak and Osman (2015) utilized self-micro emulsifying drug delivery systems (SMEDDS) to address solubility and bioavailability-associated problems of AVT. Initially, SMEDDS was developed using isopropyl myristate (oil phase), biocompatible surfactants (lecithin and TPGS), and co-surfactant (ethanol) by aqueous titration method followed by magnetic stirring. AVT was added to the oil phase. The SMEDDS was screened by pseudo-ternary phase diagram and optimized microemulsion was acquired with lecithin/TPGS (1:3), lecithin/oil (1:1), and surfactant/co-surfactant (1:1) ratios. Moreover, a combination of dextran/PEG/leucine (1:1:2) and optimized SMEDDS was added into 25% hydroalcoholic solution by magnetic stirring and subsequently subjected to spray drying to obtain solid SMEDDS. Microemulsion showed pH, viscosity, and zeta potential of 6.48, 44.37 cps, and − 28.6 mV, respectively. Solidified SMEDDS showed good powder yield (89%) with an acceptable mean particle size (3.21 μm). Uniform, wrinkled-shaped solid SMEDDS showed a good respirable fraction (>70%) during twin-stage glass impinger analysis using Aerolizer® at 60 L/min. During an in vitro cytotoxicity study using A549 cells, solid SMEDDS showed a marked 7.22 and 1.90-fold decrease in IC_{50} values compared to free drug solution and microemulsion, respectively. Briefly, the delivery of AVT in presence of TPGS in solid SMEDDS greatly enhanced the aerodynamic properties and cytotoxic activity on lung cancer cells. However, the upcoming studies should focus on the in vivo antitumor model to judge safety and efficacy of this novel platform and affirm the capacity of this system in lung cancer therapy [94].

9.2.6 TPGS Decorated Nebulized Particles

A range of nanometer and micrometer-sized TPGS carrier systems, for example, NPs, MPs, liposomes, micelles, mixed micelles, emulsion, and prodrug have been thoroughly explored as potential pulmonary delivery systems. Thus, Laouini and co-workers aimed to compare the four different commonly used TPGS-based carrier systems to probe aerodynamic characteristics and the rate of TPGS reach at broncho-alveolar level. Particularly, four different carrier systems containing TPGS such as nano emulsion (5%), liposomes (0.1%), micelles (0.05%), and solid lipid particles (0.5%) were developed using four different membrane contactor such as shirasu-porous-glass (0.9 μm), microsieve (10 μm), microsieve (10 μm), and ceramic (10.2 μm) membrane, respectively. Nano emulsion (−16.5 mV), liposomes (−28.5 mV), micelles (−19.3 mV), and solid lipid particles showed mean particle sizes of 106, 96, 154, and 2510 nm, respectively. Additionally, nano emulsion, liposomes, and micelles displayed encapsulation efficiency of 99.7, 99.9, and 87.4%, respectively. During TPGS delivery rate assessment using electronic nebulizer eflow rapid (Pari, Germany) equipped with a breathing simulator presented highest delivery rate for nano emulsion (13.43 mg/mL) followed by solid lipid particles (1 mg/mL), liposomes (0.42 mg/mL), and micelles (0.23 mg/mL). Aerodynamic assessment of developed carriers was done by two different methods namely, laser diffraction analyzer and cascade impaction. Laser diffraction revealed the highest FPD for liposomes (71%) than nano emulsion (38%), micelles (29%), and solid lipid particles (19%) with lowest volume mean distribution for liposomes (3.8 μm) compared to nano emulsion (5.6 μm), micelles (5.8 μm), and solid lipid particles (7.7 μm). Likewise, cascade impaction exhibited highest FPD for liposomes (87%) followed by nano emulsion (82%), micelles (78%), and solid lipid particles (40%) with lowest MMAD for liposomes (1.7 μm) followed by nano emulsion (2.8 μm), micelles (3.2 μm), and solid lipid particles (6.1 μm). Both aerodynamic assessment methods presented that during the nebulization, the highest FPD and lowest MMAD values were attained for liposomes. Additionally, a mathematical model, i.e., the multiple-path particle dosimetry (MPPD) was used to predict the pulmonary deposition of each carrier. MPPD study showed highest exhalation rate (60%) for liposomes while lowest one (23.7%) for solid lipid particles. Likewise, nano emulsion and micelles presented exhaled doses of 44.9 and 40.6%, respectively. Moreover, nano emulsion and micelles disclosed inhaled doses of 47.1 and 51.6%, respectively. It is apparent from the present analysis that the appropriate formulation systems for TPGS carriers are micelles and nano emulsion. In summary, three aerodynamic assessment studies suggested that proper balance between used excipients, carrier mean size, and nebulization system is important to achieve productive aerosolization outcomes [95].

Coenzyme Q10 (CQ10) is a naturally occurring 1,4-benzoquinone and is known as ubiquinone. CQ10 is a key element in proton transfer and electron transport in the mitochondrial membranes. It is a vital natural antioxidant presents in the human body with highest concentrations in liver, kidney, heart, and pancreas. CQ10 has immunostimulatory and cardio-preventive effects. CQ10 is marketed in the USA as a dietary supplement for the management of cardiovascular ailments, for example, hypertension, angina pectoris, and congestive heart failure. Deficiency of this endogenous enzyme was observed in many patients suffering from cancer. Additionally, recently published scientific data advise that a low CQ10 level is well connected to the progression of lung cancer. Considering these biological consequences pulmonary delivery of CQ10 is one of the most rational approaches for treatment of lung cancer. However, owing to CQ10 tricky physicochemical attributes, i.e., high molecular weight (863.34 g/mol), low water solubility, high lipophilicity (Log P 10), photo-degradability, the formulation development of CQ10 for pulmonary delivery is very challenging. Therefore, the design of a stable colloidal suspension with the ability of producing a satisfactory respirable fraction upon actuation can act as a valuable approach in active management of lung cancer patients. With this objective Rossi et al. (2017) developed and characterized the colloidal suspension of CQ10 with various physicochemical and aerodynamic parameters. CQ10 nanosuspension was developed using three different surfactant stabilizers, viz. lecithin, PEG_{32} stearate, and TPGS. The total solid content (0.5% w/w) and ratio between CQ10 and surfactant stabilizers (1:1) was kept constant. Initially, alcoholic phase of CQ10 (10 mL/min) was added into the aqueous phase of surfactant stabilizers under continuous magnetic stirring (600 rpm at 50 °C) and the resultant system was subjected to high-pressure homogenizer (40 cycles at 1500 bar pressure) to obtained CQ10 nanosuspension. The PEG_{32} stearate (−20.2 mV), TPGS (−20.6 mV), and lecithin (−24.6 mV) CQ10 nanosuspension presented hydrodynamic diameter of 58.9, 37.8, and 46.6 nm, respectively, suitable for pulmonary delivery. PEG_{32} stearate and TPGS CQ10 nanosuspension displayed a globular shape with a spherical homogeneous lipid core walled by the stabilizers while lecithin CQ10 nanosuspension exhibited larger and more polydisperse closed bilayer vesicles. During aerodynamic analysis using LC Sprint nebulizer, nanosuspension showed drug delivery rate in the range of 0.23–0.25 mg/min. The nebulized 4 mL PEG_{32} stearate, TPGS and lecithin CQ10 nanosuspension showed respirable fraction of 70.6, 59.7, and 55.9%, respectively. Additionally, CQ10 nanosuspension demonstrated appropriate aerosol production and aerodynamic performance via vibrating mesh (eFlow® rapid) and ultrasound (AirProjet®) nebulizers. Moreover, CQ10 nanosuspension demonstrated stable properties for 90 days during stability testing. During in vitro cellular toxicity study using individual stabilizer (lecithin and PEG_{32} stearate) all three CQ10 nanosuspension presented no cytotoxicity against A549. However, TPGS alone showed damage to the plasma membrane since TPGS supports the TOS provoked apoptosis but A549 cell damage was notably reduced for TPGS CQ10 nanosuspension because of the strong protective (antioxidant) effect of CQ10 in the nano formulation [96].

Burkholderia cenocepacia is the high-risk pathogen responsible for cystic fibrosis, and filamenting temperature-sensitive mutant Z (FtsZ) is the essential protein involved in the bacterial cell division process. Recently, benzothiadiazole derivative (C109) showed promising activity against FtsZ protein, and thus Costabile et al. (2019) judiciously developed TPGS stabilized C109 nanocrystals embedded within hydroxypropyl-β-cyclodextrin dry powders for cystic fibrosis treatment. C109: TPGS (1:2 w/w) nanosuspension was developed using the nanoprecipitation method while dry powder was obtained using HP-β-CD (1% w/v) by freeze-drying method (C109-TPGS-CD). Elongated rod-shaped C109-TPGS-CD particles showed particle size and zeta potential of 842 nm and − 20.1 mV, respectively. In in vitro aerosolization study, C109-TPGS-CD, dry powder inhalation suspension showed a % FPF of 41.9%, 39.9%, and 47.7% using commercial nebulizers, i.e., Aeroneb® Go (Aerogen, Ireland), PARI TurboBOY® (PARI GmbH, Germany) and eFlow® rapid (PARI GmbH, Germany), respectively. In in vitro antimicrobial activity using *B. cenocepacia*, C109-TPGS-CD particles showed a marked 200-fold improvement in biofilm reduction compared to C109 alone. During cytotoxicity study using wild type (16HBE) and CF bronchial epithelial (CFBE41o), C109-TPGS-CD particles displayed no significant toxicity. In an in vivo antimicrobial investigation using *Galleria mellonella* infection model, C109-TPGS-CD particles (20 mg/kg) combined with piperacillin (100 mg/kg) showed a synergistic therapeutic effect against *B. cenocepacia* lung infections [97].

9.3 Future Outlook

We have thoroughly depicted the role of TPGS in designing and developing pulmonary drug delivery systems. It appears that its unique amphiphilic structure mainly governs all the attractive applications of TPGS in inhaled formulations. Table 9.1 shows various TPGS-based micro-and nano-carrier systems with their key findings. The research studies reviewed here explored TPGS as a surfactant, co-surfactant, emulsifier, stabilizer, solubilizer, absorption enhancer, permeation enhancer, and additive to fabricate a wide range of carrier systems for effective pulmonary drug delivery. TPGS was used in delivery systems to formulate polymeric micelles, mixed micelles, mesoporous NPs, microcrystal, SMEDDS, NPs agglomerates, nanosuspension, liposomes, nanocrystals, and prodrug to improve aerodynamic performance and therapeutic efficacy of drug substances. In a few investigations, TPGS was also explored as a coating agent, coupling moiety and surface functionalization agent to develop stimuli-responsive controlled release formulations.

Notably, researchers explored the TPGS hydrophilic PEG 1000 head-to-engineered micellar structures. TPGS-based micellar structures are competent

Table 9.1 TPGS-based micro- and nano-carrier systems with their key findings

Formulation	Particle properties	Outcome	References
MPEG-PCL-TPGS-PTX micelles	Spherical and positively charged micelles (26 mV) and 110 nm	Micelles indicated pH driven sustained release profile with fourfold improvement in AUC and a twofold reduction in tumor volume compared to PTX alone.	[46]
PLA-TPGS CZT micelles	162 nm	Micelles exhibited sustained release for 96 h and typical endocytosis-driven cellular uptake with a 1.6-fold improvement in apoptosis than CZT alone.	[47]
TfR functionalized PLA-TPGS DTX micelles	Positively charged micelles (2.88 mV) and 184.8 nm	Micelles revealed sustained release for 72 h due to the unique micellar arrangement with continued interaction between micelles and DTX particles. Higher cytotoxicity for A549 cells owing to transferrin receptor refereed endocytosis mechanism.	[50]
DSPE-PEG_{2000}-TPGS LUT micelles	Spherical shape and 42.34 nm	Micelles showed sustained release (72 h) with a significant threefold lower IC_{50} value as compared to LUT alone using A549 cells and prolonged in-vivo circulating (> 24 h) properties and tumor size less than 500 mm^3 using male BALB/c nude mice.	[51]
BI-TPGS- solutol HS 15 mixed micelles	Spherical shaped and 60.52 nm	Mixed micelles showed sustained release properties and 2.60-fold lower IC_{50} compared to BI alone during the anticancer study using A549 cell lines. In contrast, the in vivo study verified the superior antitumor action and targeted delivery potential.	[55]
DDAB-TPGS BI mixed micelle	Negatively charged (1.25 mV), spherically shaped, and 31.42 nm	Mixed micelles showed sustained release properties and 4.51-fold lower IC_{50} as compared to BI alone while a marked threefold reduction in tumor weight compared to BI alone at the same dose (10 mg/kg).	[56]

Table 9.1 (continued)

Formulation	Particle properties	Outcome	References
GCK-AP-TPGS mixed micelles	Negatively charged (27.1 mV) spherically shaped and 30.28 nm	Mixed micelles showed satisfactory cellular uptake using A549 cells and efficient tumor-targeting potential due to P-gp efflux inhibition activity using A549 xenograft mouse model.	[58]
GCK-TPGS-PEG/PCL mixed micelles	Spherical shaped and 53.07 nm	Mixed micelles showed an IC_{50} value of 25.43 µg/mL using A549 cells and a 2.41-fold improved tumor inhibition rate as compared to cisplatin (2 mg/kg)	[59]
P105-TPGS OA mixed micelles	Negatively charged (8.6 mV) spherically shaped and 95.70 nm	Mixed micelles showed sustained release properties (120 h) with 1.76-fold lower IC_{50} value as compared to OA alone using A549 cells and mediate anti-metastasis effect through ERK signaling	[60]
PTX/PAR (1:1) PEG_{2000}-DSPE and TPGS-based mixed micelles	Mean particle size of 12 nm and good entrapment efficacy of 97% (PTX) and 95% (PAR)	Mixed micelles indicated sustained release (72 h) for both drugs and 79% and 70% cell death against A549 and A549-T24 cells, respectively.	[61]
PTX- PEG/ PLA/ TPGS Mixed micelles	Negatively charged (2.70 mV) spherical and 58.90 nm	Mixed micelles displayed a sustained release profile for 144 h and a marked threefold tumor growth inhibition as compared to commercial PTX formulation using an A549 xenograft model.	[62]
PIP soluplus®/ TPGS mixed micelles	Negatively charged (1.16 mV) spherically shaped and 61.90 nm	Mixed micelles revealed better cytotoxicity potential with a lower IC_{50} value and improved antitumor efficacy as compared to PIP alone against A549 cells.	[64]
HF-DEQ/TPGS/ Soluplus hybrid micelles	Negatively charged (2.34 mV) spherically shaped and 67.83 nm	Mixed micelles showed a 1.41-fold higher tumor inhibition rate as compared to HF alone using A549-xenografted nude mice and superior antitumor activity due to upregulating the Bax/Bcl-2 ratio and expression of Caspase-3.	[65]

(continued)

Table 9.1 (continued)

Formulation	Particle properties	Outcome	References
CRI-PAL-SIL mixed micelles	Spherical shape and 158.30 nm	Micelles showed a sustained release profile (120 h) and an improved cytotoxic effect against A549 cells.	[66]
CIS ester prodrug/ TPGS hybrid micelles	Spherical shaped and 38 nm	Hybrid micelles showed pH-dependent release and an evident 12-fold tumor growth inhibition in A549 multicellular spheroids compared to cisplatin alone.	[68]
iRGD mediated PTX/ shSur mixed micelles	Positively charged (30 mV) spherical and < 160 nm	Mixed micelles presented higher PTX (> 85%) release in an acidic medium. They produced substantial cell apoptosis within the tumor milieu with an apparent sixfold reduction in tumor volume compared to commercial PTX formulation in male Balb/c nude mice.	[72]
GCK-TPGS liposomes	Smooth surface, spherical and 119.3 nm	Liposomes exposed controlled release properties with 1.65-fold superior antitumor efficacy compared to GCK alone using male athymic nude mice.	[78]
LUT-TPGS liposomes	Negatively charged (23.55 mV) spherically shaped and 176.2 nm	Liposomes presented a sustained release profile (120 h) with a significant 2.34-fold reduction in tumor volume compared to LUT solution using a tumor-bearing nude mouse model.	[33]
APG-TPGS liposomes with YSV	Negatively charged (6.8 mV) spherically shaped and 118.60 nm	Liposomes displayed acceptable outcomes with >50% tumor inhibition compared to monotherapy during the in vivo antitumor study.	[82]
TPGS-functionalized PTX PLGA NPs	Negatively charged (20 mV) spherical shaped and < 150 nm	Sustained release for 200 h using PBS and selective accumulation of PTX within the tumor as compared to intravenous administration (10 mg/kg) using an A549 xenograft model.	[85]
PTX NPs with $TPGS_{2k}$ diblock aliphatic polyesters copolymer	Negatively charged (31.2 mV) spherical-shaped smooth and 202 nm	Sustained release for 30 days and marked twofold tumor volume reduction as compared to commercial PTX formulation followed by intratumoral injection using an A549 xenograft model	[86]

Table 9.1 (continued)

Formulation	Particle properties	Outcome	References
PTX-TPGS/PLGA NPs	Negatively charged (22 mV) uniform, spherical shaped and 100 nm	Showed sustained release profile for 96 h with improved PTX concentration within the tumor milieu with strong passive tumor-targeting potential owing to the EPR effect using an A549 xenograft model	[87]
DOX TPGS-functionalized poly-dopamine-modified mesoporous silica NPs	Positively charged (2.6 mV) well-ordered, porous, ellipsoidal shape, and 221.2 nm	DOX NPs showed a classical biphasic release pattern and satisfactory DOX deposition within the lung tissue. It improved antitumor efficacy due to the EPR effect	[88]
TPGS-conjugated cis-aconitic anhydride-modified DOX prodrug NPs linked with photosensitizer agent (chlorine 6)	Acid-sensitive, uniform spherical shape NPs (160 nm)	NPs showed accelerated DOX, and Ce6 release in acidic conditions (pH 5.5), and chemo-photodynamic combination therapy displayed superior apoptosis and tumor growth inhibition in the A549 lung cancer-bearing mice model.	[89]
IND TPGS nano-agglomerates	Spherical shape and 468 nm	Agglomerate powder showed satisfactory FPF (57%) using Cyclohaler® at 60 L/min	[90]
TPGS CSA porous particles	Stable, porous, and spherical particles	Spray-freeze-dried inhalable porous particles showed superior aerosol performance (FPF > 40%) using Osmohaler® at 100 L/min and suitable dissolution profiles.	[91]
CIS TPGS MPs	Spherical shaped and 2.6 μm	Spray-dried powder showed satisfactory FPD (9.84 mg) using Axahaler® at 100 L/min and controlled CIS release for 24 h in modified simulated lung fluid using a dissolution vessel containing impactor disk-assembly.	[92]
CIS/TPGS MPs	Microcrystalline solid powder	After a single pulmonary insufflation using DP-4 M dry powder insufflator™, microcrystal suspension showed a marked 10.88-, 17.17- and tenfold improvement in AUC $_{(10–48\ min)}$, targeting efficacy and targeting advantage, respectively as compared to i.v. suspension at the same dose (1.25 mg/kg).	[93]

(continued)

Table 9.1 (continued)

Formulation	Particle properties	Outcome	References
ATV SMEDDS lecithin/TPGS (1:3)	Uniform wrinkled-shaped particles	SMEDDS showed a good respirable fraction (> 70%) using Aerolizer® at 60 L/min. And a marked sevenfold decrease in IC_{50} values as compared to free drug solution during in vitro cytotoxicity study using A549 cells.	[94]
TPGS CQ10 nanosuspension	Negatively charged (20.60 mV) spherically shaped and 37.80 nm	LC Sprint nebulized particles showed respirable fraction and MMAD of 59.70% and 3.88 μm, respectively, along with 90 days of stability.	[96]
C109-TPGS-CD nanocrystals	Negatively charged (20.1 mV) elongated rod-shaped and 842 nm	Nebulized nanocrystals showed FPF of 41.9%, 39.9%, and 47.7% using commercial nebulizers, i.e., Aeroneb® Go (Aerogen, Ireland), PARI TurboBOY® (PARI GmbH, Germany) and eFlow® rapid (PARI GmbH, Germany), respectively.	[97]

enough to carry a variety of molecules, i.e., PTX, CZT, DTX, and OA, for effective airway delivery. TPGS-based micelles were found promising to achieve long circulation time, improved mean residence time, and pulmonary bioavailability, followed by pulmonary insufflation. Especially in anticancer investigations using the A549 xenograft model, TPGS-based micelles showed remarkable tumor growth inhibition, tumor volume reduction, and improved anticancer efficacy compared to the commercial anticancer formulation. TPGS is also used in various formulations to develop cocktail treatment to surpass poor response and resistance often faced with monotherapy. Additionally, a combination of TPGS with other polymers such as PEG, PLA, PCL, PEG-graft copolymer, and non-ionic triblock copolymers (Soluplus®) attained the sustained (>100 h) release profile with efficient tumor-targeting potential against A549 cells. However, further in-depth studies are needed to recognize the detailed underlying mechanisms.

Furthermore, TPGS-based nanosuspension and nanocrystals can be easily nebulized to obtain satisfactory aerosolization performance along with good stability during long-term storage. Similarly, TPGS-based spray-dried powders have also displayed better therapeutic outcomes in managing pulmonary diseases. As reviewed in the above segment, TPGS-based, dry powders are mainly formulated using coarse carriers, i.e., mannitol or lactose and anti-adherent agent leucine by the spray-drying method. Most TPGS-based dry powders showed spherical-shaped particles with mean particle size (<5 μm) suitable for deep lung deposition. These dry powders showed a good aerodynamic deposition pattern with an average respiratory fraction above 50%. Further in-depth studies would help to realize the effect of inhaler device geometry such as device resistance, air-inlet arrangement, and mouthpiece design on the aerodynamic performance of TPGS-based inhaled materials. Briefly, dynamic material scientists and clinicians have dedicated a vast amount of time and active efforts over the decades toward engineering, expansion, and clinical evaluation of TPGS-based inhaled NPs and MPs. All these energetic attempts signify a bridgehead for the clinical growth of TPGS-based inhaled particles. Even now, several preclinical and clinical investigations of TPGS-based inhaled particles are yet to be achieved for better pulmonary disease treatment.

9.4 Conclusion

In conclusion, we have thoroughly described the biophysical properties and functions of TPGS to understand the basis of designing TPGS-based particulate systems for pulmonary drug delivery in a better way. The simultaneous delivery of multiple actives using TPGS-based nanoparticulate is a helpful strategy to improve the therapeutic potential in lung cancer treatment. Shortly, this multidrug-loaded nanoparticulate carrier can be tailored with targeting moieties to further enhance selectivity toward lung cancer cells. Numerous in vitro and ex vivo investigations regarding TPGS-based delivery systems were addressed cautiously, which established the potential of TPGS to carry a variety of molecules across the pulmonary airways. Moreover, various human tumor xenograft and animal models were used, which ultimately supported the multifunctional activities of TPGS as a carrier in pulmonary drug delivery. Similarly, proof of concept studies demonstrated the potential of a TPGS-based carrier for the effective treatment of pulmonary diseases. However, there is more space for preclinical and clinical experiments to perform successful bench-to-bedside translation.

References

1. Chronic respiratory diseases. https://www.who.int/health-topics/chronic-respiratory-diseases#tab=tab_1
2. Cukic V, Lovre V, Dragisic D, Ustamujic A. Asthma and chronic obstructive pulmonary disease (COPD)–differences and similarities. Mater Sociomed. 2012;24(2):100–5.
3. Stolz D, Mkorombindo T, Schumann DM, Agusti A, Ash SY, Bafadhel M, et al. Towards the elimination of chronic obstructive pulmonary disease: a lancet commission. Lancet. 2022;S0140-6736(22):01273–9.
4. Mehta PP, Dhapte-Pawar V. Role of surfactants in pulmonary drug delivery. In: Green sustainable process for chemical and environmental engineering and science 2022 (pp. 559-577). Academic Press.
5. National Survey of Children's Health. NSCH 2011/12. Data query from the child and adolescent health measurement initiative, Data Resource Center for Child and Adolescent Health website.
6. Mehta PP, Dhapte-Pawar VS. Repurposing drug molecules for new pulmonary therapeutic interventions. Drug Deliv Transl Res. 2021;11(5):1829–48.

7. Mehta P, Bothiraja C, Kadam S, Pawar A. Potential of dry powder inhalers for tuberculosis therapy: facts, fidelity and future. Artif Cells Nanomed Biotechnol. 2018;46(sup3):S791–806.
8. Mehta PP, Dhapte-Pawar VS. Novel and evolving therapies for COVID-19 related pulmonary complications. Am J Med Sci. 2021;361(5):557–66.
9. Mehta PP. Dry powder inhalers: upcoming platform technologies for formulation development. Ther Deliv. 2019;10(9):551–4.
10. Mehta PP, Pawar AP, Mahadik KR, Kadam SS, Dhapte-Pawar V. Dry powder coating techniques and role of force controlling agents in aerosol. In: Polymer coatings: technology and applications; 2020. p. 41–74.
11. Rogers LK, Cismowski MJ. Oxidative stress in the lung–the essential paradox. Curr Opin Toxicol. 2018;7:37–43.
12. Mehta P, Bothiraja C, Mahadik K, Kadam S, Pawar A. Phytoconstituent based dry powder inhalers as biomedicine for the management of pulmonary diseases. Biomed Pharmacother. 2018;108:828–37.
13. Manisalidis I, Stavropoulou E, Stavropoulos A, Bezirtzoglou E. Environmental and health impacts of air pollution: a review. Front Public Health. 2020;8:14.
14. Gibson GJ, Loddenkemper R, Lundbäck B, Sibille Y. Respiratory health and disease in Europe: the new European lung white book. Eur Respir J. 2013 Sep;42(3):559–63.
15. Scoditti E, Massaro M, Garbarino S, Toraldo DM. Role of diet in chronic obstructive pulmonary disease prevention and treatment. Nutrients. 2019;11(6):1357.
16. Romieu I, Trenga C. Diet and obstructive lung diseases. Epidemiol Rev. 2001;23(2):268–87.
17. Keranis E, Makris D, Rodopoulou P, Martinou H, Papamakarios G, Daniil Z, Zintzaras E, Gourgoulianis KI. Impact of dietary shift to higher-antioxidant foods in COPD: a randomised trial. Eur Respir J. 2010;36(4):774–80.
18. Yang C, Wu T, Qi Y, Zhang Z. Recent advances in the application of Vitamin E TPGS for drug delivery. Theranostics. 2018;8(2):464–85.
19. Guo Y, Luo J, Tan S, Otieno BO, Zhang Z. The applications of Vitamin E TPGS in drug delivery. Eur J Pharm Sci. 2013;49(2):175–86.
20. Ribeiro AM, Estevinho BN, Rocha F. The progress and application of vitamin E encapsulation-a review. Food Hydrocoll. 2021;121:106998.
21. Liang L, Qiu L. Vitamin E succinate with multiple functions: a versatile agent in nanomedicine-based cancer therapy and its delivery strategies. Int J Pharm. 2021;600:120457.
22. Sheu MT, Chen SY, Chen LC, Ho HO. Influence of micelle solubilization by tocopheryl polyethylene glycol succinate (TPGS) on solubility enhancement and percutaneous penetration of estradiol. J Control Release. 2003;88(3):355–68.
23. Ghosh I, Michniak-Kohn B. Design and characterization of submicron formulation for a poorly soluble drug: the effect of Vitamin E TPGS and other solubilizers on skin permeability enhancement. Int J Pharm. 2012;434(1–2):90–8.
24. Kim DS, Kim DW, Kim KS, Choi JS, Seo YG, Youn YS, Oh KT, Yong CS, Kim JO, Jin SG, Choi HG. Development of a novel l-sulpiride-loaded quaternary microcapsule: effect of TPGS as an absorption enhancer on physicochemical characterization and oral bioavailability. Colloids Surf B Biointerfaces. 2016;147:250–7.
25. Zhao L, Feng SS. Enhanced oral bioavailability of paclitaxel formulated in vitamin E-TPGS emulsified nanoparticles of biodegradable polymers: in vitro and in vivo studies. J Pharm Sci. 2010;99(8):3552–60.
26. Singh H, Narang JK, Singla YP, Narang RS, Mishra V. TPGS stabilized sublingual films of frovatriptan for the management of menstrual migraine: formulation, design and antioxidant activity. J Drug Deliv Sci Technol. 2017;41:144–56.
27. Dong Y, Zhang Z, Feng SS. d-alpha-Tocopheryl polyethylene glycol 1000 succinate (TPGS) modified poly(l-lactide) (PLLA) films for localized delivery of paclitaxel. Int J Pharm. 2008;350(1–2):166–71.
28. Papas AM, Vitamin E. TPGS and its applications in nutraceuticals. In: Nutraceuticals. Academic Press; 2021. p. 991–1010.

29. Sheng X, Fan L, He C, Zhang K, Mo X, Wang H. Vitamin E-loaded silk fibroin nanofibrous mats fabricated by green process for skin care application. Int J Biol Macromol. 2013;56:49–56.
30. Zhang Z, Tan S, Feng SS. Vitamin E TPGS as a molecular biomaterial for drug delivery. Biomaterials. 2012;33(19):4889–906.
31. Rathod S, Bahadur P, Tiwari S. Nano-carriers based on vitamin E-TPGS: design principle and molecular insights into improving the efficacy of anti-cancer drugs. Int J Pharm. 2021;592:120045.
32. Fan Z, Jiang B, Shi D, Yang L, Yin W, Zheng K, Zhang X, Xin C, Su G, Hou Z. Selective antitumor activity of drug-free TPGS nanomicelles with ROS-induced mitochondrial cell death. Int J Pharm. 2021;594:120184.
33. Li J, Cheng X, Chen Y, He W, Ni L, Xiong P, Wei M. Vitamin E TPGS modified liposomes enhance cellular uptake and targeted delivery of luteolin: an in vivo/in vitro evaluation. Int J Pharm. 2016;512(1):262–72.
34. Liu Y, Ma Y, Xu J, Chen Y, Xie J, Yue P, Zheng Q, Yang M. Apolipoproteins adsorption and brain-targeting evaluation of baicalin nanocrystals modified by combination of Tween80 and TPGS. Colloids Surf B Biointerfaces. 2017;160:619–27.
35. Thant Y, Wang Q, Wei C, Liu J, Zhang K, Bao R, Zhu Q, Weng W, Yu Q, Zhu Y, Xu X. TPGS conjugated pro-liposomal nano-drug delivery system potentiate the antioxidant and hepatoprotective activity of Myricetin. J Drug Deliv Sci Technol. 2021;66:102808.
36. Xiong S, Wang Z, Liu J, Deng X, Xiong R, Cao X, Xie Z, Lei X, Chen Y, Tang G. A pH-sensitive prodrug strategy to co-deliver DOX and TOS in TPGS nanomicelles for tumor therapy. Colloids Surf B: Biointerfaces. 2019;173:346–55.
37. Luiz MT, Di Filippo LD, Alves RC, Araújo VH, Duarte JL, Marchetti JM, Chorilli M. The use of TPGS in drug delivery systems to overcome biological barriers. Eur Polym J. 2020;110129
38. Duhem N, Danhier F, Préat V. Vitamin E-based nanomedicines for anti-cancer drug delivery. J Control Release. 2014;182:33–44.
39. Alavijeh RK, Akhbari K. Vitamin E-based nanomedicines for anticancer drug delivery. In: Nanomedicine for bioactives. Singapore: Springer; 2020. p. 11–70.
40. Muddineti OS, Ghosh B, Biswas S. Current trends in the use of vitamin E-based micellar nanocarriers for anti-cancer drug delivery. Expert Opin Drug Deliv. 2017;14(6):715–26.
41. Li N, Mai Y, Liu Q, Gou G, Yang J. Docetaxel-loaded D-α-tocopheryl polyethylene glycol-1000 succinate liposomes improve lung cancer chemotherapy and reverse multi-drug resistance. Drug Deliv Transl Res. 2021;11(1):131–41.
42. Lu Y, Park K. Polymeric micelles and alternative nanonized delivery vehicles for poorly soluble drugs. Int J Pharm. 2013 Aug 30;453(1):198–214.
43. Hwang D, Ramsey JD, Kabanov AV. Polymeric micelles for the delivery of poorly soluble drugs: from nanoformulation to clinical approval. Adv Drug Deliv Rev. 2020;156:80–118.
44. Ghosh B, Biswas S. Polymeric micelles in cancer therapy: state of the art. J Control Release. 2021;332:127–47.
45. Ghezzi M, Pescina S, Padula C, Santi P, Del Favero E, Cantù L, Nicoli S. Polymeric micelles in drug delivery: an insight of the techniques for their characterization and assessment in biorelevant conditions. J Control Release. 2021;332:312–36.
46. Zhang XY, Zhang YD. Enhanced antiproliferative and apoptosis effect of paclitaxel-loaded polymeric micelles against non-small cell lung cancers. Tumour Biol. 2015;36(7):4949–59.
47. Jiang ZM, Dai SP, Xu YQ, Li T, Xie J, Li C, Zhang ZH. Crizotinib-loaded polymeric nanoparticles in lung cancer chemotherapy. Med Oncol. 2015;32(7):193.
48. Pan J, Wang Y, Feng SS. Formulation, characterization, and in vitro evaluation of quantum dots loaded in poly(lactide)-vitamin E TPGS nanoparticles for cellular and molecular imaging. Biotechnol Bioeng. 2008 Oct 15;101(3):622–33.
49. Sato Y, Watanabe S, Kodama T, Goto M, Shimosato Y. Stainability of lung cancer cells with Leu-7 and OKT-9 monoclonal antibodies. Jpn J Clin Oncol. 1985;15:537–44.
50. Singh RP, Sharma G, Sonali AP, Pandey BL, Koch B, Muthu MS. Transferrin receptor targeted PLA-TPGS micelles improved efficacy and safety in docetaxel delivery. Int J Biol Macromol. 2016;83:335–44.

51. Yan H, Wei P, Song J, Jia X, Zhang Z. Enhanced anti-cancer activity in vitro and in vivo of luteolin incorporated into long-circulating micelles based on DSPE-PEG2000 and TPGS. J Pharm Pharmacol. 2016;68(10):1290–8.
52. Cagel M, Tesan FC, Bernabeu E, Salgueiro MJ, Zubillaga MB, Moretton MA, Chiappetta DA. Polymeric mixed micelles as nanomedicines: achievements and perspectives. Eur J Pharm Biopharm. 2017 Apr;113:211–28.
53. Attia AB, Ong ZY, Hedrick JL, Lee PP, Ee PL, Hammond PT, Yang YY. Mixed micelles self-assembled from block copolymers for drug delivery. Curr Opin Colloid Interface Sci. 2011 Jun 1;16(3):182–94.
54. Naqvi AZ, Panda M. Mixed micellization: improved physicochemical behavior of different amphiphiles in presence of gemini surfactants. J Mol Liq. 2021 Dec;1(343):116876.
55. Yan H, Zhang Z, Jia X, Song J. D-α-Tocopheryl polyethylene glycol succinate/Solutol HS 15 mixed micelles for the delivery of baohuoside I against non-small-cell lung cancer: optimization and in vitro, in vivo evaluation. Int J Nanomedicine. 2016;11:4563–71.
56. Yan H, Song J, Jia X, Zhang Z. Hyaluronic acid-modified didecyldimethylammonium bromide/ d-a-tocopheryl polyethylene glycol succinate mixed micelles for delivery of baohuoside I against non-small cell lung cancer: in vitro and in vivo evaluation. Drug Deliv. 2017;24(1):30–9.
57. Li Y, Zhou T, Ma C, Song W, Zhang J, Yu Z. Ginsenoside metabolite compound K enhances the efficacy of cisplatin in lung cancer cells. J Thorac Dis. 2015;7(3):400–6.
58. Zhang Y, Tong D, Che D, Pei B, Xia X, Yuan G, Jin X. Ascorbyl palmitate/d-α-tocopheryl polyethylene glycol 1000 succinate monoester mixed micelles for prolonged circulation and targeted delivery of compound K for anti-lung cancer therapy in vitro and in vivo. Int J Nanomedicine. 2017;12:605–14.
59. Yang L, Zhang Z, Hou J, Jin X, Ke Z, Liu D, Mei D, Jia X, Lv H. Targeted delivery of ginsenoside compound K using TPGS/PEG-PCL mixed micelles for effective treatment of lung cancer. Int J Nanomedicine. 2017;12:7653–67.
60. Hao W, Zhong Q, Zhong R, Huang H, Xia Z, Ke Z, Zhang Z, Song J, Jia X. Preparation and anti-tumor evaluation of self-assembling oleanolic acid-loaded Pluronic P105/d-α-tocopheryl polyethylene glycol succinate mixed micelles for non-small-cell lung cancer treatment. Int J Nanomedicine. 2016;11:6337–52.
61. Gill KK, Kaddoumi A, Nazzal S. Mixed micelles of PEG(2000)-DSPE and vitamin-E TPGS for concurrent delivery of paclitaxel and parthenolide: enhanced chemosenstization and anti-tumor efficacy against non-small cell lung cancer (NSCLC) cell lines. Eur J Pharm Sci. 2012;46(1–2):64–71.
62. Huang G, Zang B, Wang X, Liu G, Zhao J. Encapsulated paclitaxel nanoparticles exhibit enhanced anti-tumor efficacy in A549 non-small lung cancer cells. Acta Biochim Biophys Sin Shanghai. 2015;47(12):981–7.
63. Nair HH, Alex VV, Anto RJ. Significance of nutraceuticals in cancer therapy. In: Evolutionary diversity as a source for anticancer molecules. Academic Press; 2021 Jan 1. p. 309–21.
64. Ding Y, Wang C, Wang Y, Xu Y, Zhao J, Gao M, Ding Y, Peng J, Li L. Development and evaluation of a novel drug delivery: Soluplus®/TPGS mixed micelles loaded with piperine in vitro and in vivo. Drug Dev Ind Pharm. 2018;44(9):1409–16.
65. Chen Y, Feng X, Li L, Song K, Zhang L. Preparation and anti-tumor evaluation of hinokiflavone hybrid micelles with mitochondria targeted for lung adenocarcinoma treatment. Drug Deliv. 2020;27(1):565–74.
66. de Melo-Diogo D, Gaspar VM, Costa EC, Moreira AF, Oppolzer D, Gallardo E, Correia IJ. Combinatorial delivery of Crizotinib-Palbociclib-sildenafil using TPGS-PLA micelles for improved cancer treatment. Eur J Pharm Biopharm. 2014;88(3):718–29.
67. Moura MJ, Gil MH, Figueiredo MM. Cisplatin delivery systems based on different drug encapsulation techniques. Eur Polym J. 2019 Apr 1;113:357–64.
68. He L, Xu J, Cheng X, Sun M, Wei B, Xiong N, Song J, Wang X, Tang R. Hybrid micelles based on Pt (IV) polymeric prodrug and TPGS for the enhanced cytotoxicity in drug-resistant lung cancer cells. Colloids Surf B Biointerfaces. 2020;195:111256.

69. Brooks PC, Clark RAF, Cheresh DA. Requirement of vascular integrin alpha(v) beta(3) for angiogenesis. Science. 1994;264:569–71.
70. Murphy EA, Majeti BK, Barnes LA, Makale M, Weis SM, Lutu-Fuga K, Wrasidlo W, Cheresh DA. Nanoparticle mediated drug delivery to tumor vasculature suppresses metastasis. Proc Natl Acad Sci U S A. 2008;105:9343–8.
71. Sugahara KN, Teesalu T, Karmali PP, Kotamraju VR, Agemy L, Girard OM, Hanahan D, Mattrey RF, Ruoslahti E. Tissue-penetrating delivery of compounds and nanoparticles into tumors. Cancer Cell. 2009;16:510–20.
72. Shen J, Meng Q, Sui H, Yin Q, Zhang Z, Yu H, Li Y. iRGD conjugated TPGS mediates codelivery of paclitaxel and survivin shRNA for the reversal of lung cancer resistance. Mol Pharm. 2014;11(8):2579–91.
73. Sriwidodo UAK, Wathoni N, Zothantluanga JH, Das S, Luckanagul JA. Liposome-polymer complex for drug delivery system and vaccine stabilization. Heliyon. 2022;8(2):e08934.
74. Nsairat H, Khater D, Sayed U, Odeh F, Al Bawab A, Alshaer W. Liposomes: structure, composition, types, and clinical applications. Heliyon. 2022;8(5):e09394.
75. Mehta PP, Ghoshal D, Pawar AP, Kadam SS, Dhapte-Pawar VS. Recent advances in inhalable liposomes for treatment of pulmonary diseases: concept to clinical stance. J Drug Deliv Sci Technol. 2020;56:101509.
76. Mehta P. Imagine the superiority of dry powder inhalers from carrier engineering. J Drug Deliv. 2018;2018:1.
77. Mehta P. Dry powder inhalers: a focus on advancements in novel drug delivery systems. J Drug Deliv. 2016;2016:1.
78. Yang L, Xin J, Zhang Z, Yan H, Wang J, Sun E, Hou J, Jia X, Lv H. TPGS-modified liposomes for the delivery of ginsenoside compound K against non-small cell lung cancer: formulation design and its evaluation in vitro and in vivo. J Pharm Pharmacol. 2016;68(9):1109–18.
79. Aziz N, Kim MY, Cho JY. Anti-inflammatory effects of luteolin: a review of in vitro, in vivo, and in silico studies. J Ethnopharmacol. 2018 Oct 28;225:342–58.
80. Sarawek S, Derendorf H, Butterweck V. Pharmacokinetics of luteolin and metabolites in rats. Nat Prod Commun. 2008 Dec;3(12):1934578X0800301218.
81. Fu J, Zeng W, Chen M, Huang L, Li S, Li Z, Pan Q, Lv S, Yang X, Wang Y, Yi M. Apigenin suppresses tumor angiogenesis and growth via inhibiting HIF-1α expression in non-small cell lung carcinoma. Chem Biol Interact. 2022 Jul 1;361:109966.
82. Jin X, Li M, Yin L, Zhou J, Zhang Z, Lv H. Tyroservatide-TPGS-paclitaxel liposomes: Tyroservatide as a targeting ligand for improving breast cancer treatment. Nanomedicine. 2017 Apr;13(3):1105–15.
83. Huang YT, Zhao L, Fu Z, Zhao M, Song XM, Jia J, Wang S, Li JP, Zhu ZF, Lin G, Lu R, Yao Z. Therapeutic effects of tyroservatide on metastasis of lung cancer and its mechanism affecting integrin-focal adhesion kinase signal transduction. Drug Des Devel Ther. 2016 Mar 3;10:649–63.
84. Jin X, Yang Q, Zhang Y. Synergistic apoptotic effects of apigenin TPGS liposomes and tyroservatide: implications for effective treatment of lung cancer. Int J Nanomedicine. 2017;12:5109–18.
85. Wang G, Yu B, Wu Y, Huang B, Yuan Y, Liu CS. Controlled preparation and anti-tumor efficacy of vitamin E TPGS-functionalized PLGA nanoparticles for delivery of paclitaxel. Int J Pharm. 2013;446(1–2):24–33.
86. Tiejun Zhao, Hezhong Chen, Yuchao Dong, Jiajun Zhang, Haidong Huang, Ji Zhu, Wei Zhang. Paclitaxel-loaded poly(glycolide-co-ε-caprolactone)-b-D-α-tocopheryl polyethylene glycol 2000 Succinate nanoparticles for lung cancer therapy. Int J Nanomedicine 2013:8 1947–1957.
87. Sun Y, Yu B, Wang G, Wu Y, Zhang X, Chen Y, Tang S, Yuan Y, Lee RJ, Teng L, Xu S. Enhanced anti-tumor efficacy of vitamin E TPGS-emulsified PLGA nanoparticles for delivery of paclitaxel. Colloids Surf B Biointerfaces. 2014;123:716–23.
88. Cheng W, Liang C, Xu L, Liu G, Gao N, Tao W, Luo L, Zuo Y, Wang X, Zhang X, Zeng X, Mei L. TPGS-functionalized Polydopamine-modified mesoporous silica as drug

Nanocarriers for enhanced lung cancer chemotherapy against multidrug resistance. Small. 2017;13(29):1700623.
89. Hou W, Zhao X, Qian X, Pan F, Zhang C, Yang Y, de la Fuente JM, Cui D. pH-sensitive self-assembling nanoparticles for tumor near-infrared fluorescence imaging and chemo-photodynamic combination therapy. Nanoscale. 2016;8(1):104–16.
90. Malamatari M, Somavarapu S, Bloxham M, Buckton G. Nanoparticle agglomerates of indomethacin: the role of poloxamers and matrix former on their dissolution and aerosolisation efficiency. Int J Pharm. 2015;495(1):516–26.
91. Leung SS, Wong J, Guerra HV, Samnick K, Prud'homme RK, Chan HK. Porous mannitol carrier for pulmonary delivery of cyclosporine a nanoparticles. AAPS J. 2017;19(2):578–86.
92. Levet V, Rosière R, Merlos R, Fusaro L, Berger G, Amighi K, Wauthoz N. Development of controlled-release cisplatin dry powders for inhalation against lung cancers. Int J Pharm. 2016;515(1–2):209–20.
93. Levet V, Merlos R, Rosière R, Amighi K, Wauthoz N. Platinum pharmacokinetics in mice following inhalation of cisplatin dry powders with different release and lung retention properties. Int J Pharm. 2017;517(1–2):359–72.
94. Ishak RA, Osman R. Lecithin/TPGS-based spray-dried self-micro emulsifying drug delivery systems: in vitro pulmonary deposition and cytotoxicity. Int J Pharm. 2015;485(1–2):249–60.
95. Laouini A. Characterization of different vitamin E carriers intended for pulmonary drug delivery. Int J Pharm. 2014;471:385–90.
96. Rossi I, Sonvico F, McConville JT, Rossi F, Fröhlich E, Zellnitz S, Rossi A, Del Favero E, Bettini R, Buttini F. Nebulized coenzyme Q_{10} nanosuspensions: a versatile approach for pulmonary antioxidant therapy. Eur J Pharm Sci. 2018;113:159–70.
97. Costabile G, Provenzano R, Azzalin A, Scoffone VC, Chiarelli LR, Rondelli V, Grillo I, Zinn T, Lepioshkin A, Savina S, Miro A, Quaglia F, Makarov V, Coenye T, Brocca P, Riccardi G, Buroni S, Ungaro F. PEGylated mucus-penetrating nanocrystals for lung delivery of a new FtsZ inhibitor against Burkholderia cenocepacia infection. Nanomedicine. 2020;23:102113.

Engineering of Hydrogels for Pulmonary Drug Delivery: Opportunities and Challenges

10

Vividha Dhapte-Pawar, Satish Polshettiwar, and Piyush Pradeep Mehta

Abstract

Hydrogels are one of the most widely studied materials in interdisciplinary research themes. The unique 3D polymeric network of hydrogel offers excellent swelling and mucoadhesive properties that noticeably augment the ease and efficiency of drug delivery. Additionally, biocompatibility and biodegradability properties of hydrogels are more appropriate for versatile drug delivery applications. Moreover, hydrogels are tunable into different particles ranging from nanogels, microgels to macroscopic gels for several biological applications. This chapter systematically discusses hydrogels' origin, features, detailed applications, and interdisciplinary role. Subsequently, as per the book's theme, different hydrogel systems utilized in respiratory diseases management with their design strategies, stimuli-responsive nature, and drug release kinetics are systematically summarized. Moreover, we have emphasized on the in vitro and in vivo aerodynamic outcomes of several hydrogel-based nano- and microparticulate systems. Furthermore, a separate section also illustrates some critical perspectives and precincts allied to hydrogel carriers for pulmonary drug delivery. This chapter is a detailed account of hydrogel systems reported for pulmonary diseases management in the past few years.

V. Dhapte-Pawar (✉)
Department of Pharmaceutics, Poona College of Pharmacy, Bharati Vidyapeeth (Deemed to be University), Pune, Maharashtra, India
e-mail: dhapte.vv@bharatividyapeeth.edu

S. Polshettiwar
Department of Pharmaceutical Sciences, School of Health Sciences and Technology, Dr. Vishwanath Karad MIT-World Peace University, Pune, India

P. P. Mehta
Global Respiratory R&D, Cipla R&D Center, Mumbai, Maharashtra, India
e-mail: piyu053@gmail.com

P. P. Mehta, V. Dhapte-Pawar (eds.), *Pulmonary Drug Delivery Systems: Material and Technological Advances*,
https://doi.org/10.1007/978-981-99-1923-9_10

Keywords

Hydrogels · 3D polymeric network · Nanogels · Microgels · Aerogels and hydrogel microspheres

10.1 Introduction

In 1960, Wichterle and Lim described hydrogel synthesis using biocompatible poly-2-hydroxyethyl methacrylate (PHEMA) polymer [1, 2]. Since then, hydrogels have gained great research importance because of their distinctive integrated solid and liquid properties [1]. Particularly in the past few years, research on hydrogels and hydrogel-based substances has evolved quickly [3]. Hydrogels have great demand in fundamental as well as applied sciences such as soft machines, drug delivery [4, 5], cosmetic industry [6], artificial organs fabrication [7], agriculture science [8], biotechnology industry, hygienic products, food industry [9, 10], water treatment [11, 12], electronics industry [13], and wearable electronic devices [14]. These days, hydrogel has become one of the most widely investigated materials in interdisciplinary research themes.

Hydrogel is usually described as a 3D polymer framework capable of absorbing a large quantity of water or body fluids [15]. A typical hydrogel can be synthesized using a cross-linking hydrophilic polymer chain or polymerization of water-soluble monomers with a suitable cross-linking agent [16]. Hydrogels can be synthesized using simple physical (e.g., hydrogen bonding interactions, ionic interaction using multivalent ions, heat-induced aggregation, microcrystal hydrogels), and stereocomplex formation or chemical (e.g., grafting process, condensation reaction, radical polymerization, enzymatic reaction, and high-energy radiation) cross-linking methods [17]. Additionally, many natural and synthetic biocompatible polymers such as alginate, chitosan (CH), hyaluronic acid (HA), polyesters, and poly-phosphazene are readily available to synthesize the hydrogels [16, 18]. Initially, the investigation on hydrogels was mainly focused on their fundamental attributes, such as swelling/de-swelling equilibrium [19, 20], volume phase transition [21–23], solute diffusion kinetics [24], sliding resistance [25, 26], and drug delivery [27]. The continuous progress in hydrogel research shifted from simple polymeric networks to “responsive” polymeric networks [28]. The responsive hydrogels can respond to changes in light, temperature, pH, redox potential, magnetic fields, electric fields, and ultrasound stimulation [29–31]. The engineered hydrogels have a particular interest in the biomedical domain owing to their unique physical and biochemical properties, such as multiple stimuli response [32], super-wettability [33], adhesion potential [34, 35], self-healing properties [36], and ability to mimic native biological tissues [1]. The polysaccharide-based hydrogels [37], non-cytotoxic protein-based hydrogels [38], conducting hydrogels [39], super-absorbent hybrid hydrogels [40], durable self-healing hydrogels [41], imprinted hydrogels [42] and smart stimuli driven hydrogels [28] are the well-investigated examples of engineered hydrogels. At present, hydrogels are extensively studied in tissue engineering and allied subjects to replicate the extracellular matrix [43], ease wound healing [44], targeted delivery [45], and as mechanical aid for the cells [46, 47]. The popularity of hydrogels in the biomedical sector has mainly emerged due to four fundamental properties, i.e., tailored drug release kinetics, high drug-loading capacity, biodegradability, and biocompatibility [48]. Various physicochemical properties and biological applications of the hydrogels are illustrated in Fig. 10.1.

Hydrogels and hydrogel-based materials are promising delivery cargos with good potential ranging from drug delivery, drug targeting to diagnostic applications [49, 50]. Particularly in drug delivery, hydrogels are explored to carry a variety of therapeutic moieties ranging from low molecular weight molecules to macromolecular substances [51, 52]. Moreover, hydrogels have the unique capacity to hold many water-soluble substances such as therapeutically active phytochemicals, proteins, and peptides because of their exclusive porous nature [17]. The release kinetics of the therapeutic agent can be easily modified by making minor structural changes in the hydrogels polymeric network [53, 54]. Hydrogels' mean particle size and structural morphology determine the functions and route of drug delivery for effective disease management [55]. Hydrogels can be classified into three groups as per their mean particle size, i.e., nanogels (<200 nm), microgels (0.5–10 μm), and macroscopic gels (>200 μm) [55, 56]. Several active researchers methodically utilize these different hydrogel structures for different drug delivery applications, for example, (i) macroscopic gels for in situ injection, in situ implant, and transdermal drug delivery; (ii) microgels for oral, pulmonary drug delivery, and trans-arterial chemo-embolization, and (iii) nanogels for in situ injections and intravenous injections [55]. Undoubtedly, hydrogels have shown a successful transformation in the healthcare approach, which can be probed from specific applications of the hydrogels [15, 57]. After cautiously understanding the hydrogels, various active researchers and clinicians have also tried hydrogel-based systems to effectively manage life-threatening chronic diseases such as chronic respiratory diseases (CRDs). Several recently published patents and research communications systematically studied hydrogels for pulmonary drug delivery applications. Hence in the upcoming section, we have thoroughly reviewed the different hydrogel structures with suitable examples.

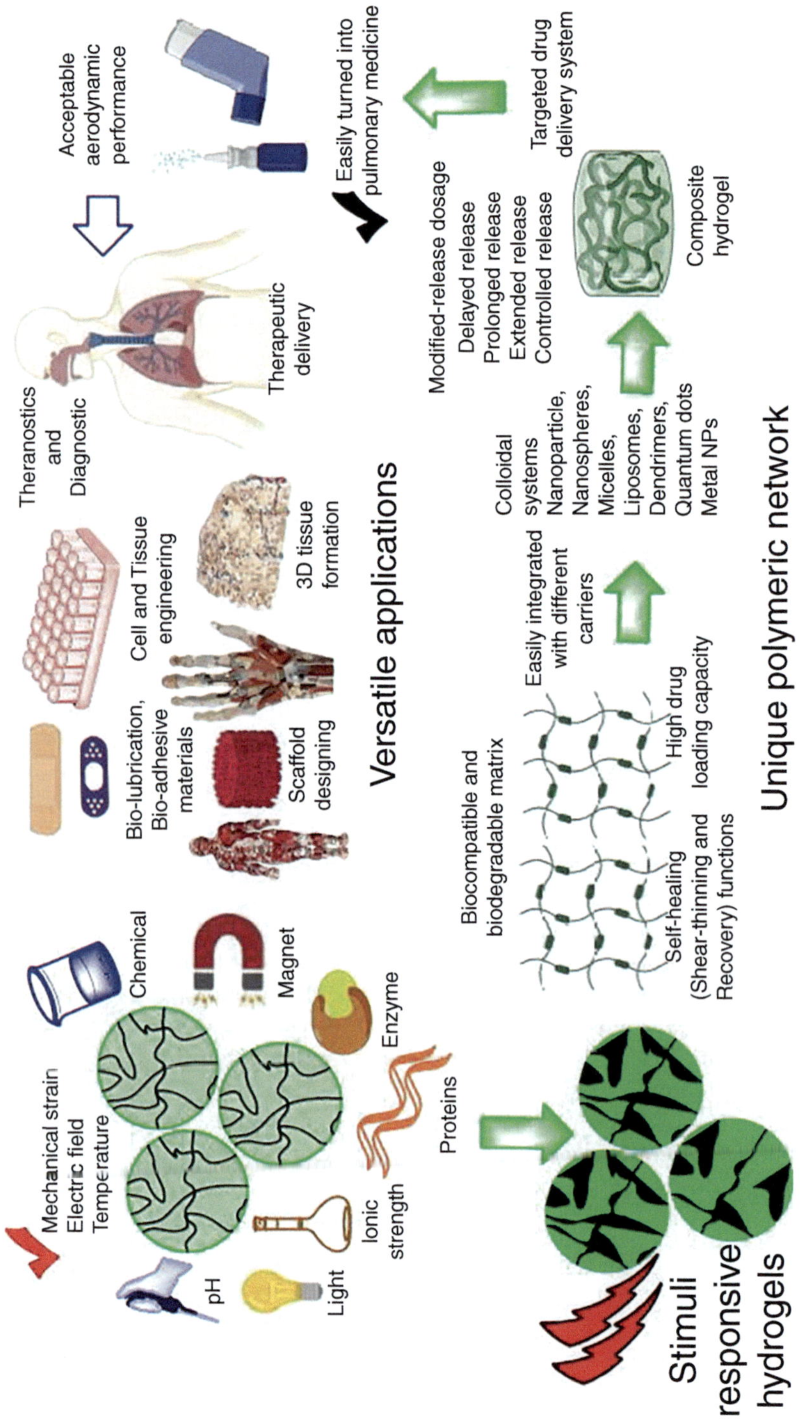

Fig. 10.1 Hydrogels: Smart approach for pulmonary drug delivery

10.2 Pulmonary Delivery

Inhalation therapies, i.e., dry powder inhalers (DPIs) [58], metered-dose inhalers [59], soft mist inhalers [60], and nebulizers [61] hold various critical advantages over other oral dosage forms in managing pulmonary diseases [62–64]. However, most inhalable drugs experience poor mean residence time and rapid lung clearance and therefore necessitate recurrent administration of the therapeutic dosage [65, 66]. The design and development of sustained-release formulation using biocompatible and mucoadhesive hydrogel carriers may overcome critical issues [67, 68]. Mucoadhesive hydrogel carriers directly interact with the mucus layer of pulmonary airways and enhance bioavailability owing to the improved dwell time of drug-loaded carriers within the pulmonary airways [50]. At present, mucoadhesive carrier hydrogels, i.e., swellable hydrogel microspheres, physically cross-linked hydrogel microparticles (HYMPs), microgel particles, and hydrogel nanoparticles (HYNPs), are widely explored in various pre-clinical and clinical studies to obtain sustained release kinetics with enhanced mean residence time [50]. Additionally, with 3D cross-linked arrangements, hydrogels possess excellent physicochemical properties and particle size (<5 μm) suitable for deep lung delivery [50]. Furthermore, they are measured as the potential carrier for controlled drug delivery because of their in vivo stability and biocompatibility [50]. Thus, hydrogel carriers are discussed in the subsequent segment, emphasizing the design, aerodynamic characteristics, and in vivo applications.

10.2.1 Hydrogel Nano- and Micro-particles

Progesterone (PGT) is a major gonadal hormone, synthesized by the ovary in the female, as well as the adrenal cortex and testes in the male [69]. Primarily, PGT is a female reproductive hormone. Chemically, PGT is a C21-steroid hormone that participates in the embryogenesis of humans and other species. Recently, various scientific communications displayed that PGT has important pleiotropic functions as a neuroprotective substance in both humans and animals [70]. PGT exerts therapeutic actions in several neurodegenerative conditions such as stroke, Parkinson's disease, and Alzheimer's disease. With current scientific indications research has been directed toward the repurposing of PGT for therapeutic management of several CNS disorders. However, the physicochemical and biological restrictions such as low solubility (<1 mg/mL), high log P (3.87), and poor bioavailability due to high intestine and first-pass metabolism rate are linked with the peripheral administration of PGT which highlight the need for more effective delivery strategy to upsurge PGT brain levels. With this aim, Cardia et al. (2019) formulated inhalable PGT HYNPs using trimethyl-CH and sodium alginate by ionotropic gelation process. Uniformly distributed, spherical-shaped PGT HYNPs exhibited particle size, zeta potential, and entrapment efficacy of 198 nm, 29 mV, and 80%, respectively. In-vitro release using water/ethanol (40:60 v/v) solution PGT HYNPs showed sustained PGT release for more than 300 min (>70%). During an in vitro biocompatibility study using a human nasal septum carcinoma cell line (RPMI 2650), PGT HYNPs showed a reduction in cell viability in a dose-dependent manner. In an in vivo pharmacokinetic study, HYNPs loaded with deuterated PGT formulation were administered via inhalation using the Nebula M2000 aerosol apparatus (1 mg/kg and 0.3 mL/min). Deuterated PGT HYNPs presented a marked fivefold enhancement in brain PGT level compared to basal PGT level following 30 min of inhalation. Moreover, PGT HYNPs showed acceptable mean particle size (<300 nm) and zeta potential (26 to 35 mV) during storage (28 days) at 4 °C. However, further detailed research will be needed to support effective clinical translation of this novel technology [71].

Stocke and co-workers developed and understood the feasibility of using HYNPs (nanogels) in pulmonary drug delivery. The pH-responsive HYNPs were formulated using acrylic acid and polyethylene glycol (PEG) 400 diacrylate (80:20) with an aqueous solution of sodium dodecyl sulfate by dialysis method. Moreover, 9-anthracenylmethyl methacrylate (1 mol %) was included based on total monomer weight. Moreover, inhalable HYNPs powder was obtained using spray drying in the presence of a mannitol carrier. Well-defined, spherical-shaped HYNPs showed a particle size of 80 nm, while swollen state HYNPs showed a particle size of 120 nm. Moreover, HYNPs showed good colloidal stability in a phosphate buffer saline (PBS) with no significant change in relative absorption performance. During aerodynamic analysis using high resistance HandiHaler® device at 60 L/min, spray-dried hydrogel composites demonstrated suitable aerosolization outcomes with emitted dose (ED) and fine particle fraction (FPF) of 78.50% and 31.70%, respectively. Briefly, this study highlighted the good prospective for these materials as smart drug delivery systems that unite the benefits of stimuli-responsive HY, nanotechnology, and effective targeting through airway delivery [72].

Tuberculosis (TB) is one of the leading causes of morbidity and mortality across the globe. The TB standard treatment regimens include 2–9 months chronic administration of 2–4 high dose (300–600 mg) actives such as isoniazid (INH) and rifampin (RFP) based on severity of diseases [73–75]. However, traditional oral or injectable treatment possesses several drawbacks such as high systemic toxicity owing to non-specific drug delivery. Therapeutic failure may occur due to painful intramuscular injections, quick drug resistance (as a result of repetitive administration of high dose of same drug), reduced bioavailability (due to first-pass metabolism), and finally poor patient compliance and adherence (due to high dose long-term therapy). In this situation, direct airway delivery of anti-TB regimens using suitable dosage form is more advantageous [73–75]. Thus, Wu et al. (2018) developed INH and RFP-loaded genipin-cross-linked deacetylated CH nanogel particles for effective treatment of TB. Initially, genipin-cross-linked deacetylated CH (2:1) nanogel particles were formulated using the emulsion cross-linking method. At the same time, INH (60 mg) and RFP (60 mg) were added to the nanogel system using the immersion method. Well-defined spherical-shaped INH/RFP nanogel showed a mean particle size of 232.20 nm and satisfactory entrapment efficacy for INZ (67.90%) and RFP (42.50%). In vitro drug release using simulated lung fluid (pH 7.2) exhibited prolonged drug release (>95 h) for INZ/RIF nanogel with no initial burst release. INH/RFP nanogel exhibited 89.90% and 73.20% release for INH and RFP, respectively. Moreover, INH/RFP nanogel system showed a minimum inhibitory concentration (MIC) of 5 mg/mL against *Mycobacterium tuberculosis* H37Rv (ATCC27294) standard strain. During in-vivo and in-vitro toxicity studies using rats and mouse embryonic fibroblast cells (NIH-3 T3), INH/RFP nanogel showed significantly lower toxicity, thereby proving safe for direct use compared to individual drug substances. An in vivo pharmacokinetic study after pulmonary administration using DP-4 M insufflator, INH/RFP nanogel system displayed a marked 1.46 and 1.64-fold (bioavailability), 3.51 and 4.21-fold (mean residence time) and 3.20 and 4.21-fold (half-life; $t_{1/2}$) improvement of INH and RFP, respectively, as compared to INH and RFP alone at the same dose. The pharmacokinetic study showed that nanogel helps to maintain INH/RFP concentration within the therapeutically effective range for a longer period of time and thus available at the site of infection above MIC to deliver the desired biological action. In short, dual-drug-loaded nanogel inhalation powder is a helpful dosage form compared to individual drug substances [76].

Pulmonary drug delivery is a better alternative to parenteral delivery of biologics such as therapeutic proteins and peptides. Recently, Graf et al. (2019) explored branched PEG-macro-monomers to develop biodegradable microgels using a photocatalyzed thiol-ene reaction to successfully deliver model proteins, i.e., lysozyme (15 kDa) and bovine serum albumin (66 kDa). Basically, branched PEG macromonomers-based microgel network was designed to entrap proteins in a cross-linked 3D network and release them by steady degradation of the 3D network. Besides, PEG-based network possesses low visibility to the immune system and is anticipated to lessen phagocytosis by macrophages (Fig. 10.2a). Moreover, freeze-drying was performed to produce microgels powder with appropriate aerodynamic properties and allow microgels particle swelling after contact with airway fluids. Furthermore, microgels powder was mixed with pre-blended lactose-containing 15% lactose fines (Fig. 10.2b). The mean particle size of microgels was 7.50 μm with satisfactory encapsulation efficiencies of 51.50% (lysozyme) and 73.60% (bovine serum albumin). Biodegradable microgels showed sustained release properties for 6–14 days using PBS (pH 7.4) media depending on the number of crosslinking moieties on the PEG macro-monomers. During in vitro aerodynamic analysis using Novolizer® device at 30 L/min, microgels loaded on pre-blended lactose carrier showed high ED and FPF of 98.08% and 48.07%, respectively. The improved aerodynamic outcome was mainly credited to added lactose fines. Lactose fines assist in saturating high-energy binding sites of coarse lactose and improving the deagglomeration performance of microgels particles. During an in vitro cytotoxicity study using mouse alveolar macrophage cells (MH-S), microgels did not show any notable cytotoxicity. In addition, PEG-microgels displayed a noticeable reduction in phagocytosis rates compared to control polystyrene particles. Briefly,

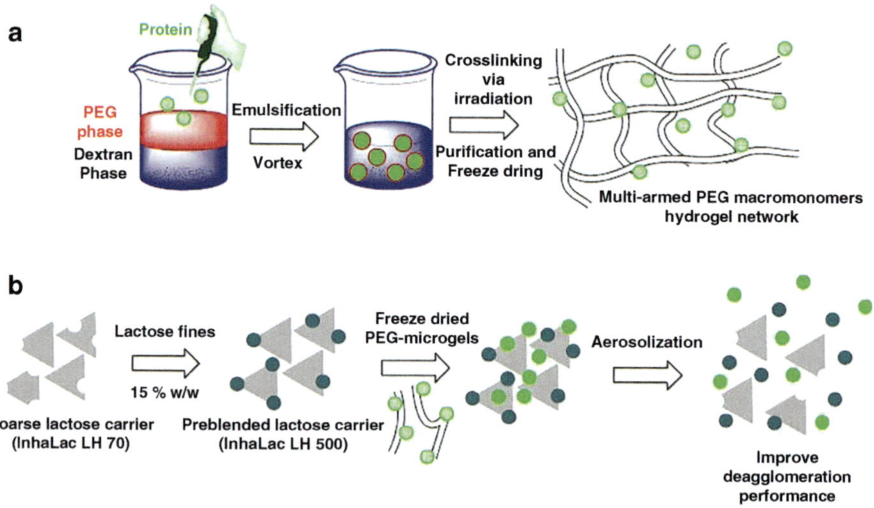

Fig. 10.2 Dry microgel particle preparation using PEG/dextran by aqueous two-phase systems (**a**) and ternary dry powder mixture using fine lactose and freeze-dried PEG-microgels (**b**)

PEG-microgels can be used to design pulmonary depot formulations to improve the local and systemic bioavailability of therapeutic proteins/peptides due to low cytotoxicity and phagocytosis rate [77].

Sildenafil is a pyrazolopyrimidinone derivative and selective phosphodiesterase-5 inhibitor frequently prescribed in the therapy of erectile dysfunction and pulmonary arterial hypertension (PAH) [78, 79]. Sildenafil is marketed as the citrate salt viz. sildenafil citrate (SC). Principally, in PAH, SC assists to increase cyclic guanosine monophosphate which results in smooth muscle relaxation, vasodilation of the pulmonary vascular bed and relieving PAH complications by regulating the blood flow in airway vasculature. Also, SC support to decrease lung inflammation and mucus production. However, due to its poor physicochemical properties such as low water solubility (4.1 mg/mL), possess ampholyte nature, i.e., two pK_a values of 8.8 (benzimidazole) and 4 (pyridinium ion) and poor oral bioavailability (40%) the formulation development of SC is very challenging. Consequently, inhalable SC-loaded HYMPs were developed to achieve enhanced lung deposition with controlled drug release properties compared to oral formulation. Particularly, SC HYMPs were formulated using biodegradable sodium carboxymethyl cellulose, sodium hyaluronate, and sodium alginate polymers by spray-drying technique. SC HYMPs were optimized using the extreme vertices mixture design, and the optimized batch showed good mean geometric particle size (2.05 μm) with entrapment efficiency (82.50%). During an in vitro drug release study in PBS (pH 7.4), SC HYMPs displayed a sustained release for over 24 h (>97.00%). In the aerodynamic analysis using low resistance mono-dose inhaler at 28.3 L/min, SC HYMPs exhibited FPF and mass median aerodynamic diameter (MMAD) of 30.04% and 4.68 μm, respectively. During swelling study using PBS (pH 7.4), SC HYMPs showed a quick upsurge in the particle size which assist to increase the HYMPs residence time within the lungs and reduce their uptake by the phagocytic cells. During the in vivo pharmacokinetic study using male albino rats, inhaled SC HYMPs showed significantly superior maximum serum concentration (C_{max}; 35-fold) and area under curve (AUC; 17.64-fold) along with prolonged elimination $t_{1/2}$ as compared to orally administered SC at the same dose (0.6 mg/kg). In agreement with swelling study SC HYMPs displayed a notable improvement in MRT of SC HYMPs compared to orally administered SC. Briefly, the inhalable spray-dried HYMPs are believed to be a potential lung delivery carriers that can be administered once daily for the treatment of PAH [80].

PAH is a potentially life-threatening ailment and is mainly treated with phosphodiesterase inhibitors, endothelin receptor antagonists, and prostacyclin. However, these treatment options are inadequate or transient to cure severely damaged pulmonary vessels. Therefore, alternative strategies such as vascular remodeling or tissue regeneration are essential for patients suffering with critical PAH. Basic fibroblast growth factor (bFGF) is an important angiogenic mitogen and a cytokine that incites tissue regeneration. Biodegradable gelatin HYs containing bFGF showed improved arteriogenesis and angiogenesis in several cardiovascular diseases. The arteriogenic and angiogenic properties of bFGF resolved PAH. However, the optimal delivery method is needed to efficiently deliver bFGF to the pulmonary airways. Consequently, Hirose et al. (2008) carefully fabricated bFGF-loaded microgelatin HY microspheres for effective management of PAH. During in vivo pharmacodynamic study using monocrotaline-induced pulmonary hypertension rats model, after a single intratracheal administration, microspheres (100 μg/0.1 mL) showed extensively lower (0.36) right ventricular/left ventricular pressure ratios at the end of 3 weeks as compared to the bFGF alone (0.58). Microspheres displayed significantly superior (threefold) vessel diameter during microanatomy analysis than the bFGF alone. Moreover, microspheres-treated rats showed an excellent (84%) survival rate after 10 weeks of intratracheal administration. Concisely, pulmonary delivery of cell signalling proteins such as bFGF was found to be a more logical and practical approach in long-term therapy of pulmonary hypertension [81].

HA is an anionic, non-sulfated glycosaminoglycan that exists in the intercellular matrix of most vertebrate epithelial, connective, and neural tissues. HA is a naturally occurring polysaccharide possessing various key applications in pharmaceutical domain such as tissue regeneration, joint lubrication, wound healing, and anti-inflammatory actions. In addition to its viscoelasticity, moisturizing properties, HA possesses unique mucoadhesive qualities which offer longer pulmonary retention. Alginate is another well-known natural product used for various applications in the pharmaceutical field. It is an anionic polysaccharide obtained from brown seaweed. It has good number of applications in the pharmaceutical and food industry owing to its biocompatibility, non-toxicity, and biodegradability. Generally, in presence of divalent cations alginate effectively undergoes an ionic inter-chain bridge process (gelation) to shape hydrogel. However, a recent study presented that a hybrid blend of alginate-HA biopolymers also successfully forms hydrogel with suitable physicochemical properties. Likewise, Athamneh et al. (2019) developed biodegradable and biocompatible alginate-HA hydrogel formulations for pulmonary drug delivery. Particularly, aerogel microspheres were formulated using alginate and HA by emulsion gelation followed by supercritical drying. The distinctive interpenetrating polymeric network structure is attributed to the small uniform and spherical shape of aerogel microspheres. Linkage of alginate-HA displayed helpful effects in terms of lowering the particles agglomeration and hydrogen bond formation between the N-acetyl-d-glucosamine moiety of HA and carboxylate groups of alginate to avoid the separation of the molecules in the aerogel microsphere matrix. Aerogel microspheres displayed good specific surface area (446–611 m^2/g), high porosity (97–98%) and lower density (0.035–0.063 g/cm^3). Aerogel microspheres (10 mg) containing alginate and HA (1:1 w/w) displayed the most suitable in vitro aerodynamic diameter using low resistance Breezhaler® device at 30 L/min. Mainly, 1:1 w/w microspheres showed similar experimental (5.0 μm) and calculated (4.0 μm) aerodynamic diameter. Lower aerodynamic diameter permits superior deposition of microspheres within the pulmonary airways. Therefore, bio-based alginate-HA microspheres showed unique physicochemical characteristics with potential for pulmonary drug delivery [82].

Idiopathic pulmonary fibrosis (IPF) is a typical chronic, progressive, and fibrotic airway disease. It is a chronic inflammatory disorder, which slowly grows to established fibrosis. IPF is the most common type of idiopathic interstitial pneumonia in which healthy lung tissue is compromised by altered extracellular matrix and alveolar structure is demolished, which leads to reduced lung compliance, disrupted gas exchange, and eventually airway failure and death [83]. The present treatment options for IPF comprise anti-fibrotic agents such as pirfenidone and nintedanib and oxygen supplements which gradually control the disease progression in patients during early stage [79]. Despite these treatment options, the average survival for IPF patients remains poor and lung transplantation remains only the therapy option for end-stage IPF patients [84]. Therefore, there is an unmet need for novel anti-fibrotic treatment options capable of slowing progression and increasing clinical results for IPF patients. Recent studies displayed that interleukin-10 (IL-10) is a cytokine with potent anti-fibrotic and anti-inflammatory properties that makes it an interesting therapeutic molecule for IPF. IL-10 exerts anti-fibrotic effects by its ability to overcome collagen I production in human skin fibroblasts. However, a key problem in IL-10 pulmonary application is very short half-life (1–2 min). Accordingly, the model approach to improve the short half-life is to use a biocompatible carrier (e.g., hydrogel) with minimal toxicity to airway tissue. Thus, Shamskhou et al. (2019) developed a novel hydrogel-based recombinant IL-10 formulation to prevent and reverse life-threatening progressive IPF. A hydrogel shaped from chemically cross-linked hyaluronan and heparin (HH-IL-10) was designed to retain both moisture (hyaluronan) as well as IL-10 (heparin). During experimental assessment HH-IL-10 hydrogel system effectively delivered the IL-10 using the capacity of heparin to bind IL-10 without bleeding reversibly or other severe obstacles. Moreover, during biological assessment after intratracheal administration, the HH-IL-10 hydrogel system significantly reduced the collagen deposition within the pulmonary parenchyma after 7 days compared to IL-10 alone. Furthermore, the HH-IL-10 hydrogel system executed the anti-fibrotic effect by suppressing transforming growth factor

(TGF-β) dependent collagen production by pulmonary fibroblasts and myofibroblasts. Hence, hydrogel-based systems are a promising and capable therapeutic alternative for progressive pulmonary diseases [85].

10.2.2 Swellable Hydrogel Carriers

The biodegradable HYMPs with sustained release properties are very helpful in pulmonary drug delivery. But, to acquire a sustained release profile in the airways, particles must first direct through aerodynamic cascade to enter the deep lung tissues (entails small particle size) where they must then circumvent rapid macrophage uptake (improved by large swollen particle size). Accordingly, to match aerodynamic characteristics and regulate macrophage clearance El-Sherbiny and coworkers logically designed swellable HYMPs as carriers for sustained lung drug delivery. Particularly, sodium fluorescein-loaded (0.2% w/w) swellable HYMPs were fabricated using 90% w/w PEG graft copolymerized CH with 10% w/w water-soluble triblock copolymer (Pluronic® F-108) by the cryo-milling method. HYMPs showed good entrapment efficacy of 65.68%. Plate-shaped HYMPs showed a particle size of 12.06 μm and lower moisture content (0.40%) with a bulk density of 0.746 g/cm^3. HYMPs were mixed with Respitose® ML001 lactose carrier (1:9) and actuated using a breath-activated Aerolizer® device at 60 L/min during aerodynamic analysis. HYMPs displayed ED and respirable fractions of 96% and 14.85%, respectively. In in vitro swelling study using PBS (pH 7.4) HYMPs showed 1101% swelling after 6 h. In an in vitro lysozyme-based biodegradation study, HYMPs exhibited faster enzymatic degradation. Pluronic® F-108 concentration played an important role in biodegradation. During in vitro release profiles using PBS (pH 7.4), HYMPs allowed more sustained release characteristics (>400 h). In a macrophage uptake study using macrophage-like, Abelson leukemia virus-transformed cell line (Raw 264.7), HYMPs displayed delayed and lower macrophage uptake due to high swelling characteristics. PEG permits stealth properties that assist in effective macrophage evasion. Briefly, new biodegradable physically cross-linked swellable HYMPs showed good potential as a carrier in controlled pulmonary drug delivery [86]. Similar research group also developed sodium fluorescein (0.2% w/w) loaded swellable HYMPs using 10% w/w Pluronic® F-108 with 90% ww PEG grafted CH [CH-HYMPs] and PEG grafted CH N-phthaloyl derivative [NPCH-HYMPs] by cryo-milling method. Physically cross-linked biodegradable swellable CH-HYMPs showed moisture content, entrapment efficacy, and volume mean diameter of 0.40%, 65.68%, and 12.06 μm, correspondingly. In contrast, NPCH-HYMPs presented moisture content, entrapment efficacy, and volume mean diameter of 0.18%, 90.63%, and 10.65 μm, respectively. The lower entrapment efficacy of CH-HYMPs was mainly attributed to the high initial swelling and thus high drug loss during polymer film washing compared to NPCH-HYMPs. An in vitro biodegradation study using lysozyme NPCH-HYMPs displayed a lower degradation outcome than CH-HYMPs due to hydrophobicity, the steric hindrance of bulky phthaloyl groups of the N-phthaloyl derivative. In swelling assessment using PBS (pH 7.4)

CH-HYMPs exhibited a fast swelling within the first 30 min while NPCH-HYMPs presented a relatively slow increase in swelling up to 1 h. This swelling profile may be attributed to the hydrophobicity of PEG-g-NPHCs copolymer resulting from the presence of lipophilic phthaloyl groups. CH-HYMPs showed a marked 1055% swelling profile after 6 h compared to initial values. HYMPs showed variation in swelling profile with respect to grafted polymer and nonionic surfactant. However, the magnitude of swelling mainly depended on the cross-linking density of HYMPs. The swelling equilibrium stage is achieved when a balance arises between the cohesive forces and osmotic driving forces. During in vitro release study using PBS (pH 7.4), swellable HYMPs presented initial burst release within the first 30 min followed by controlled release up to 20 days. Remarkably, after 20 days, NPCH-HYMPs displayed 96% sodium fluorescein release while CH-HYMPs displayed only 63% sodium fluorescein release. Briefly, respirable swellable HYMPs demonstrated promising in vitro functioning with sustained-release properties [87].

Curcumin (CUR) is the key curcuminoid of spice *Curcuma longa* (Zingiberaceae). CUR has a number of favorable therapeutic activities such as antibacterial, anti-inflammatory, antioxidant, antihyperglycemic, antimalarial, antitumor, antiviral, and anti-Alzheimer's activity [88]. Moreover, CUR exhibited notable pharmacological actions against both acute and chronic lung disorders with efficient anticancer activity against various lung cancer cell lines [89]. Furthermore, various clinical attributes of CUR have been explored methodically. CUR is safe and not toxic even at very high doses (8.0 g/day for 3 months). However, CUR is yet to touch the position of a therapeutic drug substance mostly because a typical solid dosage of CUR suffers from low oral and erratic bioavailability (0.05 μg/mL, less than 1%). The factors ascribed for low bioavailability include high molecular weight (368.39 g/mol), poor aqueous solubility (0.6 μg/mL), log P value of 2.5, high protein-binding and rapid metabolism (short elimination $t_{1/2} < 2$ h) [88, 89]. Thus, effective CUR therapy requires an appropriate formulation such as nanoparticulate system which will improve solubility and bioavailability to offer superior therapeutic efficiency. But in most cases drug-loaded NPs have shown poor airway deposition because of Brownian motion and face quick macrophage clearance [90]. Thus, El-Sherbiny et al. (2011) incorporated CUR-loaded PLGA NPs into swellable HYMPs to achieve a sustained CUR release, not only in the alveolar region where macrophage uptake prevails but mainly throughout the lumen of the airways. Basically, CUR-loaded PLGA NPs grafted with CH-grafted-PEG HYMPs were formulated by the spray-drying method. Respirable amphiphilic HYMPs showed lower moisture content (1.21%) with higher entrapment efficacy (89.63%). Highly smooth spherical-shaped HYMPs displayed acceptable mean particle size (3.08 μm) and lowered tapped density (0.149 g/mL) with integrated surface morphology suitable for superior aerodynamic performance. HYMPs (10 mg) containing 35% of CUR-loaded PLGA NPs showed the most appropriate in vitro aerodynamic diameter (1.19 μm) and FPF (30.20%) using high resistance HandiHaler® device at 60 L/min. During a dynamic swelling study using PBS (pH 7.4) HYMPs showed a prominent 12.37 and 29.54-fold improvement in the size of MPs after 6 and 20 min, respectively. HYMPs formulations with grafted copolymer and low PLGA (35%) content attained higher swelling compared to the formulations with non-grafted CH polymer and high PLGA (75%) content. This was attributed to hydrophilicity of the PEG side chains in the grafted copolymer and low concentration of hydrophobic PLGA in final HYMPs. In cytotoxicity assay using Raw 264.7 macrophage cell line, HYMPs showed low cytotoxic potential and are suitable for pulmonary drug delivery. In tumor necrosis factor-alpha (TNF-α) assay, HYMPs illustrated minimum propensity to provoke TNF-α release and thus not stimulate a local inflammatory response. Moreover, in vitro macrophage interaction investigations showed that the swellable HYMPs had good potential for sustained pulmonary release. Briefly, the present carrier system combines the benefits of swellable HYMPs and drug-loaded NPs for effective pulmonary delivery [91].

Ciprofloxacin hydrochloride (CIP) is a broad-spectrum antimicrobial agent. It is a second-generation carboxy-fluoroquinoline. CIP is a prescription medicine approved by the US FDA for the prevention and treatment of various bacterial infections such as lower respiratory tract, skin, and urinary tract infections. CIP showed slight aqueous solubility with a highly pH-dependent solubility profile due to ampholyte nature, i.e., two pKa values of 6.09 (carboxyl group) and 8.62 (N-4 in the piperazinyl group). At physiological settings CIP experiences substantial dissociation of both the carboxyl (acid) and the piperazinyl (base) groups leading to significant fractions of zwitterionic species [92]. Thus, to formulate physiologically stable and aerodynamically effective dosage form NPs integrated swellable HYMPs were developed by Du et al. (2014). CIP-loaded swellable HYMPs were formulated using the novel 3% w/v CIP-loaded PEG grafted onto phthaloyl CH (PEG-g-PHCs) nanosuspension and 1% w/v sodium alginate solution by a spray-drying method. CIP-PEG-g-PHC NPs showed a mean particle size of 218.60 nm, while spherical-shaped rough CIP-PEG-g-PHC NPs entrapped swellable HYMPs showed a mean particle size (3.90 μm) acceptable for pulmonary delivery. During a swelling study using PBS (pH 7.4) HYMPs showed a prominent 7.33 and 13.96-fold improvement in the size of MPs after 2 and 14 min, respectively. The dynamic swelling performance of HYMPs formulations was attributed to PEG side chains in the grafted copolymer and alginate chemistry. Spray-dried NPs embedded swellable HYMPs displayed sustained CIP release for more than 250 h (>50%). In a cytotoxicity study using RAW 264.7 cells, spray-dried HYMPs (320 μg/mL) showed a marked 38.40% reduction in cell viability compared to the control group. Furthermore, in vivo experimental analysis of spray-dried HYMPs showed a marked 2.5-fold improvement in plasma CIP concentration with 80% CIP release within the first 3 h compared to the micronized CIP-loaded lactose group at the same dose (15 mg/kg). Besides, spray-dried swellable HYMPs attained superior concentrations of CIP within lung tissue. Briefly, spray-dried swellable HYMPs proved to be suitable for sustained and prolonged drug release within the pulmonary airways [93].

10.2.3 Enzyme Responsive Hydrogel Carriers

Enzyme as a stimulus is very frequently explored to complete a certain piece of work to regulate material physical and chemical attributes. In comparison to physical (e.g., temperature, magnetic field, electric field, and light) or typical chemical stimuli (e.g., pH, ionic strength, and redox interactions) enzymatic stimuli are most commonly preferred since it permits sensitivity to biological signs, which are extremely selective and include catalytic amplification to enable fast response times. Thus, in the pharmaceutical field a huge range of enzyme-responsive materials are used to perform diverse activities such as drug delivery, drug targeting, and control drug release [94]. Consequently, Secret et al. (2014) judiciously designed and developed enzyme-reactive PEG HYMPs using a novel solution polymerization method. Solution polymerization method involved the production of well-defined PEG MPs from high molecular weight PEG diacrylate (PEGDA)-based precursors that include therapeutic peptides in the polymer network. Mainly, matrix metalloproteinases responsive PEG HYMPs containing a trimer peptide (Gly-Leu-Lys) were designed using a novel solution polymerization scheme. Small, uniform-sized (3 μm) and spherical-shaped PEG HYMPs were obtained from PEGDA macro-monomers with molecular weights between 2000 and 10,000 g/mol. Briefly, the solution polymerization technique permits a versatile platform to fabricate uniform and monodisperse PEG-based HYMPs for deep lung delivery [95]. A similar research group developed matrix metalloproteinase-responsive peptide-functionalized PEG-based HYMPs using the emulsion polymerization method. PEG-based HYMPs were used as a carrier for three different types of molecules, i.e., methylene blue (hydrophilic), dexamethasone (hydrophobic), and horseradish peroxidase (protein). Emulsion polymerization followed by homogenizer produced uniform and well-defined MPs. Spherical-shaped methylene blue, dexamethasone and horseradish peroxidase loaded PEG-based HYMPs showed the mean diameter of 4.5, 7.2, and 7.5 μm, respectively. Moreover, PEG-based HYMPs showed satisfactory drug loading for methylene blue (2.4 mg), dexamethasone (3.8 mg), and horseradish peroxidase (18.70 mg). Protein moiety showed higher drug loading due to the protein's retention (high mesh size) within the polymeric matrix during the rinsing and stronger adsorption on the surface of the MPs. In an in vitro release study using TESCA buffer (pH 7.4), hydrophilic drug (methylene blue) and protein (horseradish peroxidase) showed release even at small concentrations of matrix metalloproteinases while hydrophobic drug (dexamethasone) required superior concentrations of matrix metalloproteinases. Briefly, PEG HYMPs are well suitable for the enzyme-triggered release of several drugs. Thus, many lung targets can be located to attain enzyme-triggered pulmonary drug delivery by simply transforming the peptide sequence in the PEG polymeric matrix [96].

10.3 Nebulized Hydrogel Carriers

CRDs are one of the leading causes of morbidity and mortality across the globe. Current therapeutic approaches habitually fail to sufficiently treat continuously growing numbers of CRDs, indicating the need for novel therapies and drug targets [79, 89]. Thus, in the last few years clinician-scientists and physicians interest was moved toward biological interventions viz. pulmonary administration of RNA interference (RNAi) such as the use of small interfering RNA (siRNA), activation of the cytosolic RNAi pathway, regulating gene silencing effect, etc. [97–99]. Although siRNA platforms possess many key advantages over currently utilized therapies, their clinical translation is obstructed due to inefficient delivery across cellular membranes. To permit cellular internalization through endocytosis, siRNA is normally shaped into lipid- or polymer-based NPs [97–99]. But, the majority of current NPs fail to deliver encapsulated siRNA in airways. Thus, Merckx et al. (2020) formulated hybrid NPs holding siRNA-loaded nanosized hydrogel (nanogel) coated with pulmonary surfactant (Curosurf®). Of note, the coating of hybrid nanogel with pulmonary surfactant not only improved NPs stability but also enabled cytosolic siRNA delivery to treat pulmonary ailments. Particularly, siRNA (2 mg/mL)-loaded cationic dextran nanogels were formulated using an inverse mini-emulsion photopolymerization technique. Subsequently, Curosurf® dispersion (80 mg/mL) was added to siRNA nanogels under sonication to obtain pulmonary surfactant nanogel particles. Curosurf®-coated nanogels were successfully lyophilized without cryoprotectant and subsequently reconstituted for nebulization using a vibrating mesh nebulizer. Coated nanogels and lyophilized nanogels displayed hydrodynamic diameter of approx. 150 nm. Lyophilized powder formed a porous and solid superordinate structure with low moisture content (<1%), indicating the lack of collapse in the dry powder. Additionally, such powder structure permitted a rapid reconstitution and formed homogenous suspension within seconds with minimal physical processing. During the biological assessment, following lyophilization, reconstitution, and nebulization, nanogels effectively retained the physicochemical characteristics. They successfully delivered siRNA into the cytosol of a human non-small cell lung carcinoma cell line (H1299). Briefly, nanogels retained and preserved the functional integrity of Curosurf® coated siRNA nanosystems and imparted long-term storage capacity [100]. Interestingly, recent literature showed that surfactant protein B (SP-B) is a key component of the pulmonary surfactant coat to attain the improved intracytosolic siRNA delivery effect [101]. The same research group also showed that surfactant protein B (SP-B) improved the cellular interactions of proteolipid-coated hydrogel nanocomposites. Specifically, SP-B decorated hydrogel-based delivery systems are the crucial therapeutic platform for several pulmonary pathologies [101]. Lastly, to successfully push siRNA to the marketplace further preformulations and formulations efforts are needed with ample clinical evidences.

10.4 Hydrogel-Assisted Metered-Dose Inhalers

The pressurized metered-dose inhalers (pMDIs) are one of the most frequently prescribed aerosol for treatment of CRDs. MDIs are made up of three major components, i.e., (i) aluminum or stainless steel canister to hold formulation, (ii) metering valve to dispense metered quantity of the formulation after every actuation, and (iii) an actuator coupled discharge nozzle (mouthpiece) with a dust cap to deliver dose to the patient and prevent contamination during storage, respectively. The pMDIs formulation is made up of the micronized drug (0.5–5 μm), apt excipients (e.g., ethanol, oleic acid, citric acid, polyvidone K30, PEG, and glycerol), and suitable hydrofluorocarbon propellant (e.g., HFA-134a and HFA-227) [59, 102]. Fundamentally, MDIs use a propellant to deliver a fixed volume of suspension or solution to the user in the form of an aerosol. They are handy, low-priced, suitable for the patient and appropriate for a wide range of activities. Collectively knowing the advantages and limits of the pMDIS, investigators explored the same to deliver hydrogels for pulmonary ailments.

By knowing physical limitations, biological advantages of CUR and mechanical attributes of hydrogels Selvam et al. (2011) designed and developed swellable HYMPs-based pMDIs. CUR-loaded poly(lactic-co-glycolic acid) (PLGA) NPs were prepared using a single emulsion solvent evaporation technique. In contrast, swellable hydrogel microspheres were formulated using a homogenous combination of CUR-loaded PLGA NPs and CH-grafted-PEG by spray-drying method. CUR NPs showed a mean particle size of 243.40 nm, whereas HYMPs displayed a mean particle size of 3.50 μm. Moreover, spherical, smooth surface HYMPs demonstrated low moisture content (1.8%) with acceptable density. In a swelling study using PBS (pH 7.4) HYMPs showed 6.5 and 19.62-fold improvement in volume median diameter after 2 and 20 min, respectively. Swelling behavior of HYMPs was appropriate because it is not too fast to execute particle deposition in central lung and not too gradual to permit quick macrophage uptake. Swellable HYMPs demonstrated good compatibility with propellant HFA227ea. The suspension stability and re-dispersibility were primarily dependent on the concentration of swellable hydrogel microspheres. During aerosolization study at 30 L/min, CUR-PLGA NPs-SHMPs showed FPF and MMAD of 64.96% and 2.01 μm, respectively. CUR-PLGA NPs-SHMPs showed acceptable dose content uniformity when tested using USP unit spray sampling apparatus. Briefly, drug-loaded biodegradable swellable hydrogel matrices were highly apt for sustained pulmonary drug delivery. But to achieve clinical translation of these novel HYMPs or HYNPs obviously we need a rigorous preclincial and clinical experimentation [103].

10.5 Clinical Applications of Hydrogels

10.5.1 Biologic Lung Volume Reduction System

Aeris Therapeutics Inc. (US) has developed the biologic lung volume reduction (BLVR) system to attain lung volume reduction without surgery in pulmonary emphysema and COPD. In BLVR, subjects are treated using a bronchoscope to guide treatment to the most damaged parts of the airways. BLVR system contains a mixture of drugs and biologics in hydrogel scaffold, which form a biodegradable hydrogel at the site of treatment. The BLVR system was designed to support the healing process without producing inflammation. BLVR system utilizes an enzymatic primer solution to confiscate epithelial cells from the target area, and the hydrogel scaffold contains synthetic polyamine and glycosaminoglycans to endorse collagen synthesis and fibroblast attachment [104, 105]. Hydrogel scaffold helps to decrease the lung volume by permanently collapsing and closing the diseased areas of the airways. This offers space within the chest to permit the remaining healthier parts of the airways to work better [106]. Presently, BLVR is in Phase 2 of the clinical trial. Its aim is to assess the safety and efficacy of the BLVR System (20 mL) in subjects with advanced upper lobe predominant emphysema [106].

10.5.2 Amiodarone Hydrogel

University of Louisville (US) assessed the intra-operative application of amiodarone with CoSeal (amiodarone hydrogel) to avoid postoperative atrial fibrillation in subjects undergoing lung transplantation. CoSeal is a surgical sealant from Baxter Healthcare (US). CoSeal consists of two separate solutions, i.e., dilute hydrogen chloride solution, sodium phosphate/sodium carbonate solution, and synthetic PEGs. At the time of application, solutions are mixed to form a hydrogel. Thus, during the medical intervention, amiodarone hydrochloride powder will be mixed with CoSeal to form amiodarone-containing hydrogel. Amiodarone hydrogel is delivered using a device called a CO_2 driver. Following this Phase 2 trial, the sponsor proposed the randomized control trial of amiodarone hydrogel for a better understanding of the treatment [107].

10.5.3 BioSentry

Air leaks are the most frequent trouble after a lung biopsy. Thus, active clinicians and research institutes carefully assess the Bio-Seal Biopsy Track Plug (a hydrogel plug) in a clinical setting to lessen the rate of air leaks. The plug is made up of commercially available lung sealant material, i.e., desiccated polyethylene glycol, and during medical intervention, hydrogel is delivered as a solid cylinder (0.1 cm diameter and 2.5 cm length). During contact with moist lung tissue, the hydrogel plug absorbs local fluids and swells to pack the void formed by the coaxial needle puncture. The seal is airtight, and the hydrogel plug resorbs into the body over time [108]. Additionally, the Bio-Seal system is approved in the US market with an indication for use as a lung biopsy site marker [108]. Recently, Memorial Sloan Kettering Cancer Center (US) designed a study to carefully compare two techniques, i.e., injecting a very small amount of your blood called a blood patch into the biopsy hole or injecting a hydrogel-based US FDA-approved artificial plug called BioSentry to diminish the chance of air leak in lung biopsy and hoping for positive results [109].

10.6 Discussion

Hydrogels form an increasingly attractive carrier matrix owing to their unique potential applications in the area of pulmonary disease management. Table 10.1 shows a summary of hydrogel-based carriers for pulmonary delivery. As illustrated

Table 10.1 Hydrogel-based carriers for effective pulmonary delivery

Formulation	Method	Properties	Outcome	References
PGT HYNPs	Ionotropic gelation	Spherical NPs (198 nm)	HYNPs showed a sustained release profile with marked fivefold enhancement in brain progesterone level compared to basal progesterone level after 30 min of inhalation	[71]
pH-responsive HYNPs	Spray drying	Spherical-shaped NPs (80 nm) and swollen NPs (120 nm)	HYNPs showed good lung targeting potential with ED of 78.50% using HandiHaler®	[72]
INH and RIF loaded genipin-cross-linked deacetylated CH nanogel	Immersion method	Spherical nanogel (232 nm)	Inhaled spherical nanogel (232 nm) system showed a marked 1.46 (INH), and 1.64 (RIF) fold higher bioavailability values as compared to INH and RIF alone at the same dose	[76]
Lysozyme microgels	Photocatalyzed thiol-ene reaction	Swellable microgels with a mean particle size of 7.50 μm	Microgels (7.50 μm) showed good loading efficacy (51.50%) and sustained release properties for 14 days using PBS (pH 7.4) media	[77]
BSA microgels	Photocatalyzed thiol-ene reaction	Swellable microgels with a mean particle size of 7.50 μm	Microgels (7.50 μm) showed good loading efficacy (73.60%) and sustained release properties for 14 days using PBS (pH 7.4) media	[77]

(continued)

Table 10.1 (continued)

Formulation	Method	Properties	Outcome	References
Sildenafil citrate HYMPs	Spray drying	MPs with particle size of 2.05 μm and entrapment efficiency of 82.50%	HYMPs showed high FPF (30.04%) values, and inhaled HYMPs showed superior C_{max} (35-fold) and AUC (17.64-fold) values as compared to orally administered sildenafil citrate at the same dose (0.6 mg/kg)	[80]
bFGF-loaded microgelatin HY microspheres	Impregnation method	Microgelatin hydrogel microspheres showed particle size of 10 μm	Hydrogel showed a higher (84%) survival rate and significantly superior (threefold) vessels diameter as compared to the basic fibroblast growth factor-alone group in the pulmonary hypertension rats model	[81]
Aerogel microspheres	Emulsion gelation with supercritical drying	Uniform spherical microspheres	Microspheres presented lower aerodynamic diameter for superior drug deposition within the lower pulmonary airways	[82]
Recombinant IL-10	Ionotropic gelation	Commercially accessible hydrogel system (HyStem, ESI-Bio) used	After intratracheal administration effectively reduced the collagen deposition within the pulmonary parenchyma as compared to IL-10 alone	[85]
Sodium fluorescein HYMPs	Cryo-milling method	Swellable plate-shaped HYMPs (12.06 μm)	HYMPs revealed ED of 96% at 60 L/min using Aerolizer® with faster enzymatic degradation and sustained release characteristics (> 400 h)	[86]

Table 10.1 (continued)

Formulation	Method	Properties	Outcome	References
Sodium fluorescein HYMPs	Cryo-milling method	Swellable HYMPs (10.65 μm)	HYMPs presented sustained release characteristics for >20 days	[87]
CUR HYMPs	Spray drying	Spherical, smooth surface HYMPs (3.08 μm)	HYMPs controlled the TNF-α release and local inflammatory response with FPF of 30.20% at 60 L/min using HandiHaler®	[91]
CIP HYMPs	Spray drying	Swellable spherical, smooth surface HYMPs (3.90 μm)	HYMPs exhibited sustained release characteristics (> 250 h) and marked 2.5-fold improvement in plasma CIP concentration as compared to micronized CIP-loaded lactose group at the same dose (15 mg/kg)	[93]
Trimer peptide (Gly-Leu-Lys) HYMPs	Emulsion polymerization method	Spherical, monodisperse HYMPs (3 μm)	HYMPs offered a good platform for deep lung delivery	[95]
Curosurf® decorated siRNA-loaded nanogel	Inverse mini-emulsion photopolymerization method	Hollow and porous structure (<200 nm)	Nanogels effectively deliver siRNA into the cytosol of a human lung epithelial cell line	[100]
CUR HY microspheres	Single emulsion solvent evaporation technique followed by spray drying	Spherical, smooth surface microspheres (3.50 μm)	Microspheres showed FPF and MMAD of 64.96% and 2.01 μm when delivered using propellant HFA227ea at 30 L/min	[103]

in the previous segment, hydrogels can be effectively integrated into respirable nano and microparticulate systems such as nanogels, microgels, swellable microspheres, enzyme-responsive HYMPs, and HYNPs using a variety of natural polysaccharides (CH, sodium alginate and HA), pulmonary surfactant (Curosurf®) and branched/grafted PEG. Recent formulation and clinical investigations showed that the hydrogels carriers are primarily significant for airways delivery for two important reasons, i.e., (i) they can be easily size engineered for drug delivery into the secondary or ternary bronchi and (ii) they have a good ability to swell upon reaching their desired target to surpass uptake and clearance by alveolar macrophages. Additionally, various researchers showed the preparation of hybrid hydrogel delivery systems using three simple development schemes, i.e., (i) entrap a hydrogel within an NPs, (ii) fabricate 3D network within NPs and (iii) covalently link the NPs with the hydrogel matrix to achieve sustained drug release profile and superior pulmonary pharmacokinetics. Still, the types of hydrogel carriers currently investigated for pulmonary delivery are relatively few, and most of the engineered hydrogel carriers are mainly responsive to a single stimulus. Besides, the drug release from most hydrogel matrix mainly depends on traditional diffusion kinetics. Therefore, designing a multi-stimuli responsive hydrogel matrix and endogenous or exogenous triggered stimuli-responsive hydrogels would be a meaningful approach and challenging theme to scrutinize forthwith. At the same time, hydrogels would move substantially into the discipline of airway epithelium regeneration and airway remodeling to offer hypothetical and experimental basics for clinical management.

10.7 Conclusion

Hydrogels have unlocked a new promising path in pulmonary drug delivery for local and systemic applications. The above-reviewed investigations emphasized the capacity of hydrogels to deliver a broad range of therapeutic agents, which expands the scope of their function as drug carriers. Additionally, hydrogels' tunable sizes with swelling/de-swelling properties allow them to avoid the lung clearance mechanisms; mucoadhesion features raise exposure time and lessen mucociliary movement while stimuli-responsive nature permits them to attain the desired release kinetics of therapeutics, which ultimately lowers drug delivery frequencies and adverse effects. Besides, human xenograft models and in-vitro cell line assessment showed the ability of hydrogel carriers in inter/intracellular communications and potential targeting ability. All the above-stated drug delivery advantages of hydrogels enable them as excellent carriers for pulmonary drug delivery. However, a considerable gap exists between laboratory investigations and clinical applications because of the challenges regarding the safety, immunological response, and in vivo stability of these newly introduced carrier materials. Accordingly, further profound studies on both phases of in vitro and in vivo are required to support the use of hydrogels in pulmonary drug delivery.

References

1. Fan H, Gong JP. Fabrication of bioinspired hydrogels: challenges and opportunities. Macromolecules. 2020;53(8):2769–82.
2. Wichterle O, Lim D. Hydrophilic gels for biological use, vol. 185. Nature; 1960. p. 117–8.
3. Huang X, Li J, Luo J, Gao Q, Mao A, Li J. Research Progress on double-network hydrogels. Materials today. Communications. 2021;29:102757.
4. Mellati A, Hasanzadeh E, Gholipourmalekabadi M, Enderami SE. Injectable nanocomposite hydrogels as an emerging platform for biomedical applications: a review. Mater Sci Eng C. 2021;131:112489.
5. Kesharwani P, Bisht A, Alexander A, Dave V, Sharma S. Biomedical applications of hydrogels in drug delivery system: an update. J Drug Delivery Sci Technol. 2021;66:102914.
6. Mbituyimana B, Liu L, Ye W, Boni BO, Zhang K, Chen J, Thomas S, Vasilievich RV, Shi Z, Yang G. Bacterial cellulose-based composites for biomedical and cosmetic applications: research progress and existing products. Carbohydrate Polymers. 2021:118565.
7. Liu C, Xu N, Zong Q, Yu J, Zhang P. Hydrogel prepared by 3D printing technology and its applications in the medical field. Colloid and Interface Science Communications. 2021;44:100498.
8. Singh N, Agarwal S, Jain A, Khan S. 3-dimensional cross linked hydrophilic polymeric network "hydrogels": an agriculture boom. Agric Water Manag. 2021;253:106939.
9. Zha F, Rao J, Chen B. Plant-based food hydrogels: constitutive characteristics, formation, and modulation. Curr Opin Colloid Interface Sci. 2021;101505
10. Khalesi H, Lu W, Nishinari K, Fang Y. Fundamentals of composites containing fibrous materials and hydrogels: a review on design and development for food applications. Food Chemistry; 2021. p. 130329.
11. Wu Z, Zhang P, Zhang H, Li X, He Y, Qin P, Yang C. Tough porous nanocomposite hydrogel for water treatment. J Hazard Mater. 2022;421:126754.
12. Mu R, Liu B, Chen X, Wang N, Yang J. Hydrogel adsorbent in industrial wastewater treatment and ecological environment protection. Environ Technol Innov. 2020;20:101107.
13. Bashir S, Hasan K, Hina M, Soomro RA, Mujtaba MA, Ramesh S, Ramesh K, Duraisamy N. Conducting polymer/graphene hydrogel electrodes based aqueous smart supercapacitors: a review and future prospects. J Electroanal Chem. 2021;898:115626.
14. Ying B, Liu X. Skin-like hydrogel devices for wearable sensing, Soft robotics and beyond. iScience. 2021;103174
15. Caló E, Khutoryanskiy VV. Biomedical applications of hydrogels: a review of patents and commercial products. Eur Polym J. 2015;65:252–67.
16. Ahmed EM. Hydrogel: preparation, characterization, and applications: a review. J Adv Res. 2015;6:105–21.
17. Spizzirri UG, Cirillo G, editors. Functional hydrogels in drug delivery: key features and future perspectives. CRC Press; 2017. p. 1–363.
18. Fu X, Hosta-Rigau L, Chandrawati R, Cui J. Multi-stimuli-responsive polymer particles, films, and hydrogels for drug delivery. Chem. 2018;4(9):2084–107.
19. Tanaka T, Fillmore DJ. Kinetics of swelling of gels. J Chem Phys. 1979;70:1214–8.
20. Li Y, Tanaka T. Kinetics of swelling and shrinking of gels. J Chem Phys. 1990;92:1365–71.
21. Tanaka T, Sato E, Hirokawa Y, Hirotsu S, Peetermans J. Critical kinetics of volume phase transition of gels. Phys Rev Lett. 1985;55:2455–8.
22. Ilmain F, Tanaka T, Kokufuta E. Volume transition in a gel driven by hydrogen bonding. Nature. 1991;349:400–1.
23. Tong Z, Liu X. Swelling equilibria and volume phase transition in hydrogels with strongly dissociating electrolytes. Macromolecules. 1994;27:844–8.
24. Peppas NA, Khare AR. Preparation, structure and diffusional behavior of hydrogels in controlled release. Adv Drug Deliv Rev. 1993;11:1–35.

25. Gong JP, Higa M, Iwasaki Y, Katsuyama Y, Osada Y. Friction of gels. J Phys Chem B. 1997;101:5487–9.
26. Gong JP. Friction and lubrication of hydrogels -its richness and complexity. Soft Matter. 2006;2:544–52.
27. Mehta P, Mahadik K, Kadam S, Dhapte-Pawar V. Advanced applications of green hydrogels in drug delivery systems. In: Applications of advanced green materials; 2021. p. 89–130.
28. Abdollahiyan P, Baradaran B, de la Guardia M, Oroojalian F, Mokhtarzadeh A. Cutting-edge progress and challenges in stimuli responsive hydrogel microenvironment for success in tissue engineering today. J Control Release. 2020;328:514–31.
29. Bansal M, Dravid A, Aqrawe Z, Montgomery J, Wu Z, Svirskis D. Conducting polymer hydrogels for electrically responsive drug delivery. J Control Release. 2020;328:192–209.
30. Cheng Q, Hao A, Xing P. Stimulus-responsive luminescent hydrogels: design and applications. Adv Colloid Interface Sci. 2020;286:102301.
31. Ding M, Jing L, Yang H, Machnicki CE, Fu X, Li K, Wong IY, Chen PY. Multifunctional soft machines based on stimuli-responsive hydrogels: from freestanding hydrogels to smart integrated systems. Materials Today Adv. 2020;8:100088.
32. Pourjavadi A, Heydarpour R, Tehrani ZM. Multi-stimuli-responsive hydrogels and their medical applications. New J Chem. 2021;45(35):15705–17.
33. Zhang P, Zhao C, Zhao T, Liu M, Jiang L. Recent advances in bioinspired gel surfaces with superwettability and special adhesion. Adv Sci. 2019;6(18):1900996.
34. Bovone G, Dudaryeva OY, Marco-Dufort B, Tibbitt MW. Engineering hydrogel adhesion for biomedical applications via chemical Design of the Junction. ACS Biomater Sci Eng. 2021;7(9):4048–76.
35. Liu J, Lin S, Liu X, Qin Z, Yang Y, Zang J, Zhao X. Fatigue-resistant adhesion of hydrogels. Nat Commun. 2020;11(1):1–9.
36. Talebian S, Mehrali M, Taebnia N, Pennisi CP, Kadumudi FB, Foroughi J, Hasany M, Nikkhah M, Akbari M, Orive G, Dolatshahi-Pirouz A. Self-healing hydrogels: the next paradigm shift in tissue engineering? Adv Sci. 2019;6(16):1801664.
37. Zhu T, Mao J, Cheng Y, Liu H, Lv L, Ge M, Li S, Huang J, Chen Z, Li H, Yang L. Recent Progress of polysaccharide-based hydrogel interfaces for wound healing and tissue engineering. Adv Mater Interfaces. 2019;6(17):1900761.
38. Jonker AM, Löwik DW, Van Hest JC. Peptide-and protein-based hydrogels. Chem Mater. 2012;24(5):759–73.
39. Liu K, Wei S, Song L, Liu H, Wang T. Conductive hydrogels—a novel material: recent advances and future perspectives. J Agric Food Chem. 2020;68(28):7269–80.
40. Ribeiro SC, de Lima HH, Kupfer VL, da Silva CT, Veregue FR, Radovanovic E, Guilherme MR, Rinaldi AW. Synthesis of a super-absorbent hybrid hydrogel with excellent mechanical properties: water transport and methylene blue absorption profiles. J Mol Liq. 2019;294:111553.
41. Tamesue S, Saito Y, Toita R. Salinity durable self-healing hydrogels as functional biomimetic systems based on the intercalation of polymer ions into mica. Polymer. 2021;123870
42. Byrne ME, Park K, Peppas NA. Molecular imprinting within hydrogels. Adv Drug Deliv Rev. 2002;54(1):149–61.
43. Geckil H, Xu F, Zhang X, Moon S, Demirci U. Engineering hydrogels as extracellular matrix mimics. Nanomedicine (Lond). 2010;5(3):469–84.
44. Tavakoli S, Klar AS. Advanced hydrogels as wound dressings. Biomolecules. 2020;10(8):1169.
45. Narayanaswamy R, Torchilin VP. Hydrogels and their applications in targeted drug delivery. Molecules. 2019;24(3):603.
46. Ahearne M. Introduction to cell-hydrogel mechanosensing. Interface Focus. 2014;4(2):20130038.
47. Mantha S, Pillai S, Khayambashi P, Upadhyay A, Zhang Y, Tao O, Pham HM, Tran SD. Smart hydrogels in tissue engineering and regenerative medicine. Materials (Basel). 2019;12(20):3323.

48. Chai Q, Jiao Y, Yu X. Hydrogels for biomedical applications: their characteristics and the mechanisms behind them. Gels. 2017;3(1):6.
49. Correa S, Grosskopf AK, Lopez Hernandez H, Chan D, Yu AC, Stapleton LM, Appel EA. Translational applications of hydrogels. Chem Rev. 2021;121(18):11385–457.
50. Du J, Du P, Smyth HD. Hydrogels for controlled pulmonary delivery. Ther Deliv. 2013;4(10):1293–305.
51. Li J, Mooney DJ. Designing hydrogels for controlled drug delivery. Nat Rev Mater. 2016;1(12):16071.
52. Shoukat H, Buksh K, Noreen S, Pervaiz F, Maqbool I. Hydrogels as potential drug-delivery systems: network design and applications. Ther Deliv. 2021;12(5):375–96.
53. Lin Y, Dong S, Zhao W, Hu KL, Liu J, Wang S, Tu M, Du B, Zhang D. Application of hydrogel-based delivery system in endometrial repair. ACS Appl Bio Material. 2020;3(11):7278–90.
54. Bernhard S, Tibbitt MW. Supramolecular engineering of hydrogels for drug delivery. Adv Drug Deliv Rev. 2021;171:240–56.
55. Sun Z, Song C, Wang C, Hu Y, Wu J. Hydrogel-based controlled drug delivery for cancer treatment: a review. Mol Pharm. 2019;17(2):373–91.
56. Chyzy A, Tomczykowa M, Plonska-Brzezinska ME. Hydrogels as potential Nano-, micro- and macro-scale systems for controlled drug delivery. Materials (Basel). 2020;13(1):188.
57. Aswathy SH, Narendrakumar U, Manjubala I. Commercial hydrogels for biomedical applications. Heliyon. 2020;6(4):e03719.
58. Mehta PP. Dry powder inhalers: a brief overview of the drug detachment techniques. Ther Deliv. 2020;11(3):139–43.
59. Kumar R, Mehta P, Shankar KR, Rajora MA, Mishra YK, Mostafavi E, Kaushik A. Nanotechnology-assisted metered-dose inhalers (MDIs) for high-performance pulmonary drug delivery applications. Pharm Res. 2022;1-25:2831.
60. Mehta PP, Dhapte-Pawar V. Role of surfactants in pulmonary drug delivery. In: Green sustainable process for chemical and environmental engineering and science. Academic Press; 2022. p. 559–77.
61. Mehta PP, Pawar AP, Mahadik KR, Kadam SS, Dhapte-Pawar V. Dry powder coating techniques and role of force controlling agents in aerosol. In: Polymer coatings: technology and applications; 2020. p. 41–74.
62. Mehta PP. Dry powder inhalers: a concise summary of the electronic monitoring devices. Ther Deliv. 2021;12(1):1–6.
63. Mehta PP, Dhapte-Pawar VS. Novel and evolving therapies for COVID-19 related pulmonary complications. Am J Med Sci. 2021;361(5):557–66.
64. Jadhav P, Patil P, Bhagwat D, Gaikwad V, Mehta PP. Recent advances in orthogonal analytical techniques for microstructural understanding of inhalable particles: present status and future perspective. J Drug Deliv Sci Technol. 2022;68:103089.
65. Strong P, Ito K, Murray J, Rapeport G. Current approaches to the discovery of novel inhaled medicines. Drug Discov Today. 2018;23(10):1705–17.
66. Carroll W, Dhillon R. Sildenafil as a treatment for pulmonary hypertension. Arch Dis Child. 2003;88:827–8.
67. Healy AM, Amaro MI, Paluch KJ, Tajber L. Dry powders for oral inhalation free of lactose carrier particles. Adv Drug Deliv Rev. 2014;75:32–52.
68. Loira-Pastoriza C, Todoroff J, Vanbever R. Delivery strategies for sustained drug release in the lungs. Adv Drug Deliv Rev. 2014;75:81–91.
69. Singh M, Su C. Progesterone and neuroprotection. Horm Behav. 2013;63(2):284–90.
70. Wei J, Xiao GM. The neuroprotective effects of progesterone on traumatic brain injury: current status and future prospects. Acta Pharmacol Sin. 2013 Dec;34(12):1485–90.
71. Cardia MC, Carta AR, Caboni P, Maccioni AM, Erbì S, Boi L, Meloni MC, Lai F, Sinico C. Trimethyl chitosan hydrogel nanoparticles for progesterone delivery in neurodegenerative disorders. Pharmaceutics. 2019;11(12):657.

72. Stocke NA, Arnold SM, Hilt JZ. Responsive hydrogel nanoparticles for pulmonary delivery. J Drug Deliv Sci Technol. 2015;29:143–51.
73. Chaudhary KR, Puri V, Singh A, Singh C. A review on recent advances in nanomedicines for the treatment of pulmonary tuberculosis. J Drug Deliv Sci Technol. 2022;103069
74. Chae J, Choi Y, Tanaka M, Choi J. Inhalable nanoparticles delivery targeting alveolar macrophages for the treatment of pulmonary tuberculosis. J Biosci Bioeng. 2021;132(6):543–51.
75. Mehta P, Bothiraja C, Kadam S, Pawar A. Potential of dry powder inhalers for tuberculosis therapy: facts, fidelity and future. Artif Cells Nanomed Biotechnol. 2018;46(sup3):S791–806.
76. Wu T, Liao W, Wang W, Zhou J, Tan W, Xiang W, Zhang J, Guo L, Chen T, Ma D, Yu W, Cai X. Genipin-crosslinked carboxymethyl chitosan nanogel for lung-targeted delivery of isoniazid and rifampin. Carbohydr Polym. 2018;197:403–13.
77. Graf M, Ziegler CE, Gregoritza M, Goepferich AM. Hydrogel microspheres evading alveolar macrophages for sustained pulmonary protein delivery. Int J Pharm. 2019;566:652–61.
78. Pirhayati FH, Shayanfar A, Fathi-Azarbayjani A, Martinez F, Sajedi-Amin S, Jouyban A. Thermodynamic solubility and density of sildenafil citrate in ethanol and water mixtures: measurement and correlation at various temperatures. J Mol Liq. 2017 Jan 1;225:631–5.
79. Mehta PP, Dhapte-Pawar VS. Repurposing drug molecules for new pulmonary therapeutic interventions. Drug Deliv Transl Res. 2021;11(5):1829–48.
80. Shahin HI, Vinjamuri BP, Mahmoud AA, Shamma RN, Mansour SM, Ammar HO, Ghorab MM, Chougule MB, Chablani L. Design and evaluation of novel inhalable sildenafil citrate spray-dried microparticles for pulmonary arterial hypertension. J Control Release. 2019;302:126–39.
81. Hirose K, Marui A, Arai Y, Kushibiki T, Kimura Y, Sakaguchi H, Yuang H, Chandra BI, Ikeda T, Amano S, Tabata Y, Komeda M. Novel approach with intratracheal administration of microgelatin hydrogel microspheres incorporating basic fibroblast growth factor for rescue of rats with monocrotaline-induced pulmonary hypertension. J Thorac Cardiovasc Surg. 2008;136(5):1250–6.
82. Athamneh T, Amin A, Benke E, Ambrus R, Leopold CS, Gurikov P, Smirnova I. Alginate and hybrid alginate-hyaluronic acid aerogel microspheres as potential carrier for pulmonary drug delivery. J Supercrit Fluids. 2019;150:49–55.
83. Richeldi L, Collard HR, Jones MG. Idiopathic pulmonary fibrosis. Lancet. 2017;389(10082):1941–52.
84. Kumar A, Kapnadak SG, Girgis RE, Raghu G. Lung transplantation in idiopathic pulmonary fibrosis. Expert Rev Respir Med. 2018;12(5):375–85.
85. Shamskhou EA, Kratochvil MJ, Orcholski ME, Nagy N, Kaber G, Steen E, Balaji S, Yuan K, Keswani S, Danielson B, Gao M, Medina C, Nathan A, Chakraborty A, Bollyky PL, De Jesus Perez VA. Hydrogel-based delivery of Il-10 improves treatment of bleomycin-induced lung fibrosis in mice. Biomaterials. 2019;203:52–62.
86. El-Sherbiny IM, McGill S, Smyth HD. Swellable microparticles as carriers for sustained pulmonary drug delivery. J Pharm Sci. 2010;99(5):2343–56.
87. El-Sherbiny IM, Smyth HD. Novel cryomilled physically cross-linked biodegradable hydrogel microparticles as carriers for inhalation therapy. J Microencapsul. 2010;27(8):657–68.
88. Suresh K, Nangia A. Curcumin: pharmaceutical solids as a platform to improve solubility and bioavailability. CrystEngComm. 2018;20(24):3277–96.
89. Mehta P, Bothiraja C, Mahadik K, Kadam S, Pawar A. Phytoconstituent based dry powder inhalers as biomedicine for the management of pulmonary diseases. Biomed Pharmacother. 2018;108:828–37.
90. Oberdörster G. Pulmonary effects of inhaled ultrafine particles. Int Arch Occup Environ Health. 2001;74(1):1–8.
91. El-Sherbiny IM, Smyth HD. Controlled release pulmonary administration of curcumin using swellable biocompatible microparticles. Mol Pharm. 2012;9(2):269–80.
92. Sharma PC, Jain A, Jain S, Pahwa R, Yar MS. Ciprofloxacin: review on developments in synthetic, analytical, and medicinal aspects. J Enzyme Inhib Med Chem. 2010 Aug;25(4):577–89.

93. Du J, El-Sherbiny IM, Smyth HD. Swellable ciprofloxacin-loaded nano-in-micro hydrogel particles for local lung drug delivery. AAPS PharmSciTech. 2014;15(6):1535–44.
94. Billah SM, Mondal MI, Somoal SH, Pervez MN, Haque MO. Enzyme-responsive hydrogels. In: Cellulose polymers and polymeric composites. A Reference Series; 2019. p. 309–30.
95. Secret E, Kelly SJ, Crannell KE, Andrew JS. Enzyme-responsive hydrogel microparticles for pulmonary drug delivery. ACS Appl Mater Interfaces. 2014;6(13):10313–21.
96. Secret E, Crannell KE, Kelly SJ, Villancio-Wolter M, Andrew JS. Matrix metalloproteinase-sensitive hydrogel microparticles for pulmonary drug delivery of small molecule drugs or proteins. J Mater Chem B. 2015;3(27):5629–34.
97. Silva AS, Shopsowitz KE, Correa S, Morton SW, Dreaden EC, Casimiro T, Aguiar-Ricardo A, Hammond PT. Rational design of multistage drug delivery vehicles for pulmonary RNA interference therapy. Int J Pharm. 2020;591:119989.
98. Merkel OM. Can pulmonary RNA delivery improve our pandemic preparedness? J Control Release. 2022;345:549–56.
99. Zoulikha M, Xiao Q, Boafo GF, Sallam MA, Chen Z, He W. Pulmonary delivery of siRNA against acute lung injury/acute respiratory distress syndrome. Acta Pharm Sin B. 2022;12(2):600–20.
100. Merckx P, Lammens J, Nuytten G, Bogaert B, Guagliardo R, Maes T, Vervaet C, De Beer T, De Smedt SC, Raemdonck K. Lyophilization and nebulization of pulmonary surfactant-coated nanogels for siRNA inhalation therapy. Eur J Pharm Biopharm. 2020;157:191–9.
101. Merckx P, De Backer L, Van Hoecke L, Guagliardo R, Echaide M, Baatsen P, Olmeda B, Saelens X, Pérez-Gil J, De Smedt SC, Raemdonck K. Surfactant protein B (SP-B) enhances the cellular siRNA delivery of proteolipid coated nanogels for inhalation therapy. Acta Biomater. 2018;78:236–46.
102. Deb PK, Abed SN, Maher H, Al-Aboudi A, Paradkar A, Bandopadhyay S, Tekade RK. Aerosols in pharmaceutical product development. In: Drug delivery systems. Academic Press; 2020. p. 521–77.
103. Selvam P, El-Sherbiny IM, Smyth HD. Swellable hydrogel particles for controlled release pulmonary administration using propellant-driven metered dose inhalers. J Aerosol Med Pulm Drug Deliv. 2011;24(1):25–34.
104. Ingenito EP, Berger RL, Henderson AC, Reilly JJ, Tsai L, Hoffman A. Bronchoscopic lung volume reduction using tissue engineering principles. Am J Respir Crit Care Med. 2003 Mar 1;167(5):771–8.
105. Reilly J, Washko G, Pinto-Plata V, Velez E, Kenney L, Berger R, Celli B. Biological lung volume reduction: a new bronchoscopic therapy for advanced emphysema. Chest. 2007;131(4):1108–13.
106. US Biologic Lung Volume Reduction (BLVR) Phase 2 Emphysema Study. NCT00515164. https://clinicaltrials.gov/ct2/show/NCT00515164?term=hydrogels+pulmonary&draw=2&rank=4
107. Intraoperative Amiodarone to Prevent Atrial Fibrillation in Lung Transplant Patients. NCT03221764. https://clinicaltrials.gov/ct2/show/NCT03221764?term=hydrogels+pulmonary&draw=2&rank=1
108. Zaetta JM, Licht MO, Fisher JS, Avelar RL, Bio-Seal Study Group. A lung biopsy tract plug for reduction of postbiopsy pneumothorax and other complications: results of a prospective, multicenter, randomized, controlled clinical study. J Vasc Interv Radiol. 2010;21(8):1235–43.e1-3.
109. Effect of Autologous Blood Patch Injection Versus BioSentry Hydrogel Tract Plug in the Reduction of Pneumothorax Risk Following Lung Biopsy Procedures. NCT02224924. https://clinicaltrials.gov/ct2/show/NCT02224924?term=hydrogels+pulmonary&draw=2&rank=2

Resourceful Quantum Dots for Pulmonary Drug Delivery: Facts, Frontiers, and Future

11

Piyush Pradeep Mehta and Vividha Dhapte-Pawar

Abstract

Quantum dots (QDs) have captivated great interest in the scientific domain in the past few years due to their remarkable physicochemical properties. These colloidal nanomaterials have exhibited a variety of applications in different sectors. They are widely used in the medical segment for imaging purposes because of their unique optical and electronic (optoelectronic) features. Similarly, various tactics have been explored to transform the QDs for allowing the combination of biological messengers, targeting agents, and polymeric moieties to progress the biopharmaceutical properties of therapeutic agents. By knowing these key facts current chapter conscientiously summarizes the pulmonary applications of novel QDs-based nanomaterials. Respirable QDs offer good benefits for pulmonary drug delivery owing to high surface-area-to-volume ratio, in vitro/in vivo optical traceability, and ease of post-synthesis treatments such as size-selectivity and surface modifications. Initially, the chapter highlights the QDs fundamentals, synthesis schemes, and pharmaceutical applications. In the later section, the outcomes of in-vitro and in-vivo aerosolization investigations using inhaled QDs in managing microbial infections, lung cancer,

P. P. Mehta (✉)
Global Respiratory R&D, Cipla R&D Center, Mumbai, Maharashtra, India
e-mail: piyu053@gmail.com

V. Dhapte-Pawar
Department of Pharmaceutics, Poona College of Pharmacy, Bharati Vidyapeeth (Deemed to be University), Pune, Maharashtra, India
e-mail: dhapte.vv@bharatividyapeeth.edu

P. P. Mehta, V. Dhapte-Pawar (eds.), *Pulmonary Drug Delivery Systems: Material and Technological Advances*,
https://doi.org/10.1007/978-981-99-1923-9_11

and alveolar macrophage targeting are discussed in depth. Finally, the diagnostic applications of the inhaled QDs are systematically analyzed, followed by a dedicated section on the formulation and medical challenges of QDs-based nanomaterials.

Keywords

Colloidal nanomaterials · Quantum dots · Optoelectronic materials · Carbon-based quantum dots · Lung cancer and diagnostic application

11.1 Introduction

The extensive investigation of nanotechnology has concluded in developing a variety of nanomaterials with diversified physiochemical, mechanical, and biological properties, which ultimately lead to the generation of intriguing applications [1, 2]. Multifunctional bioactive nanomaterials mainly include lipid vesicles (e.g., liposomes, solid lipid nanoparticles, and nanostructured lipid carriers) [3–5], nanoparticles (NPs; e.g., polymeric NPs, magnetic NPs, and metallic NPs) [6–9], micelles [10, 11], nanospheres [12], nano-diamonds, nanotubes [13], hybrid NPs, dendrimers [14], fullerenes [15] and quantum dots (QDs) [16–19]. Among various nanomaterials, QDs have recently received major consideration owing to their practical and diverse synthesis schemes [20, 21]. Further, QDs distinctive physicochemical properties, optoelectronic features, assertive luminescent behavior, and tunable fluorescence emission attributes make them most appropriate for several essential applications [22]. Several active scientists and material chemists have meticulously investigated the significance of QDs platforms for energy storage and conversion devices (e.g., batteries and supercapacitors) [23, 24], flexible electronics (e.g., conductive films) [25, 26] and chemical/non-chemical sensors [27–29]. Additionally, QDs have been utilized for biomedical applications such as cell biology examination, medical imaging, bio-sensing, thin films, and designing of joint prostheses for a fundamental biomedical purpose [30]. Undoubtedly, QDs also have some vital applications in the drug delivery segment. QDs exhibited many potential activities in the drug delivery segment, i.e., cargo systems, modified release systems, gene therapy, photodynamic therapy, theranostics systems, and combined therapies to strengthen the novel drug delivery platform [31–33]. Figure 11.1 shows various vital applications of versatile QDs. Accordingly, in the following piece, QDs structural features, synthesis schemes, and unique properties are thoroughly described.

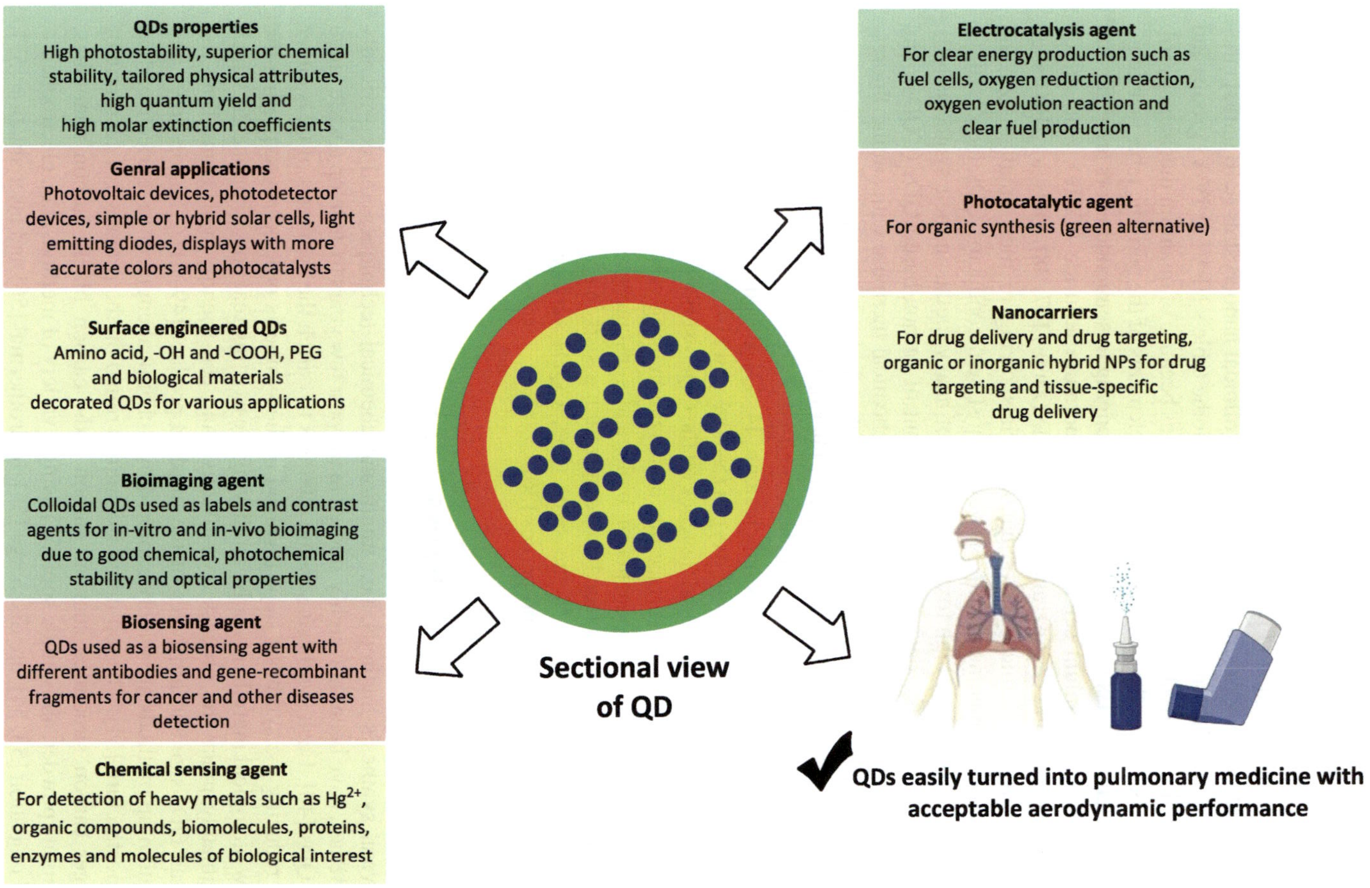

Fig. 11.1 Various vital applications of the QDs

11.2 Quantum Dots

The term "Quantum Dot" is derived from the expression quantum which means the smallest amount of any physical moiety [34]. Particularly, it is a scrutiny of small physical moieties. It flawlessly matches with the physical attributes of the QDs i.e. small particles, efficient of being shaped in to structures having large and long-lasting outcomes [35]. Fundamentally, QDs are the colloidal nanocrystals (<10 nm) synthesized using the elements belonging to the periodic table's III-VI or IV-VI groups and possess unique optical and electronic properties because of quantum mechanics [36]. QD cores belong to the III-V, II-VI, or IV-VI elements of the periodic table (semiconductors elements). Examples of elements belonging to semiconductors are cadmium sulfide (CdS), cadmium selenide (CdSe), cadmium telluride (CdTe), zinc sulfide (ZnS), zinc selenide (ZnSe), and zinc telluride (ZnTe) [37, 38]. QDs are also called zero-dimensional semiconductor nanomaterials owing to their smaller size and highly dense arrangement [39]. However, most semiconductor QDs faced critical biological issues, i.e., high toxicity because of heavy metals in their synthesis [40]. Thus, this incited the design and development of carbon-based QDs (CQDs), and CQDs displayed good biocompatibility and chemical inertness with similar fluorescence properties [41–43]. CQDs are classically quasi-spherical nanomaterials containing amorphous to the nano-crystalline interior with mainly turbo-stratic carbon, crystalline carbon (graphitic), graphene, and graphene oxide sheets merged by diamond-like sp3 hybridized carbon inclusions [44–47]. Based on the synthesis method, CQDs hold variable numbers of oxygen (5–50% w/w) [45] and carboxyl moieties at their surface [22]. The carboxyl moieties on the CQDs surface impart excellent aqueous solubility and ease the surface functionalization using various inorganic, organic materials, polymeric components or biological substances with adequate surface passivation [22]. CQDs surface functionalization assists in enhancing their physical and biological properties. At the same time, CQDs surface passivation helps to improve their fluorescence properties [30].

11.3 QDs Synthesis Schemes

Several synthesis methods, i.e., traditional chemical techniques and eco-friendly schemes, have been thoroughly studied in the past few years to develop a wide range of QDs [48]. Each synthesis scheme attempts to enrich the synthesis of QDs either by reducing the synthesis steps involved in the synthesis process, by exploring the new biological ingredients or by modifying the existing reaction components, which ultimately make the process cost-effective and eco-friendly with controlling the mean QDs size and attributes of the finished product [49–51]. The top-down and bottom-up methods are widely explored by a chemical engineer to fabricate carbon nanostructure [51]. The top-down techniques relied on the progressive cessation of larger carbon structures such as carbon rods, carbon black, carbon nanotubes, graphite powder, and candle soot [52] using a variety of methods, i.e., electrochemical oxidation [53, 54], arc discharge [55], hydrothermal [56] and laser ablation [57].

The bottom-up techniques mostly involved molecular precursors such as -OH, -COOH and NH_2 to fabricate QDs after controlled dehydration and carbonization actions [58] such as hydrothermal reactions [59], plasma-hydrothermal method [60], microwave-hydrothermal effect [61] and pyrolysis or carbonization of organic precursors. Bottom-up techniques also present better opportunities to tune the size, morphology, and physicochemical properties of QDs.

Furthermore, scientific information illustrates the lucrative use of naturally abundant resources to synthesize QDs that reveal excellent physicochemical and emission properties. These methods are also termed biological (eco-friendly/green) methods. Biological methods involved the use of eukaryotic (e.g., animals, plants, and fungi), prokaryotic (single-celled microorganisms), and isolated phytochemicals (flavonoids, alkaloids, phenols, terpenoids, and tannins) to fabricate QDs [62, 63]. Another exciting approach for synthesizing environmentally benign QDs is photocatalysts which may ultimately reduce the burden of energy utilization, product clean-up, and waste clearance [22]. Briefly, the ease of QDs synthesis and the ability to tune their physical and optical properties during synthesis make them more opportune for several vital applications, particularly in the biomedical field.

By knowing these fundamental material properties, unique photo-physical features, and ease of synthesis, several active researchers meticulously designed and developed the QDs to manage several life-threatening diseases, especially chronic respiratory disease and lung cancers. Accordingly, in the present chapter, we will discuss the QDs applications in pulmonary drug delivery in great detail. We have mainly summarized the applications of QD such as unique cargo applications, surface functionalization, bio-imaging, chemical-sensing, bio-sensing, and optical electronics properties to manage and diagnose pulmonary diseases effectively. Briefly, it is hoped that this chapter will offer a broad indication of the present status of QD research in the pulmonary drug delivery domain and open new prospects for the investigation of QDs with many improved physicochemical properties to fight the critical battle against life-threatening pulmonary diseases.

11.4 Pulmonary Applications of QDs

Diversified nanomaterials, including QDs, are extensively studied, analyzed as potential carriers for the pulmonary delivery of different drugs. According to size, shape, surface charge, and formulation type, QDs exhibited different drug distribution patterns, aerodynamic outcomes, and lung epithelium interactions. Additionally, post-functionalization of QDs with targeting substances and polymers like polyethylene glycol (PEG) significantly broadens their capability to improve their biocompatibility and in-vivo efficiency. Various essential investigations on inhalable QDs formulations are summarized in Table 11.1, and appealing case studies are briefly summarized in the subsequent sections.

The immune/inflammatory response regulation is an imperative task in several pulmonary diseases. The persistent balance between over-active immune response and suppressed immunity is highly vital for immune/inflammatory response

Table 11.1 Novel QDs-based nanoparticulate systems for effective pulmonary therapy

Formulations	Particle properties	Outcome	References
DOX-CdSe/CdS/ ZnS QDs	Negatively charged (15.42 mV) and 30.70 nm	DOX QDs showed a significant 2.25-fold improvement in in-vitro DOX uptake by alveolar macrophages compared to DOX alone and significantly low inflammatory response in C57BL/6 mice after oropharyngeal aspiration compared to DOX alone	[64]
CC-QDs	Positively charged (42.10 mV) and 19.70 nm	Dots efficiently delivered the siRNA to A549-Luc cells during the in-vitro study and after intranasal instillation to BALB/c mice showed satisfactory transgene expression with significantly lower toxicity	[65]
TAN-ZnO-QDs	Crystalline spherical and < 5 nm	QDs showed pH depended release and effectively induced the cytotoxicity against NCI-H358 cells with successful regulating of apoptosis activity	[66]
MPA-CdSe-QDs	Negatively charged (55.97 mV) and 4.69 nm	QDs showed a significant genotoxic potential with a high number of DNA strand breaks	[67]
CYST-CdSe-QDs	Positively charged (57.40 mV) and 3.68 nm	QDs showed a significant cytotoxic potential with a concentration-dependent increase in necrosis of NHBE cells	[67]
DOX-HA-PEG-ZnO-QDs	Ultra-fine NPs with 6 nm size	QDs showed 72% DOX release after 80 h at pH 5.5 and exhibited higher antitumor activity in A549 cells with higher cellular uptake due to receptor-mediated endocytosis	[68]
QDs-OSI-420	Spherical-shaped negatively charged (9.20 mV) and 48.60 nm	QDs showed significantly reduced tumor volume in 3D-spheroid cell culture of A549 as compared to erlotinib	[35]
PEGylated-FCNDs	Spherical negatively charged (12.80 mV) and 60.30 nm	Nanodots showed successful human bronchial epithelial cell membrane adherence and helpful penetration for diagnostics, imaging, and drug delivery applications	[69]
3-MPA coated CdSe-CdS/ZnS core-multishell QDs	QDs diameter without surface ligands - 5.7 nm	Core-multishell QDs did not stimulate any evident variation within the transepithelial electrical resistance and molecular morphology of the Calu-3 cell monolayers	[70]

Table 11.1 (continued)

Formulations	Particle properties	Outcome	References
CET-ETO/DNA biodots-TPGS liposomes	Spherical smooth surface negatively charged (29.30 mV) and 181.60 nm	QDs showed pH-dependent drug release and significant 180-fold lower IC_{50} values compared to ETO with augmented anticancer activity via receptor-mediated endocytosis in A549 cells	[71]
FA/GA-InP/ZnS QDs-DOX	Spherical morphology and 11 nm	QDs showed enzyme-dependent DOX release with higher cell uptake than non-functionalized QDs using A549 cells	[72]
Zn-RIF transferrin-conjugated silver QDs	Nano-crystalline and < 20 nm	Nano-systems showed a marked ten-fold superior antibacterial action against *M. smegmatis* and *M. bovis*-BCG compared to drug alone with the efficient killing of mycobacteria residing inside RAW 264.7 cells without exhibiting cytotoxicity and genotoxicity	[73]
DOX-Poly (N,N-diethyl acryl amide) functionalized graphene QDs hydrogels	<100 nm	After intraperitoneal administration, thermo-responsive hydrogels followed the Fickian diffusion release mechanism with a marked tumor volume reduction in the melanoma tumor model	[74]

regulation therapy [75, 76]. Thus, we need physically stable, site-specific therapeutic delivery platforms which deliver drug to a specific cellular target and permit a good control on therapy, persistent balance with reduced collateral damage frequency. The alveolar macrophage cells are mainly accountable for directing the host adaptive and innate immune responses involved in non-infectious and infectious pulmonary diseases. Therefore, Chakravarthy et al. (2011) fabricated physically stable, site-specific QD nanoconjugates using prototype molecule doxorubicin (DOX) for alveolar macrophage targeting. The rationale to use QD-DOX nanoconjugate is that nanoconjugate platform offers visual validation of the intracellular uptake and dissociation of DOX in the nucleus which is obligatory for drug action due to the differential photoluminescence properties of these two agents. Particularly, Chakravarthy and co-workers prepared DOX conjugated cadmium selenide/cadmium sulfide/zinc sulfide-based QDs (CdSe/CdS/ZnS QDs). CdSe/CdS/ZnS QDs were primarily synthesized by growing a CdS/ZnS graded shell on a CdSe core. Afterward, DOX was covalently coupled to CdSe/CdS/ZnS QDs using the amino group of DOX and carboxyl group of CdSe/CdS/ZnS QDs to formulate DOX-QDs nano-conjugates. CdSe/CdS/ZnS QDs showed a hydrodynamic diameter of 11.10 nm and a negative surface charge of 38.38 mV. In contrast, DOX-QDs nanoconjugates displayed a hydrodynamic diameter of 30.70 nm and a negative surface charge of 15.42 mV. The variation in the hydrodynamic diameter substantiates the surface presence of DOX, while the remarkable 2.5-fold shift in zeta potential verifies the formation of amide linkage between DOX and QDs. In an in-vitro study after 24 h, DOX-QDs nano-conjugates showed a significant 2.25-fold improvement in DOX uptake by alveolar macrophages compared to the DOX alone. During an in-vivo study using C57BL/6 mice after oropharyngeal aspiration, DOX-QDs nano-conjugates (10 nM) showed a significantly low inflammatory response and functional lung impairment compared to DOX alone (10 nM in water). Briefly, CdSe/CdS/ZnS QDs have an excellent perspective for diagnosing and treating pulmonary disorders mainly due to their selective uptake into alveolar macrophages [64].

Gene therapy is a promising therapy for many diseases including inherited genetic disorder (Parkinson's disease) or acquired diseases (e.g., cancer) [77–80]. Gene therapy involved delivery of genetic material to specific cells to inhibit the production of a target protein or increase specific gene expression. Owing to the size and the polyanionic nature of nucleic acid, its fragility toward nucleases, and the negative potential of the cell membrane surface, gene therapy needs the usage of an apt carrier for efficient and safe intracellular delivery of the therapeutic payload [77–80]. Although non-viral (synthetic) and viral carriers present good alternatives, a few fundamental difficulties such as low in-vivo transfection efficacy, hinder the US FDA / European Medicines Agency approvals and may end up in unforeseen side effects allied with these carriers. Thus, Pierrat et al. (2015) applied genuine efforts to study and understand the cationic carbon dots (CC-QDs) role in gene therapy for pulmonary disease management. Principally, Pierrat and co-workers developed a simple and economical process to formulate CC-QDs. CC-QD were formulated using one-step microwave-assisted pyrolysis of citric acid in the presence of branched polyethyleneimine (bPEI25k) for efficient pulmonary delivery of nucleic acid. CC-QDs showed particle size, zeta potential, and percent yield of 19.70 nm, 42.10 mV, and 20%, respectively. During the in-vitro study, CC-QDs efficiently delivered the siRNA to adenocarcinoma human alveolar basal epithelial cells (A549), and high specific gene knockdown was obtained without associated cytotoxicity. During an in-vivo study using BALB/c mice after intranasal instillation, CC-QDs/DNA complexes (40 mg pCMV-GLuc) showed similar transgene expression with significantly lower toxicity when compared with the golden standard formulation GL67A. Moreover, during in-vivo study CC-QDs/DNA complexes with a constant weight ratio (2.6 w/w) with smaller particle size (48.6 nm and + 50.1 mV) showed fourfold lower transfection efficacy compared to large ones (205.3 nm and + 27.3 mV). The dissimilarity in transfection efficiency is mainly due to quick engagement of small particles with periciliary layer followed by removal by mucociliary clearance while larger particles speedily pass the periciliary layer and reach the target cells. Thus, in upcoming time with control on mean particle size and engineering of CC-QDs surface with targeting moieties or PEG will extensively broaden QDs scope in pulmonary drug delivery [65].

Tangeretin (TAN) is a polymethoxyflavones found in fruits of citrus plants such as *Citrus sinensis, Citrus reticulata* (Rutaceae), and *Hura crepitans* (Euphorbiaceae). In recent years TAN has received wide scientific scrutiny due to its diverse pharmacological activites such as neuroprotection, mitochondrial biogenesis, anti-inflammatory, anti-diabetes, efflux pump inhibition, anti-microbial [81, 82]. It exhibited anti-oxidant properties via regulating several anti-oxidant enzymes, inhibits DPPH (2,2-diphenyl-1-picrylhydrazyl) radicals, hydrogen peroxides, hydroxyl radicals and scavenges superoxide anion radicals. Moreover, many pre-clinical investigations displayed that TAN shows anticancer activity in various cancer cells, arrests the cell cycle, induces apoptosis, and inhibits angiogenesis, and metastasis. It is a well-known nutraceutical molecule [81, 82]. However, its therapeutic activities are mainly hampered by poor physicochemical and biological properties such as low water solubility, poor absolute oral bioavailability (27.11%) [83], and most of the physical properties are dependent upon the substituents on molecules. To defeat these problems, researchers focused on pH-responsive inorganic nanocarriers, i.e., water soluble and stable zinc oxide QDs (ZnO QDs). Consequently, Choi and co-workers developed TAN-ZnO-QDs and thoroughly investigated them for anti-tumorigenic and anti-metastatic properties using the primary bronchoalveolar carcinoma cell (NCI-H358). Crystalline and uniformly distributed spherical-shaped TAN-ZnO-QDs showed ultra nano size (<5 nm). During an in-vitro drug release study using dialysis membranes, TAN-ZnO-QDs displayed pH-depended release of TAN. TAN-ZnO-QDs exhibited >80% TAN release within 15 h at an acidic environment (pH 5.5). However, at pH 7.4, TAN-ZnO-QDs showed <15% TAN release. During an in-vitro anticancer study using NCI-H358 cells, TAN-ZnO-QDs effectively induced cytotoxicity and morphological damage to NCI-H358 cells. TAN-ZnO-QDs also successfully induced nuclear fragmentation and imparted cell cycle arrest at the G2/M phase compared to untreated control cells. In the case of metastasis, TAN-ZnO-QDs successfully downregulated the key protein markers of metastasis, i.e., MMP2 and MMP9. Additionally, TAN-ZnO-QDs effectively regulated the apoptosis activity by escalating the expression of the pro-apoptotic protein Bax and controlling the activity of the anti-apoptotic protein Bcl-2. The pH-responsive ZnO-QDs aid in the delivery of phytoconstituent for the treatment of lung cancer but further in-vivo investigations are highly obligatory to translate this system into clinical applications [66].

Nagy et al. (2013) fabricated a typical CdSe QD (3 nm) core and functionalized the same with mercaptopropionic acid (MPA-CdSe-QDs) and cysteamine (CYST-CdSe-QDs) to obtain negatively and positively charged QDs surfaces, respectively. Functionalized QDs were studied for cytotoxic and genotoxic potential using primary normal human bronchial epithelial cells (NHBE). MPA-CdSe-QDs displayed mean particle size and zeta potential of 4.69 nm and - 55.97 mV, respectively, while CYST-CdSe-QDs showed mean particle size and zeta potential of 3.68 nm and 57.40 mV, respectively. In colloidal stability testing, CYST-CdSe-QDs (100 μg/mL) showed 27.03-fold and 8.50-fold lower leaching of Cd^{2+} ions from QDs core than the MPA-CdSe-QDs (100 μg/mL) using phosphate buffer saline (PBS) pH 4 and 7, respectively. In terms of colloidal properties, the sterically bulkier carboxyl group of the MPA allows relatively poor packing of QDs as compared to the amine end group of the CYST. CYST amine groups are competent in hydrogen bonding and thus provide a more protective layer to cover QD surface sites. During the cellular viability study, CYST-CdSe-QDs showed a significant concentration-dependent increase in NHBE necrosis during the 24 h study. In contrast, MPA-CdSe-QDs did not alter NHBE membrane permeability even after 48 h exposure. The CYST-CdSe-QDs (80 and 160 μg/mL) revealed a significant decrease ($p < 0.005$) in the mitochondrial function of NHBE. At the same time, MPA-CdSe-QDs did not show any alteration in NHBE mitochondrial function at any of the concentrations analyzed. The positively charged CYST-CdSe-QDs showed approximately two-fold higher cytotoxic potential than the negatively charged MPA-CdSe-QDs. The CYST-CdSe-QDs displayed significant cytotoxicity, reactive oxygen species formation and DNA damage. However, even in the absence of cytotoxicity and reactive oxygen species formation, MPA-CdSe-QDs caused a high number of DNA strand breaks. Additionally, the presence of p53 binding protein 1 (53BP1) in the nuclei of exposed cells and successive reduction of p53 from cytoplasmic cellular extracts confirmed the DNA damage (strand breaks and micronuclei formation) caused by MPA-CdSe-QDs. Most prominently, transcriptome sequencing showed the upregulation of the metallothionein family of genes in NHBE cells treated with MPA-CdSe-QDs but not with the CYST-CdSe-QDs. Briefly, the negatively charged QDs are relatively non-cytotoxic compared to the positively charged QDs. However, the same does not hold for their genotoxic potential. Results showed that both cytotoxic analysis and genotoxic assessment are essential for understanding cellular responses and risk assessment during the design of QDs-based pulmonary drug delivery systems [67].

Cai et al. (2016) formulated a ZnO-QDs-based pH-responsive drug delivery platform to probe intracellular controlled release of model anticancer drug, i.e., DOX. ZnO-QDs showed an ultra-small size of 3 nm. Initially, ZnO-QDs were amine-functionalized using a 3-aminopropyltriethoxysilane (APTES) solution (NH_2-ZnO-QDs). Later, dicarboxy-terminated PEG_{2000} was covalently linked to the NH_2-ZnO-QDs using amide reaction (PEG-ZnO-QDs). Moreover, to achieve better swelling and solubilization properties, hyaluronic acid was conjugated to PEG-ZnO-QDs (HA-PEG-ZnO-QDs). Furthermore, DOX molecules were loaded to PEG functionalized ZnO QDs, i.e., DOX-PEG-ZnO-QDs and DOX-HA-PEG-ZnO-QDs, forming metal-DOX complex and covalent interactions. Aggregated acid-labile luminescent ZnO-QDs showed an ultra-small size of 3 nm, while PEG-ZnO-QDs showed a stable particle size of 6 nm. The notable 37-fold reduction in zeta potential value of HA-PEG-ZnO-QDs confirms the successful conjugation of hyaluronic acid to PEG-ZnO-QDs. HA-PEG-ZnO-QDs exhibited adequately high loading efficiency (83.4%) for DOX. During the in-vitro drug release study, DOX-HA-PEG-ZnO-QDs showed 72% DOX release after 80 h in the acidic environment (pH 5.5). However, at pH 7.4, DOX-HA-PEG-ZnO-QDs showed only 36% DOX release. This variable DOX release was mainly ascribed to the fast dissolution of ZnO-QDs and quick dissociation of metal-DOX complex under acidic conditions. During an in-vitro cytotoxicity study using A549 cells, DOX-HA-PEG-ZnO-QDs exhibited higher antitumor activity than DOX-PEG-ZnO-QDs. The superior antitumor activity of DOX-HA-PEG-ZnO-QDs was primarily attributed to (i) PEG_{2000} conjugation, which offered stability in physiological milieu; (ii) targeting ligand, HA, which mainly binds to the overexpressed cell surface molecule, i.e., glycoprotein CD44. Additionally, in a cell uptake study using A549 cells,

DOX-HA-PEG-ZnO-QDs followed the receptor-mediated endocytosis pathway. pH-sensitive DOX-HA-PEG-ZnO-QDs specifically dispersed the Zn^{2+} metal ions within the acidic environment in the biological milieu. They activated the dissociation of the metal-DOX complex to achieve controlled DOX release. Briefly, PEG_{2000} and HA decorated ZnO-QDs-based pH-responsive drug delivery platform provided a satisfactory alternative for the antitumor effect [68].

Erlotinib (ERL) is a first-generation tyrosine kinase inhibitor. ERL is a quinazoline derivative applied in the treatment of metastatic and advanced non-small cell lung cancer (NSCLC) [84, 85]. However, ERL antineoplastic properties also come with adverse effects such as transient elevations in serum aminotransferase levels, clinically evident acute liver injury and over a period, majority of NSCLC subjects developed resistance to ERL. Presently, ERL is accessible as an oral coated tablet (Tarceva®, OSI Pharmaceuticals, LLC) and has poor bioavailability (<60%). With this background in consideration, the development of a drug delivery system containing QDs may be valuable in reducing the chances of developing adverse effects and resistance. During the fabrication of QD carriers a chemical linkage with an appropriate functional group, i.e., carboxylic acid (-COOH) is advantageous for effective drug conjugation. However, ERL lacks any promising functional groups for effective drug conjugation. Therefore, an active metabolite of ERL, i.e., desmethyl erlotinib; OSI-420 (free base) was explored by Kulkarni et al. (2019) for effective therapy of lung cancer (Fig. 11.2). Initially, chemical conjugation between OSI-420 and carboxylic acid functionalized CdSe/ZnS QDs was performed using carbodiimide (EDC-NHS) chemistry. Particularly, an ester linkage between carboxylic acid (-COOH) group of CdSe/ZnS QDs and free hydroxyl (-OH) group of OSI-420 was formed by preserving the reaction mixture ratio 0.4:1.

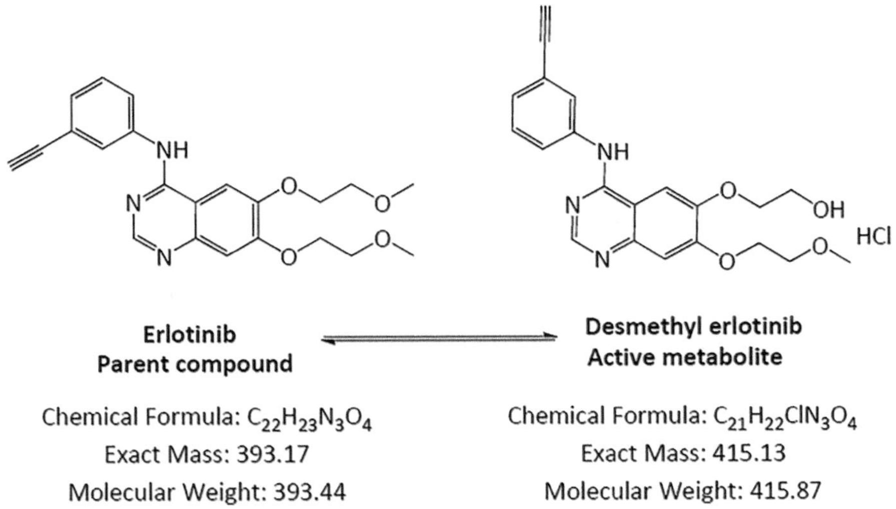

Fig. 11.2 Molecular structure and information of erlotinib and its active metabolite, i.e., desmethyl erlotinib

Spherical-shaped nanosized conjugated particles displayed mean particle size, zeta potential and percent drug conjugation of 48.60 nm, −9.20 mV and 52.20%, respectively. The intracellular porcine liver esterase enzyme-assisted drug release study was performed in lysosomal compartments. QDs-OSI-420 showed esterase enzyme concentration-dependent drug release. Mainly, with 40 U/mL esterases, QDs-OSI-420 showed 50% drug release within the first 24 h and approx. 100% drug release within 72 h while 20 U/mL esterases showed only 50% drug release within the span of 72 h. QDs improved the OSI-420 internalization during the cellular uptake study, and the cellular recovery study confirmed similar observations. As per the in-vitro cell viability analysis using A549 cell lines, QDs-OSI-420 showed half-maximal inhibitory concentration (IC_{50}) of 2.5 μM concentration while erlotinib and OSI-420 alone did not reveal >60% inhibition at 20 μM concentration. During the cytotoxic study, QDs markedly enhanced the OSI-420 toxic effect than erlotinib and OSI-420 alone. In the single and multiple-dose study using 3D-spheroid cell culture of A549, QDs-OSI 420 significantly reduced the tumor volume compared to erlotinib and OSI-420 alone. Reported QD systems can be further improved by using targeting ligands to bind particular overexpressed proteins in lung cancer cells and making the delivery system more precise [35].

PEGylated fluorescent carbon nanodots (PEGylated-FCNDs) were developed using glucose microwave synthesis. During synthesis, partially dehydrated residues improved the hydrophilicity and stability of the carbon nanodots in aqueous systems and potentially improved the further diagnostics and drug delivery applications. Well-defined spherical PEGylated-FCNDs showed a hydrodynamic diameter of 60.30 nm with a zeta-potential of −12.80 mV in PBS (0.01 M at pH 7.4). During the in-vitro cell line study, PEGylated-FCNDs were effectively located on the human bronchial epidermal cells (16HBE) surface and showed a marked three-fold improvement in lactate dehydrogenase (LDH) release compared to the control group. The improvement in LDH release and higher intercalation with cell membrane was mainly ascribed to nanodots' well-defined size and morphology. Additionally, similar cell membrane adherence and penetration properties were confirmed from the confocal imaging. During oxidative stress assays, PEGylated-FCNDs effectively decreased the cell viability due to oxidative damage mechanism, i.e., a significant increase in the intracellular O_2. Thus, PEGylated-FCNDs can be a helpful approach toward in-vitro and in-vivo diagnostics, imaging, and drug delivery applications [69].

Airways epithelial cells are widely exposed to NPs present in the environment. NPs like colloidal QDs are also judged as potential carriers for pulmonary drug delivery. Thus, to recognize the ways of distribution and intercalation of QDs with lung cell membrane Turdalieva et al. (2016) developed colloidal 3-mercaptopropionic-acid (3-MPA) coated CdSe-CdS/ZnS core-multishell QDs. Colloidal 3-MPA coated CdSe-CdS/ZnS core-multishell QDs were chemically synthesized using CdSe core with two monolayers of CdS shell, one monolayer of $Cd_{0.5}Zn_{0.5}S$ and 1.5 monolayers ZnS. Subsequently, CdSe-CdS/ZnS core-multishell QDs were covered with 3-MPA surface ligands. The diameter of the QDs without the 3-MPA surface ligands was about 5.7 nm. Using live confocal microscope imaging and QD fluorescence measurements, CdSe-CdS/ZnS core-multishell QDs showed limited QD permeation through the matured human lung cancer cell line (Calu-3) monolayers. During the QDs deposition study, a decrease in transepithelial electrical resistance of the Calu-3 cell monolayer was proportional to the QD dose. Additionally, repeated QD deposition, i.e., once every 6 days for 2 months, showed only small amounts of the QDs accumulation within the Calu-3 cell monolayer. Still, repeated QD deposition did not stimulate any evident variation within the transepithelial electrical resistance and molecular morphology of the Calu-3 cell monolayers. Briefly, colloidal 3-MPA covered CdSe-CdS/ZnS core-multishell QDs offered the potential delivery platform for effective pulmonary drug delivery. Thus, colloidal 3-MPA coated CdSe-CdS/ZnS core-multishell QDs can be used potentially for drug delivery of drugs intended to reach the surface of the lung epithelia for limited treatment time [70].

Epidermal growth factor receptor (EGFR) is a member of tyrosine kinases receptor. EGFR is accountable for controlling epithelial tissue development and homeostasis. EGFR is overexpressed in malignant cells and its overexpression was mostly related to progressive disease and poor diagnosis [86, 87]. Similarly, EGFR is also overexpressed in NSCLC cells. Cetuximab (CET) is a chimeric (mouse/human) IgG1 monoclonal antibody that efficiently binds to EGFR and effectively prevents the binding of EGF. CET (Erbitux®; Eli Lilly, USA) is indicated for the treatment of colorectal as well as head and neck cancer. Erbitux® advised initial dose is intravenous infusion (400 mg/m^2 for 2 h). To efficiently deliver CET in NSCLC therapy, Jha et al. (2020) fabricated targeted theranostic liposome carriers. Predominantly, CET conjugated, etoposide (ETO; beta-d-glucoside) and DNA biodots co-loaded D-α-tocopheryl polyethylene glycol 1000 succinate (TPGS) coated liposomes were developed for effective imaging and treatment of NSCLC. The liposomes were prepared by the solvent injection method while CET was conjugated to these liposomes using carbodiimide reaction to provide receptor-specific drug delivery. ETO: lipid (1:20 w/w) ratio was kept constant. Smooth surface liposomes showed spherical shape with particle size, zeta potential, and encapsulation efficiency of 181.60 nm, −29.30 mV, and 42.04%, respectively. Liposomes displayed satisfactory encapsulation efficiency for ETO (69.51%) and biodots (42.04%). In Bradford assay, CET showed an adequate conjugation (75%) on the liposome surface while in solid state analysis absence of specific diffraction peaks of ETO confirmed the amorphous nature of ETO and successful encapsulation within the core-shell structure of liposomes. Liposomes showed pH-dependent drug release with a significant increase in the drug release at acidic pH (5.5) compared to pH 6.4 and 7.4. Liposomes showed 50% of the drug releases within the first 16 h at lower pH (5.5), whereas liposomes required 27 and 19 h to release 50% of the drug at pH 7.4. and 6.4, respectively. Liposomes demonstrated increased fluidity of bilayer membrane at acidic pH because of the hexagonal phase transition of 1,2-distearoyl-sn-glycerol-3--phosphoethanolamine (lipid component) of liposomes at lower pH. Further, this transition raises the antitumor efficiency of liposomes and decreases the side effects caused to healthy tissues. In cytotoxicity assay using MTT in A549 cells, liposomes demonstrated significantly 180, 14.86-, 21.45-fold lower IC_{50} values than ETO alone, ETO-liposomes, and biodot/ETO-liposomes, respectively. Liposomes displayed a target-specific effect and augmented anticancer activity via receptor-mediated endocytosis. The augmented anticancer activity was proved using a blocking study during cellular uptake investigation. During an in-vivo study, theranostic liposomes demonstrated selectivity toward EGFR overexpressed cells and offered superior NSCLC targeted delivery of ETO compared to non-targeted and controlled delivery systems. In upcoming time, after suitable pre-clinical studies, theranostic biodot loaded liposomes may translate into a potential clinical alternative for concurrent imaging and therapy of lung cancer [71].

Folate receptors are the well-explored membrane-bound surface proteins. They are normally overexpressed in human malignancies (e.g., lung and nasopharyngeal cancer) and bind folates and folate conjugates with high affinity [88]. Likewise, D-glucosamine is another vital substance widely studied in targeted cancer therapy of solid tumors with high count of glucose transporters. Moreover, D-glucosamine improves the biocompatibility of NPs and worked as chemo-sensitiser in drug resistant cancer cells [89, 90]. Thus, Ranjbar-Navazi et al. (2018) developed folic acid (FA) and D-glucosamine (GA) dual functionalized indium phosphide/zinc acetate QDs (InP/ZnS QDs) for delivery of model drug DOX. InP/ZnS QDs were fabricated using hot injection method while FA/GA were linked using carbodiimide chemistry. Furthermore, carboxyl groups of InP/ZnS QDs surface were activated through carbodiimide interaction to link DOX with the system. DOX-QDs showed drug conjugation efficacy of 70.34% with spherical morphology and particle size of 11 nm. During the in-vitro drug release study, DOX-QDs showed no significant DOX release at neutral conditions (PBS pH 7.4) and minor DOX release at neutral conditions (PBS pH 5.5) because of the stable amide bonds between the DOX molecule and QDs surface carboxylic groups. Briefly, DOX-QDs showed satisfactory DOX release because of the enzymatic destruction amid bonds between DOX and QDs. During a cell uptake study using A549 cells, flow cytometry analysis of FA/GA-InP/ZnS QDs showed 87.90, 12.10, 1.88, and 1.62-fold higher uptake compared to InP/ZnS QDs, FA-InP/ZnS QDs, GA-InP/ZnS QDs and FA/GA- InP/ZnS QDs-DOX, respectively. Briefly, GA decorated QDs displayed higher uptake compared to FA decorated QDs. Higher uptake of GA decorated QDs by the A549 cells was mainly because of the overexpression of glucose transporter 2 that has evidently superior affinity to GA. Thus, synchronized imaging and delivery of higher-order GA functionalized nanostructures are deemed to offer a better means in lung cancer therapy [72].

Tuberculosis (TB), is one of the leading causes of morbidity and mortality across the globe, caused by *Mycobacterium tuberculosis*. High drug resistance, chronic dose therapy (500–600 mg/day for 6–9 months) and critical disputes allied to administration routes (low bioavailability and lack of targeted delivery) are the key issues associated to TB treatment. Rifampicin (RIF), is a semisynthetic broad-spectrum macrocyclic antibiotic, prescribed for the treatment of TB because of its effective DNA synthesis inhibition activity. Everyday oral dose of RIF (600 mg) showed serum RIF level well above the MIC (minimum inhibitory concentration) of TB bacteria. However, it may cause few adverse side effects such as renal failure, hepatotoxicity, thrombocytopenia, and antibody formation [91–94]. Thus, with the aim of improving biopharmaceutical properties and controlling adverse side effects,

Pati et al. (2016) developed novel inhaled composite therapy using RIF and QDs. In the present case, RIF is a well-known candidate to coordinate with metal cations via various chemical groups namely two aliphatic OH and two phenolic groups. Besides, flexible backbone with the company of extra oxygen and nitrogen donor atoms render RIF for developing coordination bonding with transition metal ions like zinc (Zn). Initially, Zn-RIF (18:80) complex was developed using hydrated zinc nitrate and methanol solvent. Subsequently, to synthesize Zn-RIF transferrin-conjugated silver QDs, Zn-RIF complex (500 μg/mL) and transferrin (1 mg/mL) were added concurrently to silver nitrate (1 mM) and finally, cocktail was reduced by surplus ice cold sodium borohydrate mix (Fig. 11.3). In reordering reaction, transferrin initially binds to silver QDs to form a stable complex which then eases the binding of Zn-RIF complex by covalent and non-covalent exchanges. Nano-crystalline conjugated QDs showed a mean particle size <20 nm. In antibacterial activity, conjugated QDs displayed a marked tenfold superior antibacterial action against *Mycobacterium smegmatis* and *Mycobacterium bovis*-BCG compared to the drug alone. During immuno-fluorescence assessment, conjugated QDs were dynamically endocytosed by peritoneal macrophages and dendritic cells but not by A549 lung epithelial cells, which specifies that synthesized QDs are mainly delivered to the macrophages only. In intracellular stability using mouse macrophages model, FITC tagged conjugated QDs remained stable up to 48 h with a gradual decrease of fluorescence intensity at the end of 48 h. In an intracellular killing assay using macrophage-like, Abelson leukemia virus-transformed cell line (RAW 264.7) infected with *M. smegmatis* and *M. bovis*-BCG, conjugated QDs efficiently killed mycobacteria residing inside macrophages without exhibiting cytotoxicity and genotoxicity. Sequence of invetigations presented that Zn-RIF displayed more potent antibacterial activity against both mycobacterial strains compared to RIF or

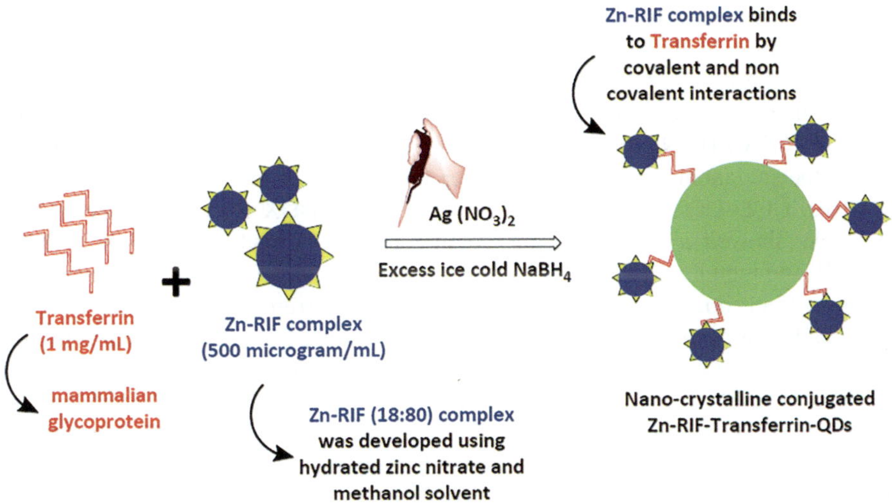

Fig. 11.3 Graphical representation of the development of Zn-RIF transferrin-conjugated silver QDs

Zn. The better penetration of Zn-RIF through the mycobacterial membrane is mainly due to its higher lipophilicity compared to RIF alone. Besides, Zn is well-known to break bacterial cell walls. This finally eases the diffusion of RIF across the mycobacterial cell walls and prevents RNA synthesis. Briefly, drug-loaded transferrin conjugated QDs showed great potential for TB therapy [73].

Polymeric nanohydrogels (HYNPs) have acquired good attention in drug delivery stream because of their unique 3D framework, wide structural flexibility, easy preparation steps and effective drug-loading capabilities [95, 96]. Similarly, graphene quantum dots (GQDs) are also one of the budding delivery vector owing to unique chemical and physical attributes, capability to house functional groups such as carbonyl, hydroxyl, epoxy, and carboxyl for rearrangement reaction and low cytotoxicity with and biocompatibility. By knowing this background Havanur et al. (2019) fabricated thermo-responsive poly(N, N-diethyl acrylamide) [PDEA]/functionalized GQDs HYNPs for effective management of metastatic lung cancer using a model drug DOX. Primarily, nanoscale polymeric drug carrier PDEA HYNPs were formulated using inverse emulsion polymerization with water:oil ratio of 1:5. GQDs were fabricated using microwave assisted pyrolysis of citric acid by a simple bottom-up approach. Afterward, synthesized GQDs were fused with HYNPs to increase the lower critical solution temperature (LCST) and anticancer functions of the delivery system. Infrared spectroscopy showed clear appearance of -OH and C-C stretch and loss of characteristic peaks of C=O and C-H. This verifies the fusion of GQDs with PDEA HYNPs. Particularly, DOX HYNPs, GQD-DOX HYNPs, leucine-GQD-DOX HYNPs, and methionine-GQD-DOX HYNPs were assessed for lung cancer activity. The particle size and encapsulation efficiency of DOX HYNPs, GQD-DOX HYNPs, leucine-GQD-DOX HYNPs, and methionine-GQD-DOX HYNPs were 59.90 nm, 68.10 nm, 91.70 nm, 99 nm and 85.87%, 41.22%, 41.69%, 47.35%, respectively. DOX HYNPs, GQD-DOX HYNPs, leucine-GQD-DOX HYNPs and methionine-GQD-DOX HYNPs showed LCST of 28 °C, 36 °C, 40 °C, and 32 °C, respectively. Further, thermo-responsive GQDs modified PDEA carriers showed acceptable DOX release when studied at different temperature conditions, i.e., 86.35% (40 °C), 68.15% (37 °C), and 49.59% (30 °C) and followed Fickian diffusion release mechanism. In cytotoxicity study using murine melanoma cell lines (B16F10) HYNPs showed IC_{50} value of 1000 μg/mL. In the in-vivo studies using the melanoma tumor model, after intraperitoneal administration of DOX HYNPs, GQD-DOX HYNPs containing 2.5 mg/kg DOX for once a week up to 4 weeks showed a marked tumor volume reduction while leucine-GQD-DOX HYNPs and methionine-GQD-DOX HYNPs were unable to show the effective metastatic effect due to superior LCST value and erratic drug release pattern. Histopathological assessment of all the HYNPs showed normal lung parenchyma with less visible melanin forming cells than the disease control group. Moreover, acceptable levels of cytokines (TNF-α and IL-6) confirmed the non-immunogenic nature of HYNPs. During in-vivo studies, i.e., tumor reduction and histopathology the whole recuperation of airways from melanoma was not attained with any of the HYNPs or GDQs-HYNPs even if it was found to be particularly lesser than the disease group supporting its potential use as a drug carrier. Briefly, in the frontier of all the above investigation, GDQs-HYNPs loaded with DOX emerged to be the best temperature-sensitive nano-drug carrier for lung cancer treatment [74].

11.5 QDs Diagnostic Application

The complexity of the physiopathology of pulmonary diseases and the significance of their early and precise diagnosis help clinicians manage the disease effectively [97–100]. From a diagnostic viewpoint, the system should identify the intended ailment with high sensitivity and differentiate the core disease from other probable disease states with similar characteristics. Furthermore, clinicians seek safer and accurate diagnosis options to overcome the adverse effects. Among the nanomaterials, QDs offer a valuable platform with low side effects, eco-friendly nature, and cost-effective synthetic methods with more efficient optronics (optical and electronic) properties than conventional diagnosis platforms to develop diagnostic materials [33]. QDs are also involved in rapid, point-of-care disease diagnostics, which offers several key benefits: equipment portability, detection time, cost, and quick performance [101].

Detection of volatile organic compounds (VOCs) based biomarkers from the breath presents an effective clinical platform for diagnosing pulmonary and extra-pulmonary diseases [102, 103]. To date, various point-of-care disease diagnostics such as solid-phase micro-extraction fibers linked with GC-MS, colorimetric assays, electrochemical detection schemes and metal functionalized elements were developed and thoroughly studied to detect VOC-based TB biomarkers. However, most point-of-care diagnostics have few limitations, such as higher detection time, on-time feasibility, and the economic side. Thus, recently, Bhattacharyya et al. (2017) thoughtfully synthesized stable colloidal suspensions containing cadmium QDs (CdSe-QDs) and CQDs as a photoluminescent platform for the effective detection of TB biomarkers. Briefly, VOCs in the breath of TB subjects were detected using CdSe-QDs and CDs-based colloidal sensing solution. Novel CdSe-QDs and CDs-based photoluminescent platforms showed remarkable variation within the excitation and emission properties, respectively, in the presence of organic compounds gathered from the breath of TB subjects. CdSe-QDs showed a shift in emission peak, while CDs displayed a reduction in peak intensity when encountered with VOCs. The transient absorbance and emission kinetics for CdSe-QDs and CD systems are distinctive because of metal-organic bonding. Besides, the typical algorithm can be quickly produced for understanding the VOCs' properties and concentration. Concisely, an intelligent, tunable fluorescent sensor system based on QDs and CDs can offer a cost-effective feasible solution for rapid pulmonary disease diagnostics [101].

11.6 Future Viewpoint

The above sections have described the synthesis and drug delivery applications of novel QDs nanomaterials. Deep theoretical and experimental investigations of QDs depicted attractive pulmonary applications and have gathered much attention as potential carriers to regular coarse sugar-based carrier systems. As reviewed in the above segment, QDs were thoroughly utilized to fabricate pH-dependent, enzyme sensitive, and thermo-responsive carrier systems for successful pulmonary drug delivery. Besides, QDs were effectively molded into various novel systems such as liposomes, polymer conjugates, drug complex, and hydrogels to improve the physicochemical and biopharmaceutical properties. Further, few active researchers engineered the QDs surface with HA, FA and GA and transferrin to attain multi-environment responsive, controlled release systems. Pegylated-QDs were also fabricated to improve anticancer efficacy for functional pulmonary diagnostics and imaging applications. Material chemists and active formulators mostly used the QDs platform to materialize anticancer drug delivery systems. QDs were adequate to carry a variety of anticancer molecules, i.e., DOX, ERL, CET, and ETO. QDs-based anticancer systems accomplished better anticancer activity primarily via the sustained (> 80 h) release profile, remarkable tumor growth inhibition and receptor-mediated endocytosis in cancerous cells. Most of these investigations also assist in comprehending the drug absorption, penetration, and drug interactions using NHBE and A549 cells. However, further in-depth active investigational efforts are needed to recognize the detailed mechanism of anticancer activity. Moreover, QDs can be easily nebulized to obtain suitable aerodynamic performance. Furthermore, in near time we need to screen the potential active molecules on basis of functional groups to develop quick coordination bond with various transition metals engaged in the synthesis of QDs as well as the micronutrients involved in human physiological functions. Such exercise with help of other advanced tools such as computational chemistry and chemical process simulation aid to design and develop a wide range of functional QDs nanomaterials with synthetic as well as potential natural active constituents.

Like pulmonary drug delivery QDs hold significant promise in disease diagnosis, imaging, and treatment specificity. QDs hold huge prospects of being explored as theranostic tools and strides [35]. Specifically, QDs are commercially accessible at diverse emission wavelengths namely emission colors such as blue, violet, red, and green which gives flexibility in picking a preferred emission wavelength for a specific application in diagnosis and imaging. Consequently, in the coming time we need to assess miscellaneous emission wavelengths to strengthen the theranostic properties of versatile QDs [35]. Surface-engineered QDs were also widely used in point-of-care diagnostics kits due to their unique dual optical and electronic properties. The QDs-based bio-imaging platform offered real-time monitoring of various diseases, including pulmonary TB and cancer. A range of organic compounds can be easily sensed and analyzed using a QDs-based imaging platform for quick and reliable diagnoses of intra- and extra-pulmonary diseases.

Despite these widespread dynamic research efforts, the number of QDs-based inhaled nanomedicines accessible to patients is significantly beneath ridges for the nanomedicines discipline, mainly due to a translational gap between animal studies and clinical investigations. Furthermore, although dry powder inhalers are more physicochemically stable, economical, and patient-friendly than nebulizers and pressurized metered-dose inhalers to date, no article is reported on QDs-based dry powder inhalers. Thus, shortly simultaneous contribution to formulation development, pre-clinical studies, and clinical progress of QDs-based inhaled nanomedicines is essential for the bench-to-bedside translational movement.

11.7 Conclusion

Although in the budding phase, QDs have revealed the considerable capability to play a big part in the design and development of drug carriers, nano-therapeutic platforms, bio-imaging agents, phototherapy, diagnostic applications, and point-of-care devices for the management of pulmonary diseases. Due to unique optoelectronics and biophysical properties, QDs carry vast potential in simultaneous disease diagnosis and drug delivery. Being extremely versatile, QDs tend to be judiciously modified into various therapeutic platforms per pulmonary drug delivery needs. As reviewed above, QD-based novel formulations show fruitful outcomes during the in-vitro release studies, in-vivo animal trials, and human xenograft models. They can overcome biochemical issues with improved bioavailability, targetability, and less cytotoxicity. QD-based systems have the vast potential to alter the fate of lung disease therapy. In a nutshell, we anticipate that this chapter will offer valuable insights into the properties, synthesis, and applications of QDs and will support active formulators soon to explore this unique class of nanomaterials in the effective management of pulmonary diseases.

References

1. Khan I, Saeed K, Khan I. Nanoparticles: properties, applications and toxicities. Arab J Chem. 2019;12(7):908–31.
2. Baig N, Kammakakam I, Falath W. Nanomaterials: a review of synthesis methods, properties, recent progress, and challenges. Mater Adv. 2021;2(6):1821–71.
3. Chacko IA, Ghate VM, Dsouza L, Lewis SA. Lipid vesicles: a versatile drug delivery platform for dermal and transdermal applications. Colloids Surf B Biointerfaces. 2020;195:111262.
4. Mehta PP, Ghoshal D, Pawar AP, Kadam SS, Dhapte-Pawar VS. Recent advances in inhalable liposomes for treatment of pulmonary diseases: concept to clinical stance. J Drug Deliv Sci Technol. 2020;56:101509.
5. Weber S, Zimmer A, Pardeike J. Solid lipid nanoparticles (SLN) and nanostructured lipid carriers (NLC) for pulmonary application: a review of the state of the art. Eur J Pharm Biopharm. 2014;86(1):7–22.
6. Grumezescu AM. Nano-and microscale drug delivery systems: design and fabrication. William Andrew; 2017.
7. Etter EL, Mei KC, Nguyen J. Delivering more for less: nanosized, minimal-carrier and pharmacoactive drug delivery systems. Adv Drug Deliv Rev. 2021;179:113994.
8. Dhapte V, Kadam S, Moghe A, Pokharkar V. Probing the wound healing potential of biogenic silver nanoparticles. J Wound Care. 2014;23(9):431–41.
9. Dhapte V, Kadam S, Pokharkar V, Khanna PK, Dhapte V. Versatile SiO 2 composites with pragmatic properties. Int Sch Res Notices. 2014;2014:170919.

10. Kesharwani P, Greish K. Polymeric micelles for drug delivery. Woodhead Publishing; 2022. p. 1–684. https://doi.org/10.1016/C2020-0-03290-1.
11. Bu X, Ji N, Dai L, Dong X, Chen M, Xiong L, Sun Q. Self-assembled micelles based on amphiphilic biopolymers for delivery of functional ingredients. Trends Food Sci Technol. 2021;114:386–98.
12. Ban Q, Zheng Y, Qin Y, Kong J. Co-assembly-driven nanocomposite formation techniques toward mesoporous nanosphere engineering: a review. Microporous Mesoporous Mater. 2021;324:111312.
13. Kiran AR, Kumari GK, Krishnamurthy PT. Carbon nanotubes in drug delivery: focus on anticancer therapies. J Drug Deliv Sci Technol. 2020:101892.
14. Mehta P, Kadam S, Pawar A, Bothiraja C. Dendrimers for pulmonary delivery: current perspectives and future challenges. New J Chem. 2019;43(22):8396–409.
15. Kazemzadeh H, Mozafari M. Fullerene-based delivery systems. Drug Discov Today. 2019;24(3):898–905.
16. Grumezescu AM, editor. Design and development of new nanocarriers. William Andrew; 2017.
17. Kamaly N, Xiao Z, Valencia PM, Radovic-Moreno AF, Farokhzad OC. Targeted polymeric therapeutic nanoparticles: design, development and clinical translation. Chem Soc Rev. 2012;41(7):2971–3010.
18. Jacob J, Haponiuk JT, Thomas S, Gopi S. Biopolymer based nanomaterials in drug delivery systems: a review. Mater Today Chem. 2018;9:43–55.
19. Noah NM, Ndangili PM. Green synthesis of nanomaterials from sustainable materials for biosensors and drug delivery. Sensors International. 2022;3:100166.
20. Jana P, Dev A. Carbon quantum dots: a promising nanocarrier for bioimaging and drug delivery in cancer. Materials today. Communications. 2022:104068.
21. Mahle R, Kumbhakar P, Nayar D, Narayanan TN, Sadasivuni KK, Tiwary CS, Banerjee R. Current advances in bio-fabricated quantum dots emphasising the study of mechanisms to diversify their catalytic and biomedical applications. Dalton Trans. 2021;
22. Lim SY, Shen W, Gao Z. Carbon quantum dots and their applications. Chem Soc Rev. 2015;44(1):362–81.
23. Liu Q, Sun J, Gao K, Chen N, Sun X, Ti D, Bai C, Cui R, Qu L. Graphene quantum dots for energy storage and conversion: from fabrication to applications. Mater Chem Front. 2020;4(2):421–36.
24. Mohanty B, Giri L, Jena BK. MXene-derived quantum dots for energy conversion and storage applications. Energy Fuel. 2021;35(18):14304–24.
25. Choi MK, Yang J, Hyeon T, et al. Flexible quantum dot light-emitting diodes for next-generation displays. npj Flex Electron. 2018;2:10.
26. Harris RD, Bettis Homan S, Kodaimati M, He C, Nepomnyashchii AB, Swenson NK, Lian S, Calzada R, Weiss EA. Electronic processes within quantum dot-molecule complexes. Chem Rev. 2016;116(21):12865–919.
27. Sharma V, Tiwari P, Mobin SM. Sustainable carbon-dots: recent advances in green carbon dots for sensing and bioimaging. J Mater Chem B. 2017;5(45):8904–24.
28. Kumar YR, Deshmukh K, Sadasivuni KK, Pasha SK. Graphene quantum dot based materials for sensing, bio-imaging and energy storage applications: a review. RSC Adv. 2020;10(40):23861–98.
29. Chen Y, Cao Y, Ma C, Zhu JJ. Carbon-based dots for electrochemiluminescence sensing. Mater Chem Front. 2020;4(2):369–85.
30. Ferrari AC, Bonaccorso F, Fal'Ko V, Novoselov KS, Roche S, Bøggild P, Borini S, Koppens FH, Palermo V, Pugno N, Garrido JA. Science and technology roadmap for graphene, related two-dimensional crystals, and hybrid systems. Nanoscale. 2015;7(11):4598–810.
31. Riyaz B, Sudhakar K, Mishra V. Quantum dot-based drug delivery for lung cancer. In Nanotechnology-Based Targeted Drug Delivery Systems for Lung Cancer 2019 (pp. 311–326). Academic Press.

32. Biju V. Chemical modifications and bioconjugate reactions of nanomaterials for sensing, imaging, drug delivery and therapy. Chem Soc Rev. 2014;43(3):744–64.
33. Dezfuli AS, Kohan E, Fateh ST, Alimirzaei N, Arzaghi H, Hamblin MR. Organic dots (O-dots) for theranostic applications: preparation and surface engineering. RSC Adv. 2021;11(4):2253–91.
34. Salehi H. Differential space, quantum systems, and prediction. 1967: 1929–1931.
35. Kulkarni NS, Parvathaneni V, Shukla SK, Barasa L, Perron JC, Yoganathan S, Muth A, Gupta V. Tyrosine kinase inhibitor conjugated quantum dots for non-small cell lung cancer (NSCLC) treatment. Eur J Pharm Sci. 2019;133:145–59.
36. Matea CT, Mocan T, Tabaran F, Pop T, Mosteanu O, Puia C, Iancu C, Mocan L. Quantum dots in imaging, drug delivery and sensor applications. Int J Nanomedicine. 2017;12:5421.
37. Nikazar S, Sivasankarapillai VS, Rahdar A, Gasmi S, Anumol PS, Shanavas MS. Revisiting the cytotoxicity of quantum dots: an in-depth overview. Biophys Rev. 2020;12(3):703–18.
38. Heath JR. Covalency in semiconductor quantum dots. Chem Soc Rev. 1998;27(1):65–71.
39. Bera D, Qian L, Tseng TK, Holloway PH. Quantum dots and their multimodal applications: a review. Materials (Basel). 2010;3(4):2260–345.
40. Zhu C, Chen Z, Gao S, Goh BL, Samsudin IB, Lwe KW, Wu Y, Wu C, Su X. Recent advances in non-toxic quantum dots and their biomedical applications. Prog Nat Sci: Mater Int. 2019;29(6):628–40.
41. Sun Y-P, Zhou B, Lin Y, Wang W, Fernando KAS, Pathak P, Meziani MJ, Harruff BA, Wang X, Wang HF, Luo PG, Yang H, Kose ME, Chen BL, Veca LM, Xie S-Y. Quantum-sized carbon dots for bright and colorful photoluminescence. J Am Chem Soc. 2006;128:7756–7.
42. Zheng LY, Chi YW, Dong YQ, Lin JP, Wang BB. Electrochemiluminescence of water-soluble carbon nanocrystals released electrochemically from graphite. J Am Chem Soc. 2009;131:4564–5.
43. Li XY, Wang HQ, Shimizu Y, Pyatenko A, Kawaguchi K, Koshizaki N. Preparation of carbon quantum dots with tunable photoluminescence by rapid laser passivation in ordinary organic solvents. Chem Commun. 2011;47:932–4.
44. Demchenko AP, Dekaliuk MO. Novel fluorescent carbonic nanomaterials for sensing and imaging. Methods Appl Fluoresc. 2013;1(4):042001.
45. Baker SN, Baker GA. Luminescent carbon nanodots: emergent nanolights. Angew Chem Int Ed. 2010;49(38):6726–44.
46. Yang ST, Cao L, Luo PG, Lu F, Wang X, Wang H, Meziani MJ, Liu Y, Qi G, Sun YP. Carbon dots for optical imaging in vivo. J Am Chem Soc. 2009;131(32):11308–9.
47. Yang ST, Wang X, Wang H, Lu F, Luo PG, Cao L, Meziani MJ, Liu JH, Liu Y, Chen M, Huang Y, Sun YP. Carbon dots as nontoxic and high-performance fluorescence imaging agents. J Phys Chem C Nanomater Interfaces. 2009;113(42):18110–4.
48. Bonilla CA, Kouznetsov VV. Green'quantum dots: basics, green synthesis, and nanotechnological applications. Green nanotechnology-overview and further. Prospects. 2016:174–92.
49. Charitidis CA, Georgiou P, Koklioti MA, Trompeta AF, Markakis V. Manufacturing nanomaterials: from research to industry. Manufacturing Rev. 2014;1:11.
50. Wang Y, Hu A. Carbon quantum dots: synthesis, properties and applications. J Mater Chem C. 2014;2(34):6921–39.
51. Valizadeh A, Mikaeili H, Samiei M, Farkhani SM, Zarghami N, Kouhi M, Akbarzadeh A, Davaran S. Quantum dots: synthesis, bioapplications, and toxicity. Nanoscale Res Lett. 2012;7(1):480.
52. Liu H, Ye T, Mao C. Fluorescent carbon nanoparticles derived from candle soot. Angew Chem Int Ed Engl. 2007;46(34):6473–5.
53. Shinde DB, Pillai VK. Electrochemical preparation of luminescent graphene quantum dots from multiwalled carbon nanotubes. Chem Eur J. 2012;18(39):12522–8.
54. Wang CI, Wu WC, Periasamy AP, Chang HT. Electrochemical synthesis of photoluminescent carbon nanodots from glycine for highly sensitive detection of hemoglobin. Green Chem. 2014;16(5):2509–14.

55. Dey S, Govindaraj A, Biswas K, Rao CN. Luminescence properties of boron and nitrogen doped graphene quantum dots prepared from arc-discharge-generated doped graphene samples. Chem Phys Lett. 2014;595:203–8.
56. Hu C, Liu Y, Yang Y, Cui J, Huang Z, Wang Y, Yang L, Wang H, Xiao Y, Rong J. One-step preparation of nitrogen-doped graphene quantum dots from oxidized debris of graphene oxide. J Mater Chem B. 2013;1(1):39–42.
57. Reyes D, Camacho M, Camacho M, Mayorga M, Weathers D, Salamo G, Wang Z, Neogi A. Laser ablated carbon nanodots for light emission. Nanoscale Res Lett. 2016;11(1):424.
58. Xu J, Tao J, Su L, Wang J, Jiao T. A critical review of carbon quantum dots: from synthesis toward applications in electrochemical biosensors for the determination of a depression-related neurotransmitter. Materials. 2021;14(14):3987.
59. Barras A, Pagneux Q, Sane F, Wang Q, Boukherroub R, Hober D, Szunerits S. High efficiency of functional carbon nanodots as entry inhibitors of herpes simplex virus type 1. ACS Appl Mater Interfaces. 2016;8(14):9004–13.
60. Wang J, Wang CF, Chen S. Amphiphilic egg-derived carbon dots: rapid plasma fabrication, pyrolysis process, and multicolor printing patterns. Angew Chem Int Ed. 2012;51(37):9297–301.
61. Tang L, Ji R, Cao X, Lin J, Jiang H, Li X, Teng KS, Luk CM, Zeng S, Hao J, Lau SP. Deep ultraviolet photoluminescence of water-soluble self-passivated graphene quantum dots. ACS Nano. 2012;6(6):5102–10.
62. Sivasankarapillai VS, Kirthi AV, Akksadha M, Indu S, Dharshini UD, Pushpamalar J, Karthik L. Recent advancements in the applications of carbon nanodots: exploring the rising star of nanotechnology. Nanoscale Adv. 2020;2(5):1760–73.
63. Rosenthal SJ, Chang JC, Kovtun O, McBride JR, Tomlinson ID. Biocompatible quantum dots for biological applications. Chem Biol. 2011;18(1):10–24.
64. Chakravarthy KV, Davidson BA, Helinski JD, Ding H, Law WC, Yong KT, Prasad PN, Knight PR. Doxorubicin-conjugated quantum dots to target alveolar macrophages and inflammation. Nanomedicine. 2011;7(1):88–96.
65. Pierrat P, Wang R, Kereselidze D, Lux M, Didier P, Kichler A, Pons F, Lebeau L. Efficient in vitro and in vivo pulmonary delivery of nucleic acid by carbon dot-based nanocarriers. Biomaterials. 2015;51:290–302.
66. Roshini A, Jagadeesan S, Arivazhagan L, Cho YJ, Lim JH, Doh YH, Kim SJ, Na J, Choi KH. pH-sensitive tangeretin-ZnO quantum dots exert apoptotic and anti-metastatic effects in metastatic lung cancer cell line. Mater Sci Eng C Mater Biol Appl. 2018;92:477–88.
67. Nagy A, Hollingsworth JA, Hu B, Steinbrück A, Stark PC, Rios Valdez C, Vuyisich M, Stewart MH, Atha DH, Nelson BC, Iyer R. Functionalization-dependent induction of cellular survival pathways by CdSe quantum dots in primary normal human bronchial epithelial cells. ACS Nano. 2013;7(10):8397–411.
68. Cai X, Luo Y, Zhang W, Du D, Lin Y. pH-sensitive ZnO quantum dots-doxorubicin nanoparticles for lung cancer targeted drug delivery. ACS Appl Mater Interfaces. 2016;8(34):22442–50.
69. Zhang X, He X, Li Y, Zhang Z, Ma Y, Li F, Liu J. A cytotoxicity study of fluorescent carbon nanodots using human bronchial epithelial cells. J Nanosci Nanotechnol. 2013;13(8):5254–9.
70. Turdalieva A, Solandt J, Shambetova N, Xu H, Blom H, Brismar H, Zelenina M, Fu Y. Bioelectric and morphological response of liquid-covered human airway epithelial Calu-3 cell monolayer to periodic deposition of colloidal 3-mercaptopropionic-acid coated CdSe-CdS/ZnS Core-multishell quantum dots. PLoS One. 2016;11(2):e0149915.
71. Jha A, Viswanadh MK, Burande AS, Mehata AK, Poddar S, Yadav K, Mahto SK, Parmar AS, Muthu MS. DNA biodots based targeted theranostic nanomedicine for the imaging and treatment of non-small cell lung cancer. Int J Biol Macromol. 2020;150:413–25.
72. Ranjbar-Navazi Z, Eskandani M, Johari-Ahar M, Nemati A, Akbari H, Davaran S, Omidi Y. Doxorubicin-conjugated D-glucosamine- and folate- bi-functionalised InP/ZnS quantum dots for cancer cells imaging and therapy. J Drug Target. 2018;26(3):267–77.

73. Pati R, Sahu R, Panda J, Sonawane A. Encapsulation of zinc-rifampicin complex into transferrin-conjugated silver quantum-dots improves its antimycobacterial activity and stability and facilitates drug delivery into macrophages. Sci Rep. 2016;6:24184.
74. Havanur S, Batish I, Cheruku SP, Gourishetti K, JagadeeshBabu PE, Kumar N. Poly(N,N-diethyl acrylamide)/functionalized graphene quantum dots hydrogels loaded with doxorubicin as a nano-drug carrier for metastatic lung cancer in mice. Mater Sci Eng C Mater Biol Appl. 2019;105:110094.
75. Rovina N, Koutsoukou A, Koulouris NG. Inflammation and immune response in COPD: where do we stand? Mediat Inflamm. 2013;2013:413735.
76. Kumar V. Pulmonary innate immune response determines the outcome of inflammation during pneumonia and sepsis-associated acute lung injury. Front Immunol. 2020;11:1722.
77. Bañuls L, Pellicer D, Castillo S, Navarro-García MM, Magallón M, González C, Dasí F. Gene therapy in rare respiratory diseases: what have we learned so far? J Clin Med. 2020;9(8):2577.
78. Munir M, Jena L, Kett VL, Dunne NJ, McCarthy HO. Spray drying: inhalable powders for pulmonary gene therapy. Mater Sci Eng C. 2021:112601.
79. Alnasser SM. Review on mechanistic strategy of gene therapy in the treatment of disease. Gene. 2021;769:145246.
80. Yahya EB, Alqadhi AM. Recent trends in cancer therapy: a review on the current state of gene delivery. Life Sci. 2021;269:119087.
81. Boye A, Ahmad I, Fakhri S, Hussain Y, Khan H. Incipient citrus polymethoxylated flavone Tangeretin as anticancer drug candidate: mechanistic insights, limitations and possible solutions. Adv Cancer Biol-Metastasis. 2021;3:100010.
82. Raza W, Luqman S, Meena A. Prospects of tangeretin as a modulator of cancer targets/pathways. Pharmacol Res. 2020;161:105202.
83. Hung WL, Chang WS, Lu WC, Wei GJ, Wang Y, Ho CT, Hwang LS. Pharmacokinetics, bioavailability, tissue distribution and excretion of tangeretin in rat. J Food Drug Anal. 2018;26(2):849–57.
84. Truong DH, Le VK, Pham TT, Dao AH, Pham TP, Tran TH. Delivery of erlotinib for enhanced cancer treatment: an update review on particulate systems. J Drug Deliv Sci Technol. 2020;55:101348.
85. Rampaka R, Ommi K, Chella N. Role of solid lipid nanoparticles as drug delivery vehicles on the pharmacokinetic variability of erlotinib HCl. J Drug Deliv Sci Technol. 2021;66:102886.
86. Sigismund S, Avanzato D, Lanzetti L. Emerging functions of the EGFR in cancer. Mol Oncol. 2018;12(1):3–20.
87. Harding J, Burtness B. Cetuximab: an epidermal growth factor receptor chemeric human-murine monoclonal antibody. Drugs Today (Barc). 2005;41(2):107–27.
88. Amreddy N, Babu A, Muralidharan R, Panneerselvam J, Srivastava A, Ahmed R, Mehta M, Munshi A, Ramesh R. Recent advances in nanoparticle-based cancer drug and gene delivery. Adv Cancer Res. 2018;137:115–70.
89. Kim DS, Park KS, Jeong KC, et al. Glucosamine is an effective chemo-sensitizer via transglutaminase 2 inhibition. Cancer Lett. 2009;273:243–9.
90. Korotcov AV, Ye Y, Chen Y, et al. Glucosamine-linked nearinfrared fluorescent probes for imaging of solid tumor xenografts. Mol Imaging Biol. 2012;14:443–51.
91. Saadat S, Rawtani D, Parikh G. Clay minerals-based drug delivery systems for anti-tuberculosis drugs. J Drug Deliv Sci Technol. 2022;31:103755.
92. Sarkar K, Kumar M, Jha A, Bharti K, Das M, Mishra B. Nanocarriers for tuberculosis therapy: design of safe and effective drug delivery strategies to overcome the therapeutic challenges. J Drug Deliv Sci Technol. 2021:102850.
93. Chae J, Choi Y, Tanaka M, Choi J. Inhalable nanoparticles delivery targeting alveolar macrophages for the treatment of pulmonary tuberculosis. J Biosci Bioeng. 2021;132(6):543–51.
94. Mehta P, Bothiraja C, Kadam S, Pawar A. Potential of dry powder inhalers for tuberculosis therapy: facts, fidelity and future. Artif Cells Nanomed Biotechnol. 2018;46(sup3):S791–806.

95. Quazi MZ, Park N. Nanohydrogels: advanced polymeric nanomaterials in the era of nanotechnology for robust functionalization and cumulative applications. Int J Mol Sci. 2022;23(4):1943.
96. Mehta P, Mahadik K, Kadam S, Dhapte-Pawar V. Advanced applications of green hydrogels in drug delivery systems. Appl Adv Green Mater. 2021:89–130.
97. Mehta PP, Dhapte-Pawar VS. Repurposing drug molecules for new pulmonary therapeutic interventions. Drug Deliv Transl Res. 2021;11(5):1829–48.
98. Mehta PP, Dhapte-Pawar V. Role of surfactants in pulmonary drug delivery. In Green Sustainable Process for Chemical and Environmental Engineering and Science 2022 (pp. 559–577). Academic Press.
99. Kumar R, Mehta P, Shankar KR, Rajora MA, Mishra YK, Mostafavi E, Kaushik A. Nanotechnology-assisted metered-dose inhalers (MDIs) for high-performance pulmonary drug delivery applications. Pharm Res. 2022;1-25
100. Mehta PP, Dhapte-Pawar VS. Novel and evolving therapies for COVID-19 related pulmonary complications. Am J Med Sci. 2021;361(5):557–66.
101. Bhattacharyya D, Sarswat PK, Free ML. Quantum dots and carbon dots based fluorescent sensors for TB biomarkers detection. Vacuum. 2017;146:606–13.
102. Lubes G, Goodarzi M. GC–MS based metabolomics used for the identification of cancer volatile organic compounds as biomarkers. J Pharm Biomed Anal. 2018;147:313–22.
103. Van Der Schee MP, Paff T, Brinkman P, Van Aalderen WM, Haarman EG, Sterk PJ. Breathomics in lung disease. Chest. 2015;147(1):224–31.

Metal-Organic Frameworks: A Toolbox for Multifunctional Pulmonary Applications

12

Piyush Pradeep Mehta and Vividha Dhapte-Pawar

Abstract

Recently, metal-organic frameworks (MOFs) have received vast awareness owing to their unique structural properties, ease of synthesis, and outstanding performance. Owing to their highly porous nature, superior surface area, precise crystal structure, and ease of surface tailoring, MOFs have become promising agents for catalysis, gas adsorption, material separation, chemo-sensors, supercapacitors, energy applications, and bio-related functions. The wide range of functionality and applications of these novel materials has led active researchers in the biomedical field to explore the drug delivery applications of MOFs. Owing to their highly tunable and adaptable nature, MOFs have emerged as a potential carrier for drug molecules, phyto-compounds, cell-imaging materials, biological substances, and diagnostic agents. Accordingly, the main goal of the present chapter is to highlight recent progress and design strategies used to materialize inhaled MOF platforms. This chapter will primarily summarize the MOF's synthesis schemes and fundamental structural properties, then a section denoting various applications ranging from energy storage to biomedical relevance. In the subsequent section, we have described the development of inhaled MOF systems focusing on particle properties, aerodynamic assessment, in-vitro human cell line studies, and xenograft models. Various MOFs-based photodynamic and pho-

P. P. Mehta (✉)
Global Respiratory R&D, Cipla R&D Center, Mumbai, Maharashtra, India
e-mail: piyu053@gmail.com

V. Dhapte-Pawar
Department of Pharmaceutics, Poona College of Pharmacy, Bharati Vidyapeeth (Deemed to be University), Pune, Maharashtra, India
e-mail: dhapte.vv@bharatividyapeeth.edu

P. P. Mehta, V. Dhapte-Pawar (eds.), *Pulmonary Drug Delivery Systems: Material and Technological Advances*,
https://doi.org/10.1007/978-981-99-1923-9_12

tothermal therapies used for simultaneous cell imaging and targeted drug delivery to treat pulmonary diseases are also reviewed. The last section of this chapter discusses opportunities and challenges in the promising field of MOFs, highlighting green synthesis methods, clinical implications, and market potential.

Keywords

Metal-organic frameworks · Novel porous materials · MIL-family MOFs · Photodynamic therapy · Targeted pulmonary delivery · Hydrothermal or solvothermal methods

12.1 Introduction

Over the past few years, porous materials have seized significant consideration and importance because of their excellent performance and wide range of applications [1]. As compared to different particulates systems such as nanoparticles (NPs), polymeric NPs, polymeric micelles [2–4], lipid vesicles (e.g., solid lipid NPs, nanostructured lipid carriers, liposomes, pro-liposomes, lipospheres, and niosomes) [5, 6], highly branched polymeric molecules (e.g., dendrimers) [7] and microparticle (MPs) (e.g., microspheres, solid lipid MPs and surface coated MPs) [8–10] porous framework materials have drawn particular attention [11]. These materials are generally referred to as porous organic frameworks (POFs), which means porous framework materials including organic units [12]. The commonly used POFs comprise covalent organic frameworks (COFs) [13], metal-organic frameworks (MOFs) [14], porous aromatic frameworks (PAFs) [15], conjugated microporous polymers (CMPs) [16] and covalent triazine frameworks (CTFs) [17–19]. These POFs share several common traits: organic-contained framework, pore structure, facile synthesis methods, design ability, and ease of surface functionalization [20]. Due to their excellent porosity, unique pore arrangement, and high specific surface area, POFs have distinguished applications in adsorption separation and catalysis [21]. Moreover, the physicochemical properties of POFs can be easily modified during their synthesis or by post-synthetic adaptation to enrich their functional characteristics [22]. As a result, POFs are promising materials for application in various fields, including adsorption, separation [23], catalysis [24], energy storage [25], biomedical discipline [26], chemical sensors [27], and magnetism [28] (Fig. 12.1). However, in the upcoming segment, we distinctively focus on one of the most popular classes of POFs, i.e., MOFs.

MOFs are principally porous 1D, 2D and 3D crystalline materials made of metal nodes coupled by organic ligands [29]. A variety of synthetic methods such as hydrothermal [30] or solvothermal [31], microwave [32] and electrochemical-assisted synthesis [33] are commonly employed in the development of MOFs wherein the dissolved metal salts and organic ligands within the solvent react at a particular temperature, causing MOF crystals' nucleation and growth [34]. However, microwave irradiation is considered the most suitable method to synthesize homogenous

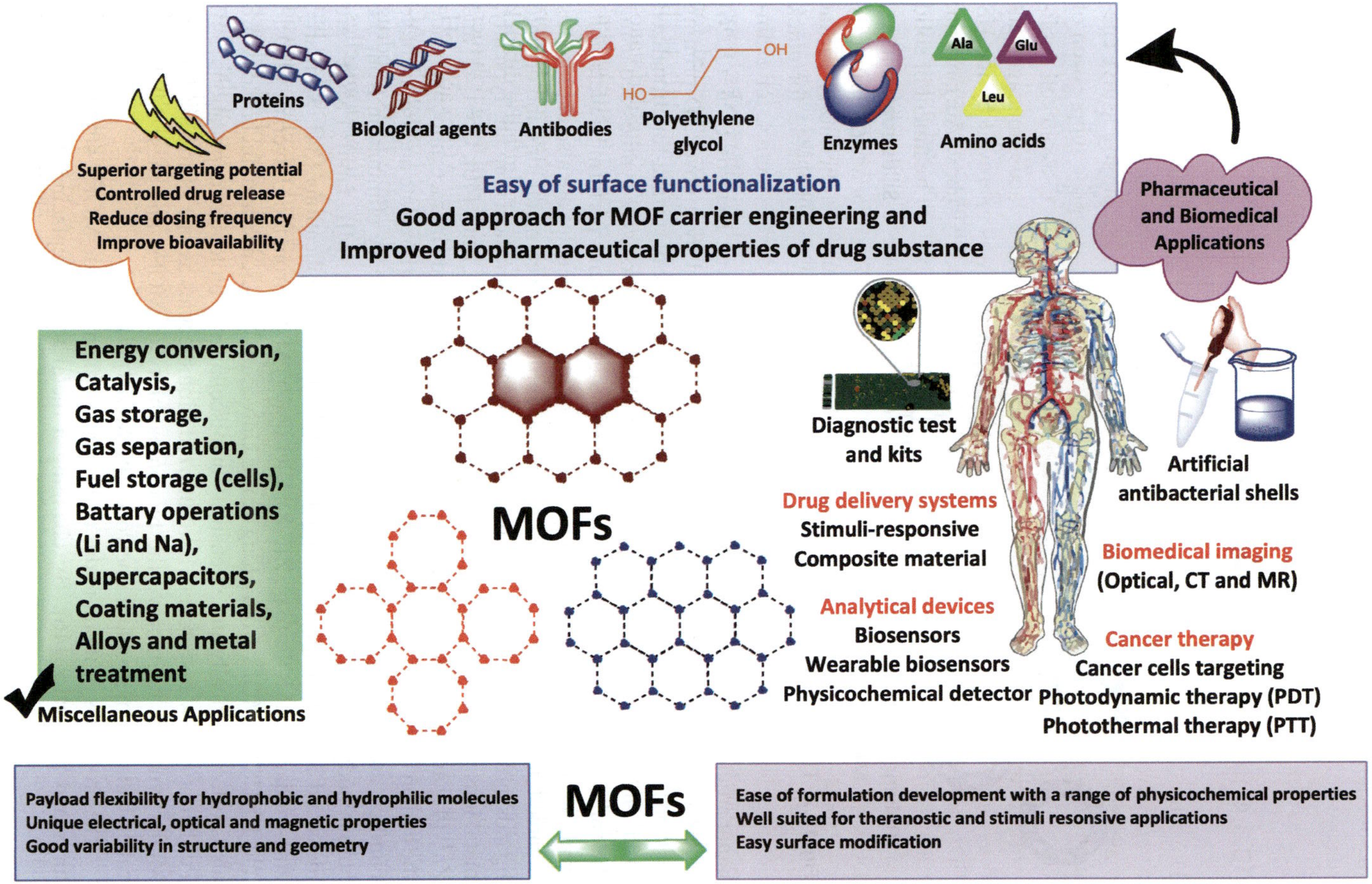

Fig. 12.1 Interdisciplinary applications of metal-organic frameworks-based nanosystems

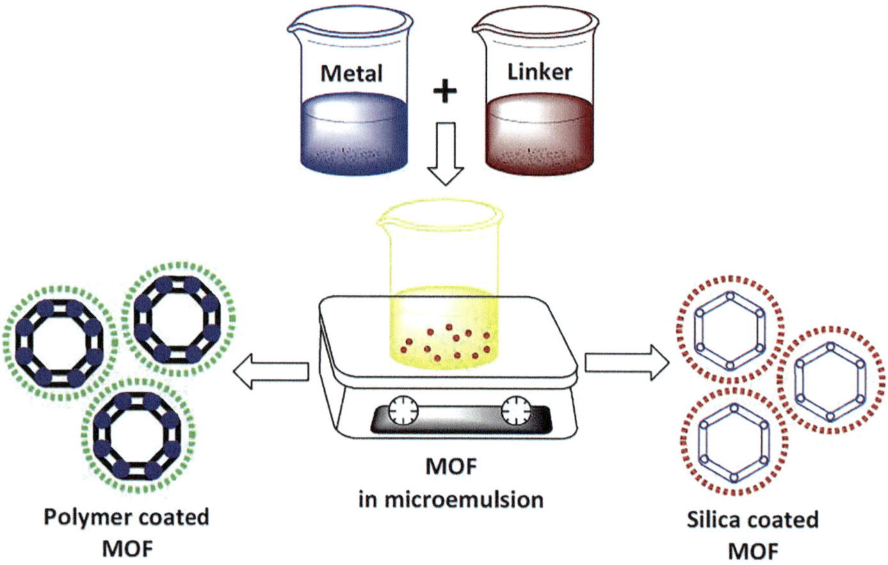

Fig. 12.2 The synthesis scheme of surface-functionalized MOF with polymers and silica

nanoparticles within a short reaction period [34]. Moreover, Denisov et al. (2020) recently fabricated MOFs using the solvothermal method by a novel 3D printing method [35]. Besides, Fig. 12.2 shows the synthesis scheme of surface-functionalized MOF with polymers and silica [11]. Different MOF carriers were synthesized using the entire width of the periodic table, wherein these elements perform as metal ions while aromatic heterocyclic or polytopic carboxylates molecules are also explored as organic ligands to couple different metal ions [36]. However, metal ions and organic ligands with minimal toxicity are most preferred in the biomedical domain. Typically, metal ions such as Ca^{2+} [37], Fe^{2} [38], Fe^{3+} [38], Zr^{2+} [39], Mn^{2+} [40], Co^{2+} [41] and Zn^{2+} [42] are used in MOF synthesis, which is present already in trace amounts inside the human body [34]. Organic ligands, particularly from carboxylate and imidazolate families, are explored due to their superior polarity and easy elimination from the human body [43, 44]. The Materials of Institute Lavoisier (MIL-family) MOFs are measured as one of the most suitable MOFs for bio-applications [34]. Similarly, zinc-imidazolate MOFs (e.g., ZIF-8), UiO-66(Zr) MOFs from the University of Oslo and HKUST-1 MOFs from Hong Kong University of Science and Technology are also explored logically and extensively in miscellaneous biological applications [34]. Furthermore, few researchers have utilized biomolecules as one of the elements in MOFs synthesis to improve the biocompatibility of MOFs, and these MOFs are called metal-biomolecule frameworks (bioMOFs) [45].

The properties of metal nodes and organic ligands decide MOFs' structural features and properties [46]. By changing a metal ion or replacing the functional moiety of an organic ligand, one can scientifically vary the properties of the MOF [46]. MOFs offer functional diversity and adaptability owing to the large specific surface area, ultra-high porosity, structural tailor ability, and crystalline nature [47, 48]. Additionally, MOFs have various open metal sites with great potential for specific adsorption and molecular identification [49]. These porous coordination polymers have significantly influenced the research disciplines and industries such as energy conversion, catalysis, gas separation, fuel storage, bioimaging [50] and biomedical section due to the above-mentioned structural diversity and properties [51, 52]. Especially in the drug delivery segment, MOFs are used as a (i) delivery vehicle for active substances, antitumor agents, antimicrobials and phyto-molecules; (ii) protective shield for enzymes, proteins, oligo-peptides, nucleic acids and cells; (iii) biomarkers detection in cancer therapy and (iv) designing of non-invasive imaging and bio-sensing platforms [53–55]. Additionally, MOFs can act as templates or precursors for several functions in the biomedical domain and quickly tune into different roles [21]. For instance, the MOF structure can be linked to specific conductive or magnetic moiety to obtain excellent photodynamic or photothermal therapy [56]. Apart from the targeted drug delivery applications of MOFs, scientific articles have shown the development of different MOF formulations in films, patches, pills, and tablets for improving patient compliance [57]. Furthermore, drug molecules can be loaded in the MOF void and tethered with the framework structure based on the delivery need [58]. Briefly, these advantageous attributes escalate the scope of MOF materials for clinical applications. In contrast to other excellent review articles, book chapters and communications that discuss the biomedical applications of MOFs from a broad perspective, this chapter differs in that it will focus on pulmonary applications of MOFs. The present chapter will thoroughly summarize the characteristics of case studies related to inhaled MOF platforms, emphasizing aerodynamic outcomes, pulmokinetics and clinical applications.

12.2 Inhaled MOF Platforms

As discussed in the above segment, initially, MOFs research was concentrated mainly on energy storage, gas separation, and catalysis owing to the distinctive benefits of MOFs such as structure tunability, high porosity, and stability [50]. Lately, a budding application area of MOFs is their deployment as drug delivery platforms owing to their intrinsic attributes of highly porous nature, high surface area, ease of surface modification, and metal ion-free nature [11]. MOFs have been widely used as pulmonary drug delivery carriers for delivering therapeutic agents to airways. It was anticipated that the networks of MOFs would offer suitable room for loading active moieties, avoiding drug leakage and thus improving the pulmonary drug delivery effectiveness. Consequently, MOF-based pulmonary drug delivery systems were thoroughly described in the present section.

12.2.1 Engineered MOFs for Phytochemicals

Curcumin (CUR) is well-known and methodically investigated polylphenol obtained from rhizome of *Curcuma longa* [59]. Checmially, CUR is diferuloylmethane with pleiotropic biological activities such as an antineoplastic agent, anti-inflammatory agent, antifungal agent, neuroprotective agent, hepatoprotective agent, radical scavenger, immunomodulator and lipoxygenase inhibitor [60, 61]. Besides, CUR is also well explored for diverse therapeutic functions such as metabolite, ligand, iron chelator, nutraceutical, biological pigment, flavoring agent and food coloring agent [62, 63]. The growing amount of scientific data from pre-clinical and clinical studies also encourage the concept that CUR plays a protective function in chronic and acute respiratory disease, acute pulmonary distress syndrome, acute lung injury and allergic asthma. CUR exerts therapeutic actions by modulating inflammation pathways and oxidative stress [61, 64]. These pre-clinical and clinical findings showed good applicability of CUR in respiratory disease management. However, CUR encounters low oral and erratic bioavailability (<1%) owing to various physicochemical restrictions such as high molecular weight (368.39 g/mol), poor aqueous solubility (0.6 μg/mL), log P of 2.5, high protein-binding and rapid metabolism (half-life <2 h) [65]. It is thus essential to design a formulation of CUR with improved solubility profile and dissolution kinetics for pulmonary administration. Accordingly, Zhou et al. (2020) developed γ-cyclodextrin (γ-CD)-based MOFs to facilitate improved solubility owing to porous nature and phase transition of CUR as well as excellent aerodynamic outcome due to uniform 3D network framework. CUR loaded γ-CD-MOFs were prepared using in-situ drug loading via a co-crystallization (vapor diffusion method) phenomenon. Typical cubic-shaped CUR-γ-CD-MOF showed a mean particle size (2.17 μm) suitable for pulmonary delivery, while micronized CUR showed a mean particle size of 2.33 μm. The surface area CUR-γ-CD-MOF using nitrogen adsorption-desorption isotherms was found to be 798 m^2/g. During in-vitro dissolution testing, CUR-γ-CD-MOF revealed a faster CUR release rate (> 98% within 5 min) as compared to the lactose-loaded micronized CUR (< 66% at the end of 90 min) using phosphate buffer saline (PBS; pH 7.4). CUR-γ-CD-MOF showed good solubility of 1.076 μg/mL with 2.02-fold lower contact angle compared to CUR alone. The lower contact angle indicated improved wettability of the CUR-γ-CD-MOF compared to CUR. Briefly, the superior wettability and unique spatial distribution of CUR within the porous interior of CD-MOFs are mainly accountable for enhanced solubility and dissolution kinetics. The Lennard-Jones (LJ) potential function-based all-atom molecular dynamic simulation study demonstrated that CUR molecules were suitably packed within the hydrophobic cavities of CD-MOFs to form nano-clusters. Particularly, single CUR molecule tend to distribute into dual γ-CD cavities with the formation of hydrogen bonds while increasing number of CUR molecules tend to scatter into different hydrophobic cavities rather than distribute into one similar hydrophobic space. Moreover, once all hydrophobic cavities were occupied then CUR molecules tried to enter into the large hydrophilic cavity of γ-CD-MOFs and shape nanoclusters. When CUR-γ-CD-MOF contacted with amphiphilic molecules, the structure of γ-CD-MOF disappeared and subsequently CUR molecules from hydrophobic cavities generated inclusion complex with γ-CD, which eventually increases its solubility and dissolution rate. In the aerodynamic analysis using the Turbospin® device (60 L/min for 4 s), CUR-γ-CD-MOF displayed higher fine particle fraction (FPF; >60%) and acceptable mass median aerodynamic diameter (MMAD; <3 μm) values as compared to the typical lactose (InhaLac® 500) based micronized CUR formulation, i.e., FPF (≤50%) and MMAD (>3 μm). The superior FPF value of CUR was primarily accredited to lower bulk density (112.9 mg/cm^3) and the unique porous structure of γ-CD-MOFs. In contrast, the lower MMAD value of CUR was mainly ascribed to the cubic shape morphology of γ-CD-MOFs which improved the powder dispersion and disaggregation during analysis. Moreover, during the in-vitro antioxidant study, using 2,2-diphenyl-1-picrylhydrazyl (DPPH) radical scavenging assay, CUR-γ-CD-MOF (0.01 mg/mL) showed a prominent 1.88-fold higher antioxidant potential than lactose-loaded micronized CUR formulation. In a few words, CD-MOFs-based carrier systems are highly suitable for the pulmonary delivery of poorly water-soluble polyphenols [66].

Paeonol (PAN) i.e., paeoniflorin or peony phenol is a phenolic phytoconstituent obtained from the *Paeonia lactiflora Pall* and *Paeonia suffruticosa Andr* [67, 68]. Chemically PAN is 2-hydroxy-4-methoxy acetophenone with low molecular weight (166.18 g/mol) and volatile compound. PAN has plentiful significant pharmacological activities such as antioxidant, analgesic, anti-inflammatory, antidiabetic, antiangiogenic, and anticancer [69]. Recent clinical studies showed that PAN efficiently prevents bleomycin and cigarette smoke stimulated pulmonary fibrosis and inflammation [70, 71], control airway vasculature in treatment of hypertension [72] and reduces acute lung injury by regulating nuclear protein [73]. The current oral delivery of PAN is mainly restricted by its low water solubility (0.54 mg/mL), poor bioavailability (<10%) and poor physicochemical stability [74]. Therefore, it is essential to formulate a suitable aerosol preparation to improve the therapeutic performance of PAN. Thus, Li et al. (2020) developed γ-CD-based MOF for successful pulmonary delivery of PAN (PAN-γ-CD-MOF) in the management of acute lung injury. Molecular simulation showed that PAN interacted strongly with γ-CD-MOF at a molar ratio of 2:1, and PAN molecules were scattered equally in the hydrophilic and hydrophobic vicinity formed by a pair of γ-CD. PAN-γ-CD-MOF displayed a 19-fold reduction in the surface area of γ-CD-MOF with 15.4% PAN loading while the absence of nano-pores during pore volume analysis confirmed that PAN molecules were effectively incorporated into cavities of γ-CD-MOF. Smooth surface and cubic morphology of PAN-γ-CD-MOF showed a mean diameter < 3 μm, whereas thermal analysis highlighted the crystalline behavior of PAN. During the in-vitro drug release study using PBS (pH 7.4), PAN-γ-CD-MOF exhibited 90% PAN release within the first 30 min while untreated PAN showed only 20% PAN release within the first 30 min. The significant increase in PAN release kinetics was mainly because of the supramolecular encapsulation of aquaphobic PAN molecules in aquaphilic γ-CD-MOF, while the lower release rate of untreated PAN was mainly attributed to its poor aqueous solubility. In an in-vitro cellular permeability study using A549 cells, PAN-γ-CD-MOF exhibited a marked five-fold improvement in

PAN cellular permeability via passive transport compared to untreated PAN at the same concentration (20 μg/mL). Further to assess the aerodynamic properties of PAN-γ-CD-MOF respirable powder, it was formulated using a combination of coarse (Inhalac® 230) and fine (Inhalac® 400) lactose by double sandwich method in a low shear mechanical blender (VH-2, V-blender). During aerodynamic analysis using Rotahaler® (50.71 L/min for 4.7 s) Inhalac® 230: Inhalac® 400: PAN-γ-CD-MOF (4:1:5) powder blend showed 1.6 and 2.33-fold enhancement in % FPF of PAN as compared to PAN-γ-CD-MOF alone and Inhalac® 230: PAN-γ-CD-MOF (3:1) powder blend, respectively. In an in-vivo pharmacokinetic study using rats after intratracheal insufflation, Inhalac® 230: Inhalac® 400: PAN-γ-CD-MOF (4:1:5) powder blend showed a marked 6.93 and 11.22-fold enhancement in the area under curve (AUC) and maximum serum concentration (C_{max}), respectively as compared to intragastric administration at the same dose (4 mg/kg). During acute lung injury study using lipopolysaccharide induced in Sprague-Dawley rat model, low dose PAN DPI (4 mg/kg) showed distinct inflammation of lung tissues while medium dose PAN DPI (8 mg/kg) showed alleviated inflammation with reduced alveolar cavity wall thicknesses and minor infiltration of inflammatory cells. High dose PAN DPI (16 mg/kg) showed completely recovered alveolar structure similar to positive control dexamethasone (3 mg/kg) group. Biological assessment of PAN-γ-CD-MOF suggested that pulmonary delivery of PAN effectively remove neutrophil infiltration and protein exudation from lung tissue. Furthermore, PAN-γ-CD-MOF showed reduced levels of inflammatory factors in histopathological examination. Overall, the targeted pulmonary delivery of PAN using γ-CD-MOF carrier appears to be a promising technique in treating acute lung injury [75].

Scutellarin (SCU) is a key phytoconstituent isolated from traditional Chinese *medicine Erigeron breviscapus* (vant) Hand-Mazz [76, 77]. SCU is also extracted from *Sempervivum ruthenicum*, *Scoparia dulcis*, and other organisms. SCU preprations were usually prescribed in China for acute cerebral infarction and paralysis provoked by cerebrovascular diseases [76, 77]. It also act as a proteasome inhibitor, antineoplastic and antioxidant agent. Moreover, SCU inhibit proliferation in non-small cell lung cancer (NSCLC) [78] and prevent idiopathic pulmonary fibrosis by modulating pulmonary vasculature [79]. However, SCU is crystalline solid with low water solubility. The oral bioavailability of SCU is remarkably poor [80]. The oral dose of 60 mg SCU in healthy subjects presented C_{max} of <5.0 ng/mL and follow uncharacteristic AUC pattern [81]. SCU displayed low absorption kinetics from the intestine and experience substantial metabolic variation before getting into the systemic circulation. Additionally, few scientific invetigations exhibited that SCU encounter a regioselective mechanism in intestine and thus mainly absorbed in the intestine in the form of its aglycone namely scutellarein [80]. In this regard, Zhao et al. (2020) formulated supramolecular γ-CD integrated MOF carriers based on non-covalent interactions for successful pulmonary delivery of SCU (SCU-γ-CD-MOF). Cubic-shaped and crystalline microparticles (MPs) of SCU-γ-CD-MOF (2:1) (<5 μm) showed a loading efficiency of 29.51%. The SCU-γ-CD-MOF showed acceptable flowability parameters, i.e., bulk density (0.20 g/mL) and Carr's index (56.08%). Moreover, thermal gravimetric analysis confirmed the protective effect of γ-CD-MOF for the SCU molecules with threefold lower weight loss compared to SCU alone. The molecular simulation displayed that polar hydroxyl and carboxyl groups of SCU created strong hydrogen bonds with the γ-CD of MOF. Further, SCU molecules showed minimal interaction and entirely symmetrical distribution within the γ-CD pair. The laryngeal deposition assessment at 65 L/min SCU-γ-CD-MOF exhibited a suitable laryngeal deposition of 57.72%. Additionally, anti-inflammatory glucocorticoid, i.e., dexamethasone (DEX), was loaded to SCU-γ-CD-MOF to comprehend the laryngeal deposition pattern of DEX and SCU. DEX-γ-CD-MOF, DEX-SCU-γ-CD-MOF (2:2:1) and DEX-SCU-γ-CD-MOF (2:1:2) exhibited 36.64, 47.35 and 54.00% laryngeal deposition at 65 L/min. The combined delivery system offered favorable laryngeal deposition of SCU. In short, the respirable γ-CD-MOF carrier has good potential in laryngeal delivery of phytoconstituent [82].

12.2.2 Functionalized MOFs for Tuberculosis

Tuberculosis (TB) is a potentially life-threatening infectious disease caused by *Mycobacterium tuberculosis* bacteria that mainly affects the lungs. TB appears in all age groups and countries. Universally, TB is one of the leading causes of death and the second leading infectious disease after coronavirus pandemic in 2020. In 2020, more than ten million people infected with TB globally [83, 84]. However, TB is curable and preventable. Isoniazid (INH) is the most reliable and commonly used antibacterial prescription approved by the U.S. Food and Drug Administration (FDA) for the prevention and treatment of TB. Chemcially, INH is a synthetic derivative of nicotinic acid and appears as white crystalline powder. INH block the synthesis of mycolic acids, key constituent of the mycobacterial cell wall to exert antibacterial action. It also interferes with mycobacterial metabolism of vitamin B6. Various INH formulations are prescribed as injection (100 mg/mL), tablets (100 and 300 mg), and syrup (10 mg/mL) for everyday as a single dose or two to three divided doses per week for treatment of TB [85]. Long-term INH therapy is frequently associated with minor, transient, and symptomless rise in serum aminotransferase levels. It is also well known to cause of acute clinically evident liver injury which can be serious and occasionally lethal [86]. Besides, INH resistance arises because of reduced bacterial wall penetration [87, 88]. In this respect, pathophysiology and natural route of TB infection with the distinctive biology of lung represents inhaled medicines as most suitable therapeutic approach for successful management of TB. Consequently, various active researchers and material chemists developed novel MOF carrier-based delivery systems for effective pulmonary delivery of INH.

The iron MOF-based theranostics and drug delivery systems are the exciting alternative for effective pulmonary therapy due to two main advantages; (i) most stable, bioactive and magnetic properties bearing MOF for theranostic activities and (ii) presence of the amine functional group to attain better-quality interaction with guest molecules. Thus, Wyszogrodzka et al. (2018) judiciously fabricated iron-based MOF (Fe-MIL-101-NH_2) as a theranostic carrier for local delivery of INH. Initially, three different Fe-MIL-101-NH_2 MOFs were developed, i.e., untreated MOF, milled MOF and milling with ultrasonication-assisted MOF. Aggregated crystalline untreated MOF and milled MOF showed mean particle sizes of 6.45 and 5.51 μm, respectively, whereas ultrasonication-assisted MOF showed a mean particle size of 3.37 μm. In morphology assessment, untreated and milled MOF presented crystal aggregates whereas ultrasonication treated MOF displayed breakage of the crystal aggregates with homogeneously dispersed crystalline structure. Infrared spectroscopy showed that INH molecules were situated within the pores of MOF via amine linkage with 12% INH loading. Particularly, infrared spectroscopy displayed substantial reduction aromatic ring breathing vibrations with phenomenon know as confinement effect, i.e., controlled breathing of the aromatic ring within the MOF pores. During an in-vitro drug release study by Franz cells using PBS (pH 7.4), Fe-MIL-101-NH_2 showed 94.20% INH release after 48 h without any initial burst effect, while crystalline INH alone showed much faster release (94.60%) within the first 3 h. The slow and steady release of INH using Fe-MIL-101-NH_2 MOF was mainly attributed to the strong interaction between INH and MOF via amine linkage. During in-vitro cytotoxicity study using fibroblasts (L929) cell lines, Fe-MIL-101-NH_2 MOF crystals formulated using micronization and ultrasonication-assisted processing successfully retained their cytotoxic potential. Fe-MIL-101-NH_2 MOF crystals were gathered in the cell cytoplasmic area and able to release drugs within the cells, making them capable carrier systems for local TB therapy. Additionally, magnetic resonance imaging (MRI) and relaxometry study of Fe-MIL-101-NH_2 MOF crystals validated that iron MOF-based carrier system can also act as the MRI contrast agent. Briefly, the unique carrier potential, imagining ability, and theranostic capacity make Fe-MIL-101-NH_2 -MOF the most suitable vehicle for INH delivery in pulmonary TB management [89].

Fernández-Paz et al. (2020) judiciously used the MIL-100-MOF-NPs template to understand the aerodynamic and intratracheal fate of INH. INH was successfully encapsulated within the MIL-100-MOF-NPs template with 29.50% encapsulation efficacy. During the in-vitro release, the equilibrium of INH release from INH-MIL-100-MOF-NPs was achieved after 2 and 4 h in water and PBS (pH 7.4), respectively, while INH-MIL-100-MOF-NPs showed slow and steady INH delivery of 14% and 80% for 3 days using water and PBS (pH 7.4) media, respectively. Afterward, INH-MIL-100-MOF-NPs loaded mannitol microspheres (1:10) were fabricated using the spray-drying method (INH-MOF-MS). During spray drying, mannitol hydroxyl groups strongly interact with MIL-100-MOF-NPs surface iron molecules and acquire effective encapsulation whereas un-interacted hydroxyl molecules offers good aqueous colloidal stability via hydrogen bonding. Well-defined, non-aggregated, spherical microspheres showed the geometric and aerodynamic diameters of 1.4 and 0.422 μm, respectively. Microspheres showed good colloidal stability in different media [e.g. MilliQ water, PBS (pH 7.4) and simulated lung fluid (SLF)] and maintain good control over particle size compared to MIL-100 MOF NPs alone. Furthermore, zeta potential stay stable and negative during the entire investigation irrespective of the test media. In in-vivo study, microspheres corresponding to 9.1 wt% of MIL-100-MOF NPs were administered intratracheally to Wistar Kyoto rats using Harvard apparatus to probe biological effects.

Intratracheally microspheres delivery showed neither pulmonary embolism sign nor toxicity for lung tissue. It retained normal parenchyma functions and thickness of alveoli wall. Additionally, after intratracheal administration (40 mg/kg) to Wistar Kyoto rats, good INH disposition in the superior lobe (16.65%), middle lobe (13.00%) and inferior lobe (18.07%) were observed. Furthermore, the dose of INH reaching the pulmonary airways can be easily modified within the minimal inhibitory concentration range (0.2–20 μg/mL). Briefly, the engineered respirable MOF-loaded microspheres platform paves the way to treat TB [90].

Wyszogrodzka-Gawel et al. (2019) developed INH-loaded Fe-MIL-101-NH_2 MOFs for dual purpose, i.e., theranostic application and local TB therapy. Initially, Fe-MIL-101-NH_2 was synthesized by solvothermal process and saturated INH solution was incorporated into the activated MOF to obtained INH-Fe-MIL-100-MOFs. INH-Fe-MIL-100-MOFs showed 12% INH content. Subsequently, INH-MOFs were loaded on hydrophobic poly(lactide-co-glycolide) (PLGA) and leucine (LEU) using spray-drying method to formulate composite particles. Furthermore, 2^3 factorial design was used to optimized the composition and spray-drying method. The INH-MOF (30%) encapsulated using PLGA/LEU (30%) ratio with 15 min blending showed good yield and aerodynamic properties with emitted dose (ED) and FPF of 90.90 and 56.70%, respectively. The INH-MOF showed irregular MPs due to bipyramidal MOF crystals, while composite particles showed a fragile and partially fragmented nature with mean particle size (<5 μm) suitable for pulmonary delivery. The cell viability and uptake studies using murine macrophages RAW 246.9 showed a marked 1.3-fold higher (21.13%) INH uptake from composite particles after 24 h than a control group. Thus, composite particles (100 μg/mL) were considered “safe” for the treated cells. In the ex-vivo analysis using rat lung tissue MRI, composite particles showed the presence of powder across the bronchi walls, including tertiary bronchi. In ex-vivo analysis, INH-MOF and composite particles showed good internalization with 13.20 and 16.66 pg/cell macrophages uptake after 24 h, respectively. Composite particles displayed improved uptake due to PLGA. PLGA assist to triggers alveolar macrophage phagocytosis and upsurge the INH concentration within cells. Additionally, Fe^{3+} ion from MOFs is well recognize to display inhibitory effect against the bacterial growth and survival. Briefly, as pulmonary TB treatment needs precise control on delivered dose and thus concurrent MRI monitoring allow tracking to develop more specific and personalized inhaled TB medicines [91].

12.2.3 Novel MOFs Platform for Anti-Inflammatory Agents

Ketoprofen (KET) is an oxo monocarboxylic acid and efficacious non-steroidal anti-inflammatory agent [92]. KET is widely used in clinical therapy owing to its anti-inflammatory, antipyretic, and analgesic actions [92]. KET exert the anti-inflammatory effect by regulating lipoxygenase and cyclooxygenase pathways by inhibiting the formation of leukotriene, prostaglandins, and thromboxane. Besides, KET also can enrich circulatory and respiratory actions by controlling endotoxic shock, stabilizing lysosomal membranes against several physiological processes such as osmotic mutilation, and release of lysosomal enzymes [93]. However, KET possess various barriers (e.g., poor solubility (51 mg/L), photolabile nature, high protein binding (> 90%), gastrointestinal tract injuries, acute renal failure after oral delivery and induce acute asthma when it is applied as the adhesive material) to its useful clinical applications [92]. But, direct lung delivery of KET is a more constructive approach. Accordingly, Zhou et al. (2020) developed γ-CD-based MOFs to facilitate pulmonary delivery of KET. KET loaded γ-CD-MOFs were prepared using in-situ drug loading via a co-crystallization method. Typical cubic-shaped KET-γ-CD-MOF particles revealed mean particle size, surface area and drug loading of 1.82 μm, 732 m^2/g, and 2.77%, respectively. In in-vitro dissolution testing, KET-γ-CD-MOF showed a faster KET release rate (> 89% within 2 h) using simulated lung fluid (SLF). The satisfactory dissolution kinetics of KET-γ-CD-MOF was primarily accredited to weak acidic nature with pH-dependent KET's solubility and SLF composition (e.g., Dipalmitoylphosphatidylcholine), which considerably improved the solubility of KET. In an aerodynamic assessment by the Turbospin® device (60 L/min for 4 sec), KET-γ-CD-MOF exhibited 2.21-fold higher FPF values and good ED (> 90%) as compared to the typical lactose (InhaLac® 230) based micronized KET formulation. The superior aerodynamic performance of KET-γ-CD-MOF was mainly endorsed to lower bulk density, homogeneous nanoscale porous structure of γ-CD-MOF and lower inter-particulate interactions. A cytocompatibility study using adenocarcinoma human alveolar basal epithelial cells (A549) and human lung cancer cell line (Calu-3), KET-γ-CD-MOF (0 to 500 μg/mL) showed good cell viability values (80%) and suitable solutions cyto-compatibility. Pulmonary exposure of γ-CD-MOF did not drastically influence lung function, i.e., lung resistance and dynamic lung compliance. As per the in-vivo inflammatory effects analysis, KET-γ-CD-MOF displayed lowered concentrations of inflammatory cytokines (interleukin-1β, interleukin-6, macrophage inflammatory protein-1 alpha and Tumour necrosis factor-α) and indicated lower pulmonary irritation compared to InhaLac® 230 based micronized KET formulation. Moreover, the ex-vivo fluorescence imaging displayed the prominent deposition rate of KET within the pulmonary airways using KET-γ-CD-MOF. Consequently, γ-CD-MOF is a capable carrier for lung drug delivery and has a wide application perspective in the therapy of lung diseases [94].

Budesonide (BUD) is a well-known glucocorticoid steroid and used for its anti-inflammatory properties. It is mainly prescribed for asthma treatment in form of aerosol. Recently, nanoporous MOFs showed promising applications in pulmonary drug delivery. Therefore, Hu et al. (2019) fabricated physically stable and nanoporous γ-CD-MOF carriers for pulmonary delivery of prototype molecule BUD. Primarily, nanoporous γ-CD-MOF particles were fabricated by vapor diffusion method. Subsequently, cholesterol and LEU-poloxamer were used to modify γ-CD-MOF powder for the improvement of physicochemical and aerodynamic properties of the powder blend. Cubic-shaped γ-CD-MOF particles showed uniform particle size (<3 μm), and cholesterol MOF powder maintained cubic morphology with mean particle size <5 μm. The γ-CD-MOF powder modified using LEU showed rough surfaces, lamellar shape particles (<3 μm) with a high degree of particle aggregation. Additionally, γ-CD-MOF, cholesterol MOF, and LEU MOF showed tapped densities of 0.45, 0.44, 0.46 g/mL, respectively. While Carr's index of 49.50, 37.60, and 32.70%, respectively. During in-vitro aerosolization study at 100 L/min (2.4 s), LEU MOF showed ED, respirable fraction and MMAD of 77.36%, 34.09%, and 5.57 μm, respectively, at the same time cholesterol MOF displayed ED and MMAD of 70.83% and 4.47 μm, respectively. Notably, cholesterol MOF demonstrated a significant 10.14-fold improvement in % FPF compared to LEU MOF. The lower FPF values of BUD using LEU MOF carrier were mainly ascribed to the aggregation of the MPs and the excessive amount of BUD deposition within the pre-separator during powder delivery. As per the in-vitro cell viability study using A549 cells, cholesterol MOF displayed >90% cell viability and good biocompatibility compared to unmodified γ-CD-MOF. In-vivo fluorescence imaging after intratracheal insufflation of Rhodamine B tagged RhB-cholesterol-MOF particles showed a marked fluorescence signal within the lung tissue compared to unmodified RhB-γ-CD-MOF at the same dose (35 mg/kg). After 12 h imaging study, RhB-cholesterol-MOF particles showed enhanced pulmonary disposition with condensed phagocytic and mucociliary clearance compared to unmodified RhB-γ-CD-MOF. Furthermore, in a pharmacokinetic study using healthy Sprague-Dawley rats after pulmonary administration by DP-4R dry insufflator device cholesterol-MOF exhibited almost comparable biological performance to Pulmicort Turbuhaler®. Both formualtions displayed approximately time to peak drug concentration (T_{max}) of 30 min and rapid clearance within 12 h. The cholesterol-MOF showed faster elimination rate and smaller AUC values compared to Pulmicort Turbuhaler® with no statistically significant difference. Furthermore, cholesterol-MOF presented higher C_{max} values compared to Pulmicort Turbuhaler® representing that MOF carrier promote the absorption of BUD effectivelty. In summary, novel MOF carriers are found to be very promising in pulmonary drug delivery. However, need to match various critical pre-clinical function to achieve successful clinical translation [95].

12.2.4 Nanostructured MOFs with Anticancer Potential

Cancer is an umbrella term for a wide range of diseases that are commonly used to state rapid formation of abnormal cells. It is a leading cause of fatalities across the world and responsible for >10 million mortalities in 2020. Moreover, the most common cause of cancer death in 2020 was lung cancer (> 1.80 million cases) [96]. Small cell lung cancer and NSCLC are the two major types of lung cancer. Now, lung cancer is a huge global dispute and figures are rapidly growing. Accordingly, the two key objectives of lung cancer treatment are to enhance patients' quality of life and survival rate [97]. Universal cancer therapy comprises chemotherapy, radiotherapy, surgical intervention, and photodynamic therapy (PDT). Radiotherapy and chemotherapy are well studied and explored for many years however are not adequate for long-term cancer rehabilitation because of their high reappearance frequency and significant adverse effects. Similarly, targeted and triggered responsive carriers were carefully studied to augment tumour drug uptake, however up to now have not revealed strong remedial benefit in human clinical trials [98]. In addition to them, a variety of nanoparticles, polymers and biological vectors are in various stages of clinical investigations [99]. The critical side effects, poor patient compliance to chronic cancer therapy and growing drug resistance are also the subject of rigorous research in present decade. In this situation, the possible advantages of cancer therapy need to be weighed alongside their potential side effects when given to patients undergoing from chronic cancer therapy [97]. Thus, compared with conventional remedies and drug carriers, active targeting or PDT are now meticulously investigated to treat cancer. In this manner, MOFs have drawn huge scientific consideration for active targeting or PDT progress owing to their structural and mechanical advantages as illustrated in above sections [100, 101]. In this regard, a recently explored MOFs vectors for cancer treatment are reviewed below.

Zhang et al. (2019) designed DNA functionalized porphyrinic PCN-224 MOFs (zirconium-based MOFs) for simultaneous cell imaging, targeted drug delivery and PDT. Specifically, MOFs were fabricated using size-controlled synthesis protocol to achieve desired surface morphology and average particle size to improve the drug loading capacity, pH-responsive release and regulate cellular uptake. Doxorubicin (DOX) was used as a model anticancer drug. Following DOX loading, the MOF was functionalized with the aptamer (DNA) of A549 cells. The aptamer was customized with fluorescein and carboxyl at the two ends to distinctively trace and target the A549 cells. The developed MOF showed monodispersed size distribution with mean particle size <250 nm and 2.35-fold shift in zeta potential compared to unmodified MOFs, which confirmed the successful implementation DNA-functionalization of PCN-224 MOFs. In an in-vitro drug release study at normal biological conditions (pH 7.4), MOF showed 5% DOX release after 48 h and confirmed the lower side effect to healthy tissues. In comparison, at the acidic environment (pH 5.0), MOF showed 30% and 45% DOX release after 12 and 48 h, respectively, representing pH-dependent release based on the capacity of functionalized MOFs. In in-vitro cytotoxicity assessment using A549 cells, MOF displayed lower cytotoxicity with >90% cell viability. Furthermore, in combined chemotherapy and PDT [650 nm laser for 15 min (100 mW/cm^2)], MOF presented a marked 1.66-fold decrease in A549 cells viability as compared to chemotherapy alone. Fluorescence tracking imaging using confocal laser scanning microscopy exhibited clearer fluorescence imaging after 6 h of incubation and stronger signals for entry of DOX in A549 cells. Concisely, this unique aptamer functionalization strategy seems potential in developing MOF based target-directed cancer therapy [102].

Guo et al. (2020) designed a multifunctional L-cysteine (CYS) decorated zirconium MOF for simultaneous delivery of chemotherapeutic agent cisplatin (CIS) and histone deacetylase inhibitor valproic acid (VAL). Uniformly dispersed and cubic-shaped crystalline MOF showed particle size of 185.90 nm and zeta potential (approx. 30 mV). In the present case, amino acid CYS was utilized to modify the surface potential and CIS and VAL loading capacity of MOF while drug loaded MOF surface was modified using polyethylene glycol (PEG) polymer to improve the biocompatibility of the system. PEGylated MOF showed a marked two-fold superior in loading capacity for CIS as compared to non-decorated MOF. During the in-vitro release study, PEGylated MOF displayed reasonable release of CIS (47.10%) under acidic conditions (pH 5.4). The in-vitro immunofluorescence assay and Western blot analysis using A549 cells PEGylated MOF (200 μg/mL) successfully down-regulated the vascular endothelial growth factor (VEGF) and PI3K/AKT protein expression, respectively along with the decrease in the drug resistance of tumor cells. During in-vitro toxicity study using A549 cells and CIS-resistant human lung adenocarcinoma cells (A549/CIS) by MTT assay, PEGylated MOF (200 μg/mL) showed 66.90% cell viability which confirmed the non-toxic nature of PEGylated MOF. Similar results were obtained when PEGylated MOF were administered to the ICR male mouse via tail vein. Further, PEGylated MOF did not induce any renal and hepatic dysfunction. An in-vivo study using A549/CDDP tumor-bearing BALB/c nude mice after combined PEGylated MOF (tail vein injection) and microwave thermal therapy showed a 100% survival rate against CIS-resistant lung cancer. Briefly, clinical experiments evidently revealed that MOFs combined with apt thermal therapy substantially enhance the remedial effect against CIS-resistant lung cancer [103].

Simon-Yarza et al. (2017) developed nanometric size crystalline biodegradable mesoporous Fe (III) polycarboxylate-based MOF with the unique pH-responsive nature for pulmonary delivery of gemcitabine monophosphate (GBM). Mainly, porous crystalline MOFs were fabricated using the coordination of cations and poly-complexing organic linkers. MOFs displayed mean particle size and zeta potential of 129 nm and – 24 mV, respectively. Further, MOFs disclosed the unique pH-responsive nature, i.e., colloidal stability of the MOF NPs remained altered under acidic conditions (from pH 4.5 to 6.5), whereas at neutral conditions (pH 7.4), MOF NPs showed aggregation (Fig. 12.3). The pH-responsive aggregation and

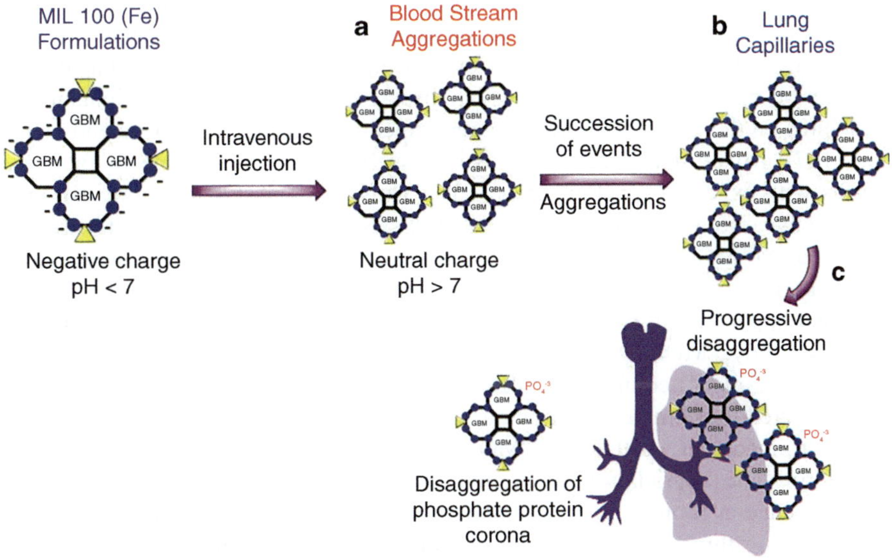

Fig. 12.3 In-vivo reversible aggregation of nanometric size crystalline biodegradable mesoporous Fe (III) polycarboxylate based MOF. (**a**) MOF Aggregation, (**b**) Retention of MOF aggregates in lung capillaries and (**c**) Progressive disaggregation of MOF aggregates

dis-aggregation were helpful for the controlled drug release within the pulmonary airways without causing toxicity during extended cancer therapy. MOFs carriers showed good drug loading (30% w/w) and encapsulation efficacy (>98%) for GBM molecules using the "green" impregnation method. Additionally, the strong interaction between the GBM phosphate group and the MOFs Lewis iron acid sites allowed controlled drug release before pulmonary particle filtration and offered satisfactory cell penetration action. In in-vivo biodistribution study using mice after i.v. administration ^{3}H-GBM (radioactive) loaded MOFs displayed a marked fivefold increase in GBM concentration within the pulmonary airways compared to GBM alone. In an in-vivo anticancer study using lung metastasis C57BL/6 J mice model after i.v. administration GBM loaded MOF displayed a notable 1.3-fold reduction in the proliferation of the metastatic nodules compared to free GBM. Moreover, the histological assessment showed a significant reduction in cancer cell proliferation and increased cell apoptosis in animals treated with GBM loaded MOF. Finally, the unique pH-dependent reversible aggregation performance and ability to target a specific tissue can be helpful to boost the therapeutic activity of the anticancer drug [104]. Optimistically, the functionalized MOFs combined with apt stimulus-response (e.g., thermal) disclosed huge budding potential for lung cancer treatment because of their positive benefits over the traditional therapy systems. However, comprehensive physical and biological scrutiny for MOF carriers is still required before surpassing the clinical cascade.

One of the appealing schemes for attaining targeted cancer therapy is to unite an established drug substance with a ligand that can carry the linked drug substance selectively to the desired site. Specifically, in anti-cancer therapy, targeted approach contains the usage of affinity ligands to access characteristically upregulated matrix proteins or biological indicators from diseased cells [105, 106]. Likewise, in past few years self-assembly of versatile organic ligands with flexible metal ions to develop MOFs is a growing theme which display huge capacity as functional nanostructures on the grounds of drug targeting, medical sensing, diagnostic and imaging [107]. Mixed-ligand technique is one such efficacious scheme to fabricate MOFs with expected functionalities [108]. Thus, Yu et al. (2019) fabricated aromatic polycarboxylic ligand containing MOFs. Specifically, mixed-ligand strategy provided two MOFs i.e. [Ni(L)0.5(4,40-bipy)0.5(H2O)2]n (MOF-A) and [Zn2(L)(bpp)2]n (H4L = 1,3-di(30,50-dicarboxylphenyl)benzene, 4,40-bipy = 4,40 ebipyridine and bpp = 1,3-di(4-pyridyl)propane) (MOF-B). In a single-crystal X-ray diffraction study, MOF A showed a four-fold interpenetrated 3D framework representing a (3,4)-connected DMD-type topological arrangement while MOF B demonstrated the complex 3D framework with the unparalleled 4-connected topological arrangement. Furthermore, MOF A exhibited excellent photocatalytic potential during the degradation of methyl blue, while MOF B showed powerful luminescence at room temperature. Additionally, during in-vitro anti-lung cancer activity using three human lung cancer cells, i.e., A549, human NSCLC derived from the lymph node (H1299) and human adenocarcinoma from lung tissue cells (PC9) by MTT assay, MOF A and B showed promising cytotoxicity action with the half-maximal inhibitory concentration (IC_{50}) values ranging from 20 to 35 mM, indicating that both MOFs revealed antitumor action against all of these cell lines in different degrees. Briefly, current assessment displayed that mixed-ligand strategy based MOFs may serve as multipurpose anti-cancer carriers, reagents and diagnostic aids in the near future [109].

12.2.5 Surface Treated MOFs for Pulmonary Hypertension

Pulmonary hypertension (PH) is one of the major causes of fatalities across the globe. It is categorized into five distinct sets according to characteristic pathophysiological conditions [110–112]. Between them pulmonary arterial hypertension (PAH) is one the most common forms of PH. PAH has many grounds but indications are analogous in all forms while their progression is differ depending on the etiology. The most ordinary indications are growing breathlessness, exhaustion, loss of consciousness, and clinical signals of heart failure. Moreover, in terms of etiology, PAH is mainly described by airway vascular endothelial dysfunction, smooth muscle proliferation, fibrosis, and in-situ thrombosis, ultimately heading to right ventricular malfunction and death [113, 114]. Five different classes of molcules such as phosphodiesterase-5 inhibitors, prostacyclin analogues, prostacyclin receptor agonists, endothelin receptor antagonists and soluble guanylate cyclase stimulators are frequently prescribed for treatment of PAH [115]. However, most of these molecules faced clinical restrictions because of systemic side effects and physicochemical barriers such as poor water solubility and low oral bioavailability. Accordingly, PAH is a rationally key disease for targeted drug delivery and Mohamed et al. (2017) put forward research efforts to develop MOFs based targeted strategy for treatment of PAH. Particularly, MIL-89 MOF NPs were made up of trans-trans muconic acid (organic linking agent) and iron-based clusters (metal-based building element) to develop PAH targeted drug delivery. Moreover, PEG coated MIL-89 MOF NPs were also developed to modulate physical and biological characteristics. Furthermore, iron based MIL-89 MOFs can act as excellent contrast agents for in-vivo imaging to track of drug distribution and progression of disease. Fine spherical MIL-89 MOF NPs showed a mean diameter of 76 nm, whereas crystalline PEGylated-MIL-89 MOF NPs displayed an average diameter of 100 nm. During thermal gravimetric analysis, both MIL-89-MOF NPs and PEGylated-MIL-89-MOF NPs were thermally stable up to 250 °C. During the macrophage viability study, both MOFs (10 mg/mL) did not affect cell viability (respiration) in the murine macrophages model or when inflammation was provoked by the addition of inflammatory lipopolysaccharide (1 mg/mL). Further, at high concentration, both MIL-89 MOF NPs (30 mg/mL) and PEGylated-MIL-89 MOF NPs (100 mg/mL) showed a significant decrease in macrophage viability. Additionally, MIL-89 MOF NPs and PEGylated-MIL-89 MOF NPs inhibited the release of the inflammatory chemokine and reserved the endothelin-1 release from endothelial cells when studied using endothelial cells cultured from normal humans and patients with PAH. During long-term storage at room temperature, PEGylated-MIL-89 MOF NPs were stable for 3 months with no evident changes in the crystallinity pattern, while MIL-89 MOF NPs showed evident changes in the crystallinity pattern after two-month storage. Therefore, both carriers MIL-89 MOF NPs and PEGylated-MIL-89 MOF NPs with the core ability to lodge PAH molecules have the additional benefit of being anti-inflammatory and reducing endothelin-1 release during the long-term PAH treatment. Briefly, data indicated that MIL-89 and PEGylated-MIL-89 are apt nanocarrier with therapeutic advantage and serves as a novel platform for treatment of PAH [116].

12.2.6 MOF as Diagnostic Platform

MOFs based nanomaterials display huge prospective in biomedical diagnosis, biosensing, and imaging. Superior potential of MOF in diagnostic field is mainly because of their unique metal nodes, huge surface area, controlled porous architecture, and exclusive catalytic action [27, 117]. Moreover, MOFs are well known to sense volatile organic compounds (VOCs) that are emitted by various metabolic pathways in human body [118]. Accordingly, material scientists tried to fabricate MOFs based diagnostic platform for concurrent detection and treatment of pulmonary aliments. Few fascinating case studies allied to section theme are listed below.

Asthma is one of the most rapidly growing non-communicable respiratory disease. It affects both adults as well as children. Approximately, it affected more than 260 million individuals in 2019 and numbers are quickly escalating across the world [119]. Recently published biological investigations have specified that hydrogen sulfide (H_2S) is linked with various pathological and physiological pathways. H_2S is also emitted in pulmonary airways tissues and reduced H_2S production in lung tissue is measured as an early detection biomarker for asthma [120]. Several analytical approaches such as gas chromatography, colorimetric assay, electrochemical assay, sulfide precipitations, and large number of fluorescent materials are normally subjected to detect and quantify H_2S from lung. However, such analytical method often require complex tissue extraction process, handing of biological fluids, tissues or cell lysates and identification of complicated fluorescence signals securely make detection task difficult and time consuming. Therefore, Zhang et al. (2019) designed nanoscale fluorescent MOF composites to transfer the complex fluorescence signals to the quickly distinguishable electronic indications for biometric detection of H_2S. Particularly, fluorescent Eu^{3+}/Ag^{+}@UiO-66-$(COOH)_2$ (EAUC-MOF) was formulated using active metal centers (Ag^{+}) to generate a strong luminescence and lanthanide-luminescence sensitizer (Eu^{3+}) for developing reactive sites for H_2S. Iso-structural UiO-66-$(COOH)_2$ exhibited a small particle size (< 100 nm) beneficial for effective dispersion. Also, UiO-66-$(COOH)_2$ showed a highly porous nature with a 1.79-fold lower surface area than UiO-66 alone because of the uncoordinated carboxyl groups. In fluorescent studies, the Eu^{3+}/Ag^{+}-based MOF sensor demonstrated quick response, high selectivity and excellent sensitivity toward in-situ detection of H_2S. Additionally, a logic network was designed using input (Ag^{+} ions and H_2S) and output (fluorescent signal;

I_{615}). In in-vitro cell viability studies using Pheochromocytoma (PC12) cells by MTT assay, Eu^{3+}/Ag^{+}-based MOF showed low cytotoxicity, excellent biocompatibility, and observations appropriate for the in-vivo detection of biomarker H_2S. Briefly, results showed that the MOF-based platforms are logically and practically applicable and opportune to detect and quantify several biomarkers involved in chronic pulmonary disease [121].

Wang et al. (2020) judiciously designed novel fluorescent MOF containing Zn(II) ions to control the augmented level of eosinophil in the pulmonary airways of asthma subjects. Mainly, novel fluorescent MOF containing Zn(II) ions as the metal nodes based {Zn3(OH)} clusters as the secondary building unit and a rigid hexacarboxylic acid linker of 5,5′,5″-(methylsilanetriyl) triisophthalic acid (H6L) prepared under solvothermal settings. Monoclinic crystalline novel fluorescent MOF containing Zn(II) ions showed the following final chemical formula {[Zn3HL(μ3-OH)(H2O)2]•3(H2O)•DMA}n, i.e., complex-I. Complex-I showed significant inhibition of eosinophil during cell viability analysis using the Cell Counting Kit-8 (CCK-8). Moreover, during ELISA assay using 5 mL atomized OVA to BALB/c mice model, complex I significantly inhibited the basic protein and eosinophil neurotoxin, causing lung epithelial cell to injure, shed with bronchospasm as compared to the control group. Briefly, results indicated that complex-I had an admirable shielding potential on asthma by reducing inflammatory precursors [122].

12.3 Future Outlook

MOFs' design, synthesis, and applications are budding hot topics in institutional and industrial organizations. As shown in the above section, substantial advances have been made in designing and developing MOFs for pulmonary drug delivery (Fig. 12.4). Table 12.1 showed different MOFs-based inhaled formulations along with the primary outcomes. A wide range of metal ions and organic building blocks are explored and purposely tailored to form hybrid materials with suitable characteristics as pulmonary drug carriers. MOFs are capable of carrying a variety of molecules, i.e., phytoconstituents (e.g., CUR, SCU, and PAN), anticancer agents (e.g., GBM, CIS, and DOX), antibiotics (e.g., INH) and anti-inflammatory molecules (e.g., KET and BUD) for successful pulmonary disease management. Additionally, PEGylated- and spray-dried mannitol-based MOFs were explored in a few investigations to improve drug deposition patterns and pulmonary biocompatibility after intratracheal administration. Furthermore, various in-vitro cell lines and in-vivo animal studies have shown optimistic outcomes with sustained and stimuli-responsive drug release using MOFs. Briefly, MOFs are biocompatible, stimuli-responsive, and target-specific novel nano-delivery systems for effective pulmonary drug delivery.

Active efforts of material chemists and researchers in the synthesis and development of MOFs have already helped translate MOFs from laboratory to commercial level. Prof-MOF (Norway) [123], NuMat Technologies (USA) [124], novoMOF (Switzerland) [125], BASF (USA) [126] and MOFapps [127] are the few key

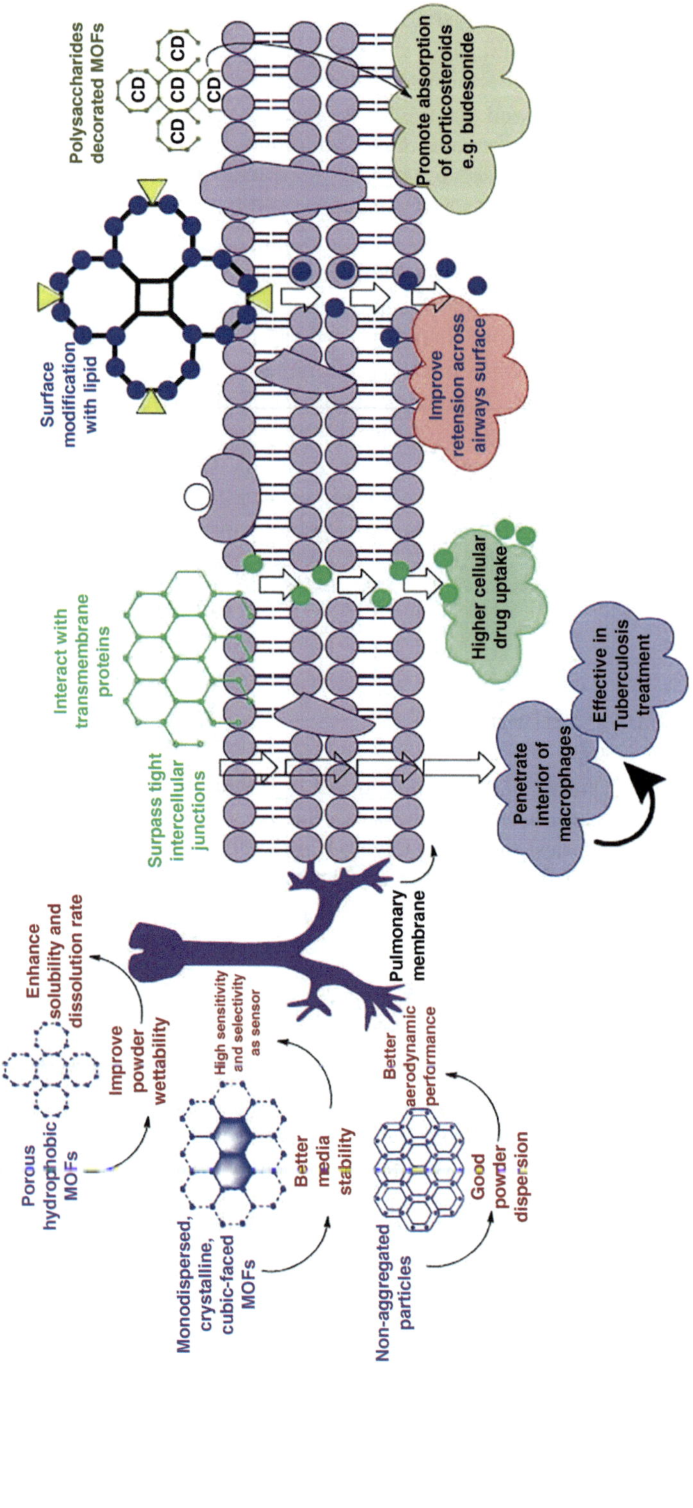

Fig. 12.4 Mechanism and applications of novel nanometric size MOFs

Table 12.1 Novel MOFs-based inhaled formulation along with the key findings

Formulation	Particle properties	Outcome	References
CUR-γ-CD-MOF	Cubic shape with 2.17 μm	MOF showed higher FPF (> 60%) using Turbospin® (60 L/min) and prominent 1.88-fold higher antioxidant potential compared to lactose based CUR formulation	[66]
PAN-γ-CD-MOF	Crystalline smooth surface and cubic shape and < 3 μm	MOF showed acceptable aerodynamic outcomes using Rotahaler® (50.71 L/min) and a marked 6.93 and 11.22-fold enhancement in AUC and C_{max}, respectively after intratracheal insufflation as compared to intragastric administration at the same dose (4 mg/kg)	[75]
SCU-γ-CD-MOF	Cubic shape and < 5 μm	MOF demonstrated acceptable laryngeal deposition (57.72%) at 65 L/min	[82]
INH-Fe-MIL-101-NH_2-MOF	Homogeneously dispersed crystalline structure with 3.37 μm	MOF successfully retained their cytotoxic potential against fibroblasts L929 cells and quickly gathered in the cell cytoplasmic area, able to release INH within the cells	[89]
INH-MIL-100-MOF-NPs	Well-defined, non-aggregated, spherical microspheres and 1.4 μm	After intratracheally administered, MOF showed higher drug deposition in the inferior lobe (18.07%), and the deposited dose is within the minimal inhibitory concentration range	[90]
INH- Fe-MIL-101-NH_2 MOFs	Irregular MPs due to bipyramidal MOF crystals and < 5 μm	MOFs MPs encapsulated using PLGA/LEU showed ED and FPF of 90.90 and 56.70%, respectively.	[90]
KET-γ-CD-MOF	Cubic shape with 1.82 μm	MOF showed higher FPF (2.21-fold) using Turbospin® (60 L/min) as compared to lactose based formulation and after intratracheal administration, displayed a lower level of inflammatory cytokines	[94]
BUD-γ-CD/cholesterol MOF	Cubic morphology with <3 μm	MOF displayed ED and MMAD of 70.83% and 4.47 μm, respectively with >90% A549 cell viability and after intratracheal insufflation showed enhanced pulmonary disposition with condensed phagocytic and mucociliary clearance as compared to unmodified MOF	[95]

Table 12.1 (continued)

Formulation	Particle properties	Outcome	References
DOX loaded DNA-functionalized porphyrinic PCN-224 MOFs	Monodispersed particles with (< 250 nm)	MOF showed pH-dependent release with PDT [650 nm laser for 15 min (100 mW/cm^2)] MOF displayed a marked 1.66-fold decrease in A549 cell viability compared to chemotherapy alone	[102]
CIS/VAL loaded CYS decorated and PEG-modified Zr-based MOF NPs	Positively charged (30 mV), cubic shape crystalline NPs with 185.90 nm	MOF showed pH-dependent release and successfully decreased the drug resistance of tumor cells with a 100% survival rate in microwave thermal combined therapy treated A549/CDDP tumor-bearing BALB/c nude mice	[103]
GBM loaded mesoporous Fe (III) polycarboxylate based MOF	Negatively charged (24 mV) crystalline porous NPs and 129 nm	pH-responsive MOF showed a five-fold increase in drug concentration within the lung compared to drug alone and a 1.3-fold reduction in the proliferation of the metastatic nodules during the in-vivo anticancer study using the C57BL/6 J mice model as compared to drug alone	[104]
MIL-89-MOF NPs	Fine spherical and 76 nm	NPs inhibited the release of the inflammatory chemokine (CXCL8) and reserved the endothelin-1 (ET-1) release in endothelial cells obtained from normal humans and patients with PAH	[116]
PEGylated-MIL-89-MOF NPs	Spherical crystalline and 100 nm	NPs inhibited the release of the inflammatory chemokine (CXCL8) and reserved the endothelin-1 (ET-1) release in endothelial cells obtained from normal human and patients with PAH and retained the crystallinity pattern during 3 months storage	[116]
Eu^{3+}/Ag^{+}@ UiO-66-$(COOH)_2$ MOF	Highly porous and < 100 nm	Biocompatible MOF sensor displayed quick response, high selectivity and excellent sensitivity towards in-situ detection of biomarker H_2S	[121]
Fluorescent MOF containing Zn(II) ions (Complex I)	Monoclinic crystalline nanomaterial	Precisely probe and control the augmented level of eosinophil in the pulmonary airways of asthma subjects	[122]

companies actively involved in synthesizing MOFs at a commercial scale. The production control and demonstration of structured hybrid nanoporous materials for industrial adsorption applications (ProDIA) is the unique theme launched by MOFapps organization to improve MOFs horizon [128]. However, the commercial applications of novel MOFs were limited to energy, catalysis, purification, gas storage and sensors. Therefore, more vigorous formulation and clinical efforts are needed to fuel the MOFs in pulmonary drug delivery and biomedical, commercial applications.

Even though the functions of MOFs varied from energy storage to drug delivery with unique applications, the toxicological and ecological effects of their metal components should not be overlooked. MOFs are frequently synthesized by solvothermal methods involving toxic solvents, long processing times and pose ecological and health issues [44]. Mainly, solvothermal synthesis schemes contain organic compounds, non-renewable polar solvents and high boiling settings [129, 130]. Additionally, numerous safety hazards that may be irrelevant at the laboratory level need to be considered for commercial-scale, such as flammability and explosion risk during material handling and asphyxiation danger of inert gas [131]. Moreover, several factors need to be judged before commercialization, such as biodegradability, cost-effectiveness, environmentally friendly production, and water stability. Thus, research institutes and industrial organizations' focus gradually shifted toward green and sustainable techniques for MOF developments. Even though using reusable and recyclable organic solvents is acceptable, removing harmful organic compounds is favored. The use of green solvents [132], biomass-based ligands as organic linkers [133], environmentally benign synthesis pathways is now thoroughly explored to design and develop green MOFs carriers. Moreover, the utility of biomolecules containing different reactive chemical moieties (e.g., carboxyl and hydroxyl) can be adopted to fabricate green MOFs [134]. Additionally, the synthesis of biomass-derived MOFs is a crucial alternative to decrease dependence on traditional hazardous routes of MOFs synthesis [135]. The judicious combination of sustainability principles and green chemistry also has substantial capability to meet commercial production's necessity. Briefly, the research focus must be switched to sustainability principles and human clinical studies for rapid expansion of MOFs in real-life biomedical applications.

12.4 Conclusion

In a nutshell, active efforts have been made in the design and development of MOFs for drug delivery, PDT applications, biosensing therapies and synergistic treatment against pulmonary diseases in the last few years. MOFs show key advantages in pulmonary therapy because of their intrinsic physicochemical properties such as tunable structure, large pore size, multifaceted nature, good chemical stability and biocompatibility. Even though some studies showed good aerodynamic properties and relevant pre-clinical results for MOFs, more detailed investigations are still necessary to resolve major MOFs' significant issues. Significantly, additional active research efforts on the controlled particle size distribution, surface functionalization, biophysical properties, pulmokinetics and biosafety are highly desirable before moving to the clinical translation phase. Finally, it is well anticipated that the analysis and assessment of MOFs for pulmonary applications in this chapter could help more interdisciplinary research at the edge of MOFs to reveal the full potential of MOFs in a crucial battle against life-threatening respiratory diseases.

References

1. Scicluna MC, Vella-Zarb L. Evolution of nanocarrier drug-delivery systems and recent advancements in covalent organic framework–drug systems. ACS Appl Nano Mater. 2020;3(4):3097–115.
2. Mehta PP, Pawar VS. Electrospun nanofiber scaffolds: technology and applications. In Applications of nanocomposite materials in drug delivery 2018 (pp. 509–573). Woodhead Publishing.
3. Mehta P, Mahadik K, Kadam S, Dhapte-Pawar V. Advanced applications of green hydrogels in drug delivery systems. Appl Adv Green Mater. 2021:89–130.
4. Kumar R, Mehta P, Shankar KR, Rajora MA, Mishra YK, Mostafavi E, Kaushik A. Nanotechnology-assisted metered-dose inhalers (MDIs) for high-performance pulmonary drug delivery applications. Pharm Res. 2022;1-25
5. Kulkarni S, Prabhakar B, Shende P. Stabilization of lipid vesicles: upcoming strategic insights for product development. J Mol Liq. 2021;31:118430.
6. Mehta PP, Ghoshal D, Pawar AP, Kadam SS, Dhapte-Pawar VS. Recent advances in inhalable liposomes for treatment of pulmonary diseases: concept to clinical stance. J Drug Deliv Sci Technol. 2020;56:101509.
7. Mehta P, Kadam S, Pawar A, Bothiraja C. Dendrimers for pulmonary delivery: current perspectives and future challenges. New J Chem. 2019;43(22):8396–409.
8. Mehta PP, Pawar AP, Mahadik KR, Kadam SS, Dhapte-Pawar V. Dry powder coating techniques and role of force controlling agents in aerosol. Polymer Coatings: Technology and Applications. 2020:41–74.
9. Mehta P. Imagine the superiority of dry powder inhalers from carrier engineering. J Drug Deliv. 2018;2018
10. Mehta P. Dry powder inhalers: a focus on advancements in novel drug delivery systems. J Drug Deliv. 2016;2016
11. Beg S, Rahman M, Jain A, Saini S, Midoux P, Pichon C, Ahmad FJ, Akhter S. Nanoporous metal organic frameworks as hybrid polymer-metal composites for drug delivery and biomedical applications. Drug Discov Today. 2017;22(4):625–37.
12. Khan NA, Hasan Z, Ahmed I, Jhung SH. Metal-organic frameworks for nanoarchitectures: nanoparticle, composite, core-shell, hierarchical, and hollow structures. In Advanced Supramolecular Nanoarchitectonics 2019 (pp. 151–194). William Andrew Publishing.
13. Geng K, He T, Liu R, Dalapati S, Tan KT, Li Z, Tao S, Gong Y, Jiang Q, Jiang D. Covalent organic frameworks: design, synthesis, and functions. Chem Rev. 2020;120(16):8814–933.
14. Zhou HC, Long JR, Yaghi OM. Introduction to metal-organic frameworks. Chem Rev. 2012;112(2):673–4.
15. Tian Y, Zhu G. Porous aromatic frameworks (PAFs). Chem Rev. 2020;120(16):8934–86.
16. Lee JS, Cooper AI. Advances in conjugated microporous polymers. Chem Rev. 2020;120(4):2171–214.

17. Liu M, Guo L, Jin S, Tan B. Covalent triazine frameworks: synthesis and applications. J Mater Chem A. 2019;7(10):5153–72.
18. Yuan Y, Yang Y, Zhu G. Multifunctional porous aromatic frameworks: state of the art and opportunities. Energy Chem. 2020;2(5):100037.
19. Yuan Y, Zhu G. Porous aromatic frameworks as a platform for multifunctional applications. ACS Central Sci. 2019;5(3):409–18.
20. Rowsell JL, Yaghi OM. Metal-organic frameworks: a new class of porous materials. Microporous Mesoporous Mater. 2004;73(1–2):3–14.
21. Zhang H, Li G, Liao C, Cai Y, Jiang G. Bio-related applications of porous organic frameworks (POFs). J Mater Chem B. 2019;7(15):2398–420.
22. Xu D, Guo J, Yan F. Porous ionic polymers: design, synthesis, and applications. Prog Polym Sci. 2018;79:121–43.
23. Wang Q, Gao T, Hao L, Guo Y, Liu W, Guo L, Wang C, Wang Z, Wu Q. Advances in magnetic porous organic frameworks for analysis and adsorption applications. TrAC Trends Anal Chem. 2020;132:116048.
24. Chen YZ, Zhang R, Jiao L, Jiang HL. Metal–organic frameork-derived porous materials for catalysis. Coord Chem Rev. 2018;362:1–23.
25. Chuhadiya S, Suthar D, Patel SL, Dhaka MS. Metal organic frameworks as hybrid porous materials for energy storage and conversion devices: a review. Coord Chem Rev. 2021;446:214115.
26. Ferreira FV, Otoni CG, Kevin J, Barud HS, Lona LM, Cranston ED, Rojas OJ. Porous nanocellulose gels and foams: breakthrough status in the development of scaffolds for tissue engineering. Mater Today. 2020;
27. Bieniek A, Terzyk AP, Wiśniewski M, Roszek K, Kowalczyk P, Sarkisov L, Keskin S, Kaneko K. MOF materials as therapeutic agents, drug carriers, imaging agents and biosensors in cancer biomedicine: recent advances and perspectives. Prog Mater Sci. 2021;117:100743.
28. He M, Liang Q, Tang L, Liu Z, Shao B, He Q, Wu T, Luo S, Pan Y, Zhao C, Niu C. Advances of covalent organic frameworks based on magnetism: classification, synthesis, properties, applications. Coord Chem Rev. 2021;449:214219.
29. Barthel S, Alexandrov EV, Proserpio DM, Smit B. Distinguishing metal-organic frameworks. Cryst Growth Des. 2018;18(3):1738–47.
30. Yaghi OM, Li H. Hydrothermal synthesis of a metal-organic framework containing large rectangular channels. J Am Chem Soc. 1995;117(41):10401–2.
31. Lee YR, Kim J, Ahn WS. Synthesis of metal-organic frameworks: a mini review. Korean J Chem Eng. 2013;30(9):1667–80.
32. Klinowski J, Paz FA, Silva P, Rocha J. Microwave-assisted synthesis of metal–organic frameworks. Dalton Trans. 2011;40(2):321–30.
33. Varsha MV, Nageswaran G. Direct electrochemical synthesis of metal organic frameworks. J Electrochem Soc. 2020;167:155527.
34. Christodoulou I, Serre C, Gref R. Metal-organic frameworks for drug delivery: degradation mechanism and in vivo fate. In Metal-organic frameworks for biomedical applications 2020 (pp. 467–489). Woodhead Publishing.
35. Denisov GL, Primakov PV, Korlyukov AA, et al. Solvothermal synthesis of the metal-organic framework MOF-5 in autoclaves prepared by 3D printing. Russ J Coord Chem. 2019;45:836–42.
36. Griffin SL, Champness NR. A periodic table of metal-organic frameworks. Coord Chem Rev. 2020;414:213295.
37. Xian S, Lin Y, Wang H, Li J. Calcium-based metal-organic frameworks and their potential applications. Small. 2021;17(22):2005165.
38. Wang D, Zhou J, Chen R, Shi R, Wang C, Lu J, Zhao G, Xia G, Zhou S, Liu Z, Wang H. Core–shell metal-organic frameworks as Fe2+ suppliers for Fe2+−mediated cancer therapy under multimodality imaging. Chem Mater. 2017;29(8):3477–89.
39. Lazaro IA, Forgan RS. Application of zirconium MOFs in drug delivery and biomedicine. Coord Chem Rev. 2019;380:230–59.

40. Taylor KM, Rieter WJ, Lin W. Manganese-based nanoscale metal–organic frameworks for magnetic resonance imaging. J Am Chem Soc. 2008;130(44):14358–9.
41. Shyngys M, Ren J, Liang X, Miao J, Blocki A, Beyer S. Metal-organic framework (MOF)-based biomaterials for tissue engineering and regenerative medicine. Front Bioeng Biotechnol. 2021;9:603608.
42. Akbarzadeh F, Motaghi M, Chauhan NP, Sargazi G. A novel synthesis of new antibacterial nanostructures based on Zn-MOF compound: design, characterization and a high performance application. Heliyon. 2020;6(1):e03231.
43. Kumar S, Jain S, Nehra M, Dilbaghi N, Marrazza G, Kim KH. Green synthesis of metal–organic frameworks: a state-of-the-art review of potential environmental and medical applications. Coord Chem Rev. 2020;420:213407.
44. Reinsch H. "Green" synthesis of metal-organic frameworks. Eur J Inorg Chem. 2016;2016(27):4290–9.
45. Tamames-Tabar C, Imbuluzqueta E, Guillou N, Serre C, Miller SR, Elkaïm E, Horcajada P, Blanco-Prieto MJ. A Zn azelate MOF: combining antibacterial effect. CrystEngComm. 2015;17(2):456–62.
46. Banerjee S, Lollar CT, Xiao Z, Fang Y, Zhou HC. Biomedical integration of metal–organic frameworks. Trends Chem. 2020;2(5):467–79.
47. Kökçam-Demir Ü, Goldman A, Esrafili L, Gharib M, Morsali A, Weingart O, Janiak C. Coordinatively unsaturated metal sites (open metal sites) in metal-organic frameworks: design and applications. Chem Soc Rev. 2020;49(9):2751–98.
48. Baumann AE, Burns DA, Liu B, et al. Metal-organic framework functionalization and design strategies for advanced electrochemical energy storage devices. Commun Chem. 2019;2:86.
49. Wang B, Xie LH, Wang X, Liu XM, Li J, Li JR. Applications of metal–organic frameworks for green energy and environment: new advances in adsorptive gas separation, storage and removal. Green Energy Environ. 2018;3(3):191–228.
50. Feng L, Qian C, Zhao Y. Recent advances in covalent organic framework-based Nanosystems for bioimaging and therapeutic applications. ACS Mater Lett. 2020;2(9):1074–92.
51. Foo ML, Matsuda R, Kitagawa S. Functional hybrid porous coordination polymers. Chem Mater. 2014;26(1):310–22.
52. Kim CR, Uemura T, Kitagawa S. Inorganic nanoparticles in porous coordination polymers. Chem Soc Rev. 2016;45(14):3828–45.
53. Karami A, Mohamed O, Ahmed A, Husseini GA, Sabouni R. Recent advances in metal-organic frameworks as anticancer drug delivery systems: a review. Anti Cancer Agents Med Chem. 2021; https://doi.org/10.2174/1871520621666210119093844.
54. Osterrieth JWM, Fairen-Jimenez D. Metal-organic framework composites for theragnostics and drug delivery applications. Biotechnol J. 2021;16(2):e2000005.
55. Lawson HD, Walton SP, Chan C. Metal-organic frameworks for drug delivery: a design perspective. ACS Appl Mater Interfaces. 2021;13(6):7004–20.
56. Bao Z, Li K, Hou P, Xiao R, Yuan Y, Sun Z. Nanoscale metal–organic framework composites for phototherapy and synergistic therapy of cancer. Mater Chem Front. 2021;5(4):1632–54.
57. Horcajada P, et al. Metal–organic frameworks in biomedicine. Chem Rev. 2012;112:1232–68.
58. Keskin S, Kizilel S. Biomedical applications of metal organic frameworks. Ind Eng Chem Res. 2011;50:1799–812.
59. Mehta P, Bothiraja C, Mahadik K, Kadam S, Pawar A. Phytoconstituent based dry powder inhalers as biomedicine for the management of pulmonary diseases. Biomed Pharmacother. 2018;108:828–37.
60. Sanphui P, Bolla G. Curcumin, a biological wonder molecule: a crystal engineering point of view. Cryst Growth Des. 2018;18(9):5690–711.
61. Lelli D, Sahebkar A, Johnston TP, Pedone C. Curcumin use in pulmonary diseases: state of the art and future perspectives. Pharmacol Res. 2017;115:133–48.
62. Amalraj A, Pius A, Gopi S, Gopi S. Biological activities of curcuminoids, other biomolecules from turmeric and their derivatives - a review. J Tradit Complement Med. 2016;7(2):205–33.

63. Priyadarsini KI. The chemistry of curcumin: from extraction to therapeutic agent. Molecules. 2014;19(12):20091–112.
64. Venkatesan N, Punithavathi D, Babu M. Protection from acute and chronic lung diseases by curcumin. Adv Exp Med Biol. 2007;595:379–405.
65. Suresh K, Nangia A. Curcumin: pharmaceutical solids as a platform to improve solubility and bioavailability. CrystEngComm. 2018;20(24):3277–96.
66. Zhou Y, Zhao Y, Niu B, Luo Q, Zhang Y, Quan G, Pan X, Wu C. Cyclodextrin-based metal-organic frameworks for pulmonary delivery of curcumin with improved solubility and fine aerodynamic performance. Int J Pharm. 2020;588:119777.
67. Mehta P, Shah R, Lohidasan S, Mahadik KR. Pharmacokinetic profile of phytoconstituent (s) isolated from medicinal plants—a comprehensive review. J Tradit Complement Med. 2015;5(4):207–27.
68. Mehta P, Lohidasan S, Mahadik KR. Pharmacokinetic behaviour of clinically important TCM prescriptions. Orient Pharm Exp Med. 2017;17(3):171–88.
69. Wang F, Shan Q, Chang X, Li Z, Gui S. Paeonol-loaded PLGA nanoparticles as an oral drug delivery system: design, optimization and evaluation. Int J Pharm. 2021;602:120617.
70. Liu MH, Lin AH, Ko HK, Perng DW, Lee TS, Kou YR. Prevention of bleomycin-induced pulmonary inflammation and fibrosis in mice by Paeonol. Front Physiol. 2017;8:193.
71. Liu MH, Lin AH, Lee HF, Ko HK, Lee TS, Kou YR. Paeonol attenuates cigarette smoke-induced lung inflammation by inhibiting ROS-sensitive inflammatory signaling. Mediat Inflamm. 2014;2014:651890.
72. Wang D, Du Y, Xu H, Pan H, Wang R. Paeonol protects mitochondrial injury and prevents pulmonary vascular remodeling in hypoxia. Respir Physiol Neurobiol. 2019;268:103252.
73. Liu X, Xu Q, Mei L, Lei H, Wen Q, Miao J, Huang H, Chen D, Du S, Zhang S, Zhou J. Paeonol attenuates acute lung injury by inhibiting HMGB1 in lipopolysaccharide-induced shock rats. Int Immunopharmacol. 2018;61:169–77.
74. Yin T, Wu J, Yuan J, Wang X. Therapeutic deep eutectic solvent based on osthole and paeonol: preparation, characterization, and permeation behavior. J Mol Liq. 2022;346:117133.
75. Li H, Zhu J, Wang C, Qin W, Hu X, Tong J, Yu L, Zhang G, Ren X, Li Z, Zhang J. Paeonol loaded cyclodextrin metal-organic framework particles for treatment of acute lung injury via inhalation. Int J Pharm. 2020;587:119649.
76. Mehta P, Dhapte V. A comprehensive review on pharmacokinetic profile of some traditional Chinese medicines. N J Sci. 2016;2016
77. Jiang RJ, Yang B, Liu ZK, Zhao YL, Liao XL, Yang J, Gao CZ, Wang F, Han B. A novel polyrotaxane-based delivery system for scutellarin: preparation, characterization, and in vitro evaluation. Carbohydr Res. 2013;380:149–55.
78. Cao P, Liu B, Du F, Li D, Wang Y, Yan X, Li X, Li Y. Scutellarin suppresses proliferation and promotes apoptosis in A549 lung adenocarcinoma cells via AKT/mTOR/4EBP1 and STAT3 pathways. Thorac Cancer. 2019;10(3):492–500.
79. Peng L, Wen L, Shi QF, Gao F, Huang B, Meng J, Hu CP, Wang CM. Scutellarin ameliorates pulmonary fibrosis through inhibiting NF-κB/NLRP3-mediated epithelial-mesenchymal transition and inflammation. Cell Death Dis. 2020;11(11):978.
80. Wang L, Ma Q. Clinical benefits and pharmacology of scutellarin: a comprehensive review. Pharmacol Ther. 2018;190:105–27.
81. Chen X, Cui L, Duan X, Ma B, Zhong D. Pharmacokinetics and metabolism of the flavonoid scutellarin in humans after a single oral administration. Drug Metab Dispos. 2006;34(8):1345–52.
82. Zhao K, Guo T, Wang C, Zhou Y, Xiong T, Wu L, Li X, Mittal P, Shi S, Gref R, Zhang J. Glycoside scutellarin enhanced CD-MOF anchoring for laryngeal delivery. Acta Pharm Sin B. 2020;10(9):1709–18.
83. Mehta P, Bothiraja C, Kadam S, Pawar A. Potential of dry powder inhalers for tuberculosis therapy: facts, fidelity and future. Artif Cells Nanomed Biotechnol. 2018;46(sup3):S791–806.
84. WHO, Tuberculosis. https://www.who.int/news-room/fact-sheets/detail/tuberculosis

85. Huaman MA, Sterling TR. Treatment of latent tuberculosis infection—An update. Clin Chest Med. 2019;40(4):839–48.
86. Saukkonen JJ, Cohn DL, Jasmer RM, Schenker S, Jereb JA, Nolan CM, Peloquin CA, Gordin FM, Nunes D, Strader DB, Bernardo J, Venkataramanan R, Sterling TR, ATS (American Thoracic Society) Hepatotoxicity of Antituberculosis Therapy Subcommittee. An official ATS statement: hepatotoxicity of antituberculosis therapy. Am J Respir Crit Care Med. 2006;174(8):935–52.
87. Moreno-Gamez S, Hill AL, Rosenbloom DI, Petrov DA, Nowak MA, Pennings PS. Imperfect drug penetration leads to spatial monotherapy and rapid evolution of multidrug resistance. Proc Natl Acad Sci. 2015;112(22):E2874-83.
88. Gygli SM, Borrell S, Trauner A, Gagneux S. Antimicrobial resistance in Mycobacterium tuberculosis: mechanistic and evolutionary perspectives. FEMS Microbiol Rev. 2017;41(3):354–73.
89. Wyszogrodzka G, Dorożyński P, Gil B, Roth WJ, Strzempek M, Marszałek B, Węglarz WP, Menaszek E, Strzempek W, Kulinowski P. Iron-based metal-organic frameworks as a Theranostic carrier for local tuberculosis therapy. Pharm Res. 2018;35(7):144.
90. Fernández-Paz C, Rojas S, Salcedo-Abraira P, Simón-Yarza T, Remuñán-López C, Horcajada P. Metal-organic framework microsphere formulation for pulmonary administration. ACS Appl Mater Interfaces. 2020;12(23):25676–82.
91. Wyszogrodzka-Gaweł G, Dorożyński P, Giovagnoli S, Strzempek W, Pesta E, Węglarz WP, Gil B, Menaszek E, Kulinowski P. An inhalable theranostic system for local tuberculosis treatment containing an isoniazid loaded metal organic framework Fe-MIL-101-NH2-from raw MOF to drug delivery system. Pharmaceutics. 2019;11(12):687.
92. Kuczyńska J, Nieradko-Iwanicka B. Future prospects of ketoprofen in improving the safety of the gastric mucosa. Biomed Pharmacother. 2021;139:111608.
93. Hu J, Liu H, Xu P, Shang Y, Liu H. Investigation of drug for pulmonary administration–model pulmonary surfactant monolayer interactions using Langmuir–Blodgett monolayer and molecular dynamics simulation: a case study of ketoprofen. Langmuir. 2019;35(41):13452–60.
94. Zhou Y, Niu B, Wu B, Luo S, Fu J, Zhao Y, Quan G, Pan X, Wu C. A homogenous nanoporous pulmonary drug delivery system based on metal-organic frameworks with fine aerosolization performance and good compatibility. Acta Pharm Sin B. 2020;10(12):2404–16.
95. Hu X, Wang C, Wang L, Liu Z, Wu L, Zhang G, Yu L, Ren X, York P, Sun L, Zhang J, Li H. Nanoporous CD-MOF particles with uniform and inhalable size for pulmonary delivery of budesonide. Int J Pharm. 2019;564:153–61.
96. WHO, Cancer. https://www.who.int/news-room/fact-sheets/detail/cancer
97. Moreau Bachelard C, Coquan E, du Rusquec P, Paoletti X, Le Tourneau C. Risks and benefits of anticancer drugs in advanced cancer patients: a systematic review and meta-analysis. EClinicalMedicine. 2021;40:101130.
98. Haemmerich D. Non-invasive image-guided targeted drug delivery. Lancet Oncol. 2018;19(8):1000–1.
99. Chen J, Zhu Y, Wu C, Shi J. Nanoplatform-based cascade engineering for cancer therapy. Chem Soc Rev. 2020;49(24):9057–94.
100. Kotzabasaki M, Froudakis GE. Review of computer simulations on anti-cancer drug delivery in MOFs. Inorganic Chem Front. 2018;5(6):1255–72.
101. Zheng Y, Zhang X, Su Z. Design of metal–organic framework composites in anti-cancer therapies. Nanoscale. 2021;13(28):12102–18.
102. Zhang Y, Wang Q, Chen G, Shi P. DNA-functionalized metal-organic framework: cell imaging, targeting drug delivery and photodynamic therapy. Inorg Chem. 2019;58(10):6593–6.
103. Guo W, Chen Z, Tan L, Wu Q, Ren X, Fu C, Du Y, Ren J, Meng X. L-cysteine decorated nanoscale metal–organic frameworks delivering valproic acid/cisplatin for drug-resistant lung cancer therapy. Chem Commun. 2020;56(27):3919–22.
104. Simon-Yarza T, Giménez-Marqués M, Mrimi R, Mielcarek A, Gref R, Horcajada P, Serre C, Couvreur P. Smart metal-organic framework nanomaterial for lung targeting. Angew Chem. 2017;129(49):15771–5.

105. Kamaly N, Xiao Z, Valencia PM, Radovic-Moreno AF, Farokhzad OC. Targeted polymeric therapeutic nanoparticles: design, development and clinical translation. Chem Soc Rev. 2012;41(7):2971–3010.
106. Srinivasarao M, Low PS. Ligand-targeted drug delivery. Chem Rev. 2017;117(19):12133–64.
107. Yoneya M, Tsuzuki S, Aoyagi M. Simulation of metal–organic framework self-assembly. Phys Chem Chem Phys. 2015;17(14):8649–52.
108. Yin Z, Zhou YL, Zeng MH, Kurmoo M. The concept of mixed organic ligands in metal–organic frameworks: design, tuning and functions. Dalton Trans. 2015;44(12):5258–75.
109. Yu LG, Sun YG, Wang ZL. Mixed-ligand strategy affording two new metal-organic frameworks: photocatalytic, luminescent and anti-lung cancer properties. J Mol Struct. 2019;1180:209–14.
110. Burger CD, DuBrock HM, Cartin-Ceba R, Moss JE, Shapiro BP, Frantz RP. Topic-based, recent literature review on pulmonary hypertension. In Mayo Clinic Proceedings 2021 (Vol. 96, No. 12, pp. 3109–3121). Elsevier.
111. Daniell H, Mangu V, Yakubov B, Park J, Habibi P, Shi Y, Gonnella PA, Fisher A, Cook T, Zeng L, Kawut SM. Investigational new drug enabling angiotensin oral-delivery studies to attenuate pulmonary hypertension. Biomaterials. 2020;233:119750.
112. Mehta PP, Dhapte-Pawar VS. Repurposing drug molecules for new pulmonary therapeutic interventions. Drug Deliv Transl Res. 2021;11(5):1829–48.
113. Deshwal H, Weinstein T, Sulica R. Advances in the management of pulmonary arterial hypertension. J Investig Med. 2021;69(7):1270–80.
114. Sommer N, Ghofrani HA, Pak O, Bonnet S, Provencher S, Sitbon O, Rosenkranz S, Hoeper MM, Kiely DG. Current and future treatments of pulmonary arterial hypertension. Br J Pharmacol. 2021;178(1):6–30.
115. Hoeper MM, McLaughlin VV, Al Dalaan AM, Satoh T, Galiè N. Treatment of pulmonary hypertension. Lancet Respir Med. 2016;4(4):323–36.
116. Mohamed NA, Davies RP, Lickiss PD, Ahmetaj-Shala B, Reed DM, Gashaw HH, Saleem H, Freeman GR, George PM, Wort SJ, Morales-Cano D, Barreira B, Tetley TD, Chester AH, Yacoub MH, Kirkby NS, Moreno L, Mitchell JA. Chemical and biological assessment of metal organic frameworks (MOFs) in pulmonary cells and in an acute in vivo model: relevance to pulmonary arterial hypertension therapy. Pulm Circ. 2017;7(3):643–53.
117. Daniel M, Mathew G, Anpo M, Neppolian B. MOF based electrochemical sensors for the detection of physiologically relevant biomolecules: An overview. Coord Chem Rev. 2022;468:214627.
118. Zhou Y, Yan B. A responsive MOF nanocomposite for decoding volatile organic compounds. Chem Commun. 2016;52(11):2265–8.
119. WHO, Asthma. https://www.who.int/news-room/fact-sheets/detail/asthma#:~:text=Asthma%20is%20a%20major%20noncommunicable,of%20breath%20and%20chest%20tightness.
120. Wang P, Zhang G, Wondimu T, Ross B, Wang R. Hydrogen sulfide and asthma. Exp Physiol. 2011;96(9):847–52.
121. Zhang X, Fang L, Jiang K, He H, Yang Y, Cui Y, Li B, Qian G. Nanoscale fluorescent metal-organic framework composites as a logic platform for potential diagnosis of asthma. Biosens Bioelectron. 2019;130:65–72.
122. Wang W, Ge A, Zhu H, Yang P. A water-stable Zn (II)-based metal-organic framework for selective detection of Fe3+ ion and treatment effect on children asthma by reducing apoptosis of eosinophil. Inorg Chim Acta. 2020;506:119526.
123. ProfMOF. https://profmof.com/
124. NuMat Technologies, Inc. https://www.numat.tech/
125. novoMOF AG. https://novomof.com/
126. BASF Corporation. https://www.basf.com/us/en/media/news-releases/2013/11/p-13-452.html
127. MOFapps AS. https://www.mofapps.com/
128. ProDIA. http://www.prodia-mof.eu/index.html

129. Raymond MJ, Slater CS, Savelski MJ. LCA approach to the analysis of solvent waste issues in the pharmaceutical industry. Green Chem. 2010;12:1826–34.
130. Babayigit A, Ethirajan A, Muller M, Conings B. Toxicity of organometal halide perovskite solar cells. Nat Mater. 2016;15:247–51.
131. Erdmenger T, Guerrero-Sanchez C, Vitz J, Hoogenboom R, Schubert US. Recent developments in the utilization of green solvents in polymer chemistry. Chem Soc Rev. 2010;39:3317–33.
132. Tao Y, Huang G, Li H, Hill MR. Magnetic metal-organic framework composites: solvent-free synthesis and regeneration driven by localized magnetic induction heat, ACS sustain. Chem Eng. 2019;7:13627–32.
133. Lomba L, Zuriaga E, Giner B. Solvents derived from biomass and theirpotential as green solvents. Curr Opin Green Sustain Chem. 2019;18:51–6.
134. Imaz I, Rubio-Martínez M, An J, Solé-Font I, Rosi NL, Maspoch D. Metalbiomolecule frameworks (MBioFs). Chem Commun. 2011;47:7287–302.
135. Werpy T, Petersen G. Top value added chemicals from biomass: volume i - results of screening for potential candidates from Sugars and Synthesis Gas, USA, 2004, https://doi.org/10.2172/15008859.

Inhalable Prodrugs for Pulmonary Therapeutics

13

Piyush Pradeep Mehta and Vividha Dhapte-Pawar

Abstract

Like the liver's major metabolic organ, pulmonary airways contain many potential metabolizing enzymes and metabolic pathways. Notably, the pulmonary metabolic network is of primary importance while exploring the inhalation route of drug delivery. The prodrug-based drug delivery system is an approach where metabolizing enzymes and metabolic pathways play a crucial role. Prodrug systems contain unique drug delivery benefits, such as enzyme-driven drug release, high drug payload, real-time controlled release properties, efficient targeting, and reduced adverse effects. Considering the lung metabolic network and typical unidirectional airway flow pattern, few dynamic formulators designed prodrug therapies to treat chronic pulmonary diseases. This chapter gives a detailed account of pulmonary metabolizing cells and enzymes, prodrug designing, and emerging prodrug integrated nanomedicine. Subsequently, this chapter touches on recent progress in prodrug-based pulmonary drug delivery systems, emphasizing biophysical properties, aerodynamic behavior, and pharmacological outcome. Moreover, the chapter also contains a devoted section regarding future experiments and critical challenges. Briefly, this chapter is a comprehensive account of inhalable prodrugs.

P. P. Mehta (✉)
Global Respiratory R&D, Cipla R&D Center, Mumbai, Maharashtra, India
e-mail: piyu053@gmail.com

V. Dhapte-Pawar
Department of Pharmaceutics, Poona College of Pharmacy, Bharati Vidyapeeth (Deemed to be University), Pune, Maharashtra, India
e-mail: dhapte.vv@bharatividyapeeth.edu

P. P. Mehta, V. Dhapte-Pawar (eds.), *Pulmonary Drug Delivery Systems: Material and Technological Advances*,
https://doi.org/10.1007/978-981-99-1923-9_13

Keywords

Prodrugs · Metabolizing enzymes · Metabolic pathways · Beclomethasone dipropionate · Ciclesonide and prodrug integrated nanomedicines

13.1 Introduction

Prodrugs are bio-reversible molecules of drug substances that undergo a chemical or enzymatic conversion in vivo to release the active drug moiety, showing the desired therapeutic action [1]. They are simple chemical derivatives which need only one to two chemical or enzymatic conversion actions to produce the active parent drug [2]. It is the well-recognized chemical approach to enhance drug substances' pharmacological, biopharmaceutical, and physicochemical properties [2]. Prodrugs offer several possibilities to defeat numerous barriers such as poor water solubility [3, 4], inadequate oral absorption [5], quick presystemic metabolism [6], insufficient penetration [7], pharmacokinetics [8], in vivo instability, toxicity [9], and local irritation [10] for successful drug formulation and delivery [11]. With progress in material sciences, biotechnology, medical sciences, and nanotechnology, prodrugs can be designed to achieve efficient drug targeting with improved pharmacological outcomes and less toxic side effects [12, 13]. Besides, the prodrug approach is well explored for almost all therapeutic categories by academic institutes, research centers, and pharmaceutical organizations. Furthermore, prodrugs offer good product life-cycle management prospects [11].

Prodrugs are pharmacologically inactive chemical moieties that characteristically consist of an active drug moiety and a pro-moiety linked by a labile chemical bond [14]. The prodrug becomes active when the pro-moiety is cleaved to release the active drug moiety [15]. The term prodrug has been utilized interchangeably with drug latentiation, bio-precursor prodrug, pro-agents, and co-drug. All these expressions are comparatively matching for relating the process that involves chemical or enzymatic conversion actions to form the active moiety with acceptable pharmacological action. Various phrases related to prodrug are reviewed in Table 13.1 for better understanding. Amine [16], carboxylic [17], hydroxyl [18], carbonyl [19], and phosphate/phosphonate groups [20] are the most common functional moieties explored to design prodrugs. Also, amides [21], carbonates [22], esters [23], carbamate [24], phosphates [25], and oximes linkages were used to formulate prodrugs [14]. Still, other unique functional moieties such as thiols [26] to form thioethers or thioesters linkages and amines to form imines or N-Mannich bases [27] have also been thoroughly studied to formulate prodrug. The prodrug platforms can be classified into different types according to the structure. The prodrug platforms can be categorized as (i) pure prodrug platform [28], (ii) combination prodrug platform [29], and (iii) small-molecule prodrug [30, 31]. Pure prodrug platform is composed of a drug substance linked to a pro-moiety using a coupling agent, usually a monomer. This platform is most commonly used as it forms a stable molecule with high drug loading and low toxicity and possesses a precise structure. However, a pure prodrug is a very challenging design process, and accurate synthesis is a significant

Table 13.1 Various terms related to prodrug

Term	Description	Example
Hard drug	It is a pharmacologically active molecule but in a form that is not susceptible to metabolic or chemical transformation	Chlorpropamide
Soft drug	It is a biologically active molecule that is biotransformed in vivo rapidly and predictably into nontoxic moieties	–
Prodrug	It is a chemically modified inert drug precursor that releases therapeutically active parent molecules upon biotransformation	Kanamycin pamoate and oleandrin acetate
Carrier-linked prodrug	It is a system in which an active molecule is covalently linked to the suitable inert carrier or transport moiety	Ketorolac and dapsone
Metabolic prodrug	It is a system obtained by chemical modification of active molecules but does not contain any inert carrier or transport moiety. It is also known as bio-precursors	Arylacetic acid
Double prodrug	A chemically modified inert drug precursor that releases a therapeutically active parent molecule upon two-step (enzymatically or chemically) biotransformation. It is also known as pro-prodrug and cascade latentiation	p-pivaloyloxybenzoate double ester prodrug of terbutaline
Mutual prodrug	It is a system in which two pharmacologically active molecules combine to form a single molecule such that each acts as a carrier for the other	Benorylate (aspirin and paracetamol)

hurdle to surpass. In combination, the prodrug platform drug substance is rationally decorated on a polymer chain or part of the polymer chain. This platform is also used to attain a specific structure with high drug loading capacity and lower systemic side effects. However, a precise multistage design process is challenging to accomplish reproducibility. Lastly, the small-molecule prodrug strategy includes the synthesis of prodrug using two or more drug substances linked by cleavable chemical bonds and quickly delivered using novel drug delivery systems [31]. The performance of such a prodrug strategy is mainly based on the properties of the individual drug substance and biological milieu.

Moreover, the advances in nanotechnology and allied fields assist prodrugs to effectively use different nanomedicines to achieve different targets such as site-specific drug delivery [32], modulated drug release [33], tumor-targeting properties [34], and simultaneous diagnosis and drug delivery [35]. Compared to other nanomedicines, prodrug integrated nanomedicines respond more precisely to biological stimuli and release the desired drug [36]. Prodrug integrated nanomedicines can be grouped into three classes, i.e., (i) prodrug-encapsulated nanomedicines, (ii) small molecular prodrug-based self-assembled nanomedicines, and (iii) prodrug polymer integrated nanomedicines [36]. Prodrug-encapsulated nanomedicines are also known as nano-assembled prodrugs [37]. Prodrug-encapsulated nanomedicines mainly utilize nanoparticles (NPs) such as inorganic NPs, lipid vesicles, polymeric micelles, dendrimers, and nanogels to carry prodrugs [38]. Remarkable dose reduction, reduced dosage frequency, superior targeting efficacy, better drug utilization, and lower side effects are the advantages allied to prodrug-encapsulated nanomedicines. At the same time, the use of suitable biological material with green processing techniques makes it a more favorable drug delivery alternative [37, 39]. The small molecular prodrug-based self-assembled nanomedicine is another appealing approach in nano-assembled prodrugs. It is composed of a low-molecular-weight nanocarrier with a small structural arrangement and an excellent ability to regulate the final conjugate's hydrophilic and hydrophobic balance [37]. These prodrugs are usually obtained via cross-linking of small molecules, and accordingly, the small molecular prodrug is self-assembled to form prodrug-based nanomedicines [40]. These prodrugs are mainly designed using cyclodextrin conjugates [41], amphiphilic peptides [42], lipid conjugates, and unsaturated fatty acids [43, 44]. Lastly, prodrug-based polymer conjugated nanomedicine is a special kind of prodrug attempt wherein drug substances are typically covalently coupled to different polymers using chemical modifications to construct prodrug-based polymer conjugated nanomedicine. A variety of polymers such as amphiphilic, hydrophobic, and hydrophilic polymers are explored to synthesize polymer–drug conjugates [45, 46]. Furthermore, conjugating drug substances to the distinctive polymer scaffold is the most commonly utilized platform to produce prodrug-based polymer nanomedicines.

Most of the prodrugs and novel prodrug integrated nanomedicines have received massive attention in cancer therapy as a practical approach to control off-target toxicity [47]. Cancer targeting prodrug usually includes four elements: (i) therapeutic moiety (e.g., drugs, biologicals, protein/peptides, and phytoconstituents), (ii) the cleavable linker (e.g., ester, amide, disulfide, and thioether); it holds drug substance and spacer molecule; quickly cleaves via reactive oxygen species, pH differential and light, (iii) the spacer (e.g., polymer and enzyme-sensitive peptide); it holds cleavable link and targeting moiety; assists in releasing therapeutic moiety in the tumor milieu, and (iv) the targeting moiety (e.g., folic acid, transferrin, and sugars); it helps to identify tumor cells to control off-target toxicity [31]. Mainly, in prodrug integrated nanomedicines, the therapeutic agent is linked to targeting moiety using cleavable linker or functional spacer and delivered to the tumor vasculature by accumulation into tumor cell and leading to the efficient delivery of a therapeutic agent into the infected cells [48, 49]. The general outline of a prodrug is displayed in Fig. 13.1.

Indisputably, prodrugs are clinically proven disease management approaches generally prescribed to treat several diseases [1]. They are widely used to treat inflammatory diseases (e.g., arthritis) [50], infectious diseases (e.g., hepatitis C and human immunodeficiency virus) [51, 52], neurological disorders (e.g., Parkinson) [53], cardiovascular diseases (e.g., arterial thrombosis) [54], pulmonary complications (e.g., pulmonary fibrosis) [55], and oncology events (e.g., tumor targeting) [56]. Prodrugs are also used to treat neglected diseases such as tuberculosis [57], leishmaniasis [58], schistosomiasis [59], sleeping sickness (African trypanosomiasis) [60], sickle cell disease [61], and Chagas' disease (American trypanosomiasis) [62, 63]. Likewise, considering the biopharmaceutical challenges, prodrugs are also

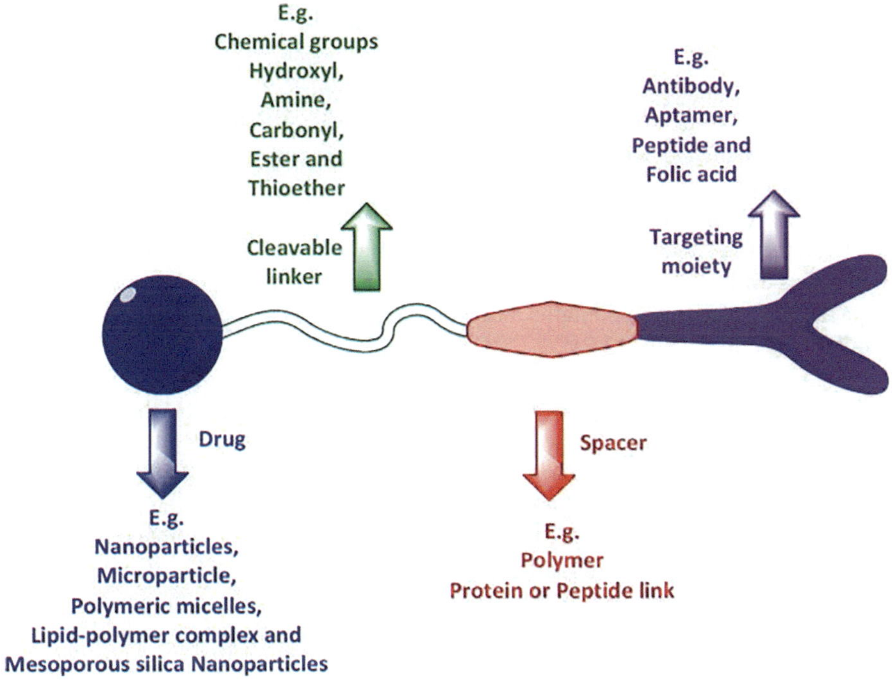

Fig. 13.1 The general outline of a prodrug

incorporated within different delivery systems such as ophthalmic [64], dermal [65], liver [66], brain [67], and pulmonary [68] to surpass critical biological membranes and milieu for better therapeutic outcomes [69, 70]. Novel tumor-targeting prodrugs were also designed and developed to manage life-threatening diseases like cancer [49]. As per World Health Organization data, lung cancer and other pulmonary complications are some of the leading causes of death in the world. Therefore, many scientific efforts have been undertaken on pulmonary delivery of prodrugs concerning their design, drug effectiveness, cancer-targeting, and biophysical properties in the past few years [71]. A good number of scientific articles have also been published on applications and clinical studies of prodrugs for pulmonary disease management [72, 73]. Thus, this chapter discusses the progress of prodrugs in lung disease management. This chapter carefully summarizes different prodrug approaches for lung therapy, including various prodrug-based nanomedicines and recent progress in disease therapy. Moreover, a dedicated discussion section provides details regarding the aerodynamic prospect, toxicological issues, and future challenges. Briefly, this chapter presents deep insights into prodrugs and novel prodrug integrated nanomedicines for the effective management of pulmonary diseases.

13.2 Drug Metabolism in the Airways

The lung drug metabolism is significant when thinking of inhalation as a route of drug delivery. Understanding the presence or absence of functional enzymes or metabolism pathways within the airways is vital in predicting the activation or deactivation of inhaled prodrugs since it significantly impacts overall pharmacokinetics action and pharmacodynamic outcomes, systemic toxicity, and clearance. Similar to the liver, the lung also contains many potential drug-metabolizing enzymes [74, 75]. It contains different types of cells, and each cell holds significant drug-metabolizing capacity [74, 75]. It also contains a diverse range of Phase I and Phase II enzymes similar to the liver (Fig. 13.2).

The regional spatial distribution of metabolizing enzymes will ultimately determine the fate of the drug within the pulmonary airways. Moreover, the delivery of drugs within the pulmonary airways due to their active transport or physicochemical properties decides the access to metabolizing enzymes, local concentrations, and exposure time [74, 75]. Briefly, drug deposition pattern, drug retention, and drug permeation significantly impact pharmacological action and lung metabolic activity. Lung metabolism is a dynamic process and is influenced by many interconnected factors. Genetic factors such as polymorphisms for specific enzymes may raise or ease enzyme activity, while environmental factors such as continuous exposure to smoking may stimulate airway cytochrome enzymes, i.e., CYP1A1/A2 expression [76, 77]. Patient-specific factors such as disease stage, gender, and age

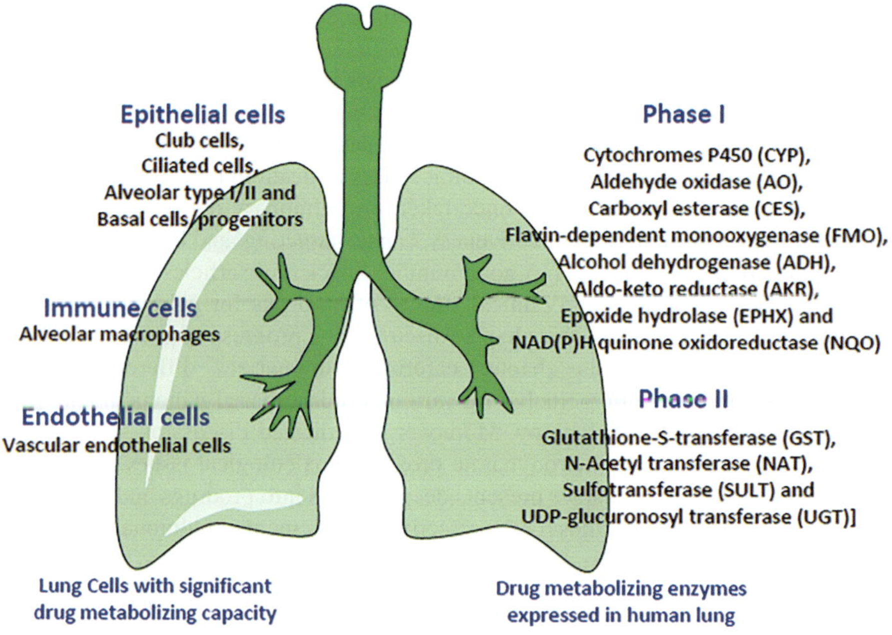

Fig. 13.2 The metabolizing cells and enzymes in the lung milieu

also noticeably impact pulmonary metabolic performance. The rate and extent of drug metabolism also differ regionally within the pulmonary airways [75]. The bronchial area typically contributes more extensively to drug metabolism than the peripheral due to non-ciliated epithelial cells, i.e., club cells. Alveolar type II pneumocytes also express a large group of vital metabolic enzymes, i.e., CYP1B1, CYP2E1, CYP3A5, GSTP1, MGST3, phenylsulfotransferase, and uridine 5'-diphospho-glucuronosyltransferase; however, it showed lower drug metabolism activity since a more significant fraction of aerosolized drugs easily perfuse to systemic circulation without being metabolized in the alveolar sac owing to higher diffusion rate within the alveolar region [78]. Likewise, vascular endothelial cells also assist in metabolizing aerosolized drugs but minorly compared to bronchiolar and alveolar epithelial cells. In a few cases, the lower metabolic capacity of the pulmonary airways causes local saturation of aerosolized drugs, which may affect the biological activity of a prodrug. Like the gastrointestinal sac, aerosolized drugs may experience the interplay between metabolism and active transport, eventually affecting their pharmacokinetics [79, 80]. In a nutshell, accurate information regarding functional enzymes and metabolism pathways is highly recommended before designing a clinically applicable inhalable prodrug.

13.3 Prodrugs Strategies for Pulmonary Therapy

Interest in inhalable prodrugs is growing due to their excellent therapeutic efficiency, lower administration frequency, and reduced systemic adverse effects. The inhaled corticosteroids, i.e., beclomethasone dipropionate (BDP) and ciclesonide (CIC), are well-known examples of prodrugs delivered by pulmonary airways. Inhaled BDP metabolic transformation is mainly influenced by the esterase enzyme (Fig. 13.3). After pulmonary delivery, BDP is metabolized via esterase enzyme to the active metabolite 17-beclomethasone monopropionate (BMP). BMP showed approximately 25–30-fold higher binding affinity to the glucocorticoid receptor than BDP. BMP was further metabolized to beclomethasone, which has a binding affinity similar to BDP. BDP showed similar metabolic transformation when tested during in vitro and in vivo studies. Thus, with a good understanding of airways, esterase activity will positively help optimize the pharmacokinetic profile and bioavailability of inhaled BDP [81]. The esterase enzyme also activates CIC after oral inhalation (Fig. 13.4). It pursues metabolic transformation similar to BDP. CIC active metabolite was further metabolized to form a "soft drug," i.e., restricted systemic exposure because of presystemic metabolic clearance. CIC inactive metabolites are fatty acid conjugates and serve as "drug pools"; i.e., the inactive conjugate may undergo hydrolysis reaction to shape the active metabolite [82, 83]. A similar metabolite effect was also found for inhaled budesonide (BUD) which forms reversible fatty acid esters, i.e., arachidonate, linoleate, oleate, palmitate, and palmitoleate conjugates [84, 85]. Briefly, the reversal metabolism mechanism has a significant impact on the pharmacokinetics of orally inhaled medications. Moreover, intracellular drug retention or

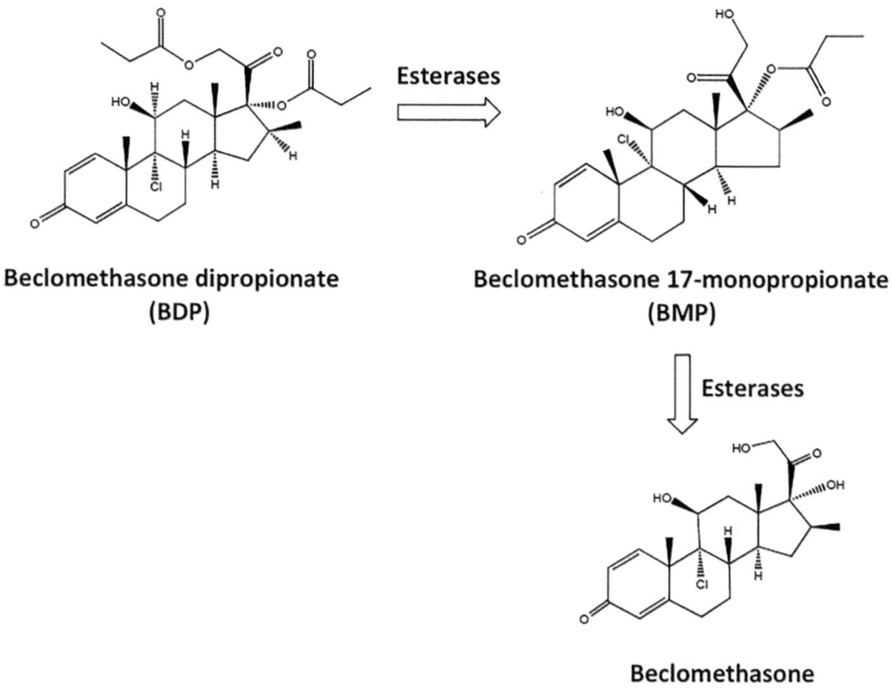

Fig. 13.3 Metabolic pathway of beclomethasone dipropionate to active metabolites beclomethasone 17-monopropionate and beclomethasone in the presence of esterase enzyme

"trapping" assists in prolonged pharmacological action with low dose frequency [84, 85]. Apart from the conventional prodrug approach, various dynamic prodrugs, i.e., nanomedicines integrated prodrugs, targeting moiety integrated prodrugs, and biological material integrated prodrugs, were also designed and developed by active researchers. Few such attractive and clinically helpful novel prodrugs are summarized in the forthcoming segments.

13.3.1 Inhalable Platinum-Based Prodrugs

Clinical applications of platinum-based anticancer substances such as carboplatin, cisplatin (CIS), and oxaliplatin are mainly limited due to dose-dependent side effects and intrinsic or acquired drug resistance. Dose-dependent side effects and resistance are generally stimulated by glutathione (GSH) and glutathione S-transferase (GSTs) overexpressed in lung cancer cells. Hence, to overcome these shortcomings and improve the therapeutic efficacy of CIS, Chen et al. (2019) logically designed Pt(IV) prodrug of CIS-containing GSTs inhibitor (NBDHEX). CIS prodrug was designed via ester linkages as tumor acidic micro-milieu enriched with esterases aids in breaking ester linkage quickly and releases the active drug in tumor cells selectively. Pt(IV) prodrug of CIS is triggered by biological reducing moieties

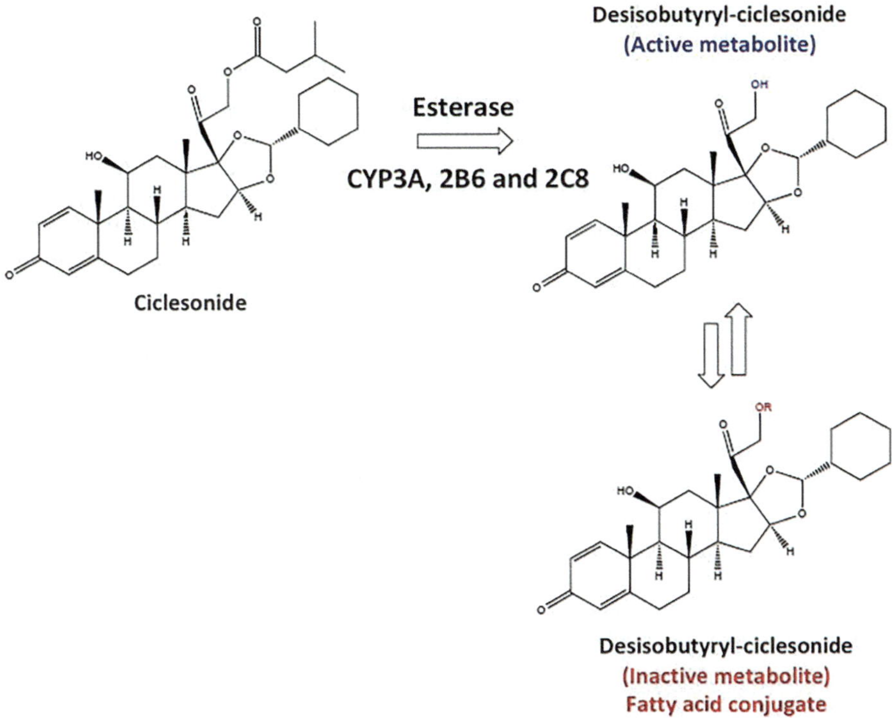

Fig. 13.4 Metabolic pathway of ciclesonide (CIC) to active metabolite desisobutyryl-ciclesonide and inactive fatty acid conjugate metabolite in the presence of esterase and CYP3A enzymes

such as GSH and ascorbic acid to discharge active Pt(II) species and axial ligands. During in vitro cytotoxicity study using human lung adenocarcinoma multidrug-resistant cells (A549/CIS), prodrug displayed more potent anticancer activity with 2.15- and 1.70-fold lower half-maximal inhibitory concentration (IC_{50}) values with resistance factor value of 0.37 as compared to CIS and NBDHEX, respectively. Moreover, the prodrug successfully restrains GSTs and promotes platinum uptake, which causes more severe DNA damage in adenocarcinoma human alveolar basal epithelial cells (A549) and A549/DDP cells than CIS. Prodrug activated cell death through an apoptotic pathway and arrested the cell at the G2/M and S phase for A549 and resistant A549/DDP cells. In an in vivo study using the A549 xenograft tumor mice model, the prodrug displayed a superior tumor-inhibiting rate and lower toxicity than CIS at the same dose (5 mg/kg i.v. once a week). In all, the Pt(IV) prodrug approach has good potential to develop as an anticancer therapy [86].

Platinum-based agents are promising anticancer agents; their therapeutic outcome is limited due to high resistance and poor intracellular uptake. To defeat these issues, Ma et al. (2021) developed novel NPs co-loaded with Pt(IV) prodrug and survivin siRNA to treat platinum-resistant lung cancer effectively. Primarily, protamine- and hyaluronic acid-based NPs were formulated using the ultrasonication method. Subsequently, survivin siRNA was added to nanocarrier with a mass ratio of 20:1 and covered with the anionic polypeptide, i.e., poly (L-glutamic acid) to control drug release with improved circulation time of NPs. Lastly, Pt(IV) prodrug developed using succinic anhydride was linked to NPs by chemical conjugation to form novel prodrug system. Spherical shaped survivin siRNA/Pt(IV) prodrug NPs showed a mean particle size of 237.60 nm with a zeta potential of –44.0 mV. Survivin siRNA/Pt(IV) prodrug NPs showed a pH-sensitive sustained-release profile, i.e., >80% drug release at the end of 120 h using phosphate buffer pH 5. During cellular uptake study using CIS-sensitive A549 and A549/DDP cells, A549 cells showed 3.32-fold higher Pt uptake than A549/DDP cells when Pt(II) was administered. At the same time, A549/DDP cells displayed much higher intracellular Pt following administration of survivin siRNA/Pt(IV) prodrug NPs. Thus, nanomaterials can successfully diminish drug efflux and overcome drug resistance. In cytotoxicity assessment, survivin siRNA/Pt(IV) prodrug NPs showed IC_{50} of 49.92 and 35.60 μM against A549 and A549/DDP, respectively, for 48 h. In pharmacokinetic studies using mice, siRNA/Pt(IV) prodrug NPs showed a marked 4.6-, 2.21-, and 2.71-fold improvement in area under curve (AUC), maximum serum concentration (C_{max}), and clearance rate as compared to Pt(II) after injection at the same dose (3.5 mg/kg). In vivo tumor inhibition ability studies using mice bearing A549/DDP cells siRNA/Pt(IV) prodrug NPs displayed the highest 82.46% tumor inhibition rate. In vivo study of prodrug NPs displayed prolonged circulation time. Thus, in tumor inhibition investigation, prodrug NPs could gradually localize at the tumor site to display antitumor action. Therefore, prodrug-based novel nanomaterials represent an excellent opportunity to improve biological efficacy with reducing systemic toxicity in treating drug-resistant lung cancer [87].

Wang et al. (2021) formulated the pH-responsive and redox-sensitive CIS prodrug and paclitaxel (PTX) co-loaded NPs for the non-small cell lung cancer (NSCLC) therapy. Specifically, tumor cells interior and exterior displayed 2- to 3-fold superior level of GSH tripeptide and highly acidic (pH 5.0–5.5) extracellular environment, respectively, compared to normal cell biology [88–90]. Moreover, pre-clinical and clinical data displayed that platinum-based combination therapy with anticancer agents such as PTX is more successful and less toxic compared to platinum-based treatments [91, 92]. Therefore, pH-responsive and redox-sensitive CIS prodrug and PTX co-loaded NPs were synthesized. Dual-responsive CIS prodrug was fabricated using polyethylene glycol (PEG) and pH-sensitive adipic acid dihydrazide and GSH responsive 3,3′-disulfanediyldipropionic acid. Furthermore, CIS prodrug and PTX co-loaded NPs were formulated using the nanoprecipitation method. Uniform spherical shaped particles showed particle size, zeta potential, and drug loading efficacy of both drugs was 112.9 nm, 10.9 mV, and > 80 %, respectively. CIS prodrug and PTX showed loading efficiency of 85.3 and 87.9 %, respectively. During the in vitro release study, co-loaded NPs showed significant pH-responsive release kinetics. Acidic pH (5.0) showed faster drug release than pH 7.4, and an increase in the GSH concentration showed improved drug release. CIS release from co-loaded NPs was 77.4%, 68.2%, and 45.7% at the GSH concentrations of 10 mM, 10 μM, and 0 mM, respectively. Co-loaded NPs displayed >70% cell uptake efficacy during in vitro cellular uptake study using A549/DDP cells.

Additionally, NPs showed remarkable cytotoxicity against A549 and A549/DDP cells with a combination index of <1. In an in vivo anticancer study using lung cancer-bearing mice, co-loaded NPs displayed a marked 3.18- and 1.61-fold higher tumor inhibition rate than CIS/PTX and CIS/PTX NPs, respectively. Moreover, in pharmacokinetics studies using BALB/c nude female mice, co-loaded NPs exhibited good C_{max} with acceptable blood circulation half-life and higher tumor accumulation than CIS/PTX combination at the same dose (5 mg/kg, i.v.). In biological assessment, co-loaded NPs offered extended circulating properties but difficult to attain controlled release properties due to injectable route of administration. Thus, in future, developed novel co-loaded NPs need to be combined with aerosol delivery for targeted and controlled drug release [93].

As discussed above to attain clinically effective anticancer therapy, Yang et al. (2020) logically developed homo-prodrug, i.e., water-soluble metformin (MET) and water-insoluble CIS co-loaded NPs cocktail therapy. Precisely, hyaluronic acid/CIS [HA/CIS] and polystyrene/MET [PY/MET] double prodrug co-assembled NPs were prepared by simple two-step self-assembly method. Cationic PY/MET prodrug micelles were initially formulated using the solvent evaporation method. Subsequently, cationic PY/MET prodrug micelles were complexed with anionic HA/CIS in a precise ratio (1:1) to obtain double prodrug co-assembled NPs. Co-assembled NPs contain 3.7 and 15.2 % of CIS and MET, respectively. Spherical shaped co-assembled NPs displayed a mean particle size of 166.5 nm with a zeta potential of -17.4 mV. During an in vitro antitumor study using Lewis lung cancer cells, HA/CIS displayed tumor cell growth inhibition while PY/MET was responsible for inhibiting tumor cell proliferation. Improved cytotoxicity of co-assembled NPs was credited to the ideal internalization of HA/CIS by CD44 receptor-mediated endocytosis and subsequent release of MET due to intracellular esterase hydrolysis. During in vivo studies using Lewis lung cancer-bearing mice, co-assembled NPs displayed quick intracellular CIS and MET cleavage with noticeably improved tumor accumulation and tumor growth inhibition and prolonged animal survival without nephrotoxicity. Co-assembled NPs also showed an upregulation of the cleaved poly-ADP-ribose polymerase protein to stimulate tumor cell apoptosis and downregulation of excision repair cross-complementation group 1 protein level to lessen the resistance to CIS. Combination therapy effectively stimulates the adenosine monophosphate-activated protein kinase-α pathway with inhibition of the mammalian target of rapamycin and exercising chemotherapeutic action. Furthermore, it shows a potent immunotherapeutic function with improved CD^{4+} and CD^{8+} T cell counts with simultaneous decrease of regulatory T cells and augmented expression of the cytokines interferon-gamma (IFN-γ) and tumor necrosis factor α (TNF-α). Briefly, double prodrug co-assembled NPs can tailor the pharmacokinetic profile and biopharmaceutical properties of drug substances used to treat lung cancer [94].

13.3.2 Inhalable Paclitaxel Prodrugs

Taxol is a tetracyclic diterpenoid extracted from the bark of the Pacific yew tree, *Taxus brevifolia*. The generic name of taxol is PTX. PTX is high valued phytoconstituent accepted by the US FDA for treatment of several types of cancers, such as Kaposi's sarcoma and lung, breast, and ovarian cancer. It binds to tubulin protein and hinders the disassembly of microtubules, thus ensuing in the reserve of cell division. In spite of high valued chemotherapeutic action PTX commercial viability is restricted because of its sources and poor physicochemical properties. Particularly, the yield of PTX is very low (milligrams) from natural sources and extremely influenced by extraction system, environmental settings, seasons, plant genotypes, and tissue nature. Moreover, insufficient supply delayed its growth into a practical chemotherapeutic agent. To surmount these issues, semisynthetic methods, novel chemical derivatives, fermentation technology, and plant cell culture approaches were adopted [95, 96]. PTX is a white to off-white crystalline powder with high molecular weight (853.9 g/mol), low water solubility (0.03 mg/mL), and lipophilic nature (log P of 3). PTX has low oral bioavailability owing to its enterocyte efflux via P-gp and hepatic presystemic metabolism, and multidrug resistance restrains its beneficial anticancer delivery [97, 98]. Many novel drug delivery strategies and nanocarrier platforms have been applied to overcome PTX poor physicochemical and biological properties [99, 100]. Similarly, few material chemists have tried to develop prodrug-based delivery carriers for lung cancer treatment. Thus, in the current section, we aim to methodically define the existing findings concerning about prodrug-based novel carriers of PTX.

Wang et al. (2018) fabricated RGD peptide-modified prodrug-based redox-sensitive nanoplatforms for lung cancer therapy. Firstly, PTX was conjugated to redox-sensitive 3,3-dithiodipropionic acid via esterification to form PTX containing a cleavable disulfide linkage. Afterward, RGD peptide was linked to PTX via amide linkage [PTX-SS-RGD]. Additionally, CIS/PLGA (1:4) mixture and PTX-SS-RGD/soybean lecithin (1:1) were used to develop core-shell lipid-polymer NPs (LNPs) via the emulsification-sonication method. LNPs displayed entrapment efficacy of 85.3 and 82.7 for PTX and CIS, respectively, with a mean particle size of 191.3 nm and zeta potential of –37.2 mV. LNPs possessed superior stability in phosphate buffer saline during 30 days storage time with no apparent change in particle sizes within the first 5 days and sustained drug release properties. During the in vitro cell viability study, LNPs exposed synergistic activities with approximately 3-fold lower IC_{50} values than free drugs combined. In in vivo studies using A549 lung tumor xenograft bearing Balb/c nude mice model, LNPs exhibited a 5.65-fold marked reduction in tumor size as compared to the initial tumor with significantly higher drug accumulation within the tumor as compared to free drug. Superior level of LNPs in the tumor tissue than normal tissues proved the favored accumulation of LPNs in the tumor tissue via enhanced permeability and retention effect (EPR) action. High drug concentrations within tumor tissue after 24 h of delivery stipulate the prolonged release properties of LNPs while extended circulating behavior was attributed to stealth effect of PEG chains present on the surface of LPNs. Lastly, superior accumulation of ligand-modified LPNs in tumor microenvironment than the LPNs implied the targeting capability of RGD that improved the deposition of the LPNs in the tumor site. Briefly, peptide-modified prodrug-based redox-sensitive systems act as a promising nanomedicine platform for lung cancer treatment [101].

Xu et al. (2018) judiciously fabricated tumor redox heterogeneity-responsive prodrug with self-induced reactive oxygen species (ROS) amplification property for achieving quick drug release with defeating multidrug resistance and lung metastasis. Practically, PTX was conjugated to D-α-tocopheryl polyethylene glycol 1000 succinate (TPGS) by redox dual-responsive linkers to achieve the prodrug conjugate. Essentially, two redox dual-sensitive PTX prodrugs, i.e., PTX-S-TPGS and PTX-SCCS-TPGS, were fabricated by coupling PTX to TPGS through single thioether and dithioether bond as a linker, individually. Moreover, ROS receptive (PTX-ROS-TPGS) and single reduction receptive (PTX-SS-TPGS) prodrugs were synthesized using arylboronic ester and disulfide bond linkage, separately. Furthermore, for non-sensitive control purposes, an ester prodrug, i.e., PTX-CC-TPGS, was developed. The self-assembled PTX prodrug micelles were formulated by the typical nanoprecipitation method using a mixture of deionized water and ethanol solution without any surfactant. The hydrophobic segment of TPGS plays a major function to shape the self-assembly of prodrugs. Specifically, phenyl moiety of TPGS eases the intermolecular hydrophobic interaction and π–π stacking with PTX planar structure to form micelles. The physicochemical and biological characterization data for developed PTX prodrugs micelles are shown in Table 13.2. Lower critical micelle concentration (CMC) numbers verified that PTX prodrugs have good capability to form stable micelles when diluted in the biological media. Spherical shaped prodrug micelles exhibited PTX loading between 28 and 33 % and

Table 13.2 The physicochemical and biological characterization data of PTX/TPGS prodrugs micelles

PTX prodrug	Prodrug micelles	CMC (μg/mL)	Drug loading (%)	Particle size (nm)	Zeta potential (mV)	4T1 IC_{50} (ng/mL)	AUC $_{0\text{-}24\,h}$ (μg/mL*h)	MRT (h)
PTX alone	–	–	–	–	–	210.23	25.09	1.85
PTX-CC-TPGS	TCP	4.27	33.20	138.4	−15.48	> 1000	09.56	19.15
PTX-S-TPGS	TSP	3.63	32.70	135.6	−16.36	62.15	44.18	08.26
PTX-SS-TPGS	TSSP	2.63	31.56	155.7	−15.32	142. 87	29.32	12.70
PTX-SCCS-TPGS	TSCSP	1.58	31.02	157.4	−18.38	80.42	41.46	11.61
PTX-ROS-TPGS	TRP	11.75	28.19	163.5	−15.10	100.15	34.78	15.21

mean particle size between 135 and 163 nm. During hemolysis assay, PTX prodrugs micelles (2 mg/mL) exhibited 4.25-fold low hemolytic potential and excellent biocompatibility compared to cremophor EL and anhydrous alcohol (1:1 v/v) mixture at 0.2 mg/mL concentration. In in vitro drug release study, TSP micelles showed cumulative higher PTX release compared to TSCSP micelles. PTX release was initiated by thiolysis of ester bond in which GSH was used. The dithioether bond from TSP micelles utilized GSH compared to TSCSP micelles' single thioether bond. Accordingly, TSP micelles displayed greater PTX release. Moreover, the hydrophobic chain length of linkage displayed a vital impact on PTX release as short chain length showed quick interaction with water molecules and redox stimuli signal. Prodrug micelles effectively stimulated tumor-specific ROS signal to replenish used ROS required for quick and complete PTX release. Prodrug micelles also assist in raised ROS to stimulate mitochondrial-dependent apoptosis through caspase-9/3 activation. TPGS moiety mainly assists in magnifying the ROS signal during the in vivo assessment. Moreover, prodrug micelles also control acquired and inherent drug resistance by varying expression of the Bcl-2 protein family and reducing mitochondria membrane potential and ATP level in lung cancer cells. During in vivo studies using the 4T1 in situ breast cancer model, prodrug micelles displayed better effectiveness for inhibiting tumor growth in the S180 sarcoma tumor model and suitably preventing lung metastasis. Prodrug micelles displayed a substantial influence on improved expression of caspase-9/3 and Bax and activity in the following order TCP < TSSP < TRP < TSCSP < TSP. The novel prodrug self-assembled polymeric micelles approach is valuable in developing stimuli-responsive drug delivery systems for multidrug-resistant and metastatic lung tumors [102].

13.3.3 Inhalable Camptothecin Prodrugs

Camptothecin (CPT) is a quinoline alkaloid extracted from the stem wood of *Camptotheca acuminata* cultivated in Southwest China [103]. CPT is well known for its variety of biological activities such as antimicrobial, antifungal, antiparasitic, antipsoriasis, antiviral, and pesticidal action [104]. CPT also displayed antitumor activity against a wide range of cancers such as NSCLC, ovarian, colorectal, and breast cancer. CPT mainly inhibits the nuclear enzyme, i.e., type I DNA topoisomerase to pertain its anticancer activity. Chemically, CPT is a planar pentacyclic quinoline alkaloid with three rings of pyrrolo-(3,4-β)-quinoline, an unsaturated pyridone group, and α-hydroxy lactone ring. Equilibrium of lactone ring plays an important role in stability and solubility of the CPT, i.e., open ring (water-soluble carboxylate form) and closed ring (water-insoluble lactone form). In acidic pH, the lactone form prevails while at biological pH, lactone ring is transformed into the less active carboxylate form [104]. The low aqueous solubility of the lactone form, fast deactivation through lactone ring hydrolysis, high protein binding, poor physiological stability, acute toxicity such as hemorrhagic cystitis of the carboxylate form, and ultimately low biological activity prohibited the clinical usage of CPT [105, 106]. A variety of investigations stated the efficient use of prodrug approach to enhance water solubility, stability of lactone form, and biological outcome of CPT [107]. Thus, in the present section the recent advances in CPT prodrug approach to surmount formulation challenges and augment pulmonary delivery of CPT through nanomedicine scheme are discussed.

Li et al. (2019) formulated a redox-sensitive hydrophobic prodrug using oleic acid via disulfanyl-ethyl carbonate linkage to address above stated issues. Moreover, redox non-sensitive hydrophobic CPT prodrug was synthesized using hexyl carbonate linker. Furthermore, prodrug micelles were fabricated using a combination of ethanol/cremophor EL (5:1). Spherical shaped redox-sensitive CPT prodrug micelles exhibited a mean hydrodynamic diameter of 14 nm and zeta potential of -7.8 mV. In the in vitro cellular investigations, redox-sensitive CPT prodrug micelles showed effective internalization within the Lewis lung carcinoma cells with an accelerated release of the active lactone moiety compared to redox non-sensitive CPT prodrug micelles at the same CPT (5 μg/mL) dose. Results specified that the presence of disulfide bond (-SS-) in CPT prodrug micelles considerably upsurge in vitro cytotoxicity because of the quicker CPT release elicited by the intracellular higher reductive GSH level. During a pharmacokinetics study, using male Sprague-Dawley rats after intravenous injection redox-sensitive and non-sensitive CPT prodrug micelles (3 mg/kg of CPT) displayed a marked 18.40 and 53.90 times higher AUC values, respectively, compared to CPT solution. This outcome confirmed that micelles displayed a much longer circulating time than CPT solution, which ultimately improves CPT accumulation within the tumor through the EPR effect. In ex vivo fluorescence imaging analysis using murine CT26 colon carcinoma bearing BALB/C mice redox-sensitive and non-sensitive CPT prodrug micelles indicated 5- to 13-folds higher fluorescence intensities in tumor tissues compared to free fluorescence imaging agent. The ex vivo fluorescence imaging analysis also supports the EPR effect model. In antitumor activity assessment using Lewis lung carcinoma bearing C57 mice redox-sensitive and non-sensitive CPT prodrug micelles indicated 86.50 and 68.90 % tumor reduction, respectively. The notably enhanced antitumor activity of CPT prodrug micelles was mainly attributed to the prolonged blood circulation, efficient cellular uptake, and GSH-triggered selective CPT release. Briefly, such redox-sensitive hydrophobic prodrug micelles strategy presents a fruitful approach for the pulmonary delivery of phytochemicals containing unstable aromatic rings to enhance antitumor efficacy [108].

Similarly, Deshmukh et al. (2010) also fabricated CPT ester prodrugs using a series of amino acids and analyzed them for in vitro anticancer potential using A549 cell lines. The esterification of the 20-position hydroxyl moiety of CPT was performed using amino acid with increasing aliphatic chain length, i.e., glycine, alanine, aminobutyric acid, and norvaline. The hydrolysis studies were performed using phosphate buffer 6.6 (tumor pH), 7.0 (lung pH), and 7.4 (physiological pH). The hydrolysis of prodrug ester bond was pH dependent and directly proportional to amino acid side chain length. CPT-norvaline prodrug showed the slower ester bond hydrolysis rate, i.e., 120 h and 144 h at lung, physiological, and tumor pH. Norvaline showed a lower hydrolysis rate due to the high steric hindrance of aliphatic chain length. In an in vitro cytotoxicity study using A549 cells,

CPT-norvaline prodrug showed a significantly lower slope (-0.443) value with a higher cell death rate at lower concentrations than CPT and other prodrugs. Briefly, the norvaline-based prodrug is the best approach for a passively targeted sustained-release pulmonary delivery system [109].

13.3.4 Inhalable Doxorubicin Prodrugs

Doxorubicin (DOX) is an anthracycline antibiotic derived from the bacterium *Streptomyces peucetius* var. caesius. DOX is a high molecular weight (543.5 g/mol), water-soluble (log P 1.27), at neutral pH orange to red-colored, photosensitive chemotherapeutic agent. The anticancer mechanisms of DOX comprise suppression of enzyme topoisomerase II and intercalation with DNA base pairs to avoid cell duplication and generation of free radicals. DOX is well known for cancer management because of its usage against several types of cancers. It was approved by the US FDA in 1974 for cancer therapy. DOX formulations (e.g., Rubex, Adriamycin, Caelyx, and Doxil® [liposomal formulation]) were prescribed for treatment of solid and metastatic breast, lung, bladder, gastric, ovarian, bone, and thyroid tumors. DOX is a versatile therapeutic agent; however, its clinical potential was primarily restrained by quick development of resistance in patients. Furthermore, partial intracellular accumulation of DOX in tumor cell is also equally liable for quick development of DOX resistance. Accordingly, different delivery tactics, for example, combination therapy, gene therapy, nanocarriers, ligand-modified NPs, polymer–drug conjugates, and prodrug assembled nanomedicine, are projected for increasing the therapeutic effectiveness of DOX in cancer rehabilitation [110, 111]. Specifically, this section aims to review the function of prodrug assembled nanoarchitectures in the delivery of DOX and successful use of this tactic in strengthening the existing cancer therapy.

Liu et al. (2018) developed hybrid micelles containing DOX prodrug and tumor-suppressor p53 genes for synergistic action against lung cancer. Initially, diblock copolymer precursor (mPEG-b-PBYP) was developed using ring-opening polymerization of 2-(but-3-yn-1-yloxy)-2-oxo-1,3,2-dioxaphospholane. Afterward, pH-sensitive DOX prodrug and polycation gene vectors were fabricated using CuAAC cycloaddition and thiol-yne "click" chemistry. Later, acid-cleavable DOX prodrug and p53 gene vector were mixed in an aqueous solution to form hybrid micelles. Spherical shaped hybrid micelles showed particle sizes between 170 and 190 nm with positive zeta potential. In the in vitro release study, hybrid micelles showed superior stability under physiological pH (7.4) and exhibited controlled release (80%) up to 90 h under acidic conditions (pH 5.0) due to the breakage of the acid-sensitive hydrazone bond. In cytotoxicity study using MTT assay showed that the diblock copolymer precursor mPEG-b-PBYP possesses biocompatibility up to 250 mg/L concentration, while hybrid micelles successfully restrain the growth of human NSCLC (H1299) and A549 cells. In live-cell imaging investigation, hybrid micelles improved the transfection efficiency of the p53 gene and successfully co-deliver anticancer agent DOX and tumor-suppressor p53 gene within A549 cells by endocytosis. In a nutshell, combination therapy of prodrug anticancer agent with appropriate tumor-suppressor genes offers a capable approach for treating lung cancer [112].

Urokinase plasminogen activator (uPA) receptors (uPAR) are overexpressed in lung tumor cells and tissues. The peptide composition resulting from uPA amino-terminal protein fraction is an effective way of targeting the uPAR on lung cancer cells. Thus, uPAR targeting U11 peptide conjugated DOX prodrug (U11-DOX) and curcumin loaded pH-sensitive NPs (U11-DOX/CUR NPs) were formulated to surpass the adverse effects and multidrug resistance issues during lung cancer treatment. Uniform, spherical U11-DOX/CUR NPs showed mean particle size and zeta potential of 121.3 nm and -33.5 mV, respectively. U11-DOX/CUR NPs displayed satisfactory entrapment efficacy for DOX (81.7%) and CUR (90.5%). Core-shell U11-DOX/CUR NPs exhibited a pH-dependent release profile with 32.8, 55.2, and 83.5 % DOX release using phosphate buffer pH 7.4, 6.0, and 5.0, respectively. At lower pH values U11-DOX/CUR NPs presented higher and faster CUR release compared to DOX. Variation in release pattern was due to physically entrapped CUR within the core of NPs, whereas the DOX was chemically coupled to the U11 peptide via PEG ligand. Thus, CUR was rapidly released by physical diffusion compared to DOX. Moreover, U11-DOX/CUR NPs displayed good serum stability for 72 h with a slight change in mean particle size. During an in vitro uptake study using DOX-resistant human NSCLC cell line (A549/ADR), U11-DOX/CUR NPs showed significantly higher cellular uptake than non-conjugated DOX NPs. In an in vitro cytotoxicity study using A549 and A549/ADR cells, U11-DOX/CUR NPs demonstrated synergistic effects with a combination index of < 1. During in vivo studies using tumor-bearing BALB/c nude mice model, U11-DOX/CUR NPs revealed evident synergetic action with good tumor tissue accumulation efficiency and 85 % tumor growth inhibition. In terms of safety, the U11-DOX/CUR NPs-treated mice model presented no difference in the physical activity level and body weight while non-conjugated DOX NPs-treated mice model exhibited body weight loss similar to control group (0.9% saline). The safety assessment reports specified that U11-DOX/CUR NPs were well tolerated. However, the key limit of the present investigation is lack of scalability which needs to improved in the succeeding research [113].

13.3.5 Inhalable Antibiotic Prodrugs

Antimicrobial treatment is the standard therapy for patients suffering from pneumonia diseases. The second-generation broad-spectrum fluoroquinoline molecule, i.e., ciprofloxacin (CIP), disclosed superb antimicrobial action and clinical success in treating pneumonia infections in both adults and children. Oral delivery of CIP is a favored remedy in the treatment of pneumonia fatality. However, CIP oral therapy has some limits due to poor water solubility with highly pH-dependent solubility profile of CIP as well as its ampholyte nature and short circulation times. Besides, repetitive parenteral delivery is painful and does not facilitate CIP delivery at the target site, i.e., alveolar macrophages, thus lessening the drug efficacy at the site of infection. Thus, in the case of pneumonia infections pulmonary delivery of CIP is a balanced approach to surmount the limits of oral and systemic delivery by directly accessing the site of infection [114, 115].

To treat alveolar macrophage infections, Jasmin et al. (2019) fabricated biologically stable and effective dosage form, i.e., glycan-targeted mannosylated CIP prodrug. Targeted mannosylated CIP prodrug was synthesized using an aqueous reversible addition-fragmentation chain transfer (RAFT) polymerization method with an 8:1 ratio of polymer to CIP. It contains mannosylated metal phenolic ester CIP prodrug. Prodrug contains 88 % of mannose and 12 % of CIP. Prodrug displayed approximately 5-fold improved intracellular delivery of CIP during time-dependent uptake study using non-transformed murine macrophages compared to non-targeted polymeric CIP prodrugs. After intratracheal administration in wild-type C57bl/6 mice using microsprayer aerosolizer-targeted CIP prodrug exhibited a significant 1.50-fold higher lung AUC than non-targeting glycan control polymer at the same dose (40 mg/kg). Moreover, targeted CIP prodrug displayed a 4.5-fold higher AUC and improved retention time by 64 h than CIP alone. Superior pharmacokinetic outcomes of CIP prodrug were mainly attributed to three aspects, i.e., (i) high molecular weight of the system compared to CIP alone; (ii) mannosylated moiety of prodrug intermingles with lung cells by targeting to mannose receptors or non-specific adhesion to epithelial cells, and (iii) mannosylated polymers have good capability to interact with the surfactant protein D and A present in the alveolar pockets. Pulmonary delivery of the targeted CIP prodrug to lethally infected mice exhibited protection against intracellular *F. novicida* infections. Additionally, CIP prodrug presented 50% and 87.5% survival rates when intratracheally administered to mice in a prophylactic and treatment regimen, respectively. Briefly, multivalent targeted CIP prodrug is the potential pulmonary delivery platform to fight against intracellular infections [116].

Su et al. (2018) demonstrate a novel intracellular enzyme-cleavable polymeric prodrug approach to modulate CIP release kinetics in the pulmonary airways. The novel targeted enzyme-cleavable polymeric prodrug approach was termed as "drugamers." It includes (i) polar mannose residues to solubilize the CIP molecule and target CIP for improved intracellular delivery and (ii) enzyme-responsible linkage to offer intracellular sustained CIP release. Initially, CIP was linked to protease-cleavable hydrolytic phenyl ester moiety or dipeptide portion and, subsequently, co-polymerized with mannose monomer by RAFT polymerization to produce well-defined drugamers without any post-polymerization coupling stage. Specifically, methacrylate-based monomer was coupled to CIP via intracellular protease-cleavable valine-citrulline dipeptide linker and a self-immolative spacer; i.e., para-aminobenzyl alcohol (lysosomal protease enzyme-sensitive spacer) hold CIP from the enzymatic cleavage position. CIP drugamer displayed molecular weight, molar mass dispersity, and drug content of 25.8 kDa, 1.10, and 9.3 %, respectively. During in vitro lysosomal protease enzyme (human liver cathepsin B)-mediated release study, CIP drugamers presented a rapid release of CIP (>50%) within 30 min while control drugamer (without cleavable peptide link) displayed no CIP release over a period of 4 h. During ex vivo release study using monocyte/macrophage-like cells

(RAW 264.7), CIP- drugamers presented time-dependent intracellular CIP release with a marked 7- to 8-fold higher intracellular CIP levels compared to control drugamer after 4 h incubation at the same dose of CIP (250 μg/mL CIP). In cell viability assay by MTT method, CIP drugamers exhibited limited cytotoxicity to RAW 264.7 cells. In an in vitro macrophage–bacteria co-culture assay using *F. novicida*-infected RAW 264.7 cells, CIP drugamer demonstrated effective 4- to 5-fold lower intracellular bacterial concentration compared to control drugamer after 24 h incubation at the same dose of CIP. In in vivo lung safety study using C57Bl/6 mice after intratracheal aerosolization (20 mg/kg once daily for 3 successive days) by Penn-Century MicroSprayer® aerosolizer model, CIP drugamer showed steadily increased mice body weight with neutrophil level matching with healthy mice (~3%). During in vivo biodistribution and pharmacokinetic investigations (20 mg/kg CIP single administration), CIP drugamer showed 1.78- and 8.44-fold higher CIP levels within the lung tissues compared to control drugamer and CIP alone, respectively. Additionally, CIP drugamer showed a marked 25.61- and 74.82-fold superior CIP levels within the alveolar macrophages compared to control drugamer and CIP alone, respectively. Moreover, sustained-release property of CIP drugamer resulted in >3-fold higher intracellular CIP levels compared to minimum inhibitory concentration (MIC) during 48 h study. In in vivo antibiotic challenge studies using lethal *F. novicida* airborne infection model, CIP drugamer and control drugamer (20 mg/kg of CIP) showed 100% survival rate while low dose (10 mg/kg of CIP) showed 62.5% (CIP drugamer) and 50% (control drugamer) survival rate. Taken together, distinctive pharmacokinetics profiles (C_{max} and AUC > MIC) and highly enzyme-sensitive nature with unique mannose targeted delivery system will be best fit for prophylactic usage of several antibiotic agents against pulmonary intracellular infections [117].

Similar research group also tested CIP-drugamer against pulmonary melioidosis. In in vivo pharmacokinetic studies using C57BL/6 mice after intratracheal aerosolization of single-dose CIP drugamer (20 mg/kg) using a MicroSprayer® aerosolizer, CIP-drugamer efficiently delivered CIP to the alveolar macrophage and showed sustained release for 7 days. During an in vivo antibacterial study using C57Bl/6 mice after pulmonary administration of CIP-drugamer (20 mg/kg), total protection with a reasonable survival rate was achieved against *B. thailandensis* challenge (>10^5 colony-forming unit). Briefly, developed modular inhalable "drugamer" system assists in attaining high and significantly prolonged CIP dosing within the pulmonary airways. Modular system is competent to repotentiate CIP and attain complete prophylaxis where the CIP alone is ineffective. Additionally, drugamer platform is completely synthetic, and therefore large-scale manufacturing is less tedious compared to novel carrier synthetic scheme and can be easily explored with other antibiotic-resistant molecules [68].

Chloramphenicol (CHP) is a semisynthetic, broad-spectrum organochlorine antibiotic obtained from *Streptomyces venezuelae*. CHP possess superior bacteriostatic activity and are usually prescribed for serious and life-threatening infections for which standard antibiotic regimens are not prevailing. Thiamphenicol (THP) is a semisynthetic derivative of CHP with higher water solubility and matching antibacterial spectrum to CHP [118]. However, both amphenicol class antibiotics displayed lower lung exposure after pulmonary administration than systemic administration due to their high apparent epithelial permeability. Consequently, to enhance lung exposure, Nurbaeti et al. (2019) designed sustained-release inhalable palmitate ester prodrugs of CHP and THP. Initially, sustained-release PLGA NPs were formulated using the emulsion solvent evaporation method. At the same time, nanoembedded microparticles (NMPs) were developed using spray-drying lactose as a carrier and leucine as a dispersing enhancer. CHP-loaded PLGA NPs (160 nm) and THP-loaded PLGA NPs (178 nm) exhibited yields of 63% and 70%, respectively, while CHP and THP NMPs showed prodrug content of 13.9% and 21%, respectively. NMPs demonstrated shriveled-like raisin shape, and this typical morphology was accredited to the collapse of the microparticle structure during the spray-drying process using leucine. The thermograms of both NMPs indicated the amorphous nature of the dry powder. An in vitro release study using Hank's Balanced Salt Solution at pH 7.4 CHP and THP NMPs displayed a sustained-release profile with the complete release over 14 days. During aerodynamic assessment using Handihaler® at 41 L/min, CHP and THP NMPs showed 79% and 82% emitted dose (ED), whereas fine particle fraction (FPF) showed 27% and 36%, respectively. Additionally, lower mass median aerodynamic diameter (MMAD) values (3 μm) indicated that the NMPs were effectively de-aggregated using a high resistance inhaler device during the aerosolization process. In a nutshell, NMPs prodrug-based dry powders showed excellent sustained-release properties with satisfactory aerodynamic deposition appropriate for lung delivery. However, further in vivo analyses are highly required to verify the clinical application of such novel delivery systems [119].

13.3.6 Prodrug Policy for Lung Cancer Treatment

Upcoming facts and figures from the recent scientific publications, communications, and ongoing clinical investigations evidently disclosed that in the next few years lung cancer are not only a potential concern for individual person but also for the entire healthcare system across the globe. Lung cancer is a multifactorial disease with diverse etiology sourced from birth defect, genetic settings, chronic illness, environmental (e.g., active or passive smoking and air pollution), nutritional or diet conditions, and constantly fluctuating lifestyle. Besides, delayed diagnosis and low survival rate are also the key factors allied with the lung cancer patient. Thus, there is a serious prerequisite for early diagnosis and timely pharmacological action to enhance patient quality of life and extend survival rate. By knowing this background picture with the key benefits of respiratory route, recent prodrug integrated novel strategies may be the hopeful alternative approach for lung cancer [120–123]. Thus, this section reviews and discusses the novel prodrug formulation explored for the lung cancer treatment.

Troxacitabine (L-1,3-dioxolane-cytidine) is a potent anticancer agent used against leukemia, but it has minimal therapeutic efficacy against solid tumors mainly because of its hydrophilicity. Thus, to optimize its hydrophilicity and enhanced therapeutic efficacy, Radi et al. (2007) thoroughly explored the troxacitabine-prodrug approach. Principally, the amino moiety of the cytosine group was modified to tune lipophilicity and anticancer activity. N^4-substituted fatty acid amide prodrugs with chain length 10–16 carbon showed appreciably superior antitumor activity against two different NSCLC compared to troxacitabine. Predominantly, the troxacitabine prodrug containing $(CH_2)_{14}$-CH_3 moiety showed an acceptable IC_{50} value of 0.004 and 0.020 μM against A549 and SW1573, respectively, with good log P (5.35) value. These in vitro cell line outcomes are very satisfactory. However, detailed in vivo therapeutic investigations are much needed [124].

Metastatic cancer cells possess high reactive oxygen species, i.e., hydrogen peroxide (H_2O_2). Thus, a prodrug activated by intracellular H_2O_2 can be used in anti-metastatic therapy. Using this hypothesis, Kim et al. (2014) designed an H_2O_2-activated prodrug theranostic system containing a boronate ester as a trigger unit and fluorophore coumarin to monitor the release of chemotherapeutic component SN-38 (an active metabolite of irinotecan). In an in vivo study using mice bearing B16F10 murine melanoma cells, intratracheally delivered prodrug (250 μg/kg) using Penn-Century MicroSprayer® showed that the prodrug effectively gathered in metastasized lung tumors and released the active anticancer agent SN-38. Briefly, a novel H_2O_2-triggered prodrug platform represents the potential approach for efficient lung cancer therapy [125].

Chronic inflammation of the pulmonary airways results in lung carcinoma via the activation of nuclear factor-kappa B (NF-κB), leading to the overexpression of cyclooxygenase (COX). Thus, the combination therapy of NF-κB and COX inhibitors is a suitable therapeutic strategy against lung cancer. Recently, Suthar et al. (2014) synthesized ester prodrugs of lantadene (potent anticancer pentacyclic triterpenoids) with nonsteroidal anti-inflammatory drug diclofenac for the concurrent inhibition of NF-kB and COX-2. In luminescent kinase assay, lead compound effectively downregulated tumor necrosis factor-α-induced activation of NF-kB and suppressed the NF-kB-mediated protein expression of COX-2 and cyclin D1. In cell viability assay, lead compound inhibited A549 cell proliferation in a dose-dependent manner and showed a 50-fold superior IC_{50} value compared to CIS. Additionally, it showed long-term 12 h stability in an acidic condition (pH 2) and hydrolyzed quickly in human plasma to discharge the active pro-moieties. The lantadene-NSAID prodrugs approach is a promising anticancer therapy against lung cancer [126].

Wang et al. (2014) developed gemcitabine (GBM)-poly(methyl methacrylate) amphiphile conjugate via reversible RAFT polymerization method to improve gemcitabine tumor-targeting potential to reduce plasma degradation (Fig. 13.5). The amphiphilic conjugate was further transformed into NPs via the nanoprecipitation method. Spherical shaped GBM prodrug NPs retained the GBM in the amorphous state with mean particle size and zeta potential of 130 nm and –60 mV, respectively. GBM prodrug NPs showed a pH-dependent controlled release profile with 46.8% and 71.6% of GBM release in 72 h in the absence and presence of protease

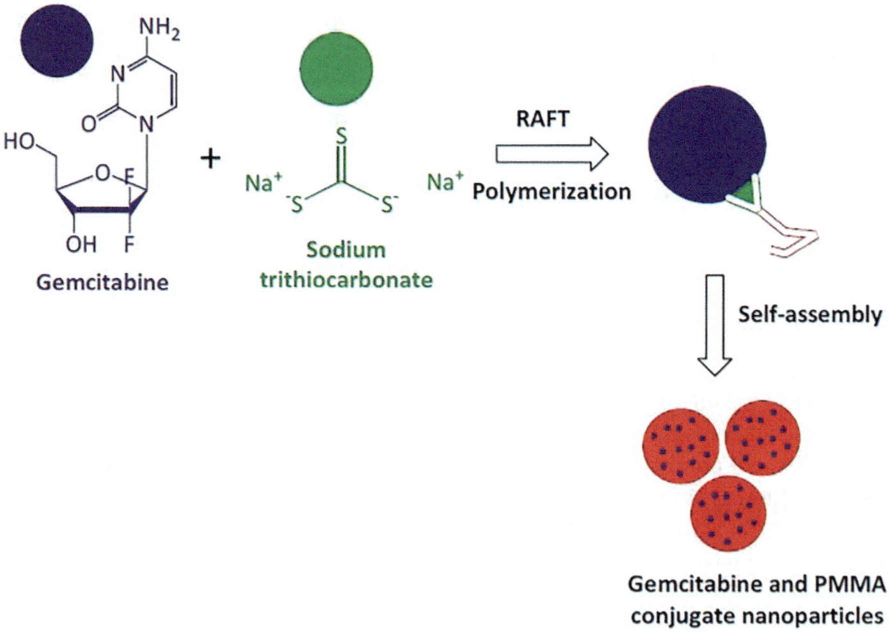

Fig. 13.5 Graphical representation of GBM-poly(methyl methacrylate) amphiphile conjugate via reversible RAFT polymerization method

cathepsin B. In an in vitro study using A549 cells, GBM prodrug NPs showed efficient anticancer activity. In an in vivo study using A549 cell-derived xenograft tumors bearing BALB/c nude mice, intravenously delivered prodrug (0.693 mg/mL) showed that the prodrug NPs effectively restrain tumor growth by 67% compared to GBM alone at the same dose without any side effects. Biocompatible, amorphous, thermoplastic polymer-based prodrug NPs in cancer chemotherapy and drug delivery denote a capable approach for efficiently delivering hydrophilic drug molecules [127].

Ruthenium-based substances are potential anticancer agents; however, biological applications are mainly restricted owing to a lack of targeting efficacy and poor release kinetics. Accordingly, Chen et al. (2020) designed novel bio-reducible ruthenium prodrugs. Prodrug with superior DNA binding affinity was used for nucleus-targeted chemotherapy and precise activation in response to the tumor microenvironment. Initially, multifunctional targeting protein was developed using green fluorescent protein (eGFP), TEV, GGG-MscL (glycine sequences and large-conductance mechano-sensitive channel protein), and H1299.2-targeting peptide and incorporated into liposomes vesicles developed using distearoyl phosphatidylcholine, cholesterol, and 1,2-distearoyl-phosphatidylethanolamine-methyl-PEG--conjugate-2000 (DSPE-PEG_{2000}) [TP-LP]. Monodisperse spherical shaped TP-LP vesicles showed mean particle size and zeta potential of 40 nm and –9.54 mV, respectively. Besides, ruthenium prodrug, i.e., mer-RuCl3(dmso)(L), was incorporated within mesoporous silica NPs [Ru(III)-MS-NPs]. Monodisperse Ru(III)-MS-NPs particles showed mean particle size and zeta potential of 50 nm and 24.57 mV, respectively. Afterward, Ru(III)-MS-NPs-TP-LP conjugate was prepared by electrostatic interaction between positively charged Ru(III)-MS-NPs and negatively charged TP-LP to attain Ru(III)-MS-NPs-mediated nuclear DNA targeting and fluorescence imaging followed by hierarchical targeting due to TP-LP. Ru(III)-MS-NPs-TP-LP conjugate retained the spherical morphology and showed a mean particle size of > 50 nm and zeta potential (–10.55 mV). Ru(III)-MS-NPs-TP-LP conjugate is stable at physiological conditions and showed a pH-dependent release profile with higher drug release at pH 5.0. During in vitro cell line study using NSCLC cell (H1299) Ru(III)-MS-NPs-TP-LP conjugate showed higher cellular uptake with 4.18-fold higher IC_{50} values 420-fold superior antitumor activity as compared to other tested cell lines due to unique hierarchical targeting potential. In an in vivo study using H1299 tumor cells bearing BALB/c nude mice model, Ru(III)-MS-NPs-TP-LP conjugate showed superior tumor accumulation. It satisfactorily inhibited tumor growth in vivo with long-term blood circulation and biocompatibility compared to ruthenium prodrug at the same dose (15 mg/kg i.v.). Thus, biocompatibility, bio-reducible, and traceable ruthenium nano-prodrugs are an efficient and safe delivery option in lung cancer therapy [128].

The lack of an efficient delivery system and poor penetration inside tumor tissues are the fundamental limitations of conventional suicide gene therapy. Recently, mesenchymal stem cells (MSCs) are materialized as a potential delivery system to suicide gene therapy. However, MSCs are genetically customized using viral gene vectors for suicide gene overexpression to stimulate therapeutic efficiency. However, such systems pose serious safety risks during clinical translation. By knowing this issue, Zhang et al. (2015) designed a novel delivery strategy that involves nonviral gene vector-based MSCs with suicide genes to lessen clinical safety risks. Furthermore, nonviral gene vector-based MSCs were co-administrated with ganciclovir prodrug-encapsulated lipid vesicles for synergistic lung cancer therapy. Ganciclovir liposomes were formulated using egg phosphatidylcholine/cholesterol/Pluronic F68 mixture (5:1:0.12) by a modified reverse-phase evaporation method. Spherical ganciclovir liposomes showed a mean particle size of 938 nm and encapsulation efficiency of 44.7 %. Liposomes displayed a sustained-release profile in physiological saline at 37 °C during the first 24 h. In pulmonary melanoma metastasis bearing C57BL6 mice model, co-administration of ganciclovir prodrug liposomes with bone marrow-derived MSCs as gene vector effectively co-target tumor tissue and show a reduction in tumor colonization with a notable increase in survival rate. Furthermore, in the in vitro 3D tumor spheroid model, bone marrow-derived MSCs showed good permeability within tumor nests. Consequently, nonviral gene vector MSCs-based systems are good alternatives for conventional suicide gene therapy with improved therapeutic efficacy [129].

Ma et al. (2015) developed β-lapachone (β-lap) prodrug nanomaterials using diester derivatives of β-lap and poly(ethylene glycol)-b-poly(D, L lactic acid) to surpass the biophysical hurdles like poor pharmacokinetics and high toxicity owing to hemolytic anemia of β-lap alone. β-lap prodrug-loaded PEG-b-PLA micelles (1:9) were developed using the film hydration method and showed high drug loading efficiency >95%. Spherical shaped micelles showed hydrodynamic diameter and zeta potential of 35 nm and –0.40 mV, respectively. In prodrug conversion studies using model enzyme porcine liver esterase, β-lap prodrug micelles showed satisfactory conversion rate with >70% β-lap release at the end of 40 h. During safety assessment regarding hemolytic anemia, β-lap prodrug micelles do not lower the hemolysis rate (5.5%) at 2.0 mg/mL compared to β-lap cyclodextrin complex. In antitumor efficacy studies using female NOD-SCID mice bearing orthotopic fire-fly luciferase-transfected A549 tumors, β-lap prodrug micelles exhibited 108 days as 50% survival time at 30 mg/kg and 139 days survival time at 70 mg/kg dose. In tumor pharmacokinetic investigations after 50 mg/kg tail vein injection, β-lap prodrug micelles were effectively converted to β-lap in the first 2 min and showed 8.5-fold superior β-lap tumor concentration than β-lap cyclodextrin complex. Moreover, during pharmacodynamic studies, a significant rise of poly [ADP-ribose] polymerase 1 activation verifies the β-lap prodrug micelles' potential in lung cancer therapy. β-lap prodrug-loaded nanomaterials showed excellent lung cancer treatment options validated by in vivo pharmacokinetic and pharmacodynamic results [130].

Boehle et al. (2001) showed the anticancer potential of combretastatin A-4 (CA4) prodrug using a NSCLC model. During in vitro studies, cell growth inhibition assay using cells derived from squamous cell carcinoma of the lung (KNS-62) and cells obtained from pleural fluid of a patient with primary adenocarcinoma of the lung (Colo-699) after 3 h exposure of CA4 (10^{-4} mol/L) showed significant inhibition of proliferation in KNS-62 (37%) and Colo-699 cells (69%). At the end of 24 h, CA4 (10^{-4} mol/L) displayed 49% and 59% inhibition of proliferation for KNS-62 and Colo-699 cells, respectively. In in vivo studies, KNS-62 and Colo-699 cells were subcutaneously transplanted into pathogen-free female SCID bg mice and treated with intraperitoneally injected CA4 (50 mg/kg). Treated mice showed a significant 19.28- and 3.17-fold reduction in tumor volume of Colo-699 and KNS-62 at the end of 42 and 38 days of treatment, respectively [131]. Similarly, ganciclovir prodrug therapy using herpes simplex virus thymidine kinase (HSVT-K30) mutant in vesicular stomatitis virus-G (VSV-G) pseudo-typed retroviral vector allows favorable treatment of NSCLC in a murine xenotransplant model [132].

13.3.7 Miscellaneous Inhalable Prodrugs

Treprostinil (TPS) is a well-known prostacyclin vasodilator prescribed in pulmonary arterial hypertension (PAH) therapy. It is obtainable in different dosage forms such as Remodulin® (i.v. and subcutaneous injection), Orenitram® (oral), and Tyvaso® (inhaled) to offer diverse routes of delivery and therapy alternative for dose titration. TPS displayed several clinical advantages in PAH therapy, but it quickly excretes from the body and entails continuous i.v. infusion or frequent oral administration to uphold its efficiency. Thus, repeated daily dosing (b.i.d. or t.i.d.), line infection with i.v. infusion, and site pain after subcutaneous injection are the key issues that are allied with oral and parenteral TPS dosage forms. Moreover, chemically, TPS is a carbotricyclic molecule with low water solubility (0.006 mg/mL) and high log P (log P=4) values. In this manner, delivery system that offers prolonged levels of TPS at desired site with extended duration of time possesses high beneficial value in the management of PAH [36, 73]. Thus, Chapman et al. (2018) designed and formulated TPS prodrug-loaded inhaled lipid NPs to surpass various biopharmaceutical and physicochemical limitations allied with TPS. Initially, TPS prodrug, i.e., hexadecyl-treprostinil (HTPS), was formulated using an esterification process with a percentage purity above 97 %. In addition, inhaled lipid NPs were formulated using HTPS, squalane, and 1,2-distearoyl-sn-glycero-3-phosphoethanolamine-N-[methoxy (PEG)-2000] (45:45:10) by ethanol injection method. Typical spherical shaped, uniform lipid NPs showed particle size in the range of 100–150 nm. Pharmacological studies were conducted using male Wistar Han rats and male beagle dogs to understand the change in pulmonary arterial pressure (PAP) after U46619 (5 μg/kg; i.v.) and right ventricular pulse pressure (RVPP) after 10 min exposure to hypoxia, respectively. In rats after nebulization using an Aeroneb® Pro nebulizer (Aerogen, Ireland) at a flow rate of 6 L/min, HTPS lipid NPs showed a notable reduction in U46619-induced PAP at 60-fold lower half-maximal effective concentration (EC_{50}) value as compared to i.v. TPS. In dogs after nebulization using a face mask attached to Aeroneb® Pro nebulizer (Aerogen, Ireland) at a flow rate between 20 and 28 L/min, HTPS lipid NPs displayed a remarkable reduction in hypoxia-driven rise in RVPP up to 6 h at a significant 550-fold lower EC_{50} value as compared to i.v. TPS. Pharmacological assessment in two different species for two different pulmonary vasoconstrictor challenges indicated that aerosolized HTPS lipid NPs could provide local and long-acting vasodilator action within the small dosage range. Briefly, aerosolized prodrug NPs represent the better prospect of effectively managing pulmonary vasodilation for a longer time at a substantially lower dose and systemic exposure than parenteral formulations [133].

Inhaled non-halogenated BUD is a key corticosteroid for maintenance treatment of asthma and other respiratory diseases. Normally, inhaled BUD is prescribed twice a day for months during maintenance therapy. Thus, reducing dosing frequency will help to improve patient compliance and adherence to the long-term medication. With this objective, Waters and Hochhaus (2018) developed a dextran-BUD prodrug to modify the release pattern and improve the pulmonary absorption of BUD. Dextran (40 kDa) BUD prodrug was formulated using a two-step synthesis process using succinate spacer. In addition, dextran-succinate-BUD prodrug was spray-dried in the presence of lactose (1:3) to obtain respirable particles. The precipitation process-based purification study showed > 99.2 % purity for the designed prodrug. Dextran-BUD prodrug containing succinate spacer showed more controlled release of BUD than dextran-BUD prodrug containing glutarate and adipate spacers in phosphate buffer (pH 7.4) at the end of 8 h. The non-enzymatic activation model showed biphasic release of BUD from dextran-succinate-BUD prodrug. The dextran-BUD prodrug activation and release of BUD increased with an increase in dextran and spacer molecular weight. Smooth spherical spray-dried particles showed normal particle size distribution with d_{50} and d_{90} of 2.74 and 6.44 μm, respectively. In aerodynamic assessment with high resistance Handihaler® device, spray-dried prodrug particles displayed high ED and FPF of 93.5% and 47.3%, respectively. Moreover, MMAD and geometric standard deviation of 4.04 µm and 3.97 showed that spray-dried prodrug particles are efficient for deep lung deposition. Accordingly prodrug approach will assist in altering the release kinetics and systemic absorption pattern of BUD. Besides, dextran-succinate-BUD prodrug has the great potential of escaping mucociliary clearance owing to its hydrophilicity that improves its pharmacological effectiveness [134].

Phosphoinositide 3-kinase (PI3K) pathway activation plays an essential function in lung inflammation and tissue remodeling. Consequently, PI3K inhibitors usually are utilized to manage non-resolving pulmonary inflammation; however, their pharmacological action is mainly restricted by surplus on-target systemic toxicity. Therefore, Campa et al. (2018) explored an inhaled prodrug pan-PI3K inhibitor (CL27c) for local therapy and asthma and pulmonary fibrosis management. In ovalbumin-induced acute asthma using BALB/c mice, 30 min aerosolization of CL27c (2 mg/mL dissolved in 0.5% tyloxapol/saline solution) demonstrated reduced insulin-induced protein kinase B phosphorylation within airways, but without any increase in blood glucose level. Additionally, after 30 min aerosolization, CL27c decreases inflammation and enhances pulmonary function in murine models of acute or glucocorticoid-resistant neutrophilic asthma. Furthermore, aerosolized CL27c micro-suspension remarkably controlled the bleomycin-induced lung fibrosis in mice and eventually enhanced survival. So, pulmonary delivery of a pan-PI3K inhibitor prodrug decreases the systemic on-target adverse effects and efficiently manages the non-resolving pulmonary inflammation [135].

As summarized in the above sections, the therapeutic results of prodrugs are helpful for respiratory disease management. In vitro, in vivo, and ex vivo trials showed that inhalable prodrugs have good potential to achieve active and passive targeting for lung cancer treatment with drug delivery potential in managing respiratory disorders. Yet, the clinical validation and marketing approval bar is set very high for pulmonary prodrugs. Thus, a few essential clinical case studies on inhalable prodrugs are listed below.

13.4 Clinical Outlook of Laninamivir Prodrug

Laninamivir octanoate hydrate is a novel antiviral agent discovered by Daiichi Sankyo Company, Ltd., used against influenza virus A and B. Laninamivir octanoate (CS-8958) is a prodrug of laninamivir (R-125489) which selectively inhibits neuraminidase of influenza virus. Laninamivir octanoate dry powder inhaler (Inavir®) is designed by Daiichi Sankyo Company, Ltd., to deliver a sufficient amount of drug to the target organ, i.e., pulmonary airways where viral growth occurs rapidly. Recently, Biota Scientific Management Pty Ltd. designed a phase 1 clinical trial (NCT02022761) to identify the safety, tolerability, and pharmacokinetics of laninamivir octanoate (40 and 80 mg) in adults with mild or moderate chronic asthma using TwinCaps® DPI device [136]. Likewise, Biota Scientific Management

Table 13.3 Outline of prodrugs utilized for pulmonary drug delivery

Prodrug	Key properties	Key outcomes	Reference
CIS/NBDHEX	Ester linkage prodrugs	More potent anticancer activity against A549/DDP cells with superior tumor growth inhibition rate in A549 xenograft tumor mice model	[86]
Survivin siRNA/ Pt(IV) NPs	Spherical NPs with particle size and negative zeta potential of 237.60 nm and 44 mV	Prodrug NPs showed a marked 4.6-fold improvement in AUC compared to Pt(II) and 82.46 % tumor growth inhibition rate in mice bearing A549/DDP cells	[87]
CIS prodrug/ PTX NPs	Uniform spherical NPs with particle size and zeta potential of 112.9 nm and 10.9 mV	Prodrug NPs displayed a 3.18-fold higher tumor growth inhibition rate than CIS/PTX with suitable blood circulation half-life	[93]
Co-assembled CIS/MET NPs	Spherical NPs with particle size and negative zeta potential of 166.5 nm and 17.4 mV	Prodrug co-assembled NPs displayed quick intracellular drug release with a reasonable tumor growth inhibition rate and prolonged animal survival without nephrotoxicity	[94]
RGD peptide/ PTX/CIS-core-shell lipid-polymer NPs	NPs showed >80% entrapment efficacy for both actives with particle size and negative zeta potential of 191.3 nm and 37.2 mV	Lipid-polymer NPs displayed 3-fold lower IC_{50} values than free drugs combined against cancer cell line and notably higher drug accumulation within the tumor with a 5.65-fold tumor growth inhibition rate than free drug	[101]
PTX/TPGS prodrug micelles	Self-assembled spherical with particle size less than 200 nm	Prodrug micelles prevent lung metastasis with reasonable control over acquired and inherent drug resistance in lung cancer cells	[102]
Redox-sensitive CPT/oleic acid prodrug micelles	Spherical with the hydrodynamic diameter and negative zeta potential of 14 nm and 7.8 mV	Prodrug micelles showed effective internalization within the Lewis lung carcinoma cells, and redox-sensitive reaction accelerated release of drug with 18.40 times higher plasma level of the drug compared to pure drug solution in in vivo studies	[108]
CPT-norvaline prodrug	pH-susceptible ester linkage prodrugs	Prodrug showed the most slower ester bond hydrolysis rate of 144 h at tumor pH (6.6) with significant cytotoxicity action against A549 cells	[109]
DOX prodrug/ tumor-suppressor p53 genes loaded hybrid micelles	Spherical and particle size between 170 and 190 nm	Micelles showed pH-dependent controlled release with an improved transfection efficiency of p53 gene and co-delivery of DOX within A549 cells by endocytosis	[112]

(continued)

Table 13.3 (continued)

Prodrug	Key properties	Key outcomes	Reference
U11-DOX/CUR NPs	Uniform spherical with particle size and negative zeta potential of 121.3 nm and 33.5 mV	pH-responsive NPs displayed good serum stability with an in vitro combination index of < 1 and 85 % tumor growth inhibition rate in tumor-bearing BALB/c nude mice model	[113]
Glycan-targeted mannosylated CIP prodrug	Aqueous RAFT polymerized prodrug	Prodrug showed approximately 5-fold improved intracellular delivery of drug-using non-transformed murine macrophages with 4.5-fold higher AUC and improved retention time up to 64 h after intratracheal administration compared to drug alone	[116]
Macrophage-targeted CIP drugamers	Intracellular enzyme-cleavable polymeric prodrug	Drugamer showed sustained CIP release with 3-fold higher intracellular CIP levels and 100 % survival rate in lethal *F. novicida* airborne infection model	[117]
Macrophage-targeted CIP polymeric prodrug	Intracellular enzyme-cleavable polymeric prodrug	Aerosolized drugamer (20 mg/kg) efficiently reached alveolar macrophage and showed sustained release for 7 days with total protection and a reasonable survival rate against the *B. thailandensis* challenge test	[68]
Chloramphenicol palmitate ester prodrug NMPs	NPs showed 160 nm particle size and prodrug content of 13.9%	Sustained-release NMPs showed ED and FPF of 79 % and 27 % using Handihaler® at 41 L/min	[119]
Thiamphenicol palmitate ester prodrug NMPs	NPs showed 178 nm particle size and prodrug content of 21%	Sustained-release NMPs showed ED and FPF of 82% and 36% using Handihaler® at 41 L/min	[119]
Troxacitabine acid prodrug	Amide linkage prodrug	During the in vitro cell line, the prodrug showed significantly lower IC_{50} values of 0.004 and 0.020 μM against A549 and SW1573, respectively	[124]
SN-38 prodrug	Boronate ester theranostic prodrug	Aerosolized prodrug exhibited effective accumulation of SN-38 within metastasized lung tumors and potent anticancer activity	[125]
Lantadene prodrug/ diclofenac	Ester linkage prodrug	Highly acid-stable prodrug showed 50-fold superior IC_{50} values against A549 cells compared to cisplatin, and luminescent kinase assay showed downregulation of NF-kB-mediated COX-2 and cyclin D1 protein expression	[126]

Table 13.3 (continued)

Prodrug	Key properties	Key outcomes	Reference
GBM prodrug NPs	Amorphous spherical NPs with particle size and negative zeta potential of 130 nm and >60 mV	Biocompatible NPs showed a 67 % tumor growth inhibition rate compared to GBM alone in A549 cell-derived xenograft tumors bearing BALB/c nude mice	[127]
Ruthenium prodrug-loaded mesoporous silica NPs	Spherical with particle size and negative zeta potential of >50 nm and 10.55 mV	Biocompatible prodrug NPs showed 4.18 and 420-fold superior IC_{50} values and antitumor activity against H1299 cells compared to other tested cell lines with prolonged blood circulation and higher tumor growth inhibition rate during in vivo studies	[128]
Nonviral gene vector-based MSCs/ ganciclovir prodrug liposomes	Spherical with a particle size of 938 nm and encapsulation efficiency of 44.7%	Prodrug liposomes exhibited a significant reduction in tumor colonization with a notable increase in survival rate in pulmonary melanoma metastasis bearing the C57BL6 mice model	[129]
β-lapachone prodrug micelles	Spherical (>95% drug loading) with a hydrodynamic diameter and negative zeta potential of 35 nm and 0.40 mV	Prodrug micelles showed satisfactory prodrug conversion with drug release using enzyme porcine liver esterase and 8.5-fold superior drug tumor concentration compared to β-lap cyclodextrin complex in female NOD-SCID mice bearing orthotopic fire-fly luciferase-transfected A549 tumors	[130]
Combretastatin A-4 prodrug	Phosphate prodrug	During the in vivo study, prodrug-treated mice showed significant 19.28- and 3.17-fold tumor volume reduction in mice bearing Colo-699 and KNS-62 cells	[131]
Hexadecyl-treprostinil prodrug lipid NPs	Uniform spherical and particle size <150 nm	In rats, Aeroneb® Pro nebulized (6 L/min) showed a notable reduction in U46619-induced PAP at a 60-fold lower EC_{50} value than i.v. drug alone. In dogs, Aeroneb® Pro nebulized (20–28 L/min) equipped face mask showed a significant 550-fold lower EC_{50} value as compared to i.v. drug alone	[133]
Spray-dried dextran (40 KDa)-succinate-BUD prodrug	Smooth spherical microparticles with d_{50} and d_{90} of 2.74 and 6.44 μm	Controlled release respirable particles showed ED and FPF of 93.5% and 47.3% using HandiHaler® device	[134]
Prodrug pan-PI3K inhibitor (CL27c)	Carboxylic acid linkage prodrug	Aerosolized CL27c micro-suspension remarkably controlled the bleomycin-induced lung fibrosis in mice and eventually enhanced survival	[135]

Pty Ltd. also studied the efficacy and safety of laninamivir octanoate delivered using TwinCaps® DPI device in adults with influenza (NCT01793883) [137]. Both clinical trials were completed with positive results. Moreover, Inavir® is now available in Japan to treat the influenza virus.

13.5 Future Outlook

This chapter carefully summarized the applications and achievements of prodrugs in the treatment of pulmonary diseases. Table 13.3 represents an outline of prodrugs utilized for pulmonary drug delivery. Respirable prodrug is a versatile and practical platform to achieve suitable aerosolization performance with a satisfactory pharmacological outcome. Most of the reported articles displayed two important motives, i.e., (i) to regulate and direct the metabolism of a drug substance by integrating easily labile moiety into the structure and (ii) to extend and improve the therapeutic value through overcoming the physiological and biological barriers. In line with various scientific reports summarized above, prodrug approach was utilized to deliver a variety of synthetic drugs, phyto-compounds, and biological substances. Moreover, it can be easily incorporated with novel delivery systems to improve therapeutic performance and drug targeting efficacy in lung cancer therapy.

Prodrug integrated nanomedicines are hopeful systems in treating lung cancer and other pulmonary complications. The primary aim of prodrug integrated nanomedicines is to increase anticancer drug efficacy and improve specific cell targeting with lower toxicity. Besides, in most of the prodrug polymer integrated nanomedicines, successful therapeutic action was achieved owing to the capability of the polymer to accumulate in tumor tissues on account of the EPR effect. In polymer-based prodrug, monomer incorporated prodrug is more practical due to ease of design and synthesis. Polymers can also be easily coated using hydrophilic, biocompatible polymer like PEG to impart "stealth" properties to the prodrug delivery system. Surely such a de novo design (novel structural frameworks) principle can be helpful to conceive metabolically active and pharmacologically sensitive inhaled prodrugs. Moreover, novel prodrug integrated nanomedicines are used to simultaneously deliver anticancer agents and biological substances to attain synergistic action with specific cell targeting to treat life threatening lung cancer. Polymeric micelles, hybrid NPs, and lipid vesicles are most commonly used to deliver prodrug at the desired site of action within pulmonary airways. Among the different prodrug integrated nanomedicines, polymeric micelles showed few advantages of good solubility, high drug content, modified drug half-life, and improved antitumor action. Besides, such nanomedicines can be quickly delivered to pulmonary airways using different delivery techniques.

Pulmonary drug delivery involves many mechanisms and devices [138–141] such as metered-dose inhalers [142], soft mist inhalers [143], nebulizers [144], and dry powder inhalers [145] to achieve desirable aerosolization performance [146–148]. Inhalation products are made up of three equally important elements, i.e., micronized drug (<5 μm), carrier (60–150 μm), and device [149–151]. The carrier molecules hold high importance in handling and actuating a drug substance due to its low concentration (4.5–500 μg) in the final formulation [152]. Accordingly, during the in vitro and clinical assessment of prodrug, adequate attention on the metabolism of carrier is also needed. In most cases, formulation and device properties conclude aerosol therapy's outcome [153]. Therefore, in vitro aerosolization analysis using different inhaler devices is highly recommended. Additionally, delivered dose uniformity testing is also essential to carefully assess the uniformity of each inhaled dose [154]. The expression of drug-metabolizing enzymes in pulmonary airways is much lower than the liver; therefore, lung epithelial tissue or sub-cellular tissue fractions treated with appropriate drug-metabolizing enzymes can be used to understand and reproduce in vivo prodrug activation. In addition, the in vitro bronchial epithelial cells integrated impactor deposition method can be adopted for a more realistic understanding of aerodynamic performance and cellular interactions of the inhaled prodrug.

Furthermore, most researchers have addressed prodrug's novel manufacturing methods, dual drug loading techniques, tumor-targeting approaches, and tumor xenograft models to study in vivo drug action. However, they have merely discussed critical issues that pertain to pulmokinetics, lung clearance mechanism, and other toxicological problems. Moreover, the interaction between pulmonary surfactants and drug substances is of specific significance as it strongly affects the in vivo activity and distribution of drug substances. Essentially, understanding various pulmonary drug-metabolizing enzyme and pathways with pulmokinetics helps to promote the clinical application of inhaled prodrugs. Therefore, in the coming times, systematic and inclusive in vitro and in vivo testing is the practical way to explore the full clinical potential of the novel pulmonary prodrugs.

13.6 Conclusion

In summary, the prodrug approach is a versatile scheme that can be applied to a wide range of formulations and many types of parent drug molecules. This chapter has carefully reviewed and analyzed the various pulmonary prodrug strategies, including prodrug integrated nanomedicines, polymer prodrug approaches, and carrier-linked prodrugs. Prodrug integrated nanomedicines and polymer conjugates appear to be promising in pulmonary drug delivery for different reasons such as physicochemical properties, water solubility, drug delivery efficiency, and drug targeting. The data summarized in this chapter indicate that several prodrugs have shown promising outcomes in the early clinical stage and achieved good remedial results; however, more detailed investigations regarding pulmonary drug metabolism, biosafety, and stability are still required to optimize prodrug in clinical application. Lastly, we feel that much more extensive experimental and clinical investigations of pulmonary prodrugs are valuable to chase.

References

1. Markovic M, Ben-Shabat S, Dahan A. Prodrugs for improved drug delivery: lessons learned from recently developed and marketed products. Pharmaceutics. 2020;12(11):1031.
2. Redasani VK, Bari SB. Prodrug design: perspectives, approaches and applications in medicinal chemistry. Academic Press; 2015.
3. Stella VJ, Nti-Addae KW. Prodrug strategies to overcome poor water solubility. Adv Drug Deliv Rev. 2007;59(7):677–94.
4. Piskáčková HB, Jansová H, Kubeš J, Karabanovich G, Váňová N, Kollárová-Brázdová P, Melnikova I, Jirkovská A, Lenčová-Popelová O, Chládek J, Roh J. Development of water-soluble prodrugs of the bisdioxopiperazine topoisomerase IIβ inhibitor ICRF-193 as potential cardioprotective agents against anthracycline cardiotoxicity. Sci Rep. 2021;11(1):1–4.
5. Peterson LW, McKenna CE. Prodrug approaches to improving the oral absorption of antiviral nucleotide analogues. Expert Opin Drug Deliv. 2009;6(4):405–20.
6. Aungst BJ, Matz N. Prodrugs to reduce presystemic metabolism. In: Stella VJ, Borchardt RT, Hageman MJ, Oliyai R, Maag H, Tilley JW, editors. Prodrugs. biotechnology: pharmaceutical aspects, vol V. New York, NY: Springer; 2007. p. 339–55.
7. N'Da DD. Prodrug strategies for enhancing the percutaneous absorption of drugs. Molecules. 2014;19(12):20780–807.
8. Wu S, Wang L, Huang X, Cao M, Hu J, Li H, Zhang H, Sun X, Meng QH, Hofstetter WL, Roth JA, Swisher SG, Fang B. Prodrug oncrasin-266 improves the stability, pharmacokinetics, and safety of NSC-743380. Bioorg Med Chem. 2014;22(19):5234–40.
9. Lin M, Guo W, Zhang Z, Zhou Y, Chen J, Wang T, Zhong X, Lu Y, Yang Q, Wei Q, Han M, Xu D, Gao J. Reduced toxicity of liposomal nitrogen mustard prodrug formulation activated by an intracellular ROS feedback mechanism in hematological neoplasm models. Mol Pharm. 2020;17(2):499–506.
10. Qandil AM. Prodrugs of non-steroidal anti-inflammatory drugs (NSAIDs), more than meets the eye: a critical review. Int J Mol Sci. 2012;13(12):17244–74.
11. Rautio J, Laine K, Gynther M, Savolainen J. Prodrug approaches for CNS delivery. AAPS J. 2008;10(1):92–102.
12. Li S, Shan X, Wang Y, Chen Q, Sun J, He Z, Sun B, Luo C. Dimeric prodrug-based nanomedicines for cancer therapy. J Control Release. 2020;326:510–22.
13. Ates-Alagoz Z, Adejare A. Prodrugs. In Remington 2021 (pp. 169-186). Academic Press.
14. Karaman R. Prodrugs design based on inter- and intramolecular chemical processes. Chem Biol Drug Des. 2013;82(6):643–68.
15. Choudhary D, Goykar H, Kalyane D, Sreeharsha N, Tekade RK. Prodrug design for improving the biopharmaceutical properties of therapeutic drugs. In The Future of Pharmaceutical Product Development and Research 2020 (pp. 179-226). Academic Press.
16. Silverman RB, Lawton GR, Ralay Ranaivo H, Chico LK, Seo J, Watterson DM. Effect of potential amine prodrugs of selective neuronal nitric oxide synthase inhibitors on blood-brain barrier penetration. Bioorg Med Chem. 2009;17(21):7593–605.
17. Majumdar S, Spaeth MM, Sivendran S, Juntunen J, Thomas JD, Sloan KB. α-(1H-Imidazol-1-yl) alkyl (IMIDA) carboxylic acid esters as prodrugs of carboxylic acid containing drugs. Tetrahedron Lett. 2007;48(26):4609–11.
18. Jensen E, Bundgaard H. Peptide esters as water-soluble prodrugs for hydroxyl containing agents: chemical stability and enzymatic hydrolysis of benzyl esters of glycine, diglycine and triglycine. Int J Pharm. 1991;71(1-2):117–25.
19. Subbaiah MA, Ramar T, Subramani L, Desai SD, Sinha S, Mandlekar S, Jenkins SM, Krystal MR, Subramanian M, Sridhar S, Padmanabhan S. (Carbonyl) oxyalkyl linker-based amino acid prodrugs of the HIV-1 protease inhibitor atazanavir that enhance oral bioavailability and plasma trough concentration. Eur J Med Chem. 2020;207:112749.
20. Wiemer AJ, Wiemer DF. Prodrugs of phosphonates and phosphates: crossing the membrane barrier. Top Curr Chem. 2015;360:115–60.

21. Guarino VR, Olson RE, Everlof JG, Wang N, McDonald I, Haskell R, Clarke W, Lentz KA. An amide-based sulfenamide prodrug of gamma secretase inhibitor BMS-708163 delivers parent drug from an oral conventional solid dosage form in male beagle dog. Bioorg Med Chem Lett. 2020;30(3):126856.
22. Huang B, Liu X, Tian Y, Kang D, Zhou Z, Daelemans D, De Clercq E, Pannecouque C, Zhan P, Liu X. First discovery of a potential carbonate prodrug of NNRTI drug candidate RDEA427 with submicromolar inhibitory activity against HIV-1 K103N/Y181C double mutant strain. Bioorg Med Chem Lett. 2018;28(8):1348–51.
23. Qin C, Chu Y, Feng W, Fromont C, He S, Ali J, Lee JB, Zgair A, Berton M, Bettonte S, Liu R. Targeted delivery of lopinavir to HIV reservoirs in the mesenteric lymphatic system by lipophilic ester prodrug approach. J Control Release. 2021;329:1077–89.
24. Xu X, Lv Y, Tang K, Song B, Jiang Q, Sun L, He Z, Zhang T. The simultaneous determination of naringenin and its valine carbamate prodrug in rat plasma using supercritical fluid chromatography -tandem mass spectrometric method. J Pharm Biomed Anal. 2021;195:113848.
25. Fu L, Jiang Z, Cai Z, Liu X, He H, Yang Y. Water-soluble phosphate prodrugs of pleuromutilin analogues with potent in vivo antibacterial activity against Gram-positive pathogens. Bioorg Med Chem Lett. 2009;19(18):5407–10.
26. Li S, Li X, Lu Y, Hou M, Xu Z, Li B. A thiol-responsive and self-immolative podophyllotoxin prodrug for cancer therapy. Tetrahedron Lett. 2021;71:153044.
27. Zhao YJ, Wei W, Su ZG, Ma GH. Poly (ethylene glycol) prodrug for anthracyclines via N-Mannich base linker: design, synthesis and biological evaluation. Int J Pharm. 2009;379(1):90–9.
28. Kapoor M, Cheryala N, Rautiola D, Georg GI, Cloyd JC, Siegel RA. Chirally pure prodrugs and their converting enzymes lead to high supersaturation and rapid transcellular permeation of benzodiazepines. J Pharm Sci. 2016;105(8):2365–71.
29. Mistry IN, Thomas M, Calder EDD, Conway SJ, Hammond EM. Clinical advances of hypoxia-activated prodrugs in combination with radiation therapy. Int J Radiat Oncol Biol Phys. 2017;98(5):1183–96.
30. Li Y, Yang M, Zhao Y, Li L, Xu W. Preparation and in vitro evaluation of amphiphilic paclitaxel small molecule prodrugs and enhancement of oral absorption. Eur J Med Chem. 2021;215:113276.
31. Meng Q, Hu H, Zhou L, Zhang Y, Yu B, Shen Y, Cong H. Logical design and application of prodrug platforms. Polymer Chem. 2019;10(3):306–24.
32. Yin W, Ke W, Lu N, Wang Y, Japir AA, Mohammed F, Wang Y, Pan Y, Ge Z. Glutathione and reactive oxygen species dual-responsive block copolymer prodrugs for boosting tumor site-specific drug release and enhanced antitumor efficacy. Biomacromolecules. 2020;21(2):921–9.
33. Deng Z, Liu S. Controlled drug delivery with nanoassemblies of redox-responsive prodrug and polyprodrug amphiphiles. J Control Release. 2020;326:276–96.
34. Guan J, Wu Y, Wang H, Zeng H, Li Z, Yang X. DiR loaded tumor targeting theranostic cisplatin-icodextrin prodrug nanoparticle for imaging guided chemo-photothermal cancer therapy. Nanoscale. 2021;13(46):19399–411.
35. Yan D, Xu X, Ren C, Chen C, Luo J, Han C, Kong L. DT-diaphorase triggered theranostic nanoparticles induce the self-burst of reactive oxygen species for tumor diagnosis and treatment. Acta Biomaterialia. 2021;125:267–79.
36. Chen KJ, Plaunt AJ, Leifer FG, Kang JY, Cipolla D. Recent advances in prodrug-based nanoparticle therapeutics. Eur J Pharm Biopharm. 2021;165:219–43.
37. Xie A, Hanif S, Ouyang J, Tang Z, Kong N, Kim NY, Qi B, Patel D, Shi B, Tao W. Stimuli-responsive prodrug-based cancer nanomedicine. EBioMedicine. 2020;56:102821.
38. Wang Z, Chen J, Little N, Lu J. Self-assembling prodrug nanotherapeutics for synergistic tumor targeted drug delivery. Acta Biomater. 2020;111:20–8.
39. Luo C, Sun J, Sun B, He Z. Prodrug-based nanoparticulate drug delivery strategies for cancer therapy. Trends Pharmacol Sci. 2014;35(11):556–66.

40. Wang W, Fan J, Zhu G, Wang J, Qian Y, Li H, Ju J, Shan L. Targeted prodrug-based self-assembled nanoparticles for cancer therapy. Int J Nanomed. 2020;15:2921–33.
41. Bai Y, Liu CP, Song X, et al. Photo- and pH- dual-responsive b-cyclodextrin-based supramolecular prodrug complex self-assemblies for programmed drug delivery. Chem Asian J. 2018;13(24):3903–11.
42. Peng M, Qin S, Jia H, Zheng D, Rong L, Zhang X. Self-delivery of a peptide-based prodrug for tumor-targeting therapy. Nano Res. 2016;9(3):663–73.
43. Caron J, Maksimenko A, Wack S, et al. Improving the antitumor activity of squalenoyl-paclitaxel conjugate nanoassemblies by manipulating the linker between paclitaxel and squalene. Adv Healthc Mater. 2013;2(1):172–85.
44. Yue Z, Wang H, Li Y, et al. Coordination-driven self-assembly of a Pt(IV) prodrug-conjugated supramolecular hexagon. Chem Commun. 2018;54(7):731–4.
45. Cheetham AG, Chakroun RW, Ma W, Cui H. Self-assembling prodrugs. Chem Soc Rev. 2017;46(21):6638–63.
46. Teng W, Jia F, Han H, et al. Polyamino acid-based gemcitabine nanocarriers for targeted intracellular drug delivery. Polym Chem. 2017;8(16):2490–8.
47. Wang Q, Guan J, Wan J, Li Z. Disulfide based prodrugs for cancer therapy. RSC Adv. 2020;10(41):24397–409.
48. Zhang Y, Cui H, Zhang R, Zhang H, Huang W. Nanoparticulation of prodrug into medicines for cancer therapy. Adv Sci (Weinh). 2021;8(18):e2101454.
49. Mahato R, Tai W, Cheng K. Prodrugs for improving tumor targetability and efficiency. Adv Drug Deliv Rev. 2011;63(8):659–70.
50. Zhao G, Ren R, Wei X, Jia Z, Chen N, Sun Y, Zhao Z, Lele SM, Zhong HA, Goldring MB, Goldring SR, Wang D. Thermoresponsive polymeric dexamethasone prodrug for arthritis pain. J Control Release. 2021;339:484–97.
51. Stephenson AA, Cao S, Taggart DJ, Vyavahare VP, Suo Z. Design, synthesis, and evaluation of liver-specific gemcitabine prodrugs for potential treatment of hepatitis C virus infection and hepatocellular carcinoma. Eur J Med Chem. 2021;213:113135.
52. Hillaireau H, Dereuddre-Bosquet N, Skanji R, Bekkara-Aounallah F, Caron J, Lepêtre S, Argote S, Bauduin L, Yousfi R, Rogez-Kreuz C, Desmaële D, Rousseau B, Gref R, Andrieux K, Clayette P, Couvreur P. Anti-HIV efficacy and biodistribution of nucleoside reverse transcriptase inhibitors delivered as squalenoylated prodrug nanoassemblies. Biomaterials. 2013;34(20):4831–8.
53. Dholkawala F, Voshavar C, Dutta AK. Synthesis and characterization of brain penetrant prodrug of neuroprotective D-264: potential therapeutic application in the treatment of Parkinson's disease. Eur J Pharm Biopharm. 2016;103:62–70.
54. Wei Z, Xin G, Wang H, Zheng H, Ji C, Gu J, Ma L, Qin C, Xing Z, Niu H, Huang W. The diosgenin prodrug nanoparticles with pH-responsive as a drug delivery system uniquely prevents thrombosis without increased bleeding risk. Nanomedicine. 2018;14(3):673–84.
55. Corboz MR, Zhang J, LaSala D, DiPetrillo K, Li Z, Malinin V, Brower J, Kuehl PJ, Barrett TE, Perkins WR, Chapman RW. Therapeutic administration of inhaled INS1009, a treprostinil prodrug formulation, inhibits bleomycin-induced pulmonary fibrosis in rats. Pulm Pharmacol Ther. 2018;49:95–103.
56. Quagliariello V, Gennari A, Jain SA, Rosso F, Iaffaioli RV, Barbarisi A, Barbarisi M, Tirelli N. Double-responsive hyaluronic acid-based prodrugs for efficient tumour targeting. Mater Sci Eng: C. 2021;131:112475.
57. Jackson M. An antitubercular prodrug leaves Mycobacterium tuberculosis facing a difficult choice, poisoning or starvation? Cell Chem Biol. 2021;28(8):1101–3.
58. Voak AA, Gobalakrishnapillai V, Seifert K, Balczo E, Hu L, Hall BS, Wilkinson SR. An essential type I nitroreductase from Leishmania major can be used to activate leishmanicidal prodrugs. J Biol Chem. 2013;288(40):28466–76.
59. Wolfe AR, Neitz RJ, Burlingame M, Suzuki BM, Lim KC, Scheideler M, Nelson DL, Benet LZ, Caffrey CR. TPT sulfonate, a single, oral dose schistosomicidal prodrug: in vivo efficacy, disposition and metabolic profiling. Int J Parasitol Drugs Drug Resist. 2018;8(3):571–86.

60. Thuita JK, Karanja SM, Wenzler T, Mdachi RE, Ngotho JM, Kagira JM, Tidwell R, Brun R. Efficacy of the diamidine DB75 and its prodrug DB289, against murine models of human African trypanosomiasis. Acta Trop. 2008;108(1):6–10.
61. Oseghale AR, Zhu X, Li B, Peterson KR, Nudelman A, Rephaeli A, Xu H, Pace BS. Conjugate prodrug AN-233 induces fetal hemoglobin expression in sickle erythroid progenitors and β-YAC transgenic mice. Blood Cells Mol Dis. 2019;79:102345.
62. Serafim EO, Silva AT, Moreno Ade H, Vizioli Ede O, Ferreira EI, Peccinini RG, Ribeiro ML, Chung MC. Pharmacokinetics of hydroxymethylnitrofurazone, a promising new prodrug for Chagas' disease treatment. Antimicrob Agents Chemother. 2013;57(12):6106–9.
63. Chung MC, Ferreira EI, Santos JL, Giarolla J, Rando DG, Almeida AE, Bosquesi PL, Menegon RF, Blau L. Prodrugs for the treatment of neglected diseases. Molecules. 2007;13(3):616–77.
64. Kour J, Kumari N, Sapra B. Ocular prodrugs: attributes and challenges. Asian J Pharm Sci. 2021;16(2):175–91.
65. Bonina FP, Puglia C, Barbuzzi T, de Caprariis P, Palagiano F, Rimoli MG, Saija A. In vitro and in vivo evaluation of polyoxyethylene esters as dermal prodrugs of ketoprofen, naproxen and diclofenac. Eur J Pharm Sci. 2001;14(2):123–34.
66. Srinivasan S, Roy D, Chavas TEJ, Vlaskin V, Ho DK, Pottenger A, et al. Liver-targeted polymeric prodrugs of 8-aminoquinolines for malaria radical cure. J Control Release. 2021;331:213–27.
67. Xiang H, Zhang Q, Han Y, Yang L, Zhang Y, Liu Q, Zhang Z, Zhang L. Novel brain-targeting 3-n-butylphthalide prodrugs for ischemic stroke treatment. J Control Release. 2021;335:498–514.
68. Chavas TEJ, Su FY, Srinivasan S, Roy D, Lee B, Lovelace-Macon L, et al. A macrophage-targeted platform for extending drug dosing with polymer prodrugs for pulmonary infection prophylaxis. J Control Release. 2021;330:284–92.
69. Jornada DH, dos Santos Fernandes GF, Chiba DE, de Melo TR, dos Santos JL, Chung MC. The prodrug approach: a successful tool for improving drug solubility. Molecules. 2015;21(1):42.
70. da Silva Santos S, Igne Ferreira E, Giarolla J. Dendrimer prodrugs. Molecules. 2016;21(6):686.
71. Najjar A, Najjar A, Karaman R. Newly developed prodrugs and prodrugs in development; an insight of the recent years. Molecules. 2020;25(4):884.
72. Tendolkar MS, Tyagi R, Handa A. Review of advances in diagnosis and treatment of pulmonary tuberculosis. Indian J Tuberc. 2021;68(4):510–5.
73. Chapman RW, Corboz MR, Malinin VS, Plaunt AJ, Konicek DM, Li Z, Perkins WR. An overview of the biology of a long-acting inhaled treprostinil prodrug. Pulm Pharmacol Ther. 2020;65:102002.
74. Rubin K, Ewing P, Bäckström E, Abrahamsson A, Bonn B, Kamata S, Grime K. Pulmonary metabolism of substrates for key drug-metabolizing enzymes by human alveolar type II cells, human and rat lung microsomes, and the isolated perfused rat lung model. Pharmaceutics. 2020;12(2):117.
75. Enlo-Scott Z, Bäckström E, Mudway I, Forbes B. Drug metabolism in the lungs: opportunities for optimising inhaled medicines. Expert Opin Drug Metab Toxicol. 2021;17(5):611–25.
76. Jamaludin J, Marlin N, Wood H, et al. Evaluating the impact of genotype on the relationship between impaired lung growth and chronic exposure to traffic derived pollutants. Eur Respir J. 2013;42(Suppl 57):P3623.
77. Tekpli X, Zienolddiny S, Skaug V, et al. DNA methylation of the CYP1A1 enhancer is associated with smoking-induced genetic alterations in human lung. Int J Cancer. 2012;131(7):1509–16.
78. Gundert-Remy U, Bernauer U, Blömeke B, et al. Extrahepatic metabolism at the body's internal–external interfaces. Drug Metab Rev. 2014;46(3):291–324.
79. Zeldin DC, Foley J, Ma J, et al. CYP2J subfamily P450s in the lung: expression, localization, and potential functional significance. Mol Pharmacol. 1996;50(5):1111–7.
80. Oesch F, Fabian E, Landsiedel R. Xenobiotica-metabolizing enzymes in the lung of experimental animals, man and in human lung models. Arch Toxicol. 2019;93(12):3419–89.

81. Winkler J, Hochhaus G, Derendorf H. How the lung handles drugs: pharmacokinetics and pharmacodynamics of inhaled corticosteroids. Proc Am Thorac Soc. 2004;1(4):356–63.
82. Mukker JK, Singh RSP, Derendorf H. Ciclesonide: a pro-soft drug approach for mitigation of side effects of inhaled corticosteroids. J Pharm Sci. 2016;105(9):2509–14.
83. Nave R, Meyer W, Fuhst R, et al. Formation of fatty acid conjugates of ciclesonide active metabolite in the rat lung after 4-week inhalation of ciclesonide. Pulm Pharmacol Ther. 2005;18(6):390–6.
84. Wieslander E, Delander E-L, Järkelid L, et al. Pharmacologic importance of the reversible fatty acid conjugation of budesonide studied in a rat cell line in vitro. Am J Respir Cell Mol Biol. 1998;19(3):477–84.
85. Borchard G, Cassará ML, Roemelé PE, et al. Transport and local metabolism of budesonide and fluticasone propionate in a human bronchial epithelial cell line (Calu-3). J Pharm Sci. 2002;91(6):1561–7.58
86. Chen H, Wang X, Gou S. A cisplatin-based platinum (IV) prodrug containing a glutathione s-transferase inhibitor to reverse cisplatin-resistance in non-small cell lung cancer. J Inorganic Biochem. 2019;193:133–42.
87. Ma S, Li X, Ran M, Ji M, Gou J, Yin T, He H, Wang Y, Zhang Y, Tang X. Fabricating nanoparticles co-loaded with survivin siRNA and Pt(IV) prodrug for the treatment of platinum-resistant lung cancer. Int J Pharm. 2021;601:120577. https://doi.org/10.1016/j.ijpharm.2021.120577.
88. Huo Q, Liang Y, Lu W, Peng X, Du C, Sun Y, Han S, Cao J, Sun Y, He B. Integrated metalloproteinase, pH and glutathione responsive prodrug-based nanomedicine for efficient target chemotherapy. J Biomed Nanotechnol. 2019;15(8):1673–87.
89. Wojtkowiak JW, Verduzco D, Schramm KJ, Gillies RJ. Drug resistance and cellular adaptation to tumor acidic pH microenvironment. Mol Pharm. 2011;8(6):2032–8.
90. Ling X, Tu J, Wang J, Shajii A, Kong N, Feng C, Zhang Y, Yu M, Xie T, Bharwani Z, Aljaeid BM. Glutathione-responsive prodrug nanoparticles for effective drug delivery and cancer therapy. Acs Nano. 2018;13(1):357–70.
91. Schiller JH, Harrington D, Belani CP, Langer C, Sandler A, Krook J, Zhu J, Johnson DH. Comparison of four chemotherapy regimens for advanced non–small-cell lung cancer. N Engl J Med. 2002;346(2):92–8.
92. Ohe Y, Ohashi Y, Kubota K, Tamura T, Nakagawa K, Negoro S, et al. Randomized phase III study of cisplatin plus irinotecan versus carboplatin plus paclitaxel, cisplatin plus gemcitabine, and cisplatin plus vinorelbine for advanced non-small-cell lung cancer: Four-Arm Cooperative Study in Japan. Ann Oncol. 2007;18(2):317–23.
93. Wang B, Hu W, Yan H, Chen G, Zhang Y, Mao J, Wang L. Lung cancer chemotherapy using nanoparticles: enhanced target ability of redox-responsive and pH-sensitive cisplatin prodrug and paclitaxel. Biomed Pharmacother. 2021;136:111249.
94. Yang T, Yu S, Liu L, Sun Y, Lan Y, Ma X, Zhu R, Li L, Hou Y, Liu Y. Dual polymeric prodrug co-assembled nanoparticles with precise ratiometric co-delivery of cisplatin and metformin for lung cancer chemo immunotherapy. Biomater Sci. 2020;8(20):5698–714.
95. Swamy MK, Vasamsetti BM. Taxol: occurrence, chemistry, and understanding its molecular mechanisms. In Paclitaxel 2022 Jan 1 (pp. 29–45). Academic Press.
96. Isah T. Anticancer alkaloids from trees: development into drugs. Pharmacogn Rev. 2016;10(20):90–9.
97. Khalifa AM, Elsheikh MA, Khalifa AM, Elnaggar YSR. Current strategies for different paclitaxel-loaded Nano-delivery Systems towards therapeutic applications for ovarian carcinoma: a review article. J Control Release. 2019;311-312:125–37.
98. Zhao B, Gu Z, Zhang Y, Li Z, Cheng L, Li C, Hong Y. Starch-based carriers of paclitaxel: a systematic review of carriers, interactions, and mechanisms. Carbohydr Polym. 2022;119628
99. Sohn JS, Jin JI, Hess M, Jo BW. Polymer prodrug approaches applied to paclitaxel. Polymer Chem. 2010;1(6):778–92.
100. Raza F, Zafar H, Khan MW, Ullah A, Khan A, Baseer A, Fareed R, Muhammad S. Recent advances in targeted delivery of paclitaxel nanomedicine for cancer therapy. Mater Adv. 2022;

101. Wang G, Wang Z, Li C, Duan G, Wang K, Li Q, Tao T. RGD peptide-modified, paclitaxel prodrug-based, dual-drugs loaded, and redox-sensitive lipid-polymer nanoparticles for the enhanced lung cancer therapy. Biomed Pharmacother. 2018;106:275–84.
102. Xu C, Sun Y, Qi Y, Yu Y, He Y, Hu M, et al. Selective self-induced stimulus amplification prodrug platform for inhibiting multidrug resistance and lung metastasis. J Control Release. 2018;284:224–39.
103. Fan X, Lin X, Ruan Q, Wang J, Yang Y, Sheng M, Zhou W, Kai G, Hao X. Research progress on the biosynthesis and metabolic engineering of the anti-cancer drug camptothecin in Camptotheca acuminate. Ind Crops Prod. 2022;186:115270.
104. Botella P, Rivero-Buceta E. Safe approaches for camptothecin delivery: structural analogues and nanomedicines. J Control Release. 2017;247:28–54.
105. Bondì ML, Di Gesù R, Craparo EF. Lipid nanoparticles for drug targeting to the brain. In Methods in enzymology 2012 Jan 1 (Vol. 508, pp. 229–251). Academic Press.
106. Yurkovetskiy AV, Fram RJ. XMT-1001, a novel polymeric camptothecin pro-drug in clinical development for patients with advanced cancer. Adv Drug Deliv Rev. 2009;61(13):1193–202.
107. Bala V, Rao S, Boyd BJ, Prestidge CA. Prodrug and nanomedicine approaches for the delivery of the camptothecin analogue SN38. J Control Release. 2013;172(1):48–61.
108. Li F, Huang Z, Chen H, Yan L, Li J, Su Y, Zhang Q, Huang Z, Zheng Y. Redox-sensitive lipophilic prodrugs: delivering unstable chemotherapeutant for improved cancer therapy. Drug Deliv. 2019;26(1):1068–79. https://doi.org/10.1080/10717544.2019.1678696.
109. Deshmukh M, Chao P, Kutscher HL, Gao D, Sinko PJ. A series of alpha-amino acid ester prodrugs of camptothecin: in vitro hydrolysis and A549 human lung carcinoma cell cytotoxicity. J Med Chem. 2010;53(3):1038–47. https://doi.org/10.1021/jm901029n.
110. Ashrafizadeh M, Mirzaei S, Gholami MH, Hashemi F, Zabolian A, Raei M, et al. Hyaluronic acid-based nanoplatforms for Doxorubicin: a review of stimuli-responsive carriers, co-delivery and resistance suppression. Carbohydr Polym. 2021;272:118491.
111. Sritharan S, Sivalingam N. A comprehensive review on time-tested anticancer drug doxorubicin. Life Sci. 2021;278:119527.
112. Liu J, He J, Zhang M, Xu G, Ni P. A synergistic polyphosphoester-based co-delivery system of the anticancer drug doxorubicin and the tumor suppressor gene p53 for lung cancer therapy. J Mater Chem B. 2018;6(20):3262–73.
113. Hong Y, Che S, Hui B, Yang Y, Wang X, Zhang X, Qiang Y, Ma H. Lung cancer therapy using doxorubicin and curcumin combination: targeted prodrug based, pH sensitive nanomedicine. Biomed Pharmacother. 2019;112:108614.
114. Alhajj N, O'Reilly NJ, Cathcart H. Developing ciprofloxacin dry powder for inhalation: a story of challenges and rational design in the treatment of cystic fibrosis lung infection. Int J Pharm. 2021:121388.
115. Nwabuife JC, Omolo CA, Govender T. Nano delivery systems to the rescue of ciprofloxacin against resistant bacteria "E. coli; P. aeruginosa; Saureus; and MRSA" and their infections. J Control Release. 2022;349:338–53.
116. Chen J, Su FY, Das D, Srinivasan S, Son HN, Lee B, Radella F 2nd, Whittington D, Monroe-Jones T, West TE, Convertine AJ, Skerrett SJ, Stayton PS, Ratner DM. Glycan targeted polymeric antibiotic prodrugs for alveolar macrophage infections. Biomaterials. 2019;195:38–50.
117. Su FY, Srinivasan S, Lee B, Chen J, Convertine AJ, West TE, Ratner DM, Skerrett SJ, Stayton PS. Macrophage-targeted drugamers with enzyme-cleavable linkers deliver high intracellular drug dosing and sustained drug pharmacokinetics against alveolar pulmonary infections. J Control Release. 2018;287:1–11.
118. Dowling PM. Chloramphenicol, thiamphenicol, and florfenicol. Antimicrobial Ther Vet Med. 2013:269–77.
119. Nurbaeti SN, Brillault J, Tewes F, Olivier JC. Sustained-release microparticle dry powders of chloramphenicol palmitate or thiamphenicol palmitate prodrugs for lung delivery as aerosols. Eur J Pharm Sci. 2019;138:105028.
120. Thai AA, Solomon BJ, Sequist LV, Gainor JF, Heist RS. Lung cancer. Lancet. 2021;398(10299):535–54.

121. Duruisseaux M, Esteller M. Lung cancer epigenetics: from knowledge to applications. In Seminars in cancer biology 2018 (Vol. 51, pp. 116–128). Academic Press.
122. Yin X, Li Y, Wang H, Jia T, Wang E, Luo Y, Wei Y, Qin Z, Ma X. Small cell lung cancer transformation: from pathogenesis to treatment. In Seminars in Cancer Biology 2022. Academic Press.
123. Yin X, Liao H, Yun H, Lin N, Li S, Xiang Y, Ma X. Artificial intelligence-based prediction of clinical outcome in immunotherapy and targeted therapy of lung cancer. In Seminars in cancer biology 2022. Academic Press.
124. Radi M, Adema AD, Daft JR, Cho JH, Hoebe EK, Alexander LE, Peters GJ, Chu CK. In vitro optimization of non-small cell lung cancer activity with troxacitabine, L-1,3-dioxolane-cytidine, prodrugs. J Med Chem. 2007;50(9):2249–53.
125. Kim EJ, Bhuniya S, Lee H, Kim HM, Cheong C, Maiti S, Hong KS, Kim JS. An activatable prodrug for the treatment of metastatic tumors. J Am Chem Soc. 2014;136(39):13888–94.
126. Suthar SK, Lee HB, Sharma M. The synthesis of non-steroidal anti-inflammatory drug (NSAID)–lantadene prodrugs as novel lung adenocarcinoma inhibitors via the inhibition of cyclooxygenase-2 (COX-2), cyclin D1 and TNF-α-induced NF-κB activation. RSC Adv. 2014;4(37):19283–93.
127. Wang W, Li C, Zhang J, Dong A, Kong D. Correction: Tailor-made gemcitabine prodrug nanoparticles from well-defined drug–polymer amphiphiles prepared by controlled living radical polymerization for cancer chemotherapy. J Mater Chem B. 2015;3(10):2229.
128. Chen F, Zhang F, Shao D, Zhang W, Zheng L, Wang W, Yang W, Wang Z, Chen J, Dong WF, Xiao F. Bioreducible and traceable Ru (III) prodrug-loaded mesoporous silica nanoparticles for sequentially targeted non-small cell lung cancer chemotherapy. Appl Mater Today. 2020;19:100558.
129. Zhang TY, Huang B, Wu HB, Wu JH, Li LM, Li YX, Hu YL, Han M, Shen YQ, Tabata Y, Gao JQ. Synergistic effects of co-administration of suicide gene expressing mesenchymal stem cells and prodrug-encapsulated liposome on aggressive lung melanoma metastases in mice. J Control Release. 2015;209:260–71.
130. Ma X, Huang X, Moore Z, Huang G, Kilgore JA, Wang Y, Hammer S, Williams NS, Boothman DA, Gao J. Esterase-activatable β-lapachone prodrug micelles for NQO1-targeted lung cancer therapy. J Control Release. 2015;200:201–11.
131. Boehle AS, Sipos B, Kliche U, Kalthoff H, Dohrmann P. Combretastatin A-4 prodrug inhibits the growth of human non-small cell lung cancer in a murine xenotransplant model. Ann Thorac Surg. 2001;71(5):1657–65.
132. Kurdow R, Boehle AS, Haye S, Boenicke L, Schniewind B, Dohrmann P, Kalthoff H. Ganciclovir prodrug therapy is effective in a murine xenotransplant model of human lung cancer. Ann Thorac Surg. 2002;73(3):905–10.
133. Chapman RW, Li Z, Corboz MR, Gauani H, Plaunt AJ, Konicek DM, Leifer FG, Laurent CE, Yin H, Salvail D, Dziak C, Perkins WR, Malinin V. Inhaled hexadecyl-treprostinil provides pulmonary vasodilator activity at significantly lower plasma concentrations than infused treprostinil. Pulm Pharmacol Ther. 2018;49:104–11.
134. Waters RC, Hochhaus G. Characterization of a dextran-budesonide prodrug for inhalation therapy. Eur J Pharm Sci. 2019;129:58–67.
135. Campa CC, Silva RL, Margaria JP, Pirali T, Mattos MS, Kraemer LR, et al. Inhalation of the prodrug PI3K inhibitor CL27c improves lung function in asthma and fibrosis. Nat Commun. 2018;9(1):5232.
136. Safety, Tolerability and Pharmacokinetics of Inhaled Laninamivir Octanoate TwinCaps® DPI in Adults With Chronic Asthma. NCT02022761 https://clinicaltrials.gov/ct2/show/NCT02022761?term=Laninamivir+Octanoate&draw=2&rank=1
137. Efficacy and Safety Study of Laninamivir Octanoate TwinCaps® Dry Powder Inhaler in Adults With Influenza (Igloo). NCT01793883 https://clinicaltrials.gov/ct2/show/NCT01793883?term=Laninamivir+Octanoate&draw=2&rank=3

138. Jadhav P, Patil P, Bhagwat D, Gaikwad V, Mehta PP. Recent advances in orthogonal analytical techniques for microstructural understanding of inhalable particles: present status and future perspective. J Drug Deliv Sci Technol. 2022:103089.
139. Mehta PP. Dry powder inhalers: a brief overview of the drug detachment techniques. Ther Deliv. 2020;11(3):139–43.
140. Mehta PP. Dry powder inhalers: a concise summary of the electronic monitoring devices. Ther Deliv. 2021;12(1):1–6.
141. Mehta PP, Dhapte-Pawar VS. Novel and evolving therapies for COVID-19 related pulmonary complications. Am J Med Sci. 2021;361(5):557–66.
142. Kumar R, Mehta P, Shankar KR, Rajora MA, Mishra YK, Mostafavi E, Kaushik A. Nanotechnology-assisted metered-dose inhalers (MDIs) for high-performance pulmonary drug delivery applications. Pharm Res. 2022;1-25
143. Mehta PP, Dhapte-Pawar V. Role of surfactants in pulmonary drug delivery. In Green sustainable process for chemical and environmental engineering and science 2022 (pp. 559-577). Academic Press.
144. Mehta PP, Dhapte-Pawar VS. Repurposing drug molecules for new pulmonary therapeutic interventions. Drug Deliv Transl Res. 2021;11(5):1829–48.
145. Mehta PP. Dry powder inhalers: upcoming platform technologies for formulation development. Ther Deliv. 2019;10(9):551–4.
146. Mehta PP, Pawar AP, Mahadik KR, Kadam SS, Dhapte-Pawar V. Dry powder coating techniques and role of force controlling agents in aerosol. Polymer coatings: technology and applications. 2020:41-74.
147. Mehta PP, Ghoshal D, Pawar AP, Kadam SS, Dhapte-Pawar VS. Recent advances in inhalable liposomes for treatment of pulmonary diseases: concept to clinical stance. J Drug Deliv Sci Technol. 2020;56:101509.
148. Mehta PP, Kadam SS, Pawar AP. Exploring the impact of extrinsic lactose fines, a USP modified sampling device and modified centrifuge tube on the delivered dose uniformity and drug detachment performance of a fluticasone propionate dry powder inhaler. J Drug Deliv Sci Technol. 2020;57:101681.
149. Mehta PP, Kadam SS, Pawar AP. Effect of USP induction ports and modified glass sampling apparatus on aerosolization performance of lactose carrier-based fluticasone propionate dry powder inhaler. J Drug Deliv Sci Technol. 2020;58:101794.
150. Mehta P, Bothiraja C, Kadam S, Pawar A. Probing the influence of lactose fines, a USP modified induction port and modified DDU apparatus on the aerodynamic behavior of a fluticasone propionate dry powder inhaler. N J Chem. 2019;43(44):17327–39.
151. Mehta PP, Kadam SS, Pawar AP. Influence of modified induction port, modified DUSA assembly and device air-inlet geometry on the aerosolization pattern of a dry powder inhaler. J Drug Deliv Sci Technol. 2020;55:101416.
152. Mehta P. Imagine the superiority of dry powder inhalers from carrier engineering. J Drug Deliv. 2018;2018
153. Mehta P. Multi-dose dry powder inhaler: advance technology for drug delivery to airways. Indian Drugs. 2018;56(11):68–71.
154. Mehta P, Bothiraja C, Kadam S, Pawar A. Effect of USP induction ports, glass sampling apparatus, and inhaler device resistance on aerodynamic patterns of fluticasone propionate–loaded engineered mannitol microparticles. AAPS Pharm Sci Tech. 2019;20(5):1–3.

Nucleic Acid Pulmonary Therapy: From Concept to Clinical Stance

14

Diana A. Fernandes

Abstract

The great advent of nucleic acids as therapeutic agents is their ability to act on the underlying transcription and translational levels of a given disease, ranging from gene knockdown to induced expression of an implicated protein. This feature holds great promise as it extends the therapy reach to a wider range of diseases, not only those considered undruggable but also the ones lacking current treatment options. Additionally, the repurpose of chemical synthesis combined with the availability of reproducible high-throughput production processes also renders these molecules easier to scale up in a cost-effective manner. The potential of nucleic acids can be further enhanced when coupled with the pulmonary route owing to its ability to tackle local and systemic treatment of diseases and reduced nuclease activity which allows for lower therapeutic target doses. Despite the overall benefits of combining nucleic acid administration with the pulmonary route, these can still be overshadowed by the inherent challenges to this type of molecules, delivery route, delivery platform, and target disease. In this chapter, the types of nucleic acids relevant to pulmonary therapy are described alongside their underlying mechanisms of action, the challenges posed by this administration route, and the delivery platform employed. A current clinical outlook pertaining to inhaled nucleic acids is also presented.

D. A. Fernandes (✉)
OnResp, Fruebjergvej 3, Copenhagen Ø, Denmark
e-mail: dianafernandes@onresp.com

P. P. Mehta, V. Dhapte-Pawar (eds.), *Pulmonary Drug Delivery Systems: Material and Technological Advances*,
https://doi.org/10.1007/978-981-99-1923-9_14

Keywords

Nucleic acid therapy · Pulmonary drug delivery · Nebulizer · Pressurized metered dose inhaler · Dry powder inhaler · Soft mist inhaler · Lipid-based nonviral vectors · Polymer-based nonviral vectors

14.1 Introduction

Nucleic acids are chemical compounds mainly responsible for the storage of genetic material, gene expression, regulation, and protein synthesis [1]. From a therapeutic standpoint, nucleic acids may include antisense oligonucleotides (ASOs), small interfering RNAs (siRNAs), microRNAs mimics (miRNAs), anti-microRNAs (Antagomirs), and aptamers (Apts) [1–3]. The great advent of these molecules as therapeutic agents is their ability to act on the underlying transcription and translational levels of a given disease, ranging from gene knockdown to induced expression of an implicated protein [1]. This feature holds great promise as it extends the therapy reach to a wider range of diseases, not only those considered undruggable but also the ones lacking current treatment options [2]. Additionally, the repurpose of chemical synthesis combined with the availability of reproducible high-throughput production processes also makes these molecules easier to scale up in a cost-effective manner.

The potential of nucleic acids can be enhanced when coupled with the pulmonary route. The pulmonary route enables the drug to be deposited whether at the site of action in the lungs for topically acting drugs or at the site of absorption for systemically acting drugs. For topically acting drugs, the advantages of pulmonary delivery entail the possibility to reduce the delivered target dose, ultimately mitigating the risk of related adverse side effects while favoring a faster onset of the drug action. For systemically acting drugs, the pulmonary route may pose an alternative to the oral or parenteral route despite requiring a device to be administered. The pulmonary epithelium displaying an area > 100 m^2 with an epithelial cell layer <1 μm in thickness could address the reduced absorption observed in the gastrointestinal tract [4, 5]. Despite the overall benefits of combining nucleic acid administration with the pulmonary route, these can still be overshadowed by the inherent challenges to this type of molecules, delivery route, delivery platform, and target disease.

Nucleic acids are susceptible to degradation and when administered via the pulmonary route are naturally exposed to its anatomy and physiology yielding different deposition profiles and featuring clearance mechanisms that hinder its effective delivery. For this reason, these must be administered and formulated in a way that allows these molecules to be retrieved on target without its function being compromised and an immunogenic response triggered. The delivery platform, herein defined as the combo between device and formulation, becomes then pivotal to meet such requirements. Nebulizers, pressurized metered dose inhalers (pMDIs), dry powder inhalers (DPIs), and soft mist inhalers (SMIs) are the devices commonly used to attain pulmonary drug delivery [6, 7]. Depending on the device, these can be filled with liquid—nebulizers, pMDIs, and SMIs—or dry powder—DPIs—dosage forms of nucleic acids. Regardless of its dosage solid state, the nucleic acids can be in their naked form while chemically modified or encapsulated in lipid- and/or polymer-based nonviral delivery vectors [1, 8] to circumvent any stress induced upon device actuation, to avoid clearance mechanisms in the lungs, or simply to be directed toward its target. In this chapter, the types of nucleic acids relevant to pulmonary therapy are described alongside their underlying mechanisms of action, the challenges posed by this administration route, and the delivery platform employed. A current clinical outlook pertaining inhaled nucleic acids is also presented.

14.2 Nucleic Acid Types

Nucleic acids are derived from deoxyribonucleic acid (DNA) and ribonucleic acid (RNA). These consist in single- (ss) or double-stranded (ds), short or long chainlike molecules made of building blocks called nucleotides (nt) [1]. Each nt is formed by three components: a nitrogen-containing aromatic base, a five-carbon sugar (pentose), and a phosphate group. Each nucleic acid contains four out of five possible nitrogen-containing bases: adenine (A), guanine (G), cytosine (C), thymine (T), and uracil (U). T, however, is only found in DNA, while U, in RNA. A, C, and G are present in all nucleic acids. DNA and RNA also differ in their pentose; in DNA, this is deoxyribose, while in RNA, this is ribose, displaying a hydroxyl group on the 2′ carbon of the sugar ring. Nucleic acids are negatively charged due to their pentose-phosphate backbone, with sizes ranging from ss chains of 20 nucleotides (~6,6 kDa) to ds chains with more than 1000 base pairs (~ >660 kDa). In this section, the types of therapeutic nucleic acids relevant to pulmonary therapy and respective underlying mechanisms are briefly overviewed. Special focus will be given to ASOs, miRNAs, siRNAs, antagomirs, mRNA, and Apts [2, 3, 9, 10].

14.2.1 Antisense Oligonucleotides (ASOs)

ASOs are ss oligonucleotides ranging from 13 to 25 nt in length (~4.3–8.3 kDa). These hybridize with mRNAs and block the translation of the corresponding proteins by the ribosomes. ASOs can also trigger the catalytic action of ribonuclease H (RNase H) that specifically cleaves the mRNA within the mRNA–ASO duplex freeing the ASO to bind to a new mRNA strand. Both mechanisms occur in the cell cytoplasm. However, ASOs can also act within the cell nucleus by forming triple helixes within the DNA, when a sequence composed entirely of purines is base paired with one composed of pyrimidines. The formation of this triple helix structure by the ASO insertion leads to the cleavage of the DNA, inhibiting gene expression at the transcription level [2, 9].

14.2.2 MicroRNAs (miRNAs)

The miRNAs are non-coding, ss chains of ~22 nucleotides (~7.3 kDa) in length. These are derived from genes in the genome of plants and animals. The miRNAs start as primary transcripts or pri-miRNAs transcribed by RNA polymerase II or III. The pri-miRNA then folds into a stem-loop structure with some unpaired nucleotides preparing the substrate for ribonuclease III, Drosha, to bind. Drosha then associates with the protein DiGeorge syndrome critical region 8 (DGCR8) forming the microprocessor complex that cleaves the pri-miRNA in pre-miRNA. The pre-miRNA is then recognized by Exportin 5, which transports it into the cytoplasm where this is cleaved by the Dicer enzyme, an endonuclease, yielding the miR:miR* duplex, composed of the guide (miR) and passenger (miR*) strands. The miR:miR* duplex will then associate with the Argonaute protein. Despite regarded as not certain, a helicase is believed to remove the passenger strand. The Argonaute protein now solely charged with the miR strand will form the complex miRISC that will target the corresponding mRNA. After hybridizing with the mRNA, the miRISC complex will recruit additional factors including a protein containing multiple glycine-tryptophan repeats called TRNC6 in humans that destabilizes the mRNA, thus preventing translation [2, 8, 9].

14.2.3 Anti-microRNAs (Antagomirs)

As previously described, miRNAs hinder the translation of the mRNA into the respective protein. As such, these are also implicated in the pathogenesis of several human diseases. Anti-micro-RNAs are a novel class of chemically engineered ss oligonucleotides ranging from 17 to 22 nt in length (~5.6–7.3 kDa). These include miRNA sponges, miRNA-masking antisense oligonucleotides, and antisense oligonucleotides targeting miRNAs. The miRNA sponge technology consists in the expression of mRNA molecules displaying multiple binding sites for the target miRNA that will function as a "sponge" to trap the desired miRNAs. That way the endogenous target mRNA will be preserved and able to function normally. This approach is quite popular in vitro than in vivo. The miRNA sponges are specific to a miRNA target and not to a gene of interest, meaning that all genes affected by the target miRNA will be impacted by its knockdown, which might not be always desirable. The miRNA-masking antisense oligonucleotide technology shields the mRNA targeted by the miRNA, preventing this from binding and blocking the expression of the desired protein. Conversely to the miRNA sponge technology, miRNA-masking antisense oligonucleotides are gene-specific; i.e., the oligonucleotides are designed to still allow protein expression while leaving the miRNA intact, free to operate in other important pathways for which its action might be relevant. Finally, antisense oligonucleotides targeting the miRNA are complementary to these and cause miRNA silencing mostly by steric blocking [11, 12].

14.2.4 Messenger RNAs (mRNAs)

The mRNAs are ss chains that can range from ~300 to 4000 nucleotides (~ 99 kDa–1320 kDa). These are translated into proteins by the ribosomes present in the cytosol of the cells. Translation, in protein synthesis, mainly encompasses three stages: initiation, elongation, and termination. In the initiation, the ribosomal subunits bind to the mRNA. Then elongation takes place and the ribosome moves along the mRNA molecule linking amino acids and forming a polypeptide chain. This is followed by termination where the ribosome reaches a stop codon which terminates protein synthesis, and the ribosome is released. The newly formed polypeptide chain undergoes several modifications before becoming a fully functioning protein [2, 9, 13].

14.2.5 Aptamers

Aptamers are ss DNA or RNA molecules of 20–60 nt (6.6–19.8 kDa) that bind with high affinity and specificity to protein targets modulating its therapeutic activity. Its underlying mechanism is target dependent [9, 14, 15].

14.2.6 Short Interfering RNAs (siRNAs)

The siRNAs are non-coding, ds chains of 20–25 nt (~6.6–8.3 kDa) in length, operating in the cytoplasm. These can derive from different sources; endogenous duplex RNA can stem from the normal transcription of genomic loci that have extensive hairpin structures or from the annealing of sense and antisense RNAs that have been transcribed from a given locus; exogenous sources of dsRNA include viral RNAs and duplex structures that have been synthetically introduced into the cells for experimental purposes.

When in the intracellular compartment, dsRNAs is recognized by the cell as foreign, triggering the Dicer enzyme to cleave the dsRNAs into shorter fragments. Following the cleavage, the Dicer complexes with the short dsRNA fragment and associates with another set of proteins (Argonauts) yielding the RNA induced silencing complex (RISC). RISC will then split the short dsRNA fragment into two single strands, the sense and antisense strands, discarding the sense strand and holding on to the antisense strand of the siRNA (sorting). RISC will then help the antisense strand to recognize the complementary sequence of the mRNA. Once the antisense strand hybridizes with the mRNA, this is cleaved and no translation into the respective protein can take place [2, 16]. Despite nucleic acid-based therapeutics' unlimited potential, this can be hindered by the delivery strategy employed. This is intimately linked to the administration route and the delivery platform.

14.3 The Pulmonary Route

14.3.1 Anatomy and Physiology of the Lungs

The anatomy and physiology of airways are crucial to determine the dynamics of inhalable drug delivery systems within the lung while providing insights on possible formulation approaches.

The lung airways are commonly grouped into two main regions [17, 18]: the conducting airway and the respiratory airway. Their structure and layout in the lung are widely described by the Weibel-A mathematical model [18]. This is a symmetric tree model that divides the lung into 24 stages (0–23), each corresponding to an airway generation (or level of bifurcation) that branches symmetrically into two similar smaller branches. Such stages allow for the prediction and profiling of aerosol deposition along the respiratory tract based on luminal diameter, length, and angle of the bronchi [18]. The physical characteristics of the respiratory airway are different when compared to the conducting airway given the distinct role of this region: gas exchange. The surface area is approximately 100 m^2, allowing for much greater contact with the inspired gas or therapeutic aerosol, conversely to the conducting airway displaying 2–3 m^2 [19].

Both conducting and respiratory airways epithelium are covered by a mucus gel composed of water (95%), glycoproteins, inorganic salts, lipids, proteins, and DNA [20]. The precise composition is not entirely known but might change according to patient condition. Aerosol particles deposit in this mucus and dissolve to release the active drug for subsequent absorption and/or pharmacological action. Nevertheless, the volume of fluid in the lung for particles to dissolve is limited (10–30 mL in humans), and it is hard to envisage the volume of fluid that an inhaled aerosol particle is exposed to, after deposition. Moreover, the thickness of the lung lining fluid layer differs between central and peripheral lungs, as the airways gradually become narrower in diameter and the lung lining layer becomes thinner until reaching the alveoli [21].

However, the alveoli are lined with a pulmonary surfactant, a lipoprotein-based complex consisting of 90% lipid and 10% protein, which is synthesized, secreted, and recycled in these structures. This pulmonary surfactant plays a dual role of reducing surface tension at the air–liquid interface, preventing the alveoli from collapsing, while defending the host from inhaled pathogens and particles, enabling their movement to the upper airways. Due to these peculiar properties, lung surfactant also plays an important role in drug delivery and absorption by extending their residence time within the lung [17]. In the case of large molecules, such as biologics, the lung surfactant can cause them to aggregate, becoming insoluble, which in turn potentiates their elimination or the trigger of immunogenic events. However, the pulmonary surfactant has also been reported in the literature as an absorption enhancer for lower molecular weight biologics such as peptides [21]. A few commercially available lung surfactants to treat respiratory distress syndrome in newborns are Infasurf®, Survanta®, Surfactin®, Curosurf®, and Newfantan® [22].

14.3.2 Particle Deposition Mechanisms

Besides lung's anatomy and physiology, particle deposition underlying mechanisms are also relevant while designing a drug delivery system. The specific pattern of deposition determines the local doses in the lung and the subsequent redistribution and clearance of deposited particles [23]. The mechanisms involved in particle deposition are ruled by a combination of different factors such as particle size, shape, breathing rate, inhalation volume, and the health condition of the patient. These include inertial impaction, gravitational sedimentation, Brownian diffusion, interception, and electrostatic precipitation.

14.3.2.1 Inertial Impaction

Inertial impaction takes place in the conducting airways, when the particle momentum is too large for it to change direction in an area where there is a rapid change in the direction of the bulk airflow. Impaction increases with air velocity, particle size, and particle density.

14.3.2.2 Gravitational Sedimentation and Brownian Diffusion

Both gravitational sedimentation and Brownian diffusion take place in the respiratory airways. In gravitational sedimentation, the particles deposit due to the higher gravitational force to which these are exposed when compared to the airflow-induced dragging force, which is lower. The rate of sedimentation deposition increases with particle size and density while decreasing with flow rate. In Brownian diffusion, the particles move randomly across the streamline due to collisions with gas molecules, depositing upon contact with the airway wall. Diffusion increases with decreasing particle size and airflow rate. However, most of these particles are exhaled by the expiratory airflow.

14.3.2.3 Interception

Interception can occur in the conducting and respiratory airways. The particles deposited by interception usually present noncircular, elongated shapes (fibers) that result in immediate deposition as soon as these contact the airway wall. Interception increases as the airway diameter becomes smaller.

14.3.2.4 Electrostatic Precipitation

Like interception, electrostatic precipitation can occur in the conducting and respiratory airways. Herein, deposition results from charged particles inducing opposite sign charges onto the surfaces of the airways. Particles electrostatically attached to pulmonary airways.

14.3.3 Particle Clearance Mechanisms

Following particle deposition, nucleic acids may be subject to in vivo mechanisms such as mucociliary clearance, clearance by macrophages, and nuclease degradation (extra- and intracellularly) [24]. In mucociliary clearance, ciliated cells drive the particles trapped in the mucus—through coordinated ciliary beat—from the lower airways to the pharynx, where they can be either swallowed and degraded in the gastrointestinal tract or ejected. Macrophages migrate to the particles and phagocytize them via chemotaxis involving opsonization. Once particles are internalized in the macrophages, these are disintegrated, for instance, by enzymes in lysosomes. Another clearance mechanism in the lungs affecting nucleic acid delivery is the nuclease-induced degradation. Nucleases cleave the phosphodiester bond between nucleotides in nucleic acids at the end of their strand (exonucleases) or at the middle of their sequence (endonucleases). Despite its action being reported as negligible when compared with the parenteral route [2], nucleases have been long considered as promising cancer biomarkers where an overexpression of these is observed in the presence of cancer [25]. This could eventually hinder any nucleic acid-based treatment targeting this condition.

14.4 Delivery Platforms

Herein the author refers to delivery platform as the combo of the device with the formulation. Both liquid- and solid-based delivery platforms are available, conditioning device actuation underlying mechanism and formulation strategy.

14.4.1 Devices

Pulmonary delivery devices can be grouped into four main categories: nebulizers, pMDIs, DPIs and SMIs. None of these devices can be considered clinically superior when compared against each other as its selection is dependent upon different factors such as patient preference, target condition, therapeutic dose, deposition profile, and formulation properties (active pharmaceutical ingredient—API, excipient), among others [6].

14.4.1.1 Nebulizers

Nebulizers convert a liquid formulation into an aerosol through compressed gas (jet nebulizers), ultrasonic (ultrasonic nebulizers), or vibrating mesh technology (mesh nebulizers) [17, 18]. Jet nebulizers disperse liquid formulations into droplets by resorting to a compressed gas stream that is forced against a small orifice, yielding a pressure drop and velocity increase, ultimately leading to the uptake of liquid through a capillary from the reservoir (Bernoulli principle). The resulting aerosol impacts on the device baffles and is subsequently fragmented into smaller droplets [6]. The larger droplets fall back into the same reservoir to undergo nebulization once again, upon following actuations. Jet nebulizers generate a continuous flow of aerosol, which results in substantial losses into the surrounding air, during the expiratory phase of the patient [17].

The second class of nebulizers is the ultrasonic nebulizer. The underlying principle consists of the vibration of a piezoelectric crystal (transducer) driven by an alternating electrical field [18], allowing for higher aerosol volumes to be nebulized in a shorter time, despite presenting a higher particle size than other nebulization systems. Ultrasonic nebulizers are particularly inefficient with suspensions, viscous solutions, or solutions with a high surface tension [17]. Also, the heating of the liquid phase during ultrasonic nebulization can lead to denaturation of thermolabile drugs, especially if this is a biologic [26]. Usually, these are not used as a first-line means of administration, and there is a tendency to move from these to the vibrating mesh technology. Its underlying mechanism consists of a mesh or a plate with multiple apertures to produce an aerosol via a piezoelectric element which either vibrates a transducer horn or is annular and encircles the mesh causing it to vibrate. Regardless of the case, the liquid is pumped across the perforated mesh producing homogeneous particles [18]. The aerosol particle size and flow are governed by the diameter of the aperture holes. The nebulization time ranges from 1 to 6 min, the aerosol velocity 1 m/s, and the residual volume 0.1–0.3 mL. Furthermore, these nebulizers are portable, handheld, and noise-free [17, 18].

14.4.1.2 Pressured Metered Dose Inhalers (pMDIs)

The pMDIs comprise a canister containing a liquid-based formulation, a metering valve, and a spray actuator. A pMDI formulation consists of a propellant-based solution or suspension containing the drug and/or additives such as a surfactant or lubricant. The propellants currently used are hydrofluoroalkanes (HFAs) given the need to replace the chlorofluoroalkanes (CFCs), because of their known detrimental effect on the ozone layer [6]. However, due to the greenhouse effect, HFAs' use in the formulation has also been gradually discouraged. To address the environmental issues posed by commonly used HFAs (HFA-134a and HFA-227ea), Koura limited has identified Zephex® 152a/HFA-152a as a potential alternative to pMDIs. The US Environmental Protection Agency (EPA) has designated HFA-152a as an acceptable replacement for fluorocarbons with high ozone depleting potential and/or global warming potential in certain applications, including as an aerosol propellant [27]. Optimal lung delivery with pMDIs entails coordination between the actuation of the device and inhalation. For correct device handling, the patient should completely exhale before placing the mouthpiece of the device between the lips, and upon the start of a slow inhalation, the canister should be pressed downward. Following actuation, it is recommended that the patient holds her or his breath for 10 s to help particle sedimentation in the peripheral airways. Actuation before or at the end of inspiration, actuation in the mouth but inspiration through the nose, and inhalation stopped by the actuation constitute some of the frequent handling errors that take place while using pMDIs [6, 17, 18]. Additionally, since the aerosol is emitted with high velocity, this causes a marked deposition of 30–65% of the dose in the upper respiratory tract, decreasing in turn the dose delivered deep into the lung [17]. This can be overcome by coupling the pMDIs with a spacer or a valved holding chamber. Spacers and valved holding chambers increase the delay time between actuation and inhalation, allowing more time for propellant evaporation, and particle deceleration, resulting in an increase of pulmonary deposition.

14.4.1.3 Dry Powder Inhalers (DPIs)

DPIs deliver solid-based formulations through the patient inspiratory flow and the turbulence induced inside the device. This turbulence is determined by the resistance against the inhalation flow required to produce a pressure drop of 4 kPa through the device, a value envisaged by pharmacopeias for the evaluation of emitted dose [28]. DPIs can be classified as low-, medium-, and high-resistance devices. Breezehaler® by Novartis, Spinhaler® by Aventis, and Rotahaler by GSK/Cipla are examples of low-resistance DPIs; Novolizer® by MEDA, Revolizer® by Cipla, and Diskus® by GSK of medium-resistance DPIs; and Turbuhaler® by AstraZeneca and Handihaler® by Boehringer Ingelheim of high-resistance DPIs (high-resistance). Through low-resistance DPIs, the disaggregation and dispersion of the dry powder formulation mainly depend on the patient's inhalation airflow rate, as the role of the turbulence induced by resistance is negligible in these cases. The low-resistance DPIs require a higher inspiratory airflow rate, which often cannot be attained by patients suffering from a disease-induced breathing limitation. As such, an increased variability on the delivered dose of these DPIs is often observed.

The flow rate dependency can be reduced with a minimum regimen of turbulence as observed in medium-resistance DPIs. The disaggregation and dispersion of the dry powder are optimal in these circumstances even in the absence of a maximal inspiratory flow rate. High-resistance DPIs, on the other hand, even if working on a lower inspiratory flow rate, demonstrated to positively affect particle generation and dispersion of the drug powder [29]. Conversely to the previously described delivery systems, DPIs have no need for propellants and the solid-based nature of the formulations it carries renders it less prone to undergo liquid-mediated reactions such as aggregation and denaturation in the long run. Additionally, from a supply chain standpoint, no cold transport and storage are required resulting in a lower burden, both resource and cost-wise, an advantage if the API is a biologic [6].

Developing a solid-based dosage form entails however an additional drying and particle engineering step that can be attained employing freeze drying followed by a milling-based technology, spray drying, or most recently spray freeze drying [30]. This hinders the formulation exercise as a combination of excipients simultaneously showcasing inhalation precedence and API stabilization potential, whether this is a biologic or a small molecule; however, particle engineering technology compatibility must be determined. Nevertheless, the increasing availability of high-throughput analytical tools, such as differential scanning fluorimetry, to support formulation screening methodologies is allowing this challenge to be addressed in reduced time with μg range of sample [4].

14.4.1.4 Soft Mist Inhalers (SMIs)

SMIs deliver a propellant-free liquid metered dose of drug solution through an impingement nozzle system, where the liquid is forced against a protruding pin that breaks the fluid apart in a low-velocity fine mist. The gentler aerosolization could be beneficial for shear-stress sensitive molecules, namely biologics. Additionally, this also allows for a higher deposition profile when compared with nebulizers, pMDIs, and DPIs, with fine particle fractions between 65% and 80% [31]. Furthermore, from a patient standpoint, Richard P. et al. [32] conducted a survey concluding that patients seem to accept well the use of an SMI-based treatment when compared to other alternatives. However, as in nebulizers and multi-dose pMDIs, a preservative compatible with the API must be added to the SMI formulation and aseptic manufacturing is required. This means increased complexity and higher-grade facilities that render the development more expensive. Respimat® by Boehringer Ingelheim's was the first SMI approved and made available in the market in 2004.

14.4.2 Formulation

Nucleic acid-based therapeutics must endure their respective manufacturing process and device actuation and, following administration, evade mucociliary clearance, clearance by macrophages, and nuclease degradation. In addition, their high molecular weight and strong negative charge hinder their passive diffusion across the cellular membrane. Reproducible and low toxicity formulation strategies, including chemical modification and/or encapsulation in drug delivery vectors, can be employed to attain successful delivery of these molecules [1].

14.4.2.1 Intrinsic Stabilizing Strategies

Base and sugar-phosphate backbone modifications and conjugation are chemical modifications that can be implemented to intrinsically stabilize nucleic acid-based drugs [2].

Base Modifications

Base modifications consist of chemical alterations in the nucleoside bases improving stability of the nucleotides without compromising their binding efficacy. Some of the most prevalent modifications include:

- Replacement of uridine by pseudouridine, N1-methylpseudouridine, 5-methyluridine (m5U), 5-methoxyuridine (5 moU), or 2 thiouridine (s2U).
- Replacement of adenosine by N1-methyl adenosine or N6 methyl adenosine.
- Replacement of cytidine for N5-methylcytidine.

Vaidyanathan et al. demonstrated that depleting uridine from Cas9 mRNA while modifying it with 5-methoxyuridine or pseudouridine reduces immunogenicity and improves translation of mRNA encoding for Cas9 [33]. N1-methylpseudouridine (m1Ψ), 5-methyluridine (m5U), 5-methoxyuridine (5moU), and 2-thiouridine (s2U) are other uracil base modifications having shown efficacy [1]. Karikó et al. also demonstrated that 5′-methyl modification of cytidine (5mC) improved mRNA translation efficiency and reduced immunogenicity both in vitro and in vivo [34].

Sugar-Phosphate Backbone Modifications

Sugar-phosphate backbone modifications usually refer to their chemistry extension to enable the increase of nuclease resistance and binding affinity of the nucleic acid-based therapeutics. For instance, ASOs integrating a phosphorothioate (PS) backbone to favor sequence stability have been clinically validated for several applications [1, 35, 36]. Methyl phosphonates, phosphoramidates, peptide nucleic acids (PNA), and boranophosphates are other strategies of backbone linkages [1].

The 2′-position of the ribose sugar is another modification site where substitutions like 2'-O-methylation (2OMe), 2'-O-methoxyethyl (2MOE), and 2'-Fluro (2F) take place. Locked nucleic acids confine the conformational space of the furanose sugar to avert backbone hydrolysis, thus enhancing stability. Phosphorodiamidate morpholino oligonucleotides (PMO) elude nuclease activity, maintaining a high solubility profile. These cannot trigger RNAse degradation of the target sequence, exerting steric blocking of translation or spliceosomal factors. These backbone modifications are largely traced in oligonucleotide or siRNA sequences [1].

Chemical Conjugation

Chemical conjugation to the backbone of nucleic acid-based biologics can improve their stability and/or delivery. Sugars, cholesterol, tocopherol, vitamin E, and fatty acids are commonly used conjugation moieties that can be added to the 3′ or 5′ end of the nucleic acid backbone. For instance, conjugation of cholesterol enables an improved cellular import and intracellular uptake through lipoprotein-mediated pathways. siRNAs conjugated with N-acetylgalactosamine, a high-affinity ligand for the hepatocyte-specific asialoglycoprotein, are currently under clinical trials. Antibodies or aptamers could also be conjugated directly with nucleic acids for targeted delivery to specific tissues or cell types [1].

14.4.2.2 Delivery Vectors

Encapsulation of nucleic acids in viral or nonviral vectors [7] also constitutes a formulation strategy to prevent nuclease degradation and enhance effective target delivery. Viruses have naturally evolved to encapsulate their DNA or RNA genome and successfully deliver nucleic acids to the cytosol or nuclei of host cells [9]. Due to their higher transfection efficiency, five major classes of viral vectors, namely, adenovirus, adeno-associated virus, lentivirus, retrovirus, and herpes simplex virus [7, 9, 15, 16, 37], have been extensively explored in nucleic acid therapy. However, the limitation of payload, inherent immunogenicity, and the difficulty of large-scale production limited their clinical application. These shortcomings have led to the development of nonviral vectors for nucleic acid delivery allowing for increased payload, low immunogenicity, and toxicity while being easier to produce over their viral counterparts. Pulmonary delivery of nucleic acids has been mainly explored using lipid- or polymer-based nonviral vectors [2].

Lipid-Based Nonviral Vectors

Lipid-based nonviral vectors may include different structures, but usually spherical platforms comprise at least one lipid bilayer surrounding one or more internal aqueous compartment. Their low toxicity, amenability to synthetic chemistry yielding increased specificity, and ease of scale-up render them suitable for the pulmonary delivery of nucleic acids. In addition, its hydrophobic nature conferred by their structure (hydrophobic head, hydrophilic tail) allows them to self-assemble while improving the payload integrity for an extended period of time. Cationic lipids (DOTMA, DOTAP), ionizable cationic lipids (DODMA, DLinDMA, DLinMC3DMA), helper lipids (cholesterol, phosphatidylcholines), rigid lipids, and PEG-lipids can be used alone or in combination with each other to develop lipid-based nonviral vectors for nucleic acid delivery in the lungs like liposomes and solid lipid nanoparticles [8, 9, 38].

Liposomes

Liposomes are closed phospholipid vesicles with one or more concentric bilayers. These form spontaneously in aqueous media while allowing for hydrophilic, hydrophobic, and lipophilic drug encapsulation. Its surface may be functionalized with peptides, glycans, and antibodies to actively target its cellular delivery. Liposomes have also been reported to evade macrophage uptake via PEGylation, fine-tuning the metabolic fate of the cargo when delivered in vivo. These can also be primed to release the nucleic acid cargo with inclusions reporting pH sensitivity, photosensitivity, or temperature sensitivity [10, 39, 40].

Solid Lipid Nanoparticles (SLN)

SLN are crystallized lipid structures with a solid lipid core and an amphiphilic exterior shell. Depending on its lipid composition, the SLN may showcase different morphologies, from a platelet shape, if its composition is triglyceride-based, to a spherical shape if this is monostearate-based [8].

Polymer-Based Nonviral Vectors

Polymer-based nonviral vectors can be manufactured from natural or synthetic materials, as well as monomers or preformed polymers. Hydrophobic and hydrophilic cargos with different molecular weights can be encapsulated within the vector core or the polymer matrix, chemically conjugated to the polymer or bound to its surface. Composition, stability, responsivity, and surface charge are some of the properties that can be modulated to control loading efficacies and release kinetics of the cargo. The most common forms of polymer-based nonviral vectors are polymersomes, micelles, and dendrimers [9].

Polymersomes are artificial vesicles, with membranes made using amphiphilic block copolymers (polyethylene glycol, PEG; polydimethylsiloxane, PDMS). Polymeric micelles are self-assembled nanospheres displaying an hydrophilic core and an hydrophobic coating. These have been used for the delivery of cancer aqueous-based therapeutics in clinical trials [41]. Dendrimers are hyperbranched polymers with complex three-dimensional structures for which the mass, size, shape, and surface chemistry can be adjusted according to the therapeutic target. Charged polymers such as polyethylenimine (PEI) and polyamidoamine (PAMAM) are frequently associated with this polymer-based nonviral vector [7, 8, 13, 42]. Dendrimers can carry numerous types of cargo but are quite popular for the delivery of nucleic acids. Their surface can also be functionalized to enable conjugation of other molecules and contrast agents.

14.5 Current Outlook

Despite the potential of nucleic acids in pulmonary delivery, their presence in the clinical pipeline still looks discrete when compared to other administration routes. Mostly naked ASOs targeting the local treatment of respiratory diseases such as asthma, chronic obstructive pulmonary disease (COPD), and idiopathic pulmonary fibrosis are under clinical trials (Table 14.1).

AIR645 is a naked ASO developed by Isis Pharmaceuticals (USA) and licensed to Altair Therapeutics (USA) that acts as a dual inhibitor of the pro-inflammatory cytokines, IL-4 and IL-13, implicated in the pathogenesis of asthma, allergic rhinitis, and other inflammatory disorders. AIR645 targets the mRNA encoding for (IL-4Rα), the signaling chain shared by the IL-4 and IL-13 receptors required for cellular responses against IL-4 and IL-13 [1, 44]

The randomized, placebo-controlled trial (NCT00941577) evaluated the safety, tolerability, bioavailability, and pharmacodynamic activity of nebulized AIR645 at multiple dose levels in 32 healthy adult subjects (0.3, 3, 10, and 20 mg/dose) and 8 mild asthma subjects (20 mg/dose). Subjects were randomized (six active: two placebo) and took six doses on study days 1, 3, 5, 8, 15, and 22. AIR645 was safe and well tolerated with no dose-limiting toxicities or safety issues detected. Adverse effects were reported as mild and no subjects were discontinued. AIR645 presence in sputum was calculated to be approximately 5 days while being dose dependent and > 1000-fold higher than in plasma, suggesting low systemic bioavailability of the drug. Following repeated inhalation of AIR645, evidence of anti-inflammatory biomarker activity was observed in subjects with mild asthma [43–45].

TPI ASM8 is an inhaled modified ASO developed by Pharmaxis (Australia) for the treatment of severe asthma administered using a vibrating mesh nebulizer. ASM8 uses proprietary RNA-silencing technology entailing two modified oligonucleotides developed to inhibit the synthesis of IL-5, GM-CSF, IL-3, and CCR3, key receptors responsible for the activation, recruitment, and survival of eosinophils. The lead compound ASM8 is in Phase 2 development. The clinical safety and efficacy of ASM8 has been established in two Phase 1 and three Phase 2 trials (NCT00550797, NCT00822861, and NCT01158898). ASM8 has been administered as a nebulized aqueous solution at an effective dose of 3–8 mg, once a day. Systemic exposure following inhalation is less than 1%. Pharmaxis envisages that ASM8 will be developed for moderate and severe asthma or for patients who are refractory to steroids or who cannot tolerate high dose steroids. ASM8 could be also advantageous over other biologic-based drug candidates such as monoclonal antibodies, given the expected greater efficacy through multi-targeting, better tolerability, and the convenience of an inhaled system delivered once daily [40, 45, 46].

Table 14.1 Nucleic acid respiratory products under clinical development [43, 44]

Product name	NCT number	Status	Indication	Type of nucleic acid	Device	Delivery vector	Company
Phase 2							
AIR645	00941577	Completed	Asthma	ASO	Nebulizer	Naked, modified	Altair Therapeutics
TPI ASM8	00550797 00822861 01158898	Completed	Allergen-induced Asthma	ASO	Vibrating mesh nebulizer	Naked, modified	Pharmaxis
Excellair	Not available	Discontinued	Asthma	siRNA	Non-disclosed	Naked	ZaBeCor Pharmaceuticals
SB010	01743768	Completed	Asthma	ASO	Non-disclosed	Naked, modified	Sterna Biologicals GmbH
Phase 1							
TAKC -02	05018533	Completed	Severe asthma	ASO	Non-disclosed	Naked, modified	TAK-Circulator Co.
TRK-250	03727802	Completed	Idiopathic pulmonary fibrosis	Non-disclosed	Non-disclosed	Non-disclosed	Toray Industries, Inc
TPI 1100	00914433	Withdrawn	COPD	ASO	Non-disclosed	Naked, modified	Pharmaxis
ARO-MUC5AC	NCT05292950	Recruiting	Muco-obstructive	iRNA	Nebulizer	Non-discloed	Arrowhead Pharmaceuticals
ARO-RAGE	NCT05276570	Recruiting	Inflammatory	iRNA	Nebulizer	Non-disclosed	Arrowhead Pharmaceuticals
ARO-MMP7	NCT05537025	Recruiting	Idiopathic Pulmonary Fibrosis	iRNA	Nebulizer	Non-disclosed	Arrowhead Pharmaceuticals

Excellair was developed by Zabecor Pharmaceuticals (USA) and consists of a siRNA-based drug which targets the spleen tyrosine kinase (Syk), a gene implicated in asthma. By targeting Syk kinase, Excellair™ inhibits an initial signaling step of inflammation, thus preventing the release of several inflammatory mediators. Some of the current treatments for asthma and other inflammatory conditions inhibit only one of the mediators of inflammation. In Phase I study, patients with asthma received 21 consecutive daily doses and 100% tolerated those well, with no adverse effects described as serious. Furthermore, 75% of patients with asthma who received Excellair reported enhanced ability to breathe easily or a reduced need to use their rescue inhaler. No patients on placebo reported improvement. Although Excellair Phase I results indicated breathing improvements in patients, the phase II trials were discontinued [40, 45].

SB010, developed by Sterna Biologicals GmbH (Germany), inhibits the transcription factor GATA-3 by an inhaled synthetic DNA molecule with enzymatic action – DNAzyme. SB010 binds to the transcription factor triggering the inflammatory response, inactivating it by enzymatic cleavage [40, 45]. In the randomized, double-blind, parallel-group, multicenter trial (NCT01743768), the efficacy of SB010 agent was tested over a period of 28 days. This was proved to be safe and well tolerated while improving lung function after specific allergen activity when compared to placebo [40, 45].

TAKC-02, endeavored by TAK-Circulator Co (Japan), is an antisense oligonucleotide that inhibits the MEX3B synthesis. MEX3B is a RNA-binding protein conserved in many animal species deeply involved in the expression of various cytokines associated with the onset and/or exacerbation of inflammatory to metabolic diseases and malignant tumors. A double-blind, randomized, placebo-controlled Phase I study was conducted (NCT05018533) to assess the safety and tolerability of single and multiple inhaled doses of TAKC-02 in healthy male subjects [40, 45].

TRK-250, developed by Toray Industries, Inc. (Japan), is a nucleic acid that inhibits the progression of pulmonary fibrosis by selectively suppressing the expression of transforming growth factor-beta 1 (TGF-β1) protein, at the gene expression level. TRK-250 was studied in a double-blind, randomized, placebo-controlled Phase I study (NCT03727802). The primary objective of the study was to assess the safety and tolerability of single and multiple inhaled doses of TRK-250 in subjects with idiopathic pulmonary fibrosis [40, 45].

TPI 1100, also by Pharmaxis, is an inhaled, modified ASO. Its FANA(TM)'s key chemistry modification technology replaces the naturally occurring D-ribose sugar moiety by a D-arabinose sugar containing fluorine. This change improves RNA-targeting properties and allows the generation of RNA-targeting agents showcasing increased resistance to nuclease breakdown, increased activity, and duration of action. TPI 1100 acts by inhibiting several forms of the enzyme phosphodiesterase (PDE), known to promote inflammation in the lung in COPD. Traditional PDE inhibitors inhibit only the PDE 4 form of the enzyme [40, 45]. Also, this resulted in a favorable safety profile during preclinical studies with activity at lower doses (NCT00914433).

ARO-MUC5AC, ARO-RAGE and ARO-MMP7 by Arrowhead Pharmaceuticals (USA) are currently undergoing Phase I clinical trials (Recruting) to treat Muco-obstructive (NCT05292950), Inflammatory (NCT05276570) and Idiopathic Pulmonary Fibrosis (NCT05537025), respectively. ARO-MUC5AC targets the reduction of mucin 5AC (MUC5AC) production associated with mucoobtructive pulmonary diseases, ARO-RAGE, the reduction of the Receptor for Advanced Glycation End products (RAGE) in inflammatory lung diseases and ARO-MMP7, inhibits the expression of matrix metalloproteinase 7 (MMP7) in idiopathic pulmonary fibrosis (IPF) [43, 44].

14.6 Concluding Remarks

By acting on the transcription and translational levels, nucleic acid therapy is impacting the way diseases can be targeted, widening the range of possible treatments. When combined with the pulmonary route, its potential can be enhanced given the possibility to treat locally and systemically in the absence of significant nuclease activity. This could ultimately allow for a simple formulation strategy and lower therapeutic dose. Despite the promising combo of nucleic acid therapy and the pulmonary route, the fine-tuning of both through the delivery platform (device + formulation) designed to the target disease still needs to be analyzed in depth. SMIs and DPIs could be suitable for nucleic acid delivery given patient preference, increased deposition profile, and formulation stability (the latter in the case of the DPI) over other treatment alternatives. Several formulation strategies have been explored and these are case dependent, but the possibility to chemically modify or conjugate nucleic acids to increase its resistance against nuclease activity (lower in the lungs) could render the formulation exercise and regulatory approval simpler as there would not always be a need to internalize them in viral or nonviral vectors. If, on the contrary, delivery vectors would be required to direct the nucleic acid cargo to its target, for instance, then lipid-based nonviral vectors could be a good approach from a low toxicity risk standpoint when compared to polymer-based nonviral vectors.

References

1. Gupta A, Andresen JL, Manan RS, Langer R. Nucleic acid delivery for therapeutic applications. Adv Drug Deliv Rev. 2021;113834. https://doi.org/10.1016/j.addr.2021.113834.
2. Chen J, Tang Y, Liu Y, Dou Y. Nucleic acid-based therapeutics for pulmonary diseases. AAPS PharmSciTech. 2018;19:3670–80. https://doi.org/10.1208/s12249-018-1183-0.
3. Fernandes DA, Leandro P, Costa E, Corvo ML. Dry powder inhaler formulation of Cu, Zn-superoxide dismutase by spray drying: a proof-of-concept. Powder Technol. 2021;389:131–7. https://doi.org/10.1016/j.powtec.2021.05.008.
4. Fernandes DA, Costa E, Leandro P, Corvo ML. Formulation of spray dried enzymes for dry powder inhalers : an integrated methodology. Int J Pharm. 2022;615(January):121492. https://doi.org/10.1016/j.ijpharm.2022.121492.

5. Stein SW, Thiel CG. The history of therapeutic aerosols : a chronological review. 2016;29:1–22. https://doi.org/10.1089/jamp.2016.1297.
6. Berkenfeld K, Lamprecht A, McConville JT. Devices for dry powder drug delivery to the lung. AAPS PharmSciTech. 2015;16(3):479–90. https://doi.org/10.1208/s12249-015-0317-x.
7. Jiao Y, Xia ZL, Ze LJ, Jing H, Xin B, and Fu S. Research Progress of nucleic acid delivery vectors for gene therapy. Biomedical Microdevices. 2020;22:1–10. https://doi.org/10.1007/s10544-020-0469-7.
8. Kanvinde S, Kulkarni T, Deodhar S, Bhattacharya D, and Dasgupta A. Non-viral vectors for delivery of nucleic acid therapies for cancer. BioTech. 2022;11.1:6. https://doi.org/10.3390/biotech11010006.
9. Kulkarni JA, et al. The current landscape of nucleic acid therapeutics. Nat Nanotechnol. 2021;16(6):630–43. https://doi.org/10.1038/s41565-021-00898-0.
10. Fattal E, Fay F. Nanomedicine-based delivery strategies for nucleic acid gene inhibitors in inflammatory diseases. Adv Drug Deliv Rev. 2021; https://doi.org/10.1016/j.addr.2021.05.019.
11. Murdaca G et al. Effects of AntagomiRs on different lung diseases in human, cellular, and animal models. International Journal of Molecular Sciences. 2019.20.16:3938. https://doi.org/10.3390/ijms20163938.
12. Lima JF, Cerqueira L, Oliveira C. Anti-miRNA oligonucleotides: a comprehensive guide for design. RNA biology. 2018.15.3: 338–52. https://doi.org/10.1080/15476286.2018.1445959.
13. Di Gioia S, et al. Nanocomplexes for gene therapy of respiratory diseases: targeting and overcoming the mucus barrier. Pulm Pharmacol Ther. 2015;34:8–24. https://doi.org/10.1016/j.pupt.2015.07.003.
14. Keefe AD, Pai S, Ellington A. Aptamers as therapeutics. Nature reviews Drug discovery. 2010;9.7:537–50. https://doi.org/10.1038/nrd3141.
15. Tang J, Cai L, Xu C, Sun S, Liu Y, Rosenecker J. and Guan S. Nanotechnologies in delivery of DNA and mRNA vaccines to the nasal and pulmonary mucosa. Nanomaterials. 2022; 12.2:226. https://doi.org/10.3390/nano12020226.
16. Nikam RR, Gore KR. Journey of siRNA: Clinical developments and targeted delivery. Nucleic acid therapeutics. 2018. 28.4:209–24. https://doi.org/10.1089/nat.2017.0715.
17. Smyth HDC and Hickey AJ. (ed.). Controlled pulmonary drug delivery. Springer Science & Business Media. 2011. https://doi.org/10.1007/978-1-4419-9745-6.
18. Nokhodchi, A, Martin, GP. (ed.). Pulmonary drug delivery advances and challenges. 2015. ISBN 978-1-118-79954-3.
19. Labiris NR, Dolovich MB. Pulmonary drug delivery. Part I: physiological factors affecting therapeutic effectiveness of aerosolized medications. Br J Clin Pharmacol. 2003;56(6):588–99. https://doi.org/10.1046/j.1365-2125.2003.01892.x.
20. Huck BC, et al. Models using native tracheobronchial mucus in the context of pulmonary drug delivery research: composition, structure and barrier properties. Adv Drug Deliv Rev. 2022;183:114141. https://doi.org/10.1016/j.addr.2022.114141.
21. Liang W, Pan HW, Vllasaliu D, Lam JKW. Pulmonary delivery of biological drugs. Pharmaceutics. 2020;12(11):1–28. https://doi.org/10.3390/pharmaceutics12111025.
22. Mehta PP, Dhapte-Pawar V. Role of surfactants in pulmonary drug delivery. In Green Sustainable Process for Chemical and Environmental Engineering and Science. Academic Press. 2022; pp. 559–577. https://doi.org/10.1016/B978-0-323-85146-6.00029-2.
23. Darquenne C. Deposition mechanisms. Journal of aerosol medicine and pulmonary drug delivery. 2020;33.4:181–5. https://doi.org/10.1089/jamp.2020.29029.cd.
24. Geiser M. Update on macrophage clearance of inhaled micro- and nanoparticles. Journal of aerosol medicine and pulmonary drug delivery. 2010;23.4:207–17. https://doi.org/10.1089/jamp.2009.0797.
25. Balian A, Hernandez FJ. Nucleases as molecular targets for cancer diagnosis. Biomarker Research. 2021;9.1:1–16. https://doi.org/10.1186/s40364-021-00342-4.
26. Depreter F, Pilcer G, Amighi K. Inhaled proteins: challenges and perspectives. Int J Pharm. 2013;447(1–2):251–80. https://doi.org/10.1016/j.ijpharm.2013.02.031.
27. "Zephex-152a." https://www.zephex.com/zephex-152a/, accessed in September 2022.

28. Islam N, Gladki E. Dry powder inhalers (DPIs)-a review of device reliability and innovation. Int J Pharm. 2008;360(1–2):1–11. https://doi.org/10.1016/j.ijpharm.2008.04.044.
29. Dal Negro RW. Dry powder inhalers and the right things to remember: a concept review. Multidiscip Respir Med. 2015;10:1. BioMed Central Ltd. https://doi.org/10.1186/s40248-015-0012-5.
30. Shoyele SA, Cawthorne S. Particle engineering techniques for inhaled biopharmaceuticals. Adv Drug Deliv Rev. 2006;58(9–10):1009–29. https://doi.org/10.1016/j.addr.2006.07.010.
31. DalbyR V, Spallek M. A review of the development of Respimat® Soft Mist™ Inhaler. Int J Pharm. 2004;283(1–2):1–9. https://doi.org/10.1016/j.ijpharm.2004.06.018.
32. Nicolaas P, Dekhuijzen FL. PPA-82857-patients—perspectives-and-preferences-in-switching--inhalers. Dovepress. 2016;12(3):1561–72. [Online]. https://www.ncbi.nlm.nih.gov/pmc/articles/PMC4993394/pdf/ppa-10-1561.pdf
33. Vaidyanathan S, et al. Uridine depletion and chemical modification increase Cas9 mRNA activity and reduce immunogenicity without HPLC purification. Mol Ther Nucleic Acid. 2018;12(September):530–42. https://doi.org/10.1016/j.omtn.2018.06.010.
34. Karikó K, Buckstein M, Ni H, Weissman D. Suppression of RNA recognition by toll-like receptors: the impact of nucleoside modification and the evolutionary origin of RNA. Immunity. 2005;23(2):165–75. https://doi.org/10.1016/j.immuni.2005.06.008.
35. Crooke CFBST. Progress in antisense oligonucleotide therapeutics. Annu Rev Pharmacol Toxicol. 1996;36:107–29. https://doi.org/10.1146/annurev.pa.36.040196.000543.
36. Brad Wan W, Seth PP. The medicinal chemistry of therapeutic oligonucleotides. J Med Chem. 2016;59(21):9645–67. https://doi.org/10.1021/acs.jmedchem.6b00551.
37. Zhang Y, et al. Novel formulations and drug delivery systems to administer biological solids. Adv Drug Deliv Rev. 2021;172:183–210. https://doi.org/10.1016/j.addr.2021.02.011.
38. Conte G et al. Hybrid lipid/polymer nanoparticles to tackle the cystic fibrosis mucus barrier in siRNA delivery to the lungs: does PEGylation make the difference. 2022, https://doi.org/10.1021/acsami.1c14975.
39. Cipolla D, Gonda I, Chan H-K. Liposomal formulations for inhalation. Ther Deliv. 2013;4(8):1047–72. https://doi.org/10.4155/tde.13.71.
40. Ferreira-Silva M, Faria-Silva C, Baptista PV, Fernandes E, Fernandes AR, and Corvo ML. Liposomal nanosystems in rheumatoid arthritis. Pharmaceutics. 2021. 13.4:454. https://doi.org/10.3390/pharmaceutics13040454.
41. Zhang L, Pornpattananangkul D, Hu C-M, Huang C-M. Development of nanoparticles for antimicrobial drug delivery. Curr Med Chem. 2010;17(6):585–94. https://doi.org/10.2174/092986710790416290.
42. Ehrhardt C. Inhalation biopharmaceutics: Progress towards comprehending the fate of inhaled medicines. Pharm Res. 2017;34(12):2451–3. https://doi.org/10.1007/s11095-017-2304-2.
43. "ClinicalTrials.gov." https://clinicaltrials.gov/, accessed in September 2022.
44. "PipelineReview.com." https://pipelinereview.com/., accessed in September 2022.
45. Hodges SGMR, Castelloe E, Chen A, Geary RS, Karras JG, Shapiro D, Yeung B, Yu R. Randomized, double-blind, placebo controlled first in human study of inhaled AIR645, an IL-4Rα oligonucleotide, in healthy volunteers. Am J Respir Crit Care Med. 2009;179:A3640. https://doi.org/10.1164/ajrccm-conference.2009.179.1_MeetingAbstracts.A3640.
46. Gail PMR, Gauvreau M, Pageau R, Seguin R, Carballo D, D'Anjou H, Campbell H, Watson RM, Parry-Billings M, Killian KJ. Efficacy of increasing doses of TPI ASM8 on allergen inhalation challenges in asthmatics. Am J Respir Crit Care Med. 2010;181:A5669. https://doi.org/10.1164/ajrccm-conference.2010.181.1_MeetingAbstracts.A5669.

GPSR Compliance

The European Union's (EU) General Product Safety Regulation (GPSR) is a set of rules that requires consumer products to be safe and our obligations to ensure this.

If you have any concerns about our products, you can contact us on ProductSafety@springernature.com

In case Publisher is established outside the EU, the EU authorized representative is:

Springer Nature Customer Service Center GmbH
Europaplatz 3
69115 Heidelberg, Germany

Batch number: 10726058

Printed by Printforce, the Netherlands